MICROEMULSIONS

Properties and Applications

SURFACTANT SCIENCE SERIES

MICROEMULSIONS
Properties and Applications

Edited by

Monzer Fanun
Al-Quds University
East Jerusalem, Palestine

CRC Press
Taylor & Francis Group
Boca Raton London New York

CRC Press is an imprint of the
Taylor & Francis Group, an **informa** business

CRC Press
Taylor & Francis Group
6000 Broken Sound Parkway NW, Suite 300
Boca Raton, FL 33487-2742

First issued in paperback 2019

© 2009 by Taylor & Francis Group, LLC
CRC Press is an imprint of Taylor & Francis Group, an Informa business

No claim to original U.S. Government works

ISBN-13: 978-1-4200-8959-2 (hbk)
ISBN-13: 978-0-367-38621-4 (pbk)

Library of Congress Cataloging-in-Publication Data

Fanun, Monzer.
    Microemulsions : properties and applications / Monzer Fanun.
        p. cm. -- (Surfactant science ; 144)
        ISBN 978-1-4200-8959-2 (alk. paper)
        1. Emulsions. I. Title.

TP156.E6F36 2008
660'.294514--dc22
                                                 2008029538

Visit the Taylor & Francis Web site at
http://www.taylorandfrancis.com

and the CRC Press Web site at
http://www.crcpress.com

Contents

Foreword

MICROEMULSIONS—A VITAL FUNDAMENTAL RESEARCH AREA MOVING RAPIDLY INTO APPLICATIONS WHILE HAVING ITS SCIENTIFIC BASIS IN OTHER SURFACTANT SELF-ASSEMBLY SYSTEMS

This book focuses on the properties and applications of microemulsions and, in particular, on their interrelationship. Of late, microemulsions have become a popular subject and applications are emerging rapidly; this further stimulates the fundamental studies. Therefore, this book is very timely and I congratulate the editor, Professor Monzer Fanun, for having prepared a volume with this focus and, in particular, achieving this so well by assembling an impressive list of contributors; this list is a good mix of established leading scientists and young colleagues entering the field recently as they will be the ones who will continue to develop our research area.

The history of microemulsions has been full of ups and downs and has been involved in many heated controversies. The name "microemulsions" itself has no doubt contributed strongly to confusion. Microemulsions are not micro but nano and are not emulsions. The history of microemulsion research is complex [1] and needs to be recounted since it provides important lessons. Here I rather wish to make a few comments from my experience with the evolution of microemulsion, which have a direct bearing on this book and its relevance.

What is a microemulsion? This question was very much in focus when I first came in contact with the field by the end of the 1960s and early 1970s. The very fact that a question like this arises leads to considerable confusion and unnecessary work. Thus, had the true nature of microemulsions been understood, a resort to the basic literature on surfactant self-assembly would have given logical explanations to many observations.

Thirty years ago, when I had just been appointed to the chair of physical chemistry at Lund University, Professor Ingvar Danielsson from Åbo Akademi in Turku, Finland came for a sabbatical. Åbo Akademi was the world-famous institution for physical chemistry where the founder of the institution, Per Ekwall, along with pupils such as Ingvar Danielsson, Krister Fontell, and Leo Mandell, had developed much of our fundamental understanding of surfactant systems, including micellization, phase behavior, and liquid crystallinity. On his retirement from Åbo, Ekwall moved to Stockholm to found the Laboratory (later Institute) of Surface Chemistry, while Danielsson took over his chair in Åbo. My first contacts with surfactant science and much of my learning were with this Stockholm-Åbo research community.

Having been concerned with aspects of surfactant aggregation on the macroscopic and aggregate levels, Danielsson took interest in a deeper molecular level of understanding, using some novel nuclear magnetic resonance (NMR) approaches, which I had developed along with my colleagues in Lund.

These studies included microemulsions and, discussing the research results and reading the literature, we became more and more concerned about the fact that different authors had very different opinions on microemulsions. (It is interesting that Ekwall and Fontell refused to use this term even though they were behind some of the pioneering and still central observations on microemulsions. Since the term referred to thermodynamically stable solutions, they found it a misnomer.) Therefore, we found it timely to suggest a definition of microemulsion as the following: A microemulsion is a system of water, oil, and amphiphile, which is a single optically isotropic and thermodynamically stable liquid solution [2]. We also gave several examples of what we considered should be included in microemulsions and what should not.

Looking into the contents of this book, and contemplating Monzer's invitation to write this foreword, I found it of interest to examine a little the acceptance of our definition among colleagues. While having a general impression that, after a quite long period of questioning, it became more and more accepted, I found it of interest to examine this further by a citation analysis. Our short note is certainly not a significant scientific contribution, but it is quite well cited (and is in fact among my 10 most cited papers). However, the citations show a very unusual variation over time, the distribution being pronouncedly "bimodal." In the first years after publication, there is quite a constant modest citation frequency. Thereafter, there is a very pronounced peak in 1989, indicating that this is the year that a more general acceptance was obtained. Afterwards, citations decrease strongly and one would have expected that the paper would as usual start to become forgotten. However, a few years ago, citations started to increase in number again and, from the citations during the first half of 2008, we can guess that this year will give the largest number of citations so far. Why is that so? Some clue can be obtained from the field of the journals where the paper is cited. Thus in 1989, most citations were in journals that focused more on physical and colloid chemistry. The pattern is very different in 2008. A majority of the citations are in journals dealing with more applied aspects, in particular, in the pharmaceutical sciences.

Is there any other evidence that microemulsions are now becoming better understood in the applied sciences, like pharmaceutics? An indication can be obtained by considering textbooks. A leading textbook in pharmacy is *Physicochemical Principles of Pharmacy* by A. T. Florence and D. Attwood [3]. Both the placing of microemulsions in the book and the text devoted to this topic reveal that even in 1998, when the third edition was published, microemulsions had received very little attention in the pharmaceutical field and that, furthermore, the nature of microemulsions was misunderstood. Thus, while there were lengthy multipage descriptions of surfactant micellization, liquid crystallinity, vesicles, and solubilization, microemulsions were dealt with in a mere seven line paragraph starting

"Microemulsions, or so-called swollen micellar systems, consist of apparently homogeneous transparent systems of low viscosity which contain a high percentage of both oil and water and high concentrations (15%–25%) of emulsifier mixture." The misconception of microemulsions in the pharmaceutical field is accentuated by the fact that rather than being placed together with other thermodynamically stable surfactant self-assembly systems, it is considered as a type of dispersion and placed under the general heading "Emulsions, Suspensions, and Other Dispersions." It is indicated from the citation analysis mentioned that if a corresponding textbook is prepared today, microemulsions would receive much more attention and would be properly classified and treated in conjunction with related surfactant systems, like micelles and liquid crystals, as they have indeed been in textbooks of physical chemistry and colloid chemistry for a long time.

This book contains significant contributions regarding the applications of microemulsions for pharmaceutical formulations, as well as for other applications, and will no doubt help considerably to provide an excellent basis for applications into new fields.

Regarding the long-standing issue of the confusion of treating microemulsions as one type of emulsion, Chapter 7 by Otto Glatter and coauthors, dealing with emulsified microemulsions, is particularly enlightening as it clearly hints to this misconception.

Stig Friberg was certainly the pioneer who demonstrated that microemulsions are indeed thermodynamically stable solutions and, therefore, should be described by phase diagrams with respect to their stability. The significance of his work on the phase behavior of surfactant–oil–water systems for the development of the microemulsion field cannot be overestimated and it is indeed very appropriate that he was invited to write the first chapter of this book. I was myself very fortunate to have early contacts with Stig Friberg. In addition, I was strongly influenced and helped by the phase diagram work of two other pioneers in the field, Per Ekwall, already mentioned above, and Kozo Shinoda in Yokohama.

Several of my collaborations with Friberg, Shinoda, and Ekwall concerned microemulsion microstructure, where they provided enlightening systems for structural investigation and deep insight into the subject.

I consider my most important contribution to the field of microemulsion as being the first, together with coworkers, to demonstrate microemulsion bicontinuity. However, this work also nicely demonstrates how important it is in microemulsion research to have a broader perspective, in particular considering other surfactant phases.

My first study dealing with surfactant phase bicontinuity did not thus concern microemulsions but cubic liquid crystalline phases. In preparing a chapter dealing with applications of NMR for a book on *Liquid Crystals and Plastic Crystals* [4], I became confused when I came to the cubic phases. As we know, cubic phases can be located in different concentration ranges in a phase diagram, inter alia between the micellar solutions and the normal hexagonal phase, and between the hexagonal and the lamellar phases. I soon realized that the surfactant self-diffusion would be very different for discrete aggregates and for connected structures. This would thus be an interesting possibility for solving the problem of the structure of cubic liquid crystalline phases. A few experiments with a postdoctoral fellow, Tom Bull, at the new pulsed NMR spectrometer, giving differences in surfactant diffusion by orders of magnitude between the two cubic phases, could directly prove that one was built up of discrete micelles while the other was bicontinuous [5]. The cubic phase, which is more dilute in surfactant, was thus found to be characterized by very slow surfactant diffusion and thus must consist of (more or less stationary) discrete aggregates. In the more concentrated cubic phase, surfactant diffusion was found to be more than one order of magnitude faster. This rather surprising finding could only be understood if the surfactant molecules could diffuse freely over macroscopic distances; thus surfactant aggregates are connected.

The distinction between discrete "droplet" structures and bicontinuous ones became central in the subsequent studies on microemulsions in Lund [6–12]. This research topic became even more emphasized when Peter Stilbs introduced the Fourier transform version of the NMR technique [13–16].

That surfactant self-assembly systems, which include liquid crystalline phases and isotropic solutions, can be divided into those that have discrete self-assembly aggregates and those where the aggregates are connected in one, two, or three dimensions was very clear for the pioneers of the microemulsion field mentioned above. Regarding lamellar phases, the two-dimensional connectivity was already appreciated at a very early stage. The same holds true for the ("normal" and "reverse") hexagonal phases, although erroneous models of linearly associated spherical micelles, "pearls-on-a-string," can be found in the literature; such a linear association was also, again incorrectly, advanced to explain droplet growth in microemulsions. The general acceptance of connectivity for these anisotropic phases stood in sharp contrast to a great difficulty to get an acceptance for bicontinuity for other phases. This is partly related to the fact that contrary to these anisotropic phases, it has been much more difficult to structurally characterize the different isotropic phases found in simple and complex surfactant systems: cubic liquid crystals, solutions in binary surfactant–water systems, and microemulsions. The first verification was due to observations of molecular self-diffusion over macroscopic distances. Electrical conductivity offers a partial insight in providing information on the extension of aqueous domains. Fluorescence quenching can provide information on the growth of nonpolar domains, but a probe has to be introduced. Later cryogenic transmission electron microscopy has developed into a very important tool for imaging different surfactant phases.

Using a similar approach as for cubic phases, it was thus quite straightforward to address the problem of microemulsion structure. Thus, by measuring oil and water self-diffusion, it was quite easy to establish whether oil or water or none of them are confined to discrete domains, "droplets." In the first work on microemulsion structure by self-diffusion, using both tracer techniques and NMR spin-echo measurements, it was clearly shown that, in addition to droplet microemulsions, over wide ranges of composition they can be bicontinuous [6]; this is manifested by both oil and water diffusion being rapid, not much less than the self-diffusion of the neat liquids.

The self-diffusion approach to microstructure is not limited to cubic phases or microemulsions. An early study concerned the demonstration of micellar growth into worm-like structures for nonionic surfactants [17,18]. Parallel pioneering studies on phase behavior of nonionic surfactants by Gordon Tiddy [19] also illustrated the same feature. Another problem, soon to be tackled, was that of the microstructure of the "sponge phases," a "microemulsion analogue," for binary surfactant systems, termed L_3 by Ekwall (and identified by him in a number of systems). While isotropic solutions in simple surfactant–water mixtures were for a long time considered synonymous with solutions of discrete surfactant micelles, there were indications of a more complex situation given by the clouding and phase separation into two solutions of nonionic surfactants at elevated temperature. Here self-diffusion was again expected to provide the solution [19,20]. For the sponge phase, water diffusion was much reduced compared to classical micellar solutions. In fact, it was close to 2/3 of the value of neat water. The surfactant diffusion was, on the other hand, found to be much more rapid, and close to 2/3 of the diffusion of the neat liquid surfactant, than what was observed for previously studied micellar solutions. The solutions are thus bicontinuous. These systems are perfect illustrations of bicontinuity and in many respects useful models of bicontinuous microemulsions. Both the water and surfactant self-diffusion coefficients are close to 2/3 of the values of the neat liquids, corresponding to an ideal zero mean curvature bicontinuous structure.

While I have illustrated here, with some examples from our own research, how progress in our understanding has been dependent on understanding alternative surfactant phases, this approach is certainly not unique. Several pioneers like Friberg, Ekwall, Shinoda, Tiddy, Scriven, and Wennerström, have provided beautiful examples of such a "holistic" view. It is my firm belief that in the ongoing expansion of the microemulsion field, that the present book emphasizes and supports a broader look into surfactant self-assembly and a resort to simpler surfactant systems are mandatory.

Björn Lindman
Coimbra University and Lund University

REFERENCES

1. Lindman, B.; Friberg, S. Microemulsions—a historical overview. In *Handbook of Microemulsion Science and Technology*, P. Kumar and K. L. Mittal, eds. Marcel Dekker, New York, 1999, pp. 1–12.
2. Danielsson, I.; Lindman, B. The definition of microemulsion. *Colloids Surfaces* 3, 1981, 391–392.
3. Florence, A.T; Attwood, D. *Physicochemical Principles of Pharmacy*, 3rd edn, Pharmaceutical Press, London, 1998.
4. Johansson, Å.; Lindman, B. In *Liquid Crystals and Plastic Crystals, Nuclear Magnetic Resonance Spectroscopy of Liquid Crystals-Amphiphilic Systems*, G.W. Gray and P. A. Winsor, eds. Ellis Horwood Publishers, Chichester, 1974, Vol. 2, pp. 192–230.

5. Bull, T.; Lindman, B. Amphiphile diffusion in cubic lyotropic mesophases. *Mol. Cryst. Liquid Cryst.* 28, 1975, 155–160.

6. Lindman, B.; Kamenka, N.; Kathopoulis, T.M.; Brun, B.; Nilsson, P.G. Translational diffusion and solution structure of microemulsions. *J. Phys. Chem.* 84, 1980, 2485–2490.

7. Nilsson, P. G.; Lindman, B. Solution structure of nonionic surfactant microemulsions from NMR self-diffusion studies. *J. Phys. Chem.* 86, 1982, 271–279.

8. Guéring, P.; Lindman, B. Droplet and bicontinuous structures in cosurfactant microemulsions from multi-component self-diffusion measurements. *Langmuir* 1, 1985, 464–468.

9. Olsson, U.; Shinoda, K.; Lindman, B. Change of the structure of microemulsions with the HLB of nonionic surfactant as revealed by NMR self-diffusion studies. *J. Phys. Chem.* 90, 1986, 4083–4088.

10. Lindman, B.; Shinoda, K.; Olsson, U.; Anderson, D.; Karlström, G.; Wennerström, H. On the demonstration of bicontinuous structures in microemulsions. *Colloids Surfaces* 38, 1989, 205–224.

11. Lindman, B.; Olsson, U. Structure of microemulsions studied by NMR Ber. Bunsenges. *Phys. Chem.* 100, 1996, 344–363.

12. Shinoda, K.; Lindman, B. Organized surfactant systems: Microemulsions. *Langmuir* 3, 1987, 135–149.

13. Stilbs, P.; Moseley, M. E. Nuclear spin-echo experiments on standard Fourier-transform NMR spectrometers—Application to multi-component self-diffusion studies. *Chem. Scripta* 13, 1979, 26–28.

14. Stilbs, P. *Prog.* Fourier transform pulsed-gradient spin-echo studies of molecular diffusion. *NMR Spectrosc.* 19, 1987, 1–45.

15. Stilbs, P.; Moseley, M. E.; Lindman, B. Fourier transform NMR self-diffusion measurements on microemulsions. *J. Magn. Reson.* 40, 1980, 401–404.

16. Lindman, B.; Stilbs, P.; Moseley, M. E. Fourier transform NMR self-diffusion and microemulsion structure. *J. Colloid Interface Sci.* 83, 1981, 569–582.

17. Nilsson, P. G.; Wennerström, H.; Lindman, B. Structure of micellar solutions of nonionic surfactants. NMR self-diffusion and proton relaxation studies of poly(ethyleneoxide) alkylethers. *J. Phys. Chem.* 87, 1983, 1377–1385.

18. Lindman, B.; Wennerström, H. Nonionic micelles grow with increasing temperature. *J. Phys. Chem.* 95, 1991, 6053–6054.

19. Mitchell, D. J.; Tiddy, G. J. T.; Waring, L.; Bostock, T.; McDonald, M. P. J. Phase behaviour of polyoxyethylene surfactants with water. Mesophase structures and partial miscibility (cloud points). *Chem. Soc. Faraday Trans.* 79, 1983, 975–1000.

20. Nilsson, P. G.; Lindman, B. Nuclear magnetic resonance self-diffusion and proton relaxation studies of nonionic surfactant solutions. Aggregate shape in isotropic solutions above the clouding temperature. *J. Phys. Chem.* 88, 1984, 4764–4769.

21. Lindman, B.; Olsson, U.; Stilbs, P.; Wennerström, H. Comment on the self-diffusion in L3 and other bicontinuous surfactant solutions. *Langmuir* 9, 1993, 625–626.

Preface

Microemulsions are microheterogeneous, thermodynamically stable, spontaneously formed mixtures of oil and water under certain conditions by means of surfactants, with or without the aid of a cosurfactant. The first paper on microemulsions appeared in 1943 by Hoar et al., but it was Schulman and coworkers who first proposed the word "microemulsion" in 1959. Since then, the term "microemulsions" has been used to describe multicomponent systems comprising nonpolar, aqueous, surfactant, and cosurfactant components. The application areas of microemulsions have increased dramatically during the past decades. For example, the major industrial areas are fabricating nanoparticles, oil recovery, pollution control, and food and pharmaceutical industries. This book is a comprehensive reference that provides a complete and systematic assessment of all topics affecting microemulsion performance, discussing the fundamental characteristics, theories, and applications of these dispersions that have been developed over the last decade.

The book opens with a chapter that describes a phase diagram approach to microemulsions by two leading authorities (Friberg and Aikens) who have contributed significantly to the field of microemulsions. In the next three chapters, Moulik and Rakshit, Mehta and coworkers, and Mejuto and coworkers, respectively, advance different approaches to describe the percolation phenomenon in microemulsion systems. Theories that predict droplet clustering along with the basic conditions required for the formation and stability of these reverse micellar systems and the composition, temperature, and pressure-dependent conductance percolation and energetics of droplet clustering are reviewed. The influence of different additives on the conductance percolation of ionic microemulsions is also reviewed.

Significant progress has been made in the formulation and characterization of new microemulsion systems. Properties of microemulsions with mixed nonionic surfactants and different types of oils are reviewed in Chapter 5 by Fanun. A comprehensive review on the influence of various simple alcohols on the internal structural organization of microemulsion systems is presented in Chapter 6 by Tomšič and Jamnik. Chapter 7 by Glatter and coworkers focuses on the effect of variations in temperature and solubilizing oil on the formation and the reversible structural transitions of emulsified microemulsions that have excellent potential in applications such as nanoreactors or host systems for solubilizing active molecules in cosmetic, pharmaceutical, and food industries. The interaction of water with room temperature ionic liquids (RTILs) has been studied in RTIL/surfactant/water-containing ternary microemulsions by solvent and rotational relaxation of neutral Coumarin probes, namely Coumarin 153 and Coumarin 151, using steady-state and picosecond time-resolved emission spectroscopy, reviewed by Seth and Sarkar in Chapter 8.

Microemulsions accommodate poorly soluble drugs (both hydrophilic and lipophilic) and protect those that are vulnerable to chemical and enzymatic degradation. They have the potential to increase the solubility of poorly soluble drugs, enhance the bioavailability of drugs with poor permeability, reduce patient variability, and offer an alternative for controlled drug release. In Chapter 9, Alany and coworkers review the formulation and characterization of microemulsions intended for drug delivery applications. Recent investigations on pharmaceutically applicable microemulsions are described in Chapter 10 by Gašperlin and Bešter-Rogač. The use of emulsions and microemulsions as a delivery system for cancer therapy is described in Chapter 11 by Karasulu and coworkers.

Enzymes when hosted in reverse micelles can catalyze reactions that are not favored in aqueous media. Products of high-added value can be thus produced in these media. The potential technical and commercial applications of enzyme-containing microemulsions as microreactors are mainly linked to their unique physicochemical properties. The potential biotechnological applications of microemulsions with immobilized biocatalysts such as enzymes are described in Chapter 12 by Kunz and coworkers and in Chapter 13 by Xenakis and coworkers.

Great efforts have been made in order to replace established but harmful, corrosive, and therefore, obsolete decontamination media for chemical warfare agents and toxic industrial chemicals. Chapter 14 by Hellweg and coworkers discusses the considerable advantages of microemulsion-based decontamination systems with respect to practical boundary conditions and fundamental principles of microemulsion formation. Additionally, the authors illustrate the further development to versatile, environmentally compatible and nonharmful systems containing nanoparticles and enzymes as active components.

Several segments of the petroleum industry can be optimized with the use of microemulsions. Research has been carried out on potential microemulsified formulations for compression-ignition, cycle-diesel engines, which, in spite of bringing about a slight increase in consumption, produce less polluting emissions. In Chapter 15, Dantas Neto and coworkers summarize recent advances in microemulsions in this type of industry.

Microemulsions can be considered as true nanoreactors, which can be used to synthesize nanomaterials. The main idea behind this technique is that by appropriate control of the synthesis parameters one can use these nanoreactors to produce tailor-made products down to a nanoscale level. Chapter 16 by Tojo and coworkers describes the use of Monte Carlo simulations to study the influence of the critical nucleus size and the chemical reaction rate on the formation of nanoparticles in microemulsions. Chapter 17 by Husein and Nassar focuses on exploring ways of maximizing the concentration of stable colloidal nanoparticles, nanoparticle uptake, in single (w/o) microemulsions. Chapter 18 by Ghosh describes the photophysical and interfacial electron transfer behavior of anatase TiO_2 nanoparticles in microemulsions.

Capillary electrophoresis is a powerful technique with relevant features of performance such as simplicity, versatility, very high resolution in short time of analysis, and low cost of operation. The final chapter by Tripodi and Lucangioli describes the use of microemulsions in capillary electrophoresis as pseudostationary phases in the electrokinetic chromatography mode. This method has extensive applications in different fields of pharmaceutical analysis for the determination of drugs and their impurities in bulk material and pharmaceutical formulations for the dosage of drugs in biological fluids.

In quintessence, this book represents the collective knowledge of young and renowned researchers and engineers in the field of microemulsions. This book covers recent advances in the characterization of the properties of microemulsions; it covers new types of materials used for the formulation and stabilization of microemulsions, and it also covers new applications. An important feature of this book is that the author of each chapter has been given the freedom to present, as he/she sees fit, the spectrum of the relevant science, from pure to applied, in his/her particular topic. Of course this approach inevitably leads to some overlap and repetition in different chapters, but that does not necessarily matter. Any author has his/her own views on, and approach to, a specific topic, molded by his/her own experience. I hope that this book will familiarize the reader with the scientific and engineering aspects of microemulsions, and provides experienced researchers, scientists, and engineers in academic and industry communities with the latest developments in this field.

I would like to thank all those who contributed as chapter authors despite their busy schedules. In total, 52 individuals from 15 countries contributed to the work. All of them are recognized and respected experts in the areas they wrote about. None of them is associated with any errors or omissions that remain. I take full responsibility. Special thanks are due to the reviewers for their valuable comments as peer review is a requirement to preserve the highest standard of publication. My appreciation goes to Barbara Glunn of Taylor & Francis for her genuine interest in this project.

Monzer Fanun
Associate Professor
Al-Quds University
East Jerusalem, Palestine

Editor

Monzer Fanun is a professor in surface and colloid science, the head of the Colloids and Surfaces Research Laboratory, and a member of the Nanotechnology Research Group at Al-Quds University, East Jerusalem, Palestine. He has authored and coauthored more than 40 professional papers. He is a member of the European Colloid and Interface Society and a fellow of the Palestinian Academy for Science and Technology. In 2003, he received his PhD in applied chemistry from the Casali Institute of Applied Chemistry a part of the Institute of Chemistry at the Hebrew University of Jerusalem, Israel.

Contributors

Patricia A. Aikens
BASF Corp.
Stony Brook, New York

Raid G. Alany
School of Pharmacy
University of Auckland
Auckland, New Zealand

Maria Carlenise Paiva de Alencar Moura
Universidade Federal do Rio Grande
  do Norte Centro de Tecnologia
Departamento de Engenharia Química
  UFRN—Federal University
  of Rio Grande do Norte
Chemical Engineering Department
Campus Universitário
Natal, Brazil

Levent Alparslan
Department of Biopharmaceutics
  and Pharmacokinetics
Faculty of Pharmacy
University of Ege
Izmir, Turkey

and

Center for Drug R&D and
  Pharmacokinetic Applications
University of Ege
Izmir, Turkey

Hans-Juergen Altmann
Armed Forces Scientific Institute
  for NBC Protection
Munster, Germany

Eduardo Lins de Barros Neto
Universidade Federal do Rio Grande
  do Norte Centro de Tecnologia
Departamento de Engenharia Química
  UFRN—Federal University
  of Rio Grande do Norte
Chemical Engineering Department
Campus Universitário
Natal, Brazil

F. Barroso
Department of Physical Chemistry
Faculty of Chemistry
University of Vigo
Vigo, Spain

Pierre Bauduin
Institut de Chimie Séparative de
  Marcoule
Bagnols-sur-Cèze, France

Marija Bešter-Rogač
Faculty of Chemistry and Chemical
  Technology
University of Ljubljana
Ljubljana, Slovenia

K. K. Bhasin
Department of Chemistry and Center
  of Advanced Studies
  in Chemistry
Panjab University
Chandigarh, Panjab, India

Liliana de Campo
Department of Applied Mathematics
The Australian National University
Canberra, New South Wales,
  Australia

Tereza Neuma de Castro Dantas
Universidade Federal do Rio Grande
  do Norte Centro de Ciências
  Exatas e da Terra
Departamento de Química
UFRN—Federal University
  of Rio Grande do Norte
Chemistry Department
Campus Universitário
Natal, Brazil

A. Cid
Department of Physical Chemistry
Faculty of Sciences
University of Vigo at Ourense
Ourense, Spain

Afonso Avelino Dantas Neto
Universidade Federal do Rio Grande
  do Norte Centro de Tecnologia
Departamento de Engenharia Química
UFRN—Federal University
  of Rio Grande do Norte
Chemical Engineering Department
Campus Universitário
Natal, Brazil

M. de Dios
Department of Physical Chemistry
Faculty of Chemistry
University of Vigo
Vigo, Spain

Gamal M. M. El Maghraby
Department of Pharmaceutics
King Saud University
Riyadh, Saudi Arabia

Monzer Fanun
Faculty of Science
  and Technology
Al-Quds University
East Jerusalem, Palestine

Stig E. Friberg
Chemistry Department
University of Virginia
Charlottesville, Virginia

L. García-Río
Departamento de Química-Física
Facultad de Química
Universidad de Santiago de
  Compostela
Santiago de Compostela,
  Spain

Mirjana Gašperlin
Faculty of Pharmacy
University of Ljubljana
Ljubljana, Slovenia

Hirendra N. Ghosh
Radiation and Photochemistry
  Division
Bhabha Atomic Research
  Center
Mumbai, Maharashtra, India

Otto Glatter
Institute of Chemistry
University of Graz
Graz, Austria

D. Gómez-Díaz
Departamento de Ingeniería
  Química
Escuela Técnica Superior
Universidad de Santiago de
  Compostela
Santiago de Compostela,
  Spain

Anja Graf
School of Pharmacy
University of Otago
Dunedin, New Zealand

Alexandre Gurgel
Universidade Federal de Viçosa
Centro de Ciências Exatas e
  Tecnológicas
Departamento de Química
  UFV—Federal University of Viçosa
Chemistry Department
Campus Universitário
Viçosa, Brazil

Thomas Hellweg
Physical Chemistry I
University of Bayreuth
Bayreuth, Germany

Maen M. Husein
Department of Chemical and
  Petroleum Engineering
University of Calgary
Calgary, Alberta, Canada

Andrej Jamnik
Faculty of Chemistry and Chemical
  Technology
University of Ljubljana
Ljubljana, Slovenia

Burçak Karaca
Department of Medical Oncology
School of Medicine
University of Ege
Izmir, Turkey

Ercüment Karasulu
Department of Biopharmaceutics
  and Pharmacokinetics
Faculty of Pharmacy
University of Ege
Izmir, Turkey

and

Center for Drug R&D and
  Pharmacokinetic Applications
University of Ege
Izmir, Turkey

H. Yesim Karasulu
Department of Pharmaceutical
  Technology
Faculty of Pharmacy
University of Ege
Izmir, Turkey

Gurpreet Kaur
Department of Chemistry and Center
  of Advanced Studies
  in Chemistry
Panjab University
Chandigarh, Panjab, India

Khushwinder Kaur
Department of Chemistry and Center
  of Advanced Studies
  in Chemistry
Panjab University
Chandigarh, Panjab, India

F. N. Kolisis
Institute of Biological Research
  and Biotechnology
National Hellenic Research
  Foundation
Athens, Greece

and

School of Chemical
  Engineering
National Technical University
  of Athens
Athens, Greece

Karen Krauel-Goellner
Institute of Food Nutrition
  and Human Health
Wellington, New Zealand

Werner Kunz
Institute of Physical and Theoretical
  Chemistry
University of Regensburg
Regensburg, Germany

Silvia Lucangioli
Analytical Chemistry and
  Physicochemistry Department
Faculty of Pharmacy and
  Biochemistry
University of Buenos Aires
Buenos Aires, Argentina

S. K. Mehta
Department of Chemistry and Center
  of Advanced Studies
  in Chemistry
Panjab University
Chandigarh, Panjab, India

J. C. Mejuto
Department of Physical Chemistry
Faculty of Sciences
University of Vigo at Ourense
Ourense, Spain

Satya Priya Moulik
Center for Surface Science
Department of Chemistry
Jadavpur University
Kolkata, West Bengal, India

Nashaat N. Nassar
Department of Chemical and
  Petroleum Engineering
University of Calgary
Calgary, Alberta, Canada

V. Papadimitriou
Institute of Biological Research
  and Biotechnology
National Hellenic Research
  Foundation
Athens, Greece

Animesh Kumar Rakshit
Department of Natural Sciences
West Bengal University
  of Technology
Kolkata, West Bengal, India

André Richardt
Armed Forces Scientific Institute
  for NBC Protection
Munster, Germany

Nilmoni Sarkar
Department of Chemistry
Indian Institute of Technology
Kharagpur, West Bengal, India

Debabrata Seth
Department of Chemistry
Indian Institute of Technology
Kharagpur, West Bengal, India

H. Stamatis
Biological Applications and
  Technologies Department
University of Ioannina
Ioannina, Greece

C. Tojo
Department of Physical Chemistry
Faculty of Chemistry
University of Vigo
Vigo, Spain

Matija Tomšič
Faculty of Chemistry and Chemical
  Technology
University of Ljubljana
Ljubljana, Slovenia

Didier Touraud
Institute of Physical and Theoretical
  Chemistry
University of Regensburg
Regensburg, Germany

Valeria Tripodi
Analytical Chemistry and
  Physicochemistry Department
Faculty of Pharmacy and Biochemistry
University of Buenos Aires
Buenos Aires, Argentina

Stefan Wellert
Physical Chemistry I
University of Bayreuth
Bayreuth, Germany

A. Xenakis
Institute of Biological Research and
  Biotechnology
National Hellenic Research
  Foundation
Athens, Greece

Anan Yaghmur
Institute of Biophysics and
  Nanosystems Research
Austrian Academy
  of Sciences
Graz, Austria

1 A Phase Diagram Approach to Microemulsions

Stig E. Friberg and Patricia A. Aikens

CONTENTS

1.1 INTRODUCTION

The phase diagram approach to microemulsions was introduced decades ago by Gillberg and collaborators [1]. At that time, it was not well received by the researchers in the area, because it emphasized that microemulsions are in fact micellar systems and the traditionally simplified thermodynamic treatment was very much in vogue at that time. Unfortunately, the ensuing arguments about the "true structure of microemulsions" shrouded the advantage of the approach, and it was only after the Israelachvili–Ninham analysis of the thermodynamics of such systems [2] that attention could be directed to the essential features of the phase diagram approach. A brief history of the development has been given by Lindman and Friberg [3].

In the following sections, the phase diagram approach will be applied to three attributes of microemulsions: (a) the importance of ordering versus disordering, (b) the temperature dependence of the behavior of microemulsions stabilized by polyethylene glycol adduct surfactants, and (c) the use of phase diagrams to obtain information on the composition of the vapor leaving microemulsion during its evaporation.

1.2 DISCUSSION

1.2.1 ORDERING–DISORDERING

The phase diagrams of microemulsions have traditionally been presented in two ways. The original one was built on the results of Ekwall on the association structures of amphiphilic systems [4] and was based on the associations in the water–surfactant combination. According to this approach, the development of the microemulsion structures was a result of the structural modifications brought about by the addition of less hydrophilic amphiphiles such as alcohols. The hydrocarbons in the microemulsions were considered solubilizates in this methodology and their effect on the structure was considered to be of secondary importance. The approach was very successful for W/O microemulsions, providing a simple tool for their formulation. A generic diagram is given in Figure 1.1.

The essential feature of importance for the microemulsion is the fact that the inverse micellar solution and the aqueous solution of normal micelles are not in mutual equilibrium except for extremely low-surfactant concentrations. For higher-surfactant concentrations, the equilibrium is with the liquid crystalline phase. As a consequence, the transition from the normal micelles to inverse micelles (Figure 1.2) does not happen directly, but through a lamellar liquid crystal (Figure 1.3).

W/O microemulsions stabilized by an ionic surfactant also employ a less hydrophilic amphiphile, which is known as the cosurfactant. The original cosurfactants were alcohols [5] and Gillberg realized early on [1] that W/O microemulsions were obtained simply by adding a hydrocarbon to Ekwall's inverse micellar solution (Figure 1.4). Addition of the hydrocarbon does not imply significant

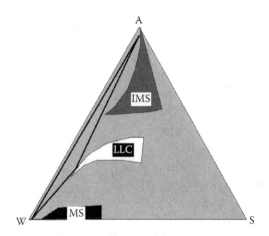

FIGURE 1.1 Partial generic phase diagram of a system water (W), surfactant (S), and medium chain length alcohol (A). (Adapted from Ekwall, P., in *Advances in Liquid Crystals*, Brown, G.H. (Ed.), Academic Press, New York, 1975, pp. 1–139. With permission.)

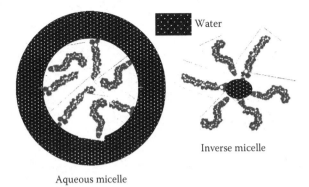

FIGURE 1.2   In the aqueous solution micelle (left), the surfactant polar groups are organized toward the surrounding water, while the hydrocarbon chains are inside the micelle. In an inverse micelle (right), the organization is opposite.

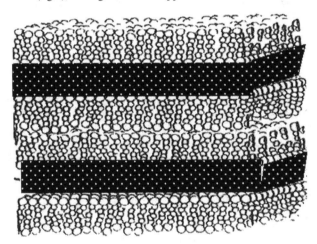

FIGURE 1.3   In a lamellar liquid crystal, water layers are separated by mirrored bilayers of surfactant.

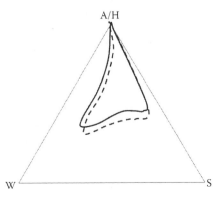

FIGURE 1.4   Addition of hydrocarbon to the inverse micellar solution (solid line) (Figure 1.1) gives a W/O microemulsion (hatched line).

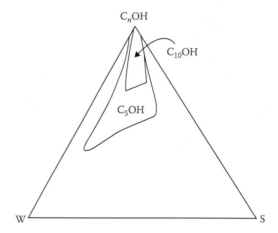

FIGURE 1.5 Comparison of the inverse micellar liquid areas for systems with pentanol and decanol.

structure changes [6] and the W/O microemulsions were hence described as inverse micellar solutions. The approach was initially not received well by Schulman's successors [7], and it is remarkable that Schulman's initial publication on the concept described these microemulsions as colloid solutions. The term "microemulsion" was coined much later [8].

The application of Ekwall's presentation of phase diagrams offers several advantages. First, it provides an explanation of the fact that when the capacity to include water in a W/O microemulsion is exceeded, the phase appearing is not an aqueous liquid, but a lamellar liquid crystal. Secondly, it provides immediate clarification of the role of the cosurfactant. As demonstrated in Figure 1.5, the effectiveness of the cosurfactant depends decisively on its chain length. The difference in the sizes of the W/O microemulsion regions in Figure 1.5 demonstrates that decanol is far less useful as a cosurfactant than pentanol (if it is even useful at all). The explanation for this fact is not, as it may appear at a first glance, the difference in the stability of the inverse micelles; it rests with the fact that the shorter pentanol chain destabilizes the lamellar liquid crystal by disordering it, and so as a result, increasing the area for the inverse micellar solution.

Following this approach, it would be logical to use butanol as a cosurfactant instead of pentanol, because its isotropic liquid region now expands continuously to the water corner (Figure 1.6).

This large continuous isotropic liquid region at first appears highly appealing, but effective utilization of shorter chain length alcohols as cosurfactants is countered by another factor. Butanol certainly destabilizes the lamellar liquid crystal efficiently (Figure 1.6), but when the hydrocarbon is added to form the microemulsion, the butanol is too water soluble and does not reach and reside at the oil/water interface sufficiently. As a result, the system forms two separate phases: a traditional macroemulsion of oil and water.

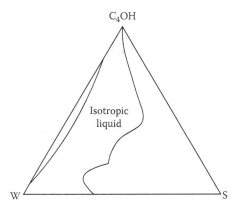

FIGURE 1.6 Isotropic liquid area for a system with butanol.

The importance of the disordering action of the cosurfactant is confirmed by a later publication concerning O/W microemulsions [9]. In this case, the pentanol per se did not provide sufficient disordering effect as demonstrated by the features in Figure 1.7.

The insufficient disordering is illustrated by the fact that the decane solubilization is limited and by the solubility gap along the sodium dodecyl sulfate (SDS)/W–C_5OH axis. The latter is caused by the lamellar liquid crystal between the aqueous and pentanol solution (Figure 1.8).

The addition of a hydrotrope, a more water soluble molecule with disordering action, supplemented the disordering and the liquid crystal range along the SDS/W–C_5OH axis disappeared (Figure 1.9) resulting in an excellent microemulsion area [10].

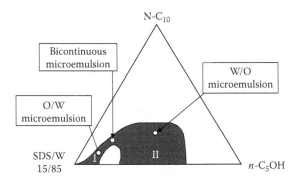

FIGURE 1.7 Isotropic liquid in the partial phase diagram of decane, n-C_{10}, pentanol, C_5OH, and a solution of 15% SDS, in water. The liquid structure passes from an O/W microemulsion to a W/O one through a bicontinuous structure without a phase separation.

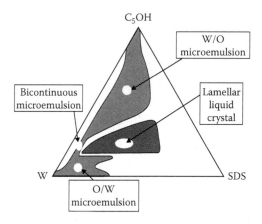

FIGURE 1.8   Part of the phase diagram water (W), SDS, and pentanol (C$_5$OH). The areas named microemulsions in this Figure 1.4 were called micellar solutions in Ekwall's terminology (From Ekwall, P., in *Advances in Liquid Crystals*, Brown, G.H. (Ed.), Academic Press, New York, 1975, pp. 1–139. With permission.)

The surfactant in this system is ionic, and hence salt has a similar action [11]. The ultimate extension of this action is amply exemplified in the early publications from the field of microemulsion-assisted petroleum recovery [12].

The phase diagram approach to microemulsions following Ekwall [1] is characterized by a section through the three-dimensional diagram according to Figure 1.10a. Alternative publications with different sectioning (Figure 1.10b) have also gained popularity [13]. Both these presentations are useful; the second one suffers from the disadvantage of not catching the strong variation in the areas with the surfactant/cosurfactant ratio as accentuated by Figure 1.8.

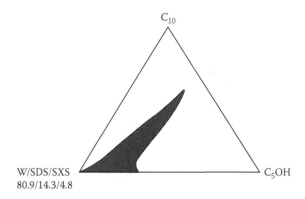

FIGURE 1.9   Microemulsion region in the system water/SDS/sodium xylene sulfonate, W/SDS/SXS, pentanol, C$_5$OH, and decane, C$_{10}$.

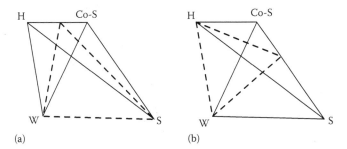

(a) (b)

FIGURE 1.10 Two main representations of the microemulsion pseudophase diagram. The left depiction (a) is the Ekwall–Gillberg approach, which treats the hydrocarbon/ cosurfactant liquid as one component, while the right model (b) combines the surfactant and cosurfactant into one component.

1.2.2 TEMPERATURE DEPENDENCE

It is seen above that the areas for microemulsions stabilized by ionic surfactants are decisively dependent on the structure of the cosurfactant to cause the necessary disorder in the system. Microemulsions stabilized by polyethylene glycol adduct nonionic surfactants, on the other hand, are characterized by the fact that cosurfactant is not used. Instead, the areas of stability now rely on temperature (Figure 1.11) although the relation with the liquid crystal structure is still the essential element [14].

The main theme of this dependence is illustrated in Figures 1.12 and 1.13 [15], which show the generic phase diagram for an alkyl ether surfactant with an aliphatic hydrocarbon of a moderate length (approx. 12 carbons), and a short polyethylene glycol chain (approx. 4 ethylene glycol units). First, the diagram is characterized by a complete disparity of the solubility of the surfactant in water

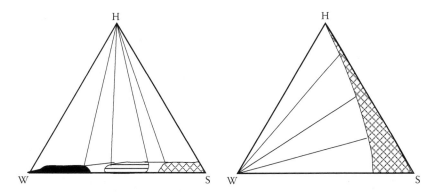

FIGURE 1.11 Phase equilibria for the system water (W), a polyethylene glycolalkyl ether (S) and an aliphatic hydrocarbon (H). Low-temperature behavior is depicted in the upper left-hand diagram, high-temperature features are depicted in the lower right-hand diagram.

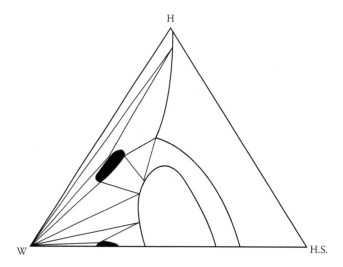

FIGURE 1.12 Features of the system at the HLB temperature.

and hydrocarbon with a moderate rise in temperature to 75°C. At lower temperature, the solubility is restricted to water transitioning into an aqueous micellar solution, followed by a lamellar liquid crystal, and finally an inverse micellar liquid with increasing surfactant content. At a higher temperature, the surfactant solubility is restricted entirely to the hydrocarbon, without any ordered structures.

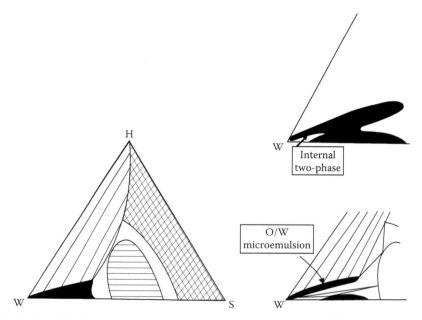

FIGURE 1.13 Three next stages after the state at left in Figure 1.11.

In between these two temperature extremes, lies the hydrophilic–lipophilic balance (HLB) temperature. Here, the aqueous micellar phase has disappeared or is severely restricted (Figure 1.13), and a liquid phase is found instead at a water/hydrocarbon ratio of approximately 1 with a moderate concentration of surfactant. This liquid is a microemulsion with a bicontinuous structure.

From a phase diagram point of view, the development of the features from those in the left-hand diagram of Figure 1.11 to those in Figure 1.12 is of interest because of the rather intricate details of the equilibria. Figure 1.13 presents the initial changes in greater detail.

The first development from the stage in the left-hand diagram of Figure 1.11 is that the hydrocarbon and the surfactant become mutually soluble and the lamellar liquid crystal is extended toward high-hydrocarbon content. These two areas remain approximately constant during the next stages. In the first of these (top right, Figure 1.13), the micellar region along the water–surfactant axis is limited with respect to the maximum water content. To reach the water corner, a certain ratio of hydrocarbon to surfactant is required. The effect is that a two-phase region is formed extending from the water corner. In the next step (bottom right, Figure 1.13), the two regions are separated and the one with highest hydrocarbon to surfactant ratio forms an O/W microemulsion, which has the specific property of being infinitely dilutable with water. It should be noted that the O/W microemulsions stabilized by an ionic surfactant such as the one in Figure 1.9 cannot be diluted with water. Any such attempt leads to phase separation and a macroemulsion is formed. The fact that there is no equilibrium between the O/W microemulsion and the remnant of the micellar solution is a remarkable feature in the diagram. Instead, the two liquids are in equilibrium with pure water and with the lamellar liquid crystal. Further progression toward the HLB temperature pattern in Figure 1.12 is depicted in Figure 1.14.

Now the O/W microemulsion area is separated from the water corner, but forms an unconnected phase in the water-rich part of the system. The ensuing phase equilibria are depicted in the enlarged partial diagram on the left-hand side of Figure 1.14. The complexity of the equilibrium conditions are well illustrated by the

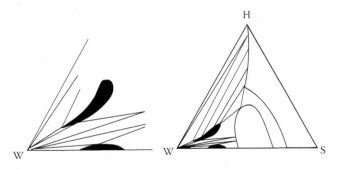

FIGURE 1.14   Subsequent step to the configurations in Figure 1.13.

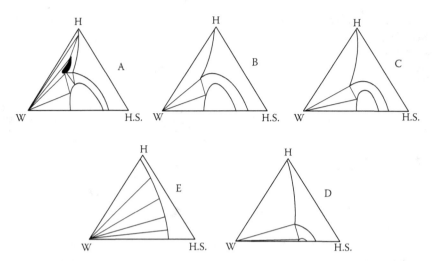

FIGURE 1.15 Subsequent patterns between that of the HLB temperature in Figure 1.12 and the final appearance to the right in Figure 1.11. (Reproduced from Shinoda, K. and Friberg, S.E., *Emulsions and Solubilization*, John Wiley & Sons, New York, 1986.)

number of phases (MS, aqueous micellar solution; O/W, O/W microemulsion; LC, liquid crystal; O, hydrocarbon–surfactant liquid) found when adding hydrocarbon to an aqueous solution of surfactant less concentrated than the separated phase. The sequence is W + MS → W + MS + LC → W + LC → W + LC + O/W → W + O/W → W + O/W + O → W + O; an extraordinary number of seven combinations for surfactant concentration less than that of any association structure phase.

The bicontinuous microemulsion region coalesces with the hydrocarbon–surfactant liquid area forming a W/O microemulsion region reaching toward greater fractions of water. The surfactant–hydrocarbon ratio for maximum water solubilization depends on the maximum solubilization of the hydrocarbon into the lamellar liquid crystal. With increasing temperature, the solubilization capacity is reduced and the maximum for water solubilization into the W/O microemulsion is shifted to greater surfactant–hydrocarbon ratios. The final result of this trend is the system of the hydrocarbon–surfactant liquid in equilibrium with water, in accordance with the right part of Figure 1.11 (Figure 1.15E).

1.2.3 VAPOR COMPOSITION FROM MICROEMULSIONS

The recently introduced algebraic method to extract information from phase diagrams [16] has been used to quantify evaporation from microemulsions [17]. The approach as such does not provide additional information to the experimentally determined phase diagram, but introduces a system to illustrate the influence of the relative humidity (RH) on the direction and volume of the evaporation and its path that is not immediately available otherwise.

The method basically relates the changes in the composition of the liquid phase to the composition of the released vapor. This information would not be of practical value if the evaporation took place in completely dry air or in vacuum; the difference in the composition of the liquid phase obviously equals the composition of the escaping vapor. In addition, if the evaporation were to take place under conditions close to equilibrium, the partial vapor pressures could be utilized to calculate the activities of the components in the liquid phase. Evaporation however, usually takes place into an atmosphere at a certain level of humidity and this fact affects the interaction between the liquid phase and its vapor. This influence is conveniently attended to using the algebraic approach as illustrated by an example [17]. The system to be discussed consists of water (W), cosurfactant (C), and surfactant (S) and is depicted in Figure 1.16.

The composition of the discharged vapor from a weight fraction composition $(W_1, C_1,$ and $S_1)$ is obtained from the tangent to the experimentally determined evaporation path, which is given a function $C(S)$.

$$C = C_1 + (S - S_1) dC_1/dS_1 \qquad (1.1)$$

Setting $S = 0$, the composition of the released vapor $(W_V, C_V,$ and 0) is found.

$$C_V = C_1 - S_1(dC_1/dS_1) \qquad (1.2)$$

and

$$W_V = 1 - C_1 + S_1(dC_1/dS_1) \qquad (1.3)$$

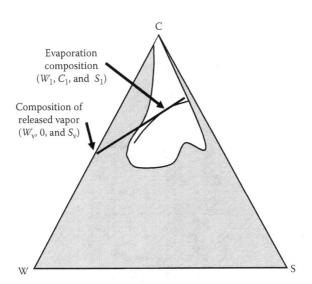

FIGURE 1.16   System used is the W/O microemulsion base in which the cosurfactant is a volatile compound, illustrating the behavior of a hydrocarbon.

This is the composition of the released vapor as obtained from experimental results. These are inherently affected by the RH of the receiving atmosphere with the actual vapor phase including the indigenous atmospheric water (W_{RH}).

$$W_V = 1 - C_1 + S_1(dC_1/dS_1) + W_{RH} \tag{1.4}$$

Realizing that the contribution from the released microemulsion vapor to the total water in the vapor is proportional to its evaporation rate and assuming ideal behavior of the vapor, the weight fractions of the two volatile compounds in the released vapor become

$$C_V = P_C M_C/[P_C M_C + M_W\{P_W(\mu Em) - 0.01RHP_W(0)\}] \tag{1.5}$$

and

$$W_V = M_W\{P_W(\mu Em) - P_W(0)\}/[P_C M_C + M_W\{P_W(\mu Em) - 0.01RHP_W(0)\}] \tag{1.6}$$

where
   P is pressure
   M is molecular weight

   As a contrast the *equilibrium values* for the receiving atmosphere are

$$C_V = P_C M_C/[P_C M_C + M_W P_W(\mu Em)] \tag{1.7}$$

and

$$W_V = M_W P_W(\mu Em)/[P_C M_C + M_W P_W(\mu Em)] \tag{1.8}$$

The combination of these equations with the phase diagram offers a great deal of insight on the evaporation [17]. A few salient points will be examined here. At first, the indigenous vapor pressure of water in the atmosphere will exceed the vapor pressure from the microemulsion for sufficiently high values of RH. As a consequence, the evaporation of water is now reversed; it is absorbed into the microemulsion liquid, as quantified by the general relation of the ratio between evaporating water and volatile organic compound, a first-order linear equation:

$$R = [P_C M_C + M_W\{P_W(\mu Em) - 0.01RHP_W(0)\}]/P_C M_C \tag{1.9}$$

The reversal of direction happens when the RH exceeds the value obtained in Equation 1.10:

$$RH > P_W(\mu Em)/0.01P_W(0) \tag{1.10}$$

Figure 1.17 illustrates this change of course in the phase diagram (negative C_V).

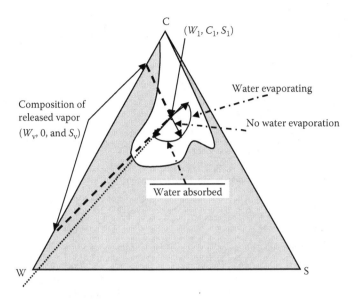

FIGURE 1.17 Phase diagram illustrating the change in water evaporation direction (arrows from the evaporation composition, W_1, C_1, and S_1) with increased RH. The evaporation changes from water evaporating to water being absorbed into the liquid from the atmosphere.

With realistic values of the vapor pressures and the molecular weights, such as

$P_C = 0.1$
$P_W(0) = 20$
$P_W(\mu Em) = 10$
$M_W = 18$
$M_C = 150$

Equation 1.9 becomes

$$W/C = 12 - 0.24 RH \tag{1.11}$$

and the reversal takes place at an RH of 50%. The RH has another critical point representing the value at which the reduction in the entire weight of the microemulsion with evaporation is reversed. The general expression is

$$RH = (M_W P_W(\mu Em) + P_C M_C)/0.01 M_W P_W(0) \tag{1.12}$$

This represents a slightly greater value than that for the reversal of the water transport direction in Equation 1.10 with a numerical value of the RH in this case being 54%. In what may be considered an oddity, this value is also the one at which the value of C_V for the released vapor instantaneously goes from positive to negative infinity with increased RH.

1.3 CONCLUSION

Phase diagrams are shown to provide valuable information on the role that structure of the surfactant, cosurfactant, and oil plays in determining the properties of the system at any composition. In addition, it is demonstrated that degree of order/disorder of lamellar liquid crystalline phases within a system stabilized by an ionic surfactant is determined by the chemical structure of the components and this in turn influences the magnitude, the nature (O/W, bicontinuous, W/O), and the location of the microemulsion phases. When nonionic polyethylene glycol ether surfactants are used instead of ionic ones, the temperature of the system is the crucial determining factor, with optimum properties at the HLB temperature of the surfactant. Finally, it is shown that straightforward algebraic analysis can be used to determine the composition of the evaporating gas from the microemulsion under practical use conditions by taking into account the atmospheric RH, utilizing the vapor pressures of the volatile components, molecular weights, and extrapolations from the phase diagram.

SYMBOLS AND TERMINOLOGIES

W	water
O	oil
C	cosurfactant
H	hydrocarbon
A	alcohol
μEm	microemulsion
RH	relative humidity
P	pressure
M	molecular weight

REFERENCES

1. Gillberg, G., Lehtinen, H., and Friberg, S. E. (1970). IR and NMR investigation of the conditions determining the stability of microemulsions, *J. Colloid Interf. Sci.* 33, 40–49.
2. Israelachvili, J. N., Mitchell, D. J., and Ninham, B. W. (1976). Thermodynamics of amphiphilic association structures, *J. Chem. Soc. Faraday Trans. II* 72, 1525–1533.
3. Lindman, B. and Friberg, S. E. (1999). Microemulsions—a historical overview, in Mittal, K., Kumar, P. (Eds), *Handbook of Microemulsions Science and Technology*, Marcel Dekker, New York, pp. 1–12.
4. Ekwall, P. (1975). Composition, properties and structures of liquid crystalline phases in systems of amphiphilic compounds, in Brown, G. H. (Ed) *Advances in Liquid Crystals*, Academic Press, New York, pp. 1–139.

5. Hoar, T. P. and Schulman, J. H. (1943). Transparent water-in-oil dispersions: The oleopathic hydro-micelle, *Nature* 152, 102–103.
6. Sjoeblom, E. and Friberg, S. (1978). Light-scattering and electron microscopy determinations of association structures in W/O microemulsions, *J. Colloid Interf. Sci.* 67, 16–30.
7. Prince, L. M. (1969). A theory of aqueous emulsions II: Mechanism of film curvature at the oil/water interface, *J. Colloid Interf. Sci.* 29, 216–221.
8. Schulman, J. H., Stoeckenius, W., and Prince, L. M. (1959). Mechanism of formation and structure of micro-emulsions by electron microscopy, *J. Phys. Chem.* 63, 1677–1680.
9. Friberg, S. E., Brancewicz, C., and Morrison, D. (1994). O/W microemulsions and hydrotropes: The coupling action of a hydrotrope, *Langmuir* 10, 2945–2949.
10. Friberg, S. E. (1997). Hydrotropes, *Curr. Opin. Surf. Colloid Sci.* 2, 490–494.
11. Friberg, S. E. and Buraczewska, I. (1978). Microemulsions in the water–potassium oleate–benzene system, *Prog. Colloid Polym. Sci.* 63, 1–9.
12. Bourrel, M. and Schechter, R. S. (1988). *Microemulsions and Related Systems,* Marcel Dekker, New York.
13. Bauduin, P., Touraud, D., and Kunz, W. (2005). Design of low-toxic and temperature-sensitive anionic microemulsions using short propyleneglycol alkyl ethers as cosurfactants, *Langmuir* 21, 8138–8145.
14. Stubenrauch, C., Frank, C., Strey, R., Burgemeister, D., and Schmidt, C. (2002). Lyotropic mesophases next to next to highly efficient microemulsions: A ^{2}H NMR study, *Langmuir* 18, 5027–5039.
15. Shinoda, K. and Friberg, S. E. (1986). *Emulsions and Solubilization,* John Wiley & Sons, New York.
16. Friberg, S. E. (2006). Weight fractions in three-phase emulsions with an Lα phase, *Colloids Surf. A.* 282/283, 369–376.
17. Friberg, S. E. and Aikens, P. A. submitted to *J. Disp. Sci. Tech.*

2 Physicochemistry of W/O Microemulsions: Formation, Stability, and Droplet Clustering

Animesh Kumar Rakshit and Satya Priya Moulik

CONTENTS

2.1 INTRODUCTION

Microemulsions are microheterogeneous, thermodynamically stable mixtures of oil and water. Here, the term "oil" means any water insoluble organic liquid. Macroscopically, they are homogeneous systems. Such oil/water disperse systems were known for a long time as there were some commercial floor cleaning products available in the American market at the turn of the twentieth century.

It was Schulman and his group at the Columbia University who first scientifically described microemulsion in 1943 [1] though the concept was there in the patent literature in mid-1930s [2,3]. The term "microemulsion" was first coined in 1959 by Schulman and his group [4]. Prior to that, different, terms like transparent emulsion, swollen micelle, micellar solution, and solubilized oil were used for such systems. Winsor [5] also separately developed such thermodynamically stable systems. Microemulsions are generally of low viscosity containing oil, water, and an amphiphile that brings down the water/oil interfacial tension (IFT), γ, to a very low value. Originally, it was thought that there exists a negative IFT which imparts stability to microemulsion [6]. Now, it is accepted that the IFT between oil and water is reduced to a very low value by the presence of an amphiphile. But there are many instances though, where the amphiphiles do not bring the IFT down to the required very low value and some short chain alcohols or amines need to be added to obtain the required IFT for the formation of a stable microemulsion. This means that in most cases the microemulsions are four component systems. These are water, oil, surfactant, and a short chain substance called a cosurfactant. Aerosol OT (AOT) is an interesting double chain surfactant, which can conveniently form a three component microemulsion system without a cosurfactant [7].

Microemulsions show a variety of structures. These are globular, bicontinuous (noodle like), cubic, or lamellar. The basic difference between emulsion and microemulsion is the fact that in the former the droplet size is in the region of micrometer whereas in the latter it is much smaller (<100 nm). There can be some systems where the droplet sizes lie in between, and such systems are called miniemulsions. Both miniemulsions and emulsions are not thermodynamically stable but they are kinetically stable. This means that the stability of a microemulsion formed under a given condition of temperature and pressure is time independent whereas the stability of a formed emulsion or miniemulsion under a given condition is a function of time. In such a case, the rate of coalescence of the droplets may be slow but is of finite magnitude and certainly is not negligible. However, there is a confusion in the literature as to the nature of microemulsion vis-à-vis emulsion, miniemulsion, nanoemulsion, submicron emulsion, etc. Broadly speaking, miniemulsion, nanoemulsion, and submicron emulsion are three names for the same system. All of them are kinetically stable. They are better stable than emulsions with less creaming as the droplet sizes are much smaller. Further, for the formation of emulsion, the requirement of surfactant is low (about 1%–2%) whereas for microemulsion the surfactant required is around 20% or more. The formation of microemulsion is a spontaneous process, requiring no energy or a very small amount of energy, whereas for emulsion and miniemulsion the energy requirement is appreciable. It should be noted that there may arise a confusion regarding the difference between microemulsion and a solution. Cosolvents like short chain alcohols, which are miscible with water, do not form microemulsion. There, the water and alcohol remain molecularly dispersed. In the microemulsion, the dispersed droplet has a definite boundary (the oil/water interface) where the amphiphile remains adsorbed to impart stability. Scattering methods like x-ray,

SANS, DLS, etc. can distinguish a microemulsion from a solution. The molecularly dispersed solutions do not have microstructure, but microemulsions have. Emulsions, on the other hand, are nontransparent.

Surfactants, particularly nonionic surfactants, can form reverse micelles in organic media. But the organic media must not be completely dry. If the medium is completely dry, then reverse micelles do not form. The reverse micelle is some what like water-in-oil (w/o) microemulsion. There, amount of water present is significantly low that satisfies the hydration of the hydrophilic head group. This water is not free, and its properties are different from normal water. If the amount of water present exceeds to that required for hydration of the hydrophilic head group of the amphiphile, then there will be both free and bound water in the water pool of the microemulsion. By NMR and calorimetric methods, presence of three types of water has been envisaged [8,9]. Such a system is considered to be microemulsion and not reverse micelles. The distinction between reverse micelle and w/o microemulsion is not sharp (Figure 2.1). A parameter ω, which is the mole ratio between water and surfactant, has been defined. Some workers suggest that systems with $\omega < 10$ are reverse micelles, microemulsions require to have $\omega > 10$. There are others who consider the threshold value of ω to be 15 [10].

Microemulsions, in general, exist in equilibrium with either excess oil or excess water or both. Winsor [5] has classified these different types of systems. When oil-in-water (o/w) microemulsion is in equilibrium with excess oil, it is

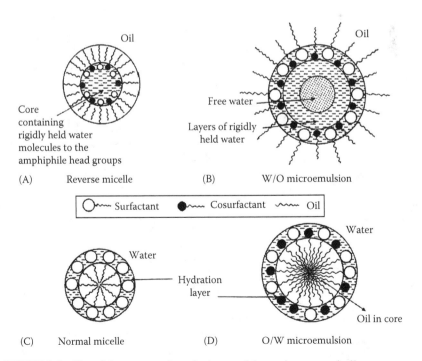

FIGURE 2.1 Pictorial representation of microemulsion and reverse micelles.

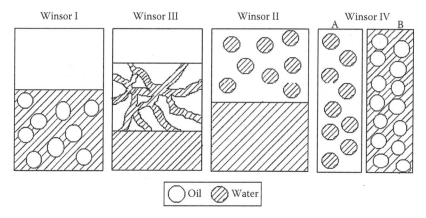

FIGURE 2.2 Different forms of Winsor structures. (From Moulik, S.P. and Rakshit, A.K., *J. Surf. Sci. Technol.*, 22, 159, 2006. With permission.)

known as Winsor I. The w/o microemulsion in equilibrium with excess water is called Winsor II. A microemulsion maintaining equilibria with both oil and water is called Winsor III. This is also termed as bicontinuous middle phase microemulsion in which both w/o and o/w dispersions remain simultaneously present. It is possible to get a microemulsion system which is not in equilibrium with either oil or water. This is known as Winsor IV system. All these four systems are illustrated in Figure 2.2. The bicontinuous structure is an extensive mutually intertwined one. There is not much of a curvature revealed from freeze–fracture electron microscopic studies (Figure 2.3) [11,12].

FIGURE 2.3 Freeze–fracture transmission electron micrographs of a biocontinuous microemulsion consisting of (A) 37.2% *n*-octane, 55.8% water, and surfactant pentathylene glycol dodecyl ether (From Vinson, P.K., Sheehan, J.G., Miller, W.G., Scriven, L.E., and Davis, H.T., *J. Phys. Chem.*, 95, 2546, 1991. With permission.) (B) 43.05% n-dodecane, 43.05% water and 13.9% didodecyl-methylammonium bromide. (From Jahn, W. and Strey, R., *J. Phys. Chem.*, 92, 2294, 1988. With permission.) In both cases 1 cm ≈ 2000 Å.

Aromatic oils are quite often used in the microemulsion preparation. Organic liquids like benzene, toluene, hexane, cyclohexane, etc. have been used [13–17]. Such preparations are useful in chemical and agrochemical industries. In the field of pharmaceutical industries, such oils are not usable. The oils useful in pharmaceutical industries are generally of higher molecular weights and molar volumes, and they are also polar. Such properties make microemulsion formation more difficult [18]. But there have been attempts to overcome such problems [19], and a good number of reports on biocompatible microemulsions can be found in literature [20–22].

In the formation of microemulsions, both ionic and nonionic surfactants are used. Cosurfactants are alcohols or amines [1,5,6,13,14]. It has been shown [23] that straight chain amines are quite different from their corresponding alkanols as cosurfactant. For example, butylamine is a more effective one on mass basis than triethylene glycol monobutyl ether. It is because the primary amine head group is more hydrophilic than alcohol, nitrile, carboxylic acid, ketone, and aldehyde head groups. In the case of amine cosurfactants, the addition of acid makes the cosurfactant more hydrophilic whereas the addition of base makes it less hydrophilic. The relative degree of hydrophilicity at the oil/water interface determines the volume of microemulsion formation. Microemulsions represent complex phase behavior, and the chemical structure of the cosurfactant has a pivotal role to play on their phase behaviors.

2.2 BASICS OF FORMATION

The otherwise immiscible oil and water are made to mix by the action of amphiphiles, and this scientific process is as well an art. The immiscibility arises due to very high IFT between water and oil. From thermodynamic consideration, at a constant pressure and temperature, the IFT, $\gamma = (\delta G/\delta A)T,P$. Since γ is positive, the Gibbs free energy change is also positive, and hence the mixing fails. Therefore to make the free energy change negative, the IFT or γ requires to be reduced to a very low value. Addition of surfactant and cosurfactant helps achieve this goal. The associated work process is given by the following energetic relation,

$$\Delta G = \Delta H - T\Delta S + \gamma \Delta A \tag{2.1}$$

where ΔG, ΔH, ΔS, T, and ΔA are the Gibbs free energy change, enthalpy change, entropy change, temperature in Kelvin, and change in interfacial area, respectively. The enthalpy change when immiscible oil and water are mixed is negligible. Now, as the droplet size decreases, there is a positive change in entropy ($T\Delta S \gg \gamma \Delta A$) causing negative ΔG for the system. Thus, the dispersion o/w or w/o becomes spontaneous and stable. It was thought earlier that the IFT becomes zero or even negative for the spontaneous formation of microemulsion [6]. This concept has been later changed [24]. By the action of surfactant, a constant but moderate IFT corresponds to a monolayer formation at the oil/water interface results. The addition of a cosurfactant brings down the IFT further to a very low value. For a multicomponent system, the change in IFT (γ) can be expressed by the relation

$$\partial \gamma = -\sum_i \Gamma_i d\mu_i = -\sum_i \Gamma_i RT d \ln C_i \qquad (2.2)$$

where γ_i, μ_i, and C_i are the Gibbs surface excess, chemical potential, and the concentration of the ith component, respectively, and R and T have their usual significance.

The integrated form of Equation 2.2 for a two component system is then [25,26]

$$\gamma - \gamma_o = -RT \left[\int_0^{C_1} \Gamma_1 d \ln C_1 - \int_0^{C_2} \Gamma_2 d \ln C_2 \right] \qquad (2.3)$$

where Γ_1 and Γ_2 are the surface excesses of the component 1 and 2, respectively at their concentrations C_1 and C_2, and γ_o is the IFT between oil and water in absence of surfactant and cosurfactant.

We have seen earlier that the microemulsion formation is a spontaneous process which is controlled by the nature of amphiphile, oil, and temperature. The mechanical agitation, heating, or even the order of component addition may affect microemulsification. The complex structured fluid may contain various aggregation patterns and morphologies known as microstructures. Methods like NMR, DLS, dielectric relaxation, SANS, TEM, time-resolved fluorescence quenching (TRFQ), viscosity, ultrasound, conductance, etc. have been used to elucidate the microstructure of microemulsions [25,26].

2.3   STABILITY OF MICROEMULSION

The thermodynamics of microemulsion discussed in the beginning of the chapter has accounted for the basic conditions required for the formation and stability of reverse micellar systems. The energetics of formation in terms of Gibbs free energy, enthalpy, and entropy need to be quantified with reference to the system composition and the droplet structures. For the formation of w/o system, a simple method called "dilution method" can extract energetic information for many combinations along with the understanding of their structural features. The method has been amply studied and presented in literature [4,27–32]. We, herein, introduce and present the method with basic theory and examples.

2.3.1   METHOD

For obtaining the information about microemulsion structure, the knowledge of concentration of surfactant and cosurfactant in oil and at the interface is a requirement. Various techniques like SAXS, SANS, IFT, DLS, conductance, viscosity etc. have been used to study the microstructure of microemulsion. The dilution experiments have been found to be very useful and convenient [13,28,33,34]. It is generally considered that the surfactants remain present at the interface between the dispersed droplets and the liquid continuum. However, the cosurfactants (generally short chain alcohols or amines) remain distributed between water, oil, and interface depending on their solubility in these regions. When surfactant and water concentrations are fixed, a threshold concentration of cosurfactant is needed for the formation of a stable microemulsion. The size of the microemulsion droplets becomes a function of the cosurfactant amount. The more the cosurfactant, the lower is the droplet size. There is always a bending stress associated with droplet formation. Short chain surface active cosurfactant molecules can easily snuggle themselves among the surfactant molecules at the droplet/liquid interface whereby releasing the bending stress. The cosurfactant, as we mentioned earlier, gets distributed between oil and the interface, and this distribution coefficient is quite difficult to obtain directly though it is an essentially physicochemical component for the formation and stability of a microemulsion. The dilution method has been developed and used to determine the composition of the interphase in a w/o microemulsion system, i.e., the concentrations of surfactant and cosurfactant therein.

In this method, fixed quantities of surfactant and water are taken in a container to which a known amount of oil is added. The contents are then mixed well and titrated with a cosurfactant until a single phase solution is obtained indicated by the total system clarity. The amount of cosurfactant added is then noted. The system is then destabilized by adding oil into it, and restabilized by tritation of it with cosurfactant for the second time and noting its amount again. This procedure of destabilization by adding oil and restabilization by titrating with cosurfactant is repeated for several times with noting their amounts at each protocol step. The whole process is performed at a constant temperature. At the point of clarity, the total number of moles (n) of the cosurfactant (an alkanol) in the system is given by

$$n_a = n_a^i + n_a^w + n_a^o \qquad (2.4)$$

where the superscripts i, w, and o represent the interphase, water, and oil phases, respectively. In terms of the experimental protocol, the ratio between the number of moles of oil and that of the cosurfactant is fixed. The droplet dimensions remain the same, only they get diluted in each step. We can then write

$$k = n_a^o/n_o \qquad (2.5)$$

where n_a^o, n_o, and k are the number of moles of alkanol in oil, the total number of moles of oil and a constant, respectively.

From Equations 2.4 and 2.5 we get

$$n_a = n_a^i + n_a^w + kn_o \qquad (2.6)$$

Dividing all through by n_s (normalization), the equation becomes

$$n_a/n_s = (n_a^i + n_a^w)/n_s + k(n_o/n_s) \qquad (2.7)$$

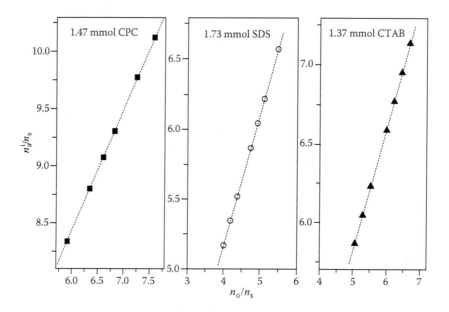

FIGURE 2.4 Plot of n_a^i/n_s versus n_o/n_s for w/o microemulsion systems with 27.8 mmol of water and 7.4 mmol of IPM stabilized by butan-1-ol and CPC, SDS, and CTAB at 305 K. (From Hait, S.K. and Moulik, S.P., *Langmuir*, 18, 6736, 2002. With permission.)

The (n_a/n_s) is thus linearly related with (n_o/n_s) with slope and intercept k and $(n_a^i + n_a^w)$, respectively. The value of n_a^i can be computed from the intercept (I) if n_a^w is known at a given temperature or determined by a separate experiment. The constant k is system specific. The nature of validity of Equation 2.7 is illustrated in Figure 2.4. In literature, such nice linear correlations are found for various systems with a single or mixed surfactants [32,35–41].

The distribution coefficient of the alkanol (cosurfactant) between oil and interphase in the case of w/o microemulsion in terms of mole fractions of the components can be written [36] as

$$K_d = X_a^i/X_a^o = [n_a^i/(n_a^i + n_s)]/[n_a^o/(n_a^o + n_o)] \qquad (2.8)$$

$$= [n_a^i(1+k)]/k[(1+I)n_s - n_a^w] \qquad (2.9)$$

the terms above have been defined earlier. The n_a^w may be taken to be zero since the solubility of alkanols in water is very small, particularly for the representatives with more than four carbon atoms. For lower alkanol, such as butanol, its aqueous solubility reported in literature can be used. Under the above conditions, Equation 2.9 takes the form

$$K_d = I(1+k)/k(1+I) \qquad (2.10)$$

The k and I being known, the value of K_d can be readily obtained. As the number of moles of surfactant in the system, n_s, the number of moles of oil in the system, n_o, and the number of moles of alcohol in oil, n_a^o are all experimentally and independently obtainable, the number of moles of alcohol at the interphase and hence the composition of the interphase can be estimated.

Digout et al. [42] have used the above procedure in determining the interfacial composition and distribution coefficient for w/o microemulsion system comprising cetylpyridinium chloride (CPC), alkanols, water, and alkanes. Some results are presented in Table 2.1.

It is obvious from Table 2.1 that the number of moles of alcohol in the oil phase and at the interphase depends upon the temperature as well as the chain length of the alkanol. In the pentane/CPC/BuOH/water system, n_a^i decreased with increase in temperature, and n_a^o increased under the same condition. However, when BuOH

TABLE 2.1
Interfacial Composition and Distribution Coefficient of Several Microemulsion Systems at Different Temperatures

System	Temperature (K)	$10^3\, n_a^i$ (mol)	$10^3\, n_a^o$ (mol)	K_d	ΔH_t^o (kJ mol^{-1})	ΔS_t^o (J mol^{-1} K^{-1})
		Pentane/CPC/Alkanol/Water				
BuOH	283	4.52	11.10	7.2	−15.8	−40.0
	288	3.98	11.54	6.6	−17.9	−46.0
	298	3.37	12.59	5.4	−22.2	−60.0
PenOH	283	3.70	6.88	11.4	9.1	52
	288	3.79	7.07	11.1	−6.5	−2
	298	4.23	9.53	7.5	−38.8	−113
HexOH	283	3.32	6.12	16.5	−63.3	−200
	288	2.75	5.14	12.2	−29.5	−82
	298	3.37	6.79	14.1	40.0	157
		Heptane/CPC/Alkanol/Water				
BuOH	298	10.85	13.18	6.3	−26.3	−73
	308	8.72	12.30	5.1	−5.7	−5
	318	9.87	15.44	5.4	15.6	63
PenOH	298	8.28	18.48	7.6	18.0	77
HexOH	308	7.50	12.37	8.6	19.5	81
	318	8.6	14.08	9.8	21.1	85
	298	7.59	10.85	10.1	−28.8	−77
	308	6.51	10.18	8.7	−12.1	−21
	318	7.12	11.64	9.4	5.1	34

Source: Adapted from Digout, L.G., Bren, K., Palepu R., and Moulik, S.P., *Colloid Polym. Sci.*, 279, 655, 2001.

was replaced by PenOH then both n_a^i and n_a^o increased with temperature. Neverthe-less, in both the cases the K_d values decreased with increase in temperature. This meant that at least in these two systems the mole fraction of alcohol in the interphase declined with increasing temperature in the narrow studied range.

It is interesting to note that by replacing octane for pentane both n_a^i and n_a^o produced minimum at 308 K when alkanols were butanol, pentanol, and hexanol (data not shown). The K_d value on the other hand exhibited minimum at 308 K for both butanol and hexanol; in the case of pentanol, there was a regular increase in K_d. Digout et al. [42] also studied microemulsion forming systems with nonane and decane (data not shown) as oil, and observed almost similar types of behavior as found for systems with pentane and octane. This simply indicated that chain length of alkanols and hydrocarbon oils have a large say on the overall microemulsion formation and their structures.

Moulik et al. [35] have made a detailed study on the compositions and energetics of formation of several w/o microemulsion-forming systems with isopropyl myristate (IPM) as oil, butanol as cosurfactant and CPC, cetyltrimethyl ammonium bromide (CTAB), and sodium dodecyl sulfate (SDS) as surfactants. The study has shown that variations of n_a^i, n_a^o, and K_d were independent of the nature (i.e., whether cationic or anionic) of the surfactant. They have also shown that by keeping surfactant concentration constant and by changing the water amount, the variation of n_a^i and n_a^o with moles of water were similar though there were some differences in their absolute magnitudes except at higher ω. Such results are depicted in Figure 2.5.

From the Equation 2.6, as explained earlier, one can obtain the intercept and the slope, and can compute the ratio β. Thus

$$\beta = \text{intercept/slope} = (n_a^i + n_a^w)/(n_s k) = (n_a^i + n_a^w)n_o/(n_s n_a^o) \qquad (2.11)$$

The β values did vary fairly linearly with temperature for the above-referred studied systems. But both the intercepts and slopes have shown exponential dependence on the concentration of surfactants. The authors have suggested that the "dependence of β with temperature is a measure of the relative adjustment between the surfactant and cosurfactant molecules at the interface and in the continuous oil phase for the sake of stability" [35].

From the knowledge of K_d, it is possible to get the standard Gibbs free energy of transfer of alkanol from the organic phase to the interphase. The K_d is essentially an equilibrium constant, and, therefore, the free energy of transfer of alkanol from oil to the interphase (ΔG_t^o) is

$$\Delta G_t^o = - RT \ln K_d \qquad (2.12)$$

By determining K_d at different temperatures, it is possible to compute the entropy of transfer, ΔS_t^o, and hence the enthalpy of transfer ΔH_t^o by the following ener-getic relations.

$$-T \Delta S_t^o = -RT \ln K_d - RT^2 d \ln K_d/dT \qquad (2.13)$$

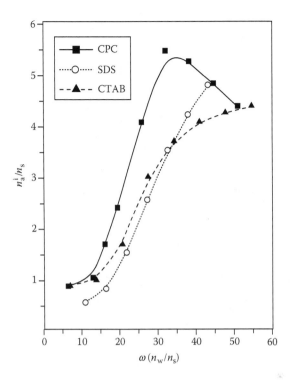

FIGURE 2.5 Plot of n_a^i/n_s versus ω for w/o microemulsion systems comprising 7.4 mmol of IPM stabilized by butan-1-ol and CPC, SDS, and CTAB at 305 K. (From Hait, S.K. and Moulik, S.P., *Langmuir* 18, 6736, 2002. With permission.)

and

$$\Delta H_t^\circ = RT^2 \, d \ln K_d / dT \qquad (2.14)$$

It has been observed that the ΔH_t° and ΔS_t° values obtained have shown striking compensation [31] correlations (Figure 2.6). The negative free energy values have suggested that the alkanol was spontaneously transferred from the oil phase to the interphase.

Since the ΔG_t° values were reasonably low, it has been considered that the mutual interaction between the surfactant and the cosurfactant was not strong at the interphase. For a large number of alkanol/surfactant systems, such small values of ΔG_t° have also been observed by Bansal et al. [43]. Moreover, for the C_4 alkanol (butanol), the transfer process was least spontaneous; higher the alkanol chain length favorable was their transfer to the interphase although the process was only marginal.

Although the dilution experiments have been reported on mixtures of a single surfactant and a cosurfactant, reports on mixed surfactants in presence of

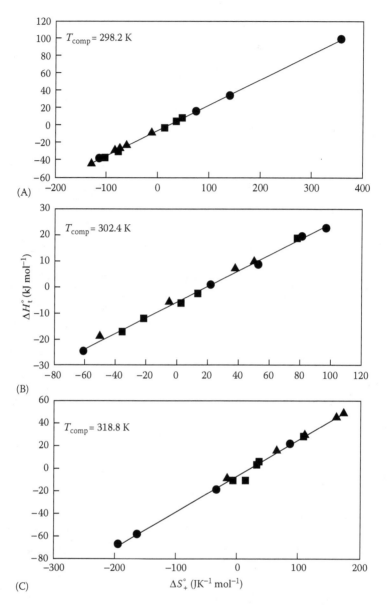

FIGURE 2.6 Enthalpy–entropy compensation plots for the water/CPC/alkane (C6, C9) microemulsions. ◆, 1-butanol; ▼, 2-butanol; ▲, 2-methyl-1-butanol; ●, 1-propanol; ■, 1-hexanol.

TABLE 2.2
Thermodynamic Parameters for the Transfer of Pentanol from Oil to the Droplet Interface at 303 K for Water/Bj58 + CTAB + Pentanol/Heptane (Decane) Systems

X_{Bj58}	ΔG_t^o (kJ mol^{-1}) Heptane (Decane)	ΔH_t^o (kJ mol^{-1}) Heptane (Decane)	ΔS_t^o (JK^{-1} mol^{-1}) Heptane (Decane)
0.0	−6.76 (−5.21)	−25.8 (−14.4)	−62.7 (−30.2)
0.2	−4.89 (−3.73)	−19.5 (−7.94)	−48.2 (−13.9)
0.4	−3.66 (−2.34)	−6.94 (−3.04)	−10.8 (−2.31)
0.6	−2.68 (−2.13)	4.23 (1.42)	22.8 (11.7)
0.8	−1.82 (−1.63)	15.1 (14.4)	55.7 (53.0)
1.0	−2.61 (−2.43)	16.1 (15.9)	61.8 (60.4)

Source: Adapted from Mitra, R.K., Pal, B.K., and Moulik, S.P., *J. Colloid Interf. Sci.*, 300, 755, 2006.

Note: Errors in ΔG_t^o, ΔH_t^o, and ΔS_t^o are ±3%, ±5%, and ±8%, respectively.

cosurfactants are very rare. In a recent report, Mitra et al. [38] have shown the use of mixed surfactants Brij-58 (Bj58) and CTAB in presence of two cosurfactants, butanol, and pentanol at different proportions of the two surfactants in the heptane and decane continuum. The analyses of the results have been made as per above theoretical rationale. A comprehensive presentation of their results is given in Tables 2.2 and 2.3. The mole fraction of Bj58 in the mixture was varied between 0 and 1.0 at different temperatures and the interfacial compositions and energetic parameters have been evaluated. The transfer of pentanol from oil to

TABLE 2.3
The n_a^i, n_a^0, and K_d Values for the Water/Bj58 + CTAB + (Butanol/Pentanol)/Heptane System at 303 K

X_{Bj58}	$10^4\ n_a^i$/mol (Bu/Pn)	$10^3\ n_a^o$/mol (Bu/Pn)	K_d (Bu/Pn)
0.0	2.36/8.16	2.52/0.68	3.32/19.3
0.2	2.56/6.03	3.38/1.70	1.83/6.95
0.4	2.56/7.32	4.18/3.22	1.95/4.27
0.6	4.52/6.21	6.22/4.72	2.00/2.90
0.8	3.59/4.14	6.86/5.62	1.63/2.06
1.0	5.59/6.53	6.04/5.01	2.27/2.82

Source: Adapted from Mitra, R.K., Pal, B.K., and Moulik, S.P., *J. Colloid Interf. Sci.*, 300, 755, 2006.

Note: Bu and Pn are butanol and pentanol, respectively. Error in K_d is ±5%.

the interphase has been observed to be more favorable than butanol, and the process declined with increasing proportion of Bj58 in the surfactant mixture. The Gibbs free energy of transfer was more spontaneous in heptane continuum than decane, and the spontaneity decreased with increasing proportion of the nonionic surfactant in the mixture. The process enthalpies were exothermic up to $X_{Bj58} = 0.5$, thereafter the enthalpy was endothermic. Its reflection on entropy yielded negative entropy of transfer up to $X_{Bj58} = 0.5$, which was afterwards positive. The magnitudes of these values were appreciably higher when heptane was the oil. The droplet surroundings' order/disorder depended on the type of the oil present in the system as well as the composition of the surfactant mixture. The difference between the oil was maximum in the absence of Bj58, which declined with increasing proportion of the nonionic surfactant, and in the absence of CTAB the differences of the energetic parameters between the two oil surroundings were minimal. Further elaborate studies are required to shed light on the significant influence of ionic surfactants in the process of w/o microemulsion formation in presence of nonionic surfactants. This appears to be a good area of further research in the understanding of basics of w/o microemulsion formation.

In addition to the understanding of droplet interfacial properties, the dilution method can as well shed light on the structural aspects of the system viz, droplets dimension, their population, amphiphile compositions at the interface, etc. Such information has been found to corroborate with results of DLS, SANS, SAXS, NMR, and other sophisticated techniques [25,26,44 and references therein]. The rationale behind such analysis along with typical results is presented in what follows.

The distribution of the amphilies at the w/o interface can lead to the determination of various structural parameters using a simplified but elegant structural model [35,37,42,45].

The w/o microemulsion droplets are assumed spherical and mono disperse with a surface monolayer (called the interphase) comprising closely arranged surfactant and cosurfactant molecules determined by the composition, temperature, and the system type. The total droplet volume V_d and their surface area A_d are then given by the relations

$$V_d = (4/3)\pi R_e^3 N_d \qquad (2.15)$$

$$A_d = 4\pi R_e^2 N_d \qquad (2.16)$$

The R_e and N_d are effective droplet radius and the total number of droplets per milliliter of solution; V_d is related to the volume of water (V_w), volumes of surfactant (V_s), and cosurfactant (V_a^i) at the interface. The V_d then can be represented as

$$V_d = V_w + V_s + V_a^i \qquad (2.17)$$

The knowledge of n_a^i gathered from the above discussed dilution protocol can be used to get V_a^i from the relation

$$V_a^i = n_a^i M_d / \rho_a \qquad (2.18)$$

where M_a and ρ_a are the molar mass and the density of the cosurfactant (alkanol), respectively.

An equivalent relation like Equation 2.16 can be used to evaluate V_s. The Equation 2.16 can also be written as

$$A_d = [n_s(A_s + IA_a)]N \qquad (2.19)$$

where A_s and A_a are the cross-sectional areas of the surfactant and the cosurfactant head groups, respectively, and N is the Avogadro number.

The R_e and N_d evaluation can be performed from the following relations:

$$R_e = 3V_d / A_d \qquad (2.20)$$

$$N_d = 3V_d / 4\pi R_e^3 \qquad (2.21)$$

The number of surfactant (N_s) and cosurfactant molecules (N_a) at the droplet interface is consequently related by the equations

$$N_s = n_s N / N_d \qquad (2.22)$$

and

$$N_a = n_a^i N / N_d \qquad (2.23)$$

The effective droplet radius (R_e) $(R_e = R_w + L$ [the thickness of the interfacial monomolecular amphiphile film]) is related with R_w by the relation

$$R_w = [(V_w + V_s^h + V_a^h)/V_d]^{1/3} R_e \qquad (2.24)$$

The head group volumes of the surfactant and cosurfactant V_s^h and V_a^h, respectively follow the relations

$$V_s^h = (4/3\pi^{1/2})A_s^{3/2} N_s \qquad (2.25)$$

and

$$V_a^h = (4/3\pi^{1/2})A_s^{3/2} N_a \qquad (2.26)$$

The structural parameters R_w, R_e, N_d, N_a, and N_s for water/CTAB-alkanol/IPM and water/SDS-alkanol/IPM systems at 305 K are presented in Table 2.4. The results

TABLE 2.4

Structural Parameters for 1/1/4 SDS/H$_2$O/IPM and CTAB/H$_2$O/ IPM Microemulsion Systems at 305 K

Alkanol	$R_w(R_e)$ (nm)	$10^{18} N_d$ (mL)	$N_s(N_a)$ Droplet	N_a/N_s
		SDS System		
Butanol	4.32 (5.61)	1.48	742 (707)	0.95
Pentanol	3.49 (4.84)	2.11	365 (833)	2.3
Hexanol	3.16 (4.70)	2.30	278 (900)	3.2
Heptanol	2.87 (4.62)	2.42	203 (988)	4.9
Octanol	2.89 (4.85)	2.09	212 (1107)	5.2
Nonanol	2.79 (5.03)	1.88	190 (1257)	6.6
		CTAB System		
Butanol	3.86 (5.19)	1.24	683 (399)	0.58
Pentanol	4.19 (5.60)	1.36	511 (682)	1.3
Hexanol	3.80 (5.30)	1.60	370 (831)	2.2
Heptanol	3.64 (5.27)	1.63	367 (802)	2.7
Octanol	4.07 (5.75)	1.26	469 (891)	1.9
Nonanol	3.89 (5.73)	1.27	421 (1000)	2.4

Source: Adapted from Mohareb, M.M., Palepu, R.M., and Moulik, S.P., *J. Disp. Sci. Technol.*, 27, 1209, 2006; for butanol from Hait, S.K. and Moulik, S.P., *Langmuir*, 18, 6736, 2002.

have evidenced that the droplet size decreases with increasing alkanol chain length because of their increasing surface activity. The N_d values for SDS derived systems were larger than CTAB-derived ones for the latter imparted better surface activity than the former. The N_a/N_s ratios were consequently more for the SDS systems than the CTAB formulations. On a comparative basis, more alkanol molecules have populated the droplet surface with their increasing size as evidenced from the results of N_a and N_s. Such structural properties for several other microemulsion systems have been presented in Table 2.5.

2.3.2 DROPLET DIMENSIONS

The composition-dependent conductance percolation of w/o microemulsion system (called volume percolation) has also been effectively studied to generate information on the internal structure of the system. Fang and Venable [46] have shown that the radius of the assumed spherical droplets of water (R_w) in w/o microemulsion can be calculated from the percolation results by the following relation.

TABLE 2.5
Structural Properties of Several W/O Microemulsion
Systems at 0.2 Weight Fraction Water at 293 K

System	$R_w(R_e)$ (nm)	$N_s(N_{cs})$	N_s/N_{cs}
H_2O/CTAB-Bu/Dc	4.45 (5.26)	241 (636)	0.38
H_2O/CTAB-Bu/Hp	4.52 (5.34)	257 (636)	0.40
H_2O/CTAB-HA/Dc	5.35 (6.27)	477 (475)	1.00
H_2O/AOT-Bu/Dc	5.68 (6.49)	493 (357)	1.38
H_2O/AOT-HA/Dc	5.56 (6.38)	474 (270)	1.76
H_2O/AOT-Bu/Hp	5.80 (6.61)	523 (340)	1.54
H_2O/SDS-HA/Dc	6.30 (7.25)	1057 (517)	2.04
H_2O/SDS-Bu/Hp	6.90 (7.85)	1360 (612)	2.22
H_2O/SDS-Bu/Dc	7.48 (8.47)	1781 (401)	4.44

Source: Adapted from Bisal, S.R., Bhattacharya, P.K., and Moulik, S.P., *J. Phys. Chem.*, 94, 350, 1990.

Note: Bu, butanol; HA, hexyl amine; Hp, heptane; Dc, decane.

$$R_w = 3[M_s + rM_{cs} + \{(W_w + W_{cs})/W_s\}M_s/\{N\rho_d(A_s + rA_{cs})\} \qquad (2.27)$$

where
  M stands for molar mass
  W is the weight fraction
  subscripts s and cs stand for surfactant and cosurfactant, respectively
  ρ_d and N are density of the dispersed phase, cross-sectional area, and
    Avogadro number, respectively
  $r(= N_{cs}/N_s)$ is called structural ratio (ratio between the number of molecules
    of s and cs at the droplet interface)

In addition, the effective radius of the water droplet (R_e), i.e., radius of the water core and interface can be computed from the following relation:

$$R_e = (3/4\pi)^{1/3}(4\pi R_w^3/3 + N_s M_s/\rho_s N + N_{cs} M_{cs}/\rho_{cs} N)^{1/3} \qquad (2.28)$$

where N_s, N_{cs}, ρ_s, and ρ_{cs} are aggregation number of surfactant per droplet, aggregation number of cosurfactant per droplet, densities of surfactant, and cosurfactant, respectively. Moulik et al. [45,47] have used the above relations to compute various quantities for w/o microemulsions. Cross-sectional areas of surfactants and cosurfactants, i.e., of CTAB and SDS. AOT, 1-butanol, and n-hexylamine required to be used are 0.505, 0.35, 0.678, 0.20, and 0.25 nm^2, respectively [45].

2.3.3 ENERGETICS OF DROPLET CLUSTERING

It is considered that the dispersed water droplets with amphiphile-coates in a w/o microemulsion come closer at a specific temperature or a specific composition and associate or cluster to augment easier transport (or transfer) of ions for the manifestation of the process of percolation. In the percolation stage, this process efficiently occurs. The underlying droplet clustering process must be controlled by the laws of energetics, whose evaluation is an important task from fundamental as well as application points of view. Considering the starting of droplet clustering as a change in phase, the standard Gibbs free energy (ΔG_{cl}^o) of the process can be written as

$$\Delta G_{cl}^o = RT \ln X_d \qquad (2.29)$$

where X_d is the mole fraction of the microdroplets corresponding to θ_{cl} at a constant ω. The X_d value can be found from the relation

$$X_d = n_d/(n_d + n_o) = A_t R_w M_o/[A_t R_w M_o + 3V_o \rho_o N] \qquad (2.30)$$

where
  n_d and n_o are the number of moles of droplet and oil, respectively
  A_t is the total cross-sectional area of the amphiphiles
  R_w is the water pool radius
  V_o, ρ_o, and M_o are the volume, density, and molar mass of the oil, respectively
  N is the Avogadro number
  R_w can be obtained from Equation 2.27

In the absence of a cosurfactant, Equation 2.27 changes to

$$R_w = 3[M_s + (W_w/W_s)]/N\rho_d a_s \qquad (2.31)$$

All the terms in Equation 2.31 are already defined. The enthalpy of clustering (ΔH_{cl}^o) can be obtained from the Gibbs–Helmholtz equation. Thus

$$d(\Delta G_{cl}^o/\theta_{cl})/d(1/\theta_{cl}) = \Delta H_{cl}^o \qquad (2.32)$$

The entropy change for the process (ΔS_{cl}^o) is then evaluated from the relation

$$\Delta S_{cl}^o = (\Delta H_{cl}^o - \Delta G_{cl}^o)/\theta_t \qquad (2.33)$$

The above energetic rationale for clustering of microemulsion droplets has been amply exploited by workers in this discipline [48–54]. To estimate ΔH_{cl}^o, microemulsion at a fixed ω was diluted with several portions of oil, and the θ_t for all the diluted preparations were determined from conductance measurements. The ΔG_{cl}^o values were calculated from the system compositions. The rest was graphical and analytical manipulations of the data. Moulik and Ray [54] have also measured the enthalpy of clustering directly by calorimetry, and have discussed the differences between these values and those evaluated by the method of percolation by conductometry working on water/AOT/heptane system. We reiterate that in the above-described rationale, it has been considered that the droplets in the clustered state retain their individual identity but exist in a different phase (a pseudophase-like micelles). The individual identities of the droplets during percolation has been supported by DLS measurements on samples before, at and above the percolation threshold [48,49,55]. In Table 2.6, energetic parameters for clustering for different systems in different environments have been presented to have a ready understanding of the basic physicochemical control over the phenomenon. For several studied systems ΔH_{cl}° has been found to have a sigmoidal dependence on ω. For water/AOT/heptane system in presence of sodium cholate (NaC) and sodium salicylate (NaSl), transitions at $35 > \omega < 40$ were observed [48,49,51].

TABLE 2.6
Energetic Parameters for the Clustering of W/O Microemulsions at 308 K

	System: H_2O/AOT/Hp		
ω	$-\Delta G_{cl}^{\circ}$ (kJ mol^{-1})	ΔH_{cl}° (kJ mol^{-1})	ΔS_{cl}° (J mol^{-1} K^{-1})
10	14.0	16.3 [5.40]	98
20	17.0	42.0 [47.0]	191
30	19.6	77.9 [71.4]	316
40	20.6	93.3 [117.4]	391

	System: H_2O/AOT/Dc; $\omega = 25$ in Presence of Additives			
Additive	θ_t/K	ΔG_{cl}° (kJ mol^{-1})	ΔH_{cl}° (kJ mol^{-1})	ΔS_{cl}° (J mol^{-1} K^{-1})
0	303.0	22.5	10.3	108
Urea (18)	299.4	22.2	13.2	118
NaC (10)	283.9	19.6	23.1	150
NaSl (10)	300.0	20.6	22.0	143
Rs (18)	291.8	21.7	14.3	123
Hq (20)	285.0	19.7	17.4	130
Pg (20)	288.7	20.0	23.1	150

Source: Adapted from Paul, B.K., Mitra, R.K., and Moulik, S.P., in *Encyclopedia of Surface and Colloid Science*, Somasundaran, P. (ed.), Taylor & Francis, Boca Raton, FL, 2006; Hait, S.K., Moulik, S.P., Rogers, M.P., Burke, S.E., and Palepu, R., *J. Phys. Chem. B.*, 105, 7145, 2001; Moulik, S.P. and Ray, S., *Pure Appl. Chem.*, 66, 521, 1994.

Note: Hq, hydroquinone; Rs, resorcinol; Pg, pyrogallol; NaSl, sodium salicylate; NaC, sodium cholate. Parentheted concentrations of additives for H_2O/AOT/Dc system were in mM. The ΔH_{cl}° values for H_2O/AOT/Hp system in third brackets were from calorimetry.

2.3.4 Dynamics of Dispersed Droplets

The nanodimensional droplets in microemulsion are in kinetic motion with a certain distribution of energy depending on the composition, and environmental conditions viz, temperature and pressure. By the TRFQ method, it has been demonstrated [56,57] that the droplets continuously associate and dissociate in a preparation. In such a sate, transfer or transport of materials may take place among the droplets. Mechanisms for such transfer process have also been proposed. This important phenomenon has been studied in detail and elaborated with theory and demonstrated by transport experiments like conductance and viscosity [58]. It may be mentioned here that the effects of temperature and water soluble polyacrylamide on the monophasic area of the pseudoternary phase diagram of the system cyclohexane/SDS/n-propanol/water have been studied. It has been observed that there exists a critical temperature below which the microemulsion zone increases with increase in temperature though at higher temperature the zone area decreases. Polyacrylamide also seems to have similar type of effect [59]. Herein, we describe and discuss at length the conductance method detailing with the microemulsion behavior and its consequences. The composition-dependent viscosity and pressure-induced conductance of microemulsions are briefly described. For their details, we refer to the literature citations [47,60].

2.3.5 Conductance of Microemulsions

Investigations on transport of ions in microemulsions can provide information on its internal dynamics. For o/w system, the continuum is aqueous and the conductance is fairly large, and very nominally depends on oil concentration. This is quite opposite for w/o systems, where conductance is appreciably dependent on over all composition or ω. Two mechanisms are under active consideration for such ion transport in w/o microemulsion systems. In the first (hopping mechanism), the droplets associate, and materials (essentially surfactant ions) hop from droplet to droplet manifesting large increase in conductance. In the second (fusion of droplets followed by their fission), droplets coalesce or fuse and transfer of ions takes place followed by fission of associated droplets [61,62]. Both the models are supported by experimental results but the second is considered more appropriate and useful. The aggregation and clustering of droplets in w/o microemulsion have been supported from conductivity changes with variation in both composition and temperature [47].

According to Winsor's classification, we have four types of microemulsion systems: Winsor I (WI), Winsor II (WII), Winsor III (WIII), and Winsor IV (WIV). The ion conductance of WI (o/w system) is reasonably high just like aqueous solution, where water is the dispersion medium whereas the conductance in WII (w/o system) is very low, where the dispersion medium is oil. In WIII systems, where both o/w and w/o dispersions are simultaneously present (the bicontinuous system) the conductance is reasonably large. It has been observed that conductance sharply rises after a threshold concentration of water in the system at a constant temperature or after a threshold temperature at a given concentration of water in the system. This phenomenon is known as "percolation." The concentration or temperature at which such a dramatic change occurs is called the "percolation threshold." Obviously, for w/o microemulsion system, there may be two percolation thresholds: one with respect to composition, and the other with respect to temperature. A pictorial presentation of the process is illustrated in Figure 2.7, wherein φ_t and θ_t are the composition- and temperature-related thresholds, respectively. Differentiation of the plots may locate the transition or threshold points in terms of peaks. Hait et al. [63] have proposed the use of Sigmoidal–Boltzmann equation for the estimation of the threshold values mathematically by way of computer fitting of the data points to the equation (Equation 2.34). The procedure is also equally applicable to get critical micellar concentration of surfactants, where the plots of physical properties of surfactant solutions are of Sigmoidal nature [47]

$$\log\sigma = (\log\sigma_i - \log\sigma_f)/[1 + \exp(\theta - \theta_t)/d\theta] + \log\sigma_f \qquad (2.34)$$

where σ and θ represent conductance and temperature, respectively, and the subscripts i, f, and t stand for initial, final and percolation state, respectively.

The conductance percolation phenomenon can be explained assuming a regular "lattice model." This lattice will have sites and edges, i.e., they will have vertices and bonds. One can randomly fill up the sites and edges and a point can arise in the system where a long range connectivity between various sites develops and a random network is formed. This point where suddenly the long range connectivity forms is known as the "percolation threshold." Both

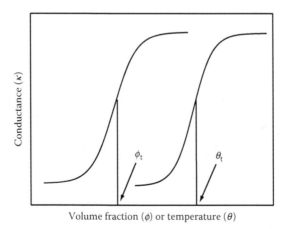

FIGURE 2.7 Nature of volume and temperature percolation of conductance. Scale: arbitrary.

the temperature and composition of the system can augment this long range connectivity and hence the "percolation threshold." The phenomenon of percolation is said to be of two types: (1) "static percolation" and (2) "dynamic percolation." By static percolation, we mean percolation in a system which is a mixture of insulators and solid conductors where the conductance is almost zero below the threshold condition (or connectivity). In the dynamic percolation (as in w/o microemulsion), the droplets are in motion with chances of collision which is very high at the threshold condition. To start with, therefore, conductance very slowly increases and shots up sharply at the threshold where the droplets efficiently associate or cluster. This clustering results in "fusion" followed by "mass transfer" and "fission." In a simplified model for w/o microemulsion, two droplets containing conducting ions are considered to fuse to form a dumbbell-shaped bigger drop. The conducting ion then can easily move from one partner into the other under the influence of the applied electric field. Subsequently, the bigger drop breaks into two individual droplets. These phenomena happen in a combined and concerted manner and hence the conductance efficiently rises as the conducting ions are getting a pathway for easier movement as has been depicted in Figure 2.8. The transfer of ions can also occur by hopping mechanism mentioned earlier [64]. But the "fusion-mass-transfer-fission" model (Figure 2.8) in a general way fits better to the dynamic percolation process [52,65]. During percolation, transient complex structures are formed. The single droplets relax with associated spatial rearrangement of clusters and charge carriers get transported along the clusters. A detailed analysis of these processes for ionic microemulsions of water/AOT/decane at $\omega = 26.3$ has been dealt with by Feldman et al. [66,67]. Uses of scaling equations for percolation, assessments of activation energy, and energetics of percolation have been studied by different workers. Also, the droplets' dimension, their polydispersity, and diffusion coefficients at the percolation stage have been determined by the light-scattering method. The results have been analyzed and rationalized in the light of physicochemistry.

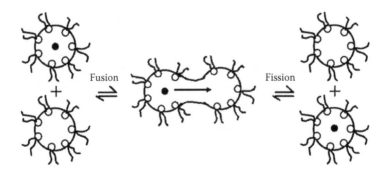

FIGURE 2.8 Dynamics of droplets in w/o microemulsion fusion, mass exchange, and fission leading to mass transfer and transport.

2.3.6 USEFUL PERCOLATION EQUATIONS

The temperature-induced percolation process has been found to obey a scaling equation of the following form

$$\sigma = p(\theta - \theta_t)^n \tag{2.35}$$

where
σ is the conductance of the system at temperature
θ, and θ_t is the threshold temperature
p and n are two-scaling constants

Equation 2.35 can be linearized in the form

$$\ln \sigma = \ln p + n \ln(\theta - \theta_t) \tag{2.36}$$

Hence, n and p values can be obtained from the linear plot of $\ln \sigma$ versus $\ln(\theta - \theta_t)$. The value of n is 1.9 for static percolation, which is expected also to hold for dynamic percolation [66,68]. However, it should be clearly noted that this value of 1.9 has been seldom found in practice.

The scaling law relating conductance and microemulsion composition at constant temperature is also similar to Equation 2.35. Thus

$$\sigma = \kappa(\phi - \phi_t)^\mu \tag{2.37}$$

where ϕ and ϕ_t are volume fraction of the dispersed phase and that at the percolation threshold, respectively. The exponent μ is a constant as is κ.

Interestingly, the value of μ has been suggested [66] to be the same as n, i.e., 1.9. Here too the experimental values rarely correspond to 1.9. In Table 2.7, the observed n and μ values obtained for different systems are recorded for a ready reference wherein much lower values than 1.9 may be found. Interestingly, the n and μ values obtained were seldom greater than the expected value. This was also found by Dogra et al. [69] where the n values were always lower than 1.9 (Table 2.8). In our opinion, microemulsions are soft systems, and inelasticity of droplet collision has a bearing on the lower values of the scaling exponents.

It may be mentioned here that the conductance of w/o microemulsions also becomes a function of the pressure. Boned et al. [70,71] have studied the effect of pressure up to 1000 bar on the conductance and viscosity of two microemulsion systems (a) water–AOT–undecane, and (b) glycerol–AOT–isooctane. The first system was an aqueous microemulsion, and the second was a nonaqueous one. The systems were also studied as a function of volume fraction, ϕ in the dispersion (ϕ = water plus AOT or glycerol plus AOT). In both systems, the percolation thresholds, ϕ_t were obtained. In aqueous microemulsion, as the pressure increased, the ϕ_t continuously decreased indicating higher droplet–droplet attractive interaction whereas in nonaqueous microemulsion there was almost no change in the ϕ_t values. It seems from the above that there was droplet cluster formation in aqueous

TABLE 2.7

Percolation Parameters and Activation Energy for Different W/O Microemulsion Systems in Presence of Additives at $W = 25$

		System: H_2O/AOT/ Hp		
Additive	θ_t (K)	ln P	N	E_p (kJ mol^{-1})
0	315.6	6.47	1.25	508
2-MP (10)	309.4	7.41	0.82	790
3-MC (10)	321.6	7.37	1.04	685
5-MR (10)	319.6	7.20	1.21	687
		System: H_2O/AOT/ Oc		
0	301.3	7.58	1.39	683
2-MP (20)	304.5	7.62	1.03	698
3-MC (20)	299.7	7.66	1.06	682
5-MR (20)	296.9	7.54	1.07	614
		System: H_2O/AOT/Dc		
0	296.3	6.07	0.93	580
2-MP (10)	291.8	7.68	0.88	1018
3-MC (10)	289.9	7.71	1.30	933
5-MR (10)	290.7	6.25	1.42	606
Hq (10)	289.3	5.16	1.51	629
Rs (10)	292.3	5.03	1.54	568
Pg (10)	291.5	5.74	1.47	648
Ct (10)	295.6	6.08	1.07	751
NaSl (10)	300.0	6.17	0.94	491
NaC (10)	283.9	5.99	1.09	491

Source: Adapted from Hait, S.K. and Moulik, S.P., *Langmuir*, 18, 6736, 2002.

Note: 2-MP, 2-methyl phenol; 3-Mc, 3-methoxy catechol; 5-MR, 5-methoxy resorcinol; Hq, hydroquinone; Rs, resorcinol; Pg, pyrogallol; Ct, catechol; NaSl, sodium salicylate; NaC, sodium cholate. Additive concentrations in millimole are shown in parenthesis.

systems whereas there was no such clustering in glycerol. They also have estimated two-scaling exponents, below and above the threshold pressure value of ϕ_t. The values were -1.2 and 2, respectively. However, these scaling exponents were independent of the system and pressure; thus they showed kind of universality.

The pressure-related conductance percolation follows the scaling equations:

$$\sigma = C_1(P)\sigma_1(P)[K(P_t)(P_t - P)]^\mu \quad \text{for } P > P_t + \delta_P$$
$$\sigma = C_2(P)\sigma_2(P)[K(P_t)(P - P_t)]^{-s} \quad \text{for } P < P_t - \delta_P'$$

(2.38)

where C_1 and C_2 are the prefactors, σ_1 and σ_2 are the conductances of the disperse phase, P_t is the threshold pressure, K is the pressure derivative of ϕ_t, and $(\delta_P + \delta_P')$

TABLE 2.8

Percolation Threshold, Scaling Law Parameters, and Activation Energies at the Percolation Threshold for Water-Induced Percolation of the Water/SDS + Myrj 45 (1:1)/Cyclohexane Microemulsion Systems in Different Conditions

Alkanol	Temperature (K)	ϕ_w^P	N	E_p (kJ mol^{-1})
n-PrOH	303	0.166 ± 0.006	0.742 ± 0.063	15.2
	313	0.113 ± 0.004	0.621 ± 0.022	
	323	0.061 ± 0.011	0.874 ± 0.069	
n-BuOH	303	0.208 ± 0.002	1.214 ± 0.079	18.0
	313	0.156 ± 0.038	1.201 ± 0.018	
	323	0.149 ± 0.041	0.998 ± 0.038	
n-PenOH	303	0.247 ± 0.008	0.642 ± 0.024	27.7
	313	0.190 ± 0.024	1.639 ± 0.038	
	323	0.172 ± 0.008	0.742 ± 0.149	
n-HexOH	303	0.257 ± 0.009	0.731 ± 0.095	35.6
	313	0.212 ± 0.048	1.221 ± 0.067	
	323	0.206 ± 0.016	0.909 ± 0.125	

Source: Adapted from Dogra, A. and Rakshit, A.K., *J. Phys. Chem. B.*, 108, 10053, 2004.

is the crossover regime or the translational interval which is of the order of $(\sigma_2 + \sigma_1)^{1/(\mu + S)}$. If K is independent of P, and if σ_1, σ_2, and C_2 are also independent of P, the equation simplifies to be a function, f of P as

$$\ln \sigma = f\left(\ln|P - P_t|\right) \qquad (2.39)$$

2.3.7 Conductance Percolation and EMT

The conductance of a microheterogeneous system like microemulsion can be quantitatively explained by "Effective Medium Theory" (EMT) [72,73]. The theory offers calculation of conductance, optical, and other properties of composites from the magnitudes of the pure components. The well-known Clausius-Mosotti equation is a good example of the use of EMT in optical properties [74]. A good amount of discussion on EMT is available [75,76]. It should be mentioned here that in general EMTs neglect correlations among various bonds. Also, the EMT could not be effectively used in microheterogeneous systems where metals and metal oxides are dispersed in a continuous medium. To overcome the problems, a modified EMT called EMTDD (EMT with dipole–dipole interaction) was suggested by Granqvist and Hunderi [77]. The dipole–dipole interaction is helpful in explaining the formation of chains and also clusters from droplets.

In a nonconducting dispersion medium, the EMT has been used by Bottcher [78]. The general form of the EMT equation is the following:

$$(\sigma - \sigma_m)/3\sigma = (\sigma_d - \sigma_m)\varphi/(\sigma_d + 2\sigma) \tag{2.40}$$

where σ, σ_m, σ_d, and φ are the conductivity of the solution, that of the continuous medium, that of the disperse phase, and the volume fraction of the disperse phase, respectively. The disperse phase has been assumed to consist of spherical entities. Assuming σ_m (conductance of the nonconducting medium) to be equal to zero, the above equation takes the scaling form.

$$\sigma = 1.5\sigma_d(\varphi - 0.333) \tag{2.41}$$

where 0.333 is termed as the threshold volume fraction, ϕ_t. It is less than 0.333 when the disperse phase is made up of nonspherical entities. Comparing with the scaling Equation 2.37, it is noticed that $\kappa = 1.5\sigma_d$, and the exponent $\mu = 1$. Bernasconi and Weismann [79] have developed and discussed various EMTs, and have calculated conductivities of site-disordered resistance networks. They have considered cluster of bonds and accommodated correlations between the values of bonds with a common site. It was observed that a general cluster EMT and a "disturbed neighborhood" treatment introduced made vast improvement over the single-bond EMT. For dilute and quasidilute systems, i.e., at low cluster concentration, EMT yields fairly accurate results.

Moulik et al. [80] further simplified the original EMT and EMTDD equations for various types of dispersed particles. They found that

For spheres,

$$\sigma = 1.5\sigma_d(\phi - 0.333) \tag{2.42}$$

For chains,

$$\sigma = -0.1519\,\sigma_d[0.1566\phi + 1.7216\phi^2 - 0.729\phi^3]\sigma_d \tag{2.43}$$

For clusters,

$$\sigma = -0.0984\sigma_d + [0.539\phi + 0.5679\phi^2]\sigma_d \tag{2.44}$$

In general, the EMTDD (cluster) equation was not valid; the EMT and the EMTDD (chain) formalisms evidenced good validity. Considering the clustering of droplets, Bernasconi and Wiesmann [79] proposed an EMT relation of the form

$$\sigma = 1.05\sigma_d(\phi - 0.157) \tag{2.45}$$

The percolation threshold of EMTDD for clusters ($\phi_c = 0.156$) is in exact agreement with Equation 2.44 which has been found to be valid up to $\phi_t \leq \frac{3}{4}$. For higher values of ϕ, the Equation 2.44 transforms into the form of EMT. It has been noticed that the systems that satisfy EMT and EMTDD (chain) have $r_{EMT} \geq 1$, while those that do not satisfy them have $r_{EMT} \leq 1$ (where r_{EMT} is the droplet radius derived from the EMT rationale). Further, in presence of a cosurfactant, the EMT and EMTDD (chain) theories are obeyed; without a cosurfactant both EMTDD (cluster) and Bernasconi–Wiesmann propositions stand valid. The data treatment has considered spherical droplets for arriving at the above conclusions. These spherical inclusions can undergo association at higher concentration forming clusters and chains with prolate and oblate geometry. The above analysis [80] has shown that water-induced percolation in w/o microemulsion may fit into both EMTDD (cluster) and BW theories, since the latter is a cluster extension of EMT. In presence of cosurfactant, the systems tend to fit into the EMT; for $r_{EMT} \gg 1$, it may also fit into the EMTDD (chain).

2.3.8 VISCOSITY PERCOLATION

The viscosity of microemulsion systems can have characteristic features. Dilute microemulsions obey Einstein's well-known viscosity relation, $\eta_r = 1 + 2.5\phi$ where η_r is the relative viscosity of the solution or dispersion, and ϕ is the volume fraction of the solute or the dispersed phase. However, at higher concentrations, this relation is not obeyed probably due to many body interactions. There is no accurate theory yet on viscosity of microemulsions but several semiempirical relations are available [81]. Salts like NaCl can increase the viscosity of nonionic amphiphile-derived microemulsions. In presence of NaCl, water SDS-propanol/cyclohexane w/o system has shown increase in viscosity with increasing amount of water; the diameter of the droplets increased leading to aggregation (and channel formation) ultimately to transform into a bicontinuous system [82]. The structural transitions from w/o to o/w in microemulsions with the addition of water may evidence maxima in the viscosity–ω profiles [82,83]. In the percolation mode, the increase in conductance should be followed with a decrease in viscosity with increase in ω of the w/o microemulsion system: also if there is a viscosity maximum there should be a conductance minimum in the respective profiles. However, such phenomena are not commonly observed. The viscosity and conductance can simultaneously increase because of efficient transport of ions through the formed channels and conduits in the existing structure. This is contrary to Walden rule (constancy of the product of viscosity and conductance) as observed by Moulik et al. and others [82–87] though there are reports where the rule seems to have been followed [88]. It should be mentioned here that the Curie–Prigogine principle [89] of irreversible thermodynamics suggests the invalidity of the Walden rule [81].

The changing of system composition affects the viscosity of microemulsions. An oil continuous system can change into a bicontinuous system and ultimately to a water continuous ensemble. Such changes yielding different structural organizations (states) become associated with distinct changes in viscosity. Therein, a percolation in viscosity obeying scaling law can be witnessed. Thus, in the prepercolation stage

$$\eta = A'(\phi_t - \phi)^{-S} \tag{2.46}$$

and in the postpercolation stage is

$$\eta = A(\phi - \phi_t)^{\mu} \tag{2.47}$$

where
  ϕ is the volume fraction of the dispersed phase
  ϕ_t is its percolation threshold
  μ and S are the scaling exponents
  A and A' are constants

These equations are obeyed, if $\phi > \phi_t + \delta'$, and $\phi < \phi_t + \delta''$, respectively where δ' and δ'' are the respective crossover regimes.

The equations in the logarithmic forms are normally used to get the constants and the exponents for the pre- and postpercolation stages. The values of the exponents, here again, do not tally with the expected value of 1.9; estimated values are on the whole lower as discussed earlier. Here, it would be interesting to note that when heptane is replaced by nonane in water/Brij-35 + propanol/heptane system, the viscosity decreases whereas in presence of NaCl the viscosity increases [82]. This is similar to the effect on cloud point. Presence of additives like polyethylene glycols (PEG) 400 increased the viscosity of microemulsion having CTAB. Frenkel–Eyring equation could be used to compute the activation energy of viscous flow [90].

2.3.9 ANTIPERCOLATION AND DOUBLE PERCOLATION

It is also possible to observe "antipercolation" phenomenon in microemulsion. In this case, the conductance is expected to decrease as the variables like temperature, pressure, and composition change. Tingey et al. [91] have shown the presence of such a phenomenon in water/DDAB (didodecylammonium bromide)/supercritical propane system as a function of pressure between 80 and 4000 bar. The conductance decreased by about three orders of magnitude. It was suggested that the interconnected channels as present in bicontinuous microemulsion broke into droplets and hence the conductance decreased. Ajith et al. [82] and John et al. [83] have reported that both nonionic [water/Brij-35/n-propanol/alkane] in presence of salt and ionic [water/SDS/n-propanol/cyclohexane] microemulsions in presence of salt showed reasonable temperature-induced percolation at high ω values at atmospheric pressure. But these systems manifested decrease in conductance at low ω as the temperature was increased though the decrease was not large. It was suggested that there was no interconnection among the droplets, and there was barrier for the ions to travel from one droplet to another.

It may be mentioned here that two stage (or double) percolation has been reported by Ray and Moulik [92] and Maiti et al. [93] for (water/AOT/decanol) and water/DTAB (octadecyltrimethylammonium bromide)-butanol/heptane microemulsion systems, respectively. The double percolation process for AOT/decane/NaCl (0.5%) was also reported by Eicke et al. by conductivity, viscosity, and electro-optical Kerr effect [94]. The two processes demarcated three structural regimes viz, o/w, w/o, and oil or water continuous.

2.3.10 ACTIVATION ENERGY FOR PERCOLATION

All physicochemical processes are temperature dependent, and require activation energies for their occurrence. According to Eyring's hypothesis, a transient complex or a state is required to attain with an energy (called the activation energy), which makes the process to proceed towards its required direction: higher the energy of activation required, lesser is the efficacy of the process. Conceptually, a barrier has to be overcome for the augmentation of a process, like a ball requiring energy to move uphill to roll down on the other side downhill, and higher the mountain height higher is the energy requirement for the ball to crossover. In the conductance percolation of w/o microemulsion, activation energy is required for the process to occur. For the "fusion-mass transfer-fission" mechanism kinetically there should be a rate determining step whose activation energy is the key-factor for the percolation process, and therein arises the need for research to probing into the transport process of percolation.

Both in the pre- and postpercolation stages, the temperature dependence of conductance can furnish with the information on the activation energies required for the conductance process to occur with some specific and involved fast and slow physicochemical courses. With the help of Arrhenius type equation, in a suitable form, the processing of data can evaluate the required activation energy of percolation, E_p. Thus

$$\sigma = a \exp(-E_p/RT) \qquad (2.48)$$

or

$$\ln \sigma = \ln a - E_p/RT \qquad (2.49)$$

In the equations a is a constant and the other terms have their usual significance. The E_p values can be conveniently evaluated from the use of the linear equation which has been done by different workers for their systems of percolating microemulsions both without and with additives. The values obtained are quite large, and are system and condition dependent. A generalization of the E_p values is yet to be found, the trends are system specific. For instance, it may not follow a trend with respect to additives either; there can be a maximum, a minimum, or a different mode of variation with changing concentration of an additive [48,49,95,96]. It has been observed that the percolation-resisting additives may produce lower E_p than percolation-assisting additives. This has led to the consideration that the "fission" of the fused droplets is the rate-determining process, whose activation energy ought to be the guiding principle for the overall conductance of a percolating microemulsion system [49,97]. The bridging or fusing of two droplets forming a doublet may be assisted by additives like bile salts, hydrotropes, etc. resulting in the transfer of ions from one to the other but it is their fission that propagates the process of the transfer to other droplets, and eventually energy of activation for fission controls the overall process. The fission of the doublets may require higher energy. It is, therefore, not how quickly the droplets associate but how strong the association that energetically guides the ion transport process during clustering. It has, however, been found that, at a fixed ω, although the E_p values do not show any trend with the chain length of the hydrocarbon oils but there may be a decreasing trend of E_p with the increasing concentration of the additive. Thus, in many instances percolation-resistive additives have yielded more or less comparable E_p values. The activation energy for the conductance of a salt solution is normally much smaller than E_p of dynamic percolation because of difference in the mechanism of ion conduction in the two distinctly separate processes. The above discussion is a gross and general assessment of the experimental results. For a quantified analysis, intricate considerations of steric and other interacting factors have to be incorporated, and adaptation of this is much wanted. A short account of results realized from different systems under varied conditions is presented in Table 2.7. Further, Dogra et al. [69] have studied the effect of change of cosurfactant in a mixed surfactant microemulsion system water/SDS + Myrj 45 + cosurfactant/cyclohexane. It can be seen from the Table 2.8 that the percolation activation energy E_p goes on increasing from 15.2 to 35.6 kJ mol^{-1} as cosurfactant was changed from n-PrOH to n-HexOH. Further, among the isomeric alcohols, the n-alcohols have lower percolation activation energy, i.e., in presence of n-alcohols the percolation is easier to happen in comparison to the other isomeric alcohols. From the data of n-alcohols, it was found that the percolation energy barrier changes by ~6.6 kJ mol^{-1} per CH$_2$ group.

2.3.11 ADDITIVE EFFECTS ON PERCOLATION

The phenomenon of conductance percolation of w/o microemulsion systems has been found to be susceptible to the presence of certain additives: both assistance and resistance to the phenomenon have been witnessed. The bile salts, cholate, and deoxycholate have been found to accelerate the process, and the former with three hydroxyl groups in the molecule was more effective than the latter with two hydroxyl groups. The salt dehydrocholate without any hydroxyl group was indifferent [50,96]. It has been proposed that the hydroxyl groups can efficiently couple two microemulsion droplets by way of making bridges through hydrogen bond formation for ready ion transport through the formed channels. In other words, cholate and deoxycholate can assist droplet-coupling (or fusion) which dehydrocholate without any hydroxyl group in the molecule cannot. This model has been further supported by the use of a number of hydrotropes which have produced distinct influencing effects on the conduction percolation of different types of w/o microemulsions [51]. It has been demonstrated that aromatic compounds with benzene ring in them viz, benzene, toluene, naphthalein, anthracin, etc. distinctly hindered or resisted the percolation process. They acted as barriers between two droplets to resist fusion or coupling and subsequent mass transfer [49,50,96,98–102]. Other additives like salts (both organic and inorganic classes) had minor effects on percolation; such influences originated from changes in the morphology, shape, and size of the dispersed phase in addition to the change in the polarity of the medium. Typical illustrations for specific additives, NaC and NaSl, on the assistance and resistance to "fusion-bridging-and-mass-transfer" model are presented in Figure 2.9. Polymers have been found to influence the percolation temperatures of different w/o microemulsions. In correlating the effect of triblock copolymer on the percolation temperature of water/AOT/isooctane system, two distinct ranges of [copolymer]/[nanodroplets] ratios were observed to be important on thermodynamic ground [102]. For nonionic systems, because of repulsive interaction between the polymer and the surfactant, only steric consideration is important. The comb-like polymer, PMOAVE [poly(maleic anhydride methyl vinyl ether)] was found to

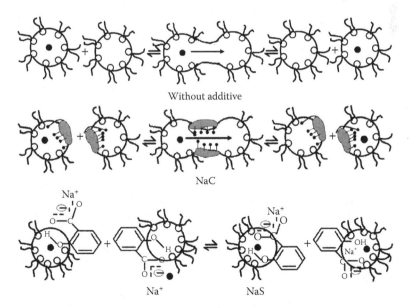

FIGURE 2.9 Normal, assisted, and resisted percolation with and without additives. normal, no additive; assisted, with NaC; resisted, with NaSl.

lower the threshold percolation temperature of water/AOT/isooctane system while a block copolymer (PBd-d-PEO) was found to increase the percolation temperature [98]. Electrolytes have been reported to retard the percolation temperature of w/o microemulsions [99].

The effects of salts, acids, alkali, guanidium chloride, etc. have also been studied [101,102]. Additives like urea, thiourea, ethylene glycol (EG), etc. have evidenced increase in micellar association and percolation of AOT-derived systems. The influence of water soluble biopolymers (proteins, enzymes, etc.) on the temperature-induced percolation of water/AOT/isooctane microemulsion has also been investigated [10,103,104]. The presence of cytochrome c favored the transition at lower volume fraction and temperature. The enzymes chymotrypsin and trypsin have increased the threshold with increasing enzyme concentration. The enzyme east alcohol dehydrogenase has increased the percolation temperature with distortion of the spherical structure of AOT-derived microemulsion system; the spherical structure was restored upon addition of bile salts [105]. The influences of EG and PEGs on the AOT- and IG (igepal)-derived microemulsion systems with different hydrocarbon oils have also been studied in detail [106] using conductometry and time-resolved electrical birefringence methods. The addition of EG induced percolation of IG systems, the effect of EG was contrary to almost all other systems studied. In the presence of PEG, the droplet structure of the systems was maintained. The dependence of concentration and molecular weight of PEGs has been explained in terms of interfacial adsorption of the polymers.

The rigidity/flexibility of the oil/water interface is a guiding factor for the connectivity of droplets and the transport or exchange of materials among them. For instance, cholesterol, a membrane stiffening agent, resists percolation whereas EG and urea assist it. Molecular packing at the interface is thus a determinant factor in the formation of aggregates and their connectivity. The presence of a second surfactant can modify the packing state of the interface. Addition of nonionic surfactant to AOT-stabilized reverse micelles decreases the interfacial rigidity to facilitate droplet clustering. Liu et al. [107]. have studied water/AOT/Brij/heptane systems and have rationalized their findings in the light of amphiphile packing. The decrease in the percolation threshold with increasing polar head group area of the nonionic surfactant has been explained on the basis of altered interfacial packing for the microemulsions comprising water/AOT/nonionics (Brijs, Spans, and Tweens)/IPM by Mitra and Paul [41,108]. For further information on the additive effects on droplet clustering and percolation, we refer to our earlier review [60]. The field is interesting and requires more exploration to frame a general rationale for the process of percolation in relation to the presence of additives under varied environmental conditions. Dogra et al. [69] studied the phase behavior and percolation of mixed surfactant microemulsion system water/SDS + Myrj 45/cyclohexane with different alcohols as cosurfactant. They showed that as NaCl concentration increases there is Winsor transition (WI $\rightarrow$ W III $\rightarrow$ W II), which follows the expectation from Sabatini equation [109]

$$C_r = c \left| \ln(m^*/m) \right| \tag{2.50}$$

where

C_r is the curvature

m^* and m are optimal salinity and electrolyte concentration, respectively

c is a proportionality constant

When $m < m^*$, Cr is positive which indicates o/w microemulsion. When $m > m^*$, C_r is negative indicating w/o microemulsion. It should be mentioned that optimal salinity is the concentration of salt in the microemulsion where amounts of oil and water in the microemulsion are exactly same, and these are experimentally determined. The optimum salinity seems to be a function of the HLB of the system, which changes with temperature.

ACKNOWLEDGMENTS

SPM thanks Jadavpur University for an Emeritus Professorship, and INSA, New Delhi for an Honorary Scientist position during the tenure of which the work was completed. AKR thankfully acknowledges the Emeritus Fellowship provided by AICTE, New Delhi and the necessary facilities extended by the authorities of WBUT, Kolkata for completion of the work. Authors thank Mr. Abhijit Dan for his help in the preparation of the manuscript.

SYMBOLS AND TERMINOLOGIES

ϕ	volume fraction
ϕ_t	threshold volume fraction
$(\delta', \delta'', \delta_p, \text{ and } \delta'_p)$	crossover regimes
A	constant
A'	constant
A_{cs}	cross-sectional area of cosurfactant head
A_d	droplet surface area in unit volume
AOT	aerosol OT (surfactant)
A_s	cross-sectional area of surfactant head
A_t	total cross-sectional area of amphiphile in unit volume
BuOH	n-butanol
C	concentration
CPC	cetylpyridinium chloride
C_r	curvature
CTAB	cetyltrimethylammonium bromide (surfactant)
I	intercept
IPM	isopropyl myristate (oil)
K	constant
K_d	distribution coefficient

L	length of surfactant molecule
m	electrolyte concentration
m^*	optimal salinity
M_a	molar mass of alkanol
M_{cs}	molar mass of cosurfactant
M_o	molar mass of oil
M_s	molar mass of surfactant
n	constant
N	Avogadro number
n_o	number of moles of oil
n_a^i	number of moles of alkanol at the interface
n_a^o	number of moles of alkanol in the oil
n_a^w	number of moles of alkanol at the interface
N_a	number of molecules of alkanol at the interface
N_{cs}	number of cosurfactant molecules at the droplet interface
n_d	number of moles of droplet
N_d	number of droplets per unit volume
N_s	number of surfactant molecules at the droplet interface
o/w	oil-in-water
OTAB	octadecyltrimethylammonium bromide
P	pressure
p	constant
PenOH	pentanol
r	N_{cs}/N_s
R	gas constant
R_e	effective radius of droplet
R_w	radius of water pool
S	surface area
SDS	sodium dodecyl sulfate
T	absolute temperature
V_a^h	volume of alkanol head group
V_a^i	volume of alkanol at the interface
V_d	droplet volume
V_o	volume of oil
V_s	volume of surfactant molecule
V_s^h	volume of surfactant head group
V_w	volume of water molecule
w/o	water-in-oil
W_{cs}	weight fraction of cosurfactant
W_s	weight fraction of surfactant
W_w	weight fraction of water
X	mole fraction
X_d	mole fraction of droplet
β	intercept/slope

Γ	Gibbs surface excess
γ	IFT
ΔG	free energy
ΔG_t^o	standard Gibbs free energy of transfer
ΔH	enthalpy
ΔH_t^o	standard enthalpy of transfer
ΔS	entropy
ΔS_t^o	standard entropy of transfer
θ	temperature
θ_{cl}	clustering temperature
θ_t	threshold temperature
μ_i	chemical potential of the ith component
ρ_a	density of alkanol
ρ_{cs}	density of cosurfactant
ρ_o	density of oil
ρ_s	density of surfactant
σ	solution conductance
σ_d	conductance of the disperse phase
σ_f	final conductance
σ_i	initial conductance
σ_m	conductance of the medium
ω	mole ratio between water and surfactant

REFERENCES

1. J.P. Hoar and J.H. Schulman 1943 Transparent (w/o) dispersions: Oleopathic hydro-micelle, *Nature* 152, 102–103.
2. V.R. Kokatnur 1935 U.S. Patent 2, 111, 000.
3. D. Bowden and J. Holmstine 1936 U.S. Patent 2, 045, 455.
4. J.H. Schulman, W. Stockenius, and L.M. Prince 1959 Mechanism of formation and structure of micro emulsions by electron microscopy, *J. Phys. Chem.* 63, 1677–1680.
5. P.A. Winsor 1948 Hydrotropy, solubilisation and related emulsification processes, *Trans. Faraday Soc.* 44, 376–398.
6. L.M. Prince 1977 *Microemulsions: Theory and Practice*, New York, Academic Press.
7. M.J. Hou, M. Kim, and D.O. Shah 1988 A light scattering study on the droplet size and interdroplet interaction in microemulsions of AOT–oil–water system, *J. Colloid Interf. Sci.* 123, 398–412.
8. A.N. Maitra 1984 Determination of size parameters of water–Aerosol OT–oil reverse micelles from their nuclear magnetic resonance data, *J. Phys. Chem.* 88, 5122–5125.
9. P.R. Majhi and S.P. Moulik 1999 Microcalorimetric investigation of AOT self asso-ciation in oil and the state of pool water in water/oil microemulsions, *J. Phys. Chem. B* 103, 5977–5983.
10. M.P. Pileni 1993 Reverse micelles as microreactors, *J. Phys. Chem.* 97, 6961–6973.
11. P.K. Vinson, J.G. Sheehan, W.G. Miller, L.E. Scriven, and H.T. Davis 1991 Viewing microemulsions with freeze–fracture transmission electron microscopy, *J. Phys. Chem.* 95, 2546–2550.

12. W. Jahn and R. Strey 1988 Microstructure of microemulsions by freeze fracture electron microscopy. *J. Phys. Chem.* 92, 2294–2301.
13. K.S. Birdi 1982 Microemulsions: Effect of alkyl chain length of alcohol and alkane, *Colloid Polym. Sci.* 260, 628–631.
14. H.N. Singh, Ch. Durga Prasad, and Sanjeev Kumar 1993 Water solubilization in micro-emulsions containing amines as cosurfactant, *J. Am. Oil Chem. Soc.* 70, 69–73.
15. C. Alba-Simionesco, J. Teixeira, and C.A. Angell 1989 Structural characterization of glass forming oil/water microemulsions by neutron scattering, *J. Chem. Phys.* 91, 395–398.
16. P. Fini, M.L. Curri, M. Castagnolo, F. Ciampi, and A. Agostiano 2003 Calorimetric study of CdS nanoparticle formation in water-in-oil microemulsion, *Mat. Sci. Eng.* C 23, 1077–1081.
17. C. Lai, S. Tang, Y. Wang, and K. Wei 2005 Formation of calcium phosphate nano-particles in reverse microemulsions, *Mat. Lett.* 59, 210–214.
18. R. Aboofazeli, N. Patel, M. Thomas, and M.J. Lawrence 1995 Investigations into the formation and characterization of phospholipids microemulsions. 4. Pseudoternary phase diagrams of systems containing water–lecithin–alcohol and oil—the influ-ence of oil. *Int. J. Pharm.* 125, 107–116.
19. M.J. Lawrence and W. Warisnoicharoen 2006 Recent advances in microemulsions as drug delivery vehicles, in *Nanoparticulates as Drug Carriers* ed. V.P. Torchilin, Imperial College Press, London, pp. 125–171.
20. S. Gupta and S.P. Moulik 2007 Biocompatible microemulsions and their prospec-tive uses in drug delivery, *J. Pharm. Sci.* 97, 22–45.
21. S. Watnasirichaikul, N.M. Davies, T. Rades, and I.G. Tucker 2000 Preparation of biodegradable insulin nanocapsules from biocompatible microemulsions, *Pharm. Res.* 17, 684–689.
22. J. Flanagan and H. Sing 2006 Recent advances in the delivery of food derived bioactives and drugs using microemulsions, in *Nanocarrier Technologies: Frontiers of Nanotherapy* ed. M.R. Mozafari, Springer, New York, pp. 95–111.
23. K.R. Wormuth and E.W. Kaler 1987 Amines as microemulsion cosurfactants, *J. Phys. Chem.* 91, 611–617.
24. K. Holmberg 1998 Quarter century progress and new horizons in microemulsions, in *Micelles, Microemulsions, and Monolayers Science and Technology* ed. D.O. Shah, Marcel Dekker Inc, New York, 161–192.
25. S.E. Friberg and P. Bothorel 1987 *Microemulsions: Structure and Dynamics*, CRC Press, Boca Raton.
26. S.P. Moulik and B.K. Pal 1998 Structure, dynamics and transport properties of microemulsions, *Adv. Colloid Interf. Sci.* 78, 99–195.
27. E. Sjoeblon and U. Henriksson 1984, *The importance of the alcohol chain length and the nature of the hydrocarbon for the properties of ionic microemulsion systems in Surfactants in Solution* ed. K.L. Mittal and B. Lindman, Plenum, New York, pp. 1867–1880.
28. S. Kumar, S. Singh, and H.N. Singh 1986 Effect of chain length of alkanes on water-in-oil microemulsions, *J. Surf. Sci. Technol.* 21, 85–91.
29. S.P. Moulik, W.M. Aylward, and R. Palepu 2001 Phase behaviours and conductivity study of water/CPC/alkan-1-ol (C_4 and C_5)/1-hexane water/oil microemulsions with reference to their structure and related thermodynamics. *Can. J. Chem.* 79, 1–12.
30. M. Giustini, S. Murgia, and G. Palazzo 2004 Does the Schulman's titration of microemulsions really provide meaningful parameters? *Langmuir* 20, 7381–7384.
31. M.M. Mohareb, R.M. Palepu, and S.P. Moulik 2006 Interfacial and thermodynamic properties of formation of water-in-oil microemulsions with surfactants (SDS & CTAB) and cosurfactants (*n*-alkanols C5-C9), *J. Disp. Sci. Technol.* 27, 1209–1216.

32. Y. Bayrak 2004 Interfacial composition and formation of w/o microemulsion with different amphiphiles and oils, *Colloids Surf. A* 247, 99–103.
33. G.J.M. Koper, W.F.C. Sager, J. Smeets, and D. Bedeaux. Aggregation in oil-continuous water/sodium bis(2-ethylhexyl)sulfosuccinate/oil microemulsions, *J. Phys. Chem.* 1995, 99 (35), 13291–13300.
34. G. Gu, W. Wang, and H. Yan 1998 Phase equilibria and thermodynamic properties in microemulsions, *J. Therm. Anal. Calori.* 51, 115–123.
35. S.K. Hait and S.P. Moulik 2002 Interfacial composition and thermodynamics of formation of water/isopropyl myristate water-in-oil microemulsions stabilized by butan-1-ol and surfactants like cetyl pyridinium chloride, cetyl trimethyl ammonium bromide, and sodium dodecyl sulfate, *Langmuir* 18, 6736–6744.
36. S.P. Moulik, L.G. Digout, W.M. Aylward, and R. Palepu 2000 Studies on the interfacial composition and thermodynamic properties of w/o microemulsions, *Langmuir* 16, 3101–3106.
37. D. Mitra, I. Chakraborty, S.C. Bhattacharya, S.P. Moulik, S. Roy, D. Das, and P.K. Das 2006 Physicochemical studies on cetylammonium bromide and its modified (mono-, di-, and trihydroxyethylated) head group analogues. Their micellization characteristics in water and thermodynamic and structural aspects of water-in-oil microemulsions formed with them along with *n*-hexanol and isooctane, *J. Phys. Chem. B* 110, 11314–11316.
38. R.K. Mitra, B.K. Pal, and S.P. Moulik 2006 Phase behavior, interfacial composition and thermodynamic properties of mixed surfactant (CTAB and Brij-58) derived w/o microemulsions with 1-butanol and 1-pentanol as cosurfactants and *n*-heptane and *n*-decane as oils, *J. Colloid Interf. Sci.* 300, 755–764.
39. X. Li, K. Ueda, and H. Kuneida 1999 Solubilization and phase behavior of microemulsions with mixed anionic–cationic surfactants and hexanol, *Langmuir* 15, 7973–7979.
40. R.K. Mitra and B.K. Paul 2005 Physicochemical investigations of microemulsification of eucalyptus oil and water using mixed surfactants (AOT + Brij-35) and butanol, *J. Colloid Interf. Sci.* 283, 565–577.
41. B.K. Paul and R.K. Mitra 2005 Water solubilization capacity of mixed reverse micelles: Effect of surfactant component, the nature of the oil, and electrolyte concentration, *J. Colloid Interf. Sci.* 288, 261–279.
42. L.G. Digout, K. Bren, R. Palepu, and S.P. Moulik 2001 Interfacial composition, structural parameters and thermodynamic properties of water-in-oil microemulsions *Colloid Polym. Sci.* 279, 655–663.
43. V.K. Bansal, D.O. Shah, and J.P. O'connell 1980 Influence of alkyl chain length compatibility on microemulsion structure and solubilization, *J. Colloid Interf. Sci.* 75, 462–475.
44. S.R. Bisal, P.K. Bhattacharya, and S.P. Moulik 1990 Conductivity study of microemulsions: Dependence of structural behavior of water/oil systems on surfactant, cosurfactant, oil and temperature, *J. Phys. Chem.* 94, 350–355.
45. I. Chakarborty and S.P. Moulik 2005 Physicochemical studies on microemulsions: 9. Conductance percolation of AOT-derived W/O microemulsion with aliphatic and aromatic hydrocarbon oils, *J. Colloid Interf. Sci.* 289, 530–541.
46. J. Fang and R.L. Venable 1987 Conductivity study of the microemulsion system sodium dodecyl sulfate-hexylamine-heptane–water, *J. Colloid Interf. Sci.* 116, 269–277.
47. B.K. Paul, R.K. Mitra, and S.P. Moulik 2006 Microemulsions: Percolation of conduction and thermodynamics of droplet clustering, in *Encyclopedia of Surface and Colloid Science* ed. P. Somasundaran, 2nd Edition, Taylor & Francis, Boca Raton, FL, pp. 3927–3956.

48. S.P. Moulik, G.C. De, B.B. Bhowmik, and A.K. Panda 1999 Physicochemical studies on microemulsions. 6. Phase behavior, dynamics of percolation, and energetics of droplet clustering in water/AOT/n-heptane system influenced by additives (sodium cholate and sodium salicylate), *J. Phys. Chem. B* 103, 7122–7129.

49. S.K. Hait, S.P. Moulik, M.P. Rogers, S.E. Burke, and R. Palepu 2001 Physicochemical studies on microemulsions. 7. Dynamics of percolation and energetics of clustering in water/aot/isooctane and water/AOT/decane w/o microemulsions in presence of hydrotopes (sodium salicylate, α-naphthol, β-naphthol, resorcinol, catechol, hydroquinone, pyrogallol and urea) and bile salt (sodium cholate), *J. Phys. Chem. B* 105, 7145–7154.

50. L. Mukhopadhyay, P.K. Bhattacharya, and S.P. Moulik 1990 Additive effects on the percolation of water/AOT/decane microemulsion with reference to the mechanism of conduction, *Colloids Surf. A* 50, 295–302.

51. L. Mukhopadhyay, P.K. Bhattacharya, and S.P. Moulik 1993 Effect of butanol and cholesterol on the conductance of AOT aided water/xylene microemulsion, *Indian J. Chem.* 32A, 485–492.

52. P. Alexandradis, J.F. Holzwarth, and T.A. Hatton 1995 Thermodynamics of droplet clustering in percolating AOT water-in-oil microemulsions, *J. Phys. Chem.* 99, 8222–8232.

53. L.M.M. Nazario, T.A. Hatton, and J.P.S.G. Crespo 1996 Nonionic cosurfactants in AOT reversed micelles: Effect on percolation, size, and solubilization site, *Langmuir* 12, 6326–6335.

54. S.P. Moulik and S. Ray 1994 Thermodynamics of clustering of droplets in water/AOT/heptane microemulsion, *Pure Appl. Chem.* 66, 521–525.

55. M. Adachi, M. Harada, A. Shioi, and Y. Sato 1991 Extraction of amino acids to microemulsion, *J. Phys. Chem.* 95, 7925–7931.

56. R. Johannsson, M. Almgren, and J. Alsins 1991 Fluorescence and phosphorescence study of AOT/water/alkane systems in the L2 reversed micellar phase, *J. Phys. Chem.* 95, 3819–3823.

57. M.J. Suarej, H. Levy, and J. Lang 1993 Effect of addition of polymer to water-in-oil microemulsions on droplet size and exchange of materials between droplets, *J. Phys. Chem.* 97, 9808–9816.

58. H. Mays and J. Ilgenfrintz 1996 Intercluster exchange rates in AOT water-in-oil microemulsions: Percolation, material transport mechanism and activation energy, *J. Chem. Soc. Faraday Trans.* 92, 3145–3150.

59. A.C. John and A.K. Rakshit 1993 Formation of microemulsion: Effect of temperature and polyacrylamide, *J. Colloid Interf. Sci.* 156, 202–206.

60. B.K. Paul and S.P. Moulik 1997 Microemulsion: An overview, *J. Disp. Sci. Technol.* 18, 301–367.

61. F. Bordi and C. Cametti 2000 Ion transport and electrical conductivity in heterogeneous systems: The case of microemulsions, in *Interfacial Dynamics* ed. N. Kallay, Marcel Dekker, New York, pp. 541–563.

62. J. Texter 2001 Microstructure effects on transport in reverse microemulsions, In *Liquid Interfaces in Chemical, Biological and Pharmaceutical Applications* ed. A.G. Volkov, Marcel Dekker, New York, p. 241.

63. S.K. Hait, S.P. Moulik, and R. Palepu 2002 Refined method of assessment of parameters of micellization of surfactants and percolation of w/o microemulsion, *Langmuir* 18, 2471–2476.

64. A.N. Maitra, C. Mathew, and M. Varshney 1990 Closed and open structure aggregates in microemulsion and mechanism of percolation conduction, *J. Phys. Chem.* 94, 5290–5292.

65. S. Ray, S. Paul, and S.P. Moulik 1996 Physicochemical studies on microemulsions: V. Additive effects on the performance of scaling equations and activation energy for percolation of conductance of water/AOT/heptane microemulsion. *J. Colloid Interf. Sci.* 183, 6–12.

66. Y. Feldman, N. Kozlovich, I. Nir, and N. Garti 1995 Dielectric relaxation in sodium bis 2(ethyl hexyl) sulfosuccinate–water–decane microemulsions near the percolation temperature threshold, *Phys. Rev. E* 51, 478–491.

67. Y. Feldman, N. Kozlovich, Y. Alexandrov, R. Nigmatullin, and Y. Ryabov 1996 Mechanism of the cooperative relaxation in microemulsions near the percolation threshold, *Phys. Rev. E* 54, 5420–5427.

68a. P.G. De Gennes and C. Taupin 1982 Microemulsions and the flexibility of oil/water interfaces, *J. Phys. Chem.* 86, 2294–2304.

68b. M. Sahimi, B.D. Hughes, L.E. Scriven, and H.T. Davis 1983 Critical exponent of percolation conductivity by finite-size scaling, *J. Phys. Chem. Solid Stat. Phys.* 16, 2521–2527.

69. A. Dogra and A.K. Rakshit 2004 Phase behavior and percolation studies on microemulsion system water/SDS + Myrj45/cyclohexane in presence of various alcohols as cosurfactants, *J. Phys. Chem. B* 108, 10053–10061.

70. C. Boned, Z. Saidi, P. Xans, and J. Peyrelasse 1994 Percolation phenomenon in ternary microemulsions: The effect of pressure, *Phys. Rev. E* 49, 5295–5302.

71. Z. Saidi, J.L. Daridon, and C. Boned 1995 The influence of pressure on the phase diagram, the phenomenon of percolation and interactions in a ternary microemulsion, *J. Phys. D: Appl. Phys.* 28, 2108–2112.

72. J. Garboczi and J.G. Berryman 2001 Elastic moduli of a material containing composite inclusions: Effective medium theory and finite element computations, *Mech. Mater.* 33, 455–470.

73. E.J. Garboczi and J.G. Berryman 2000 New effective medium theory for the diffusivity or conductivity of a multiscale concrete microstructure material, *Concreter Sci. Eng.* 2, 88–96.

74. H.-Ch. Weissker, J. Furthmiller, and F. Bechstedt 2003 Validity of effective medium theory for optical properties of embedded nanocrystallites from ab initio super cell calculations, *Phys. Rev. B* 67, 165322-1–165322-5.

75. R. Landauer 1978 Electrical conductivity in inhomogeneous media, in *Electrical Transport and Optical Properties of Inhomogeneous Media* eds. J.C. Garland and D.B. Tanner, *AIP Conf. Proc.* No. 40, AIP, New York, pp. 2–45.

76. C.G. Granqvist and O. Hunderi 1977 Optical properties of ultrafine gold particles, *Phys. Rev. B* 16, 3513–3534.

77. C.G. Granqvist, O. Hunderi 1978 Conductivity of inhomogeneous materials: Effective medium theory with dipole–dipole interaction, *Phys. Rev. B* 18, 1554–1561.

78. C.J.F. Bottcher 1945 The dielectric constant of crystalline powders, *Recl. Trav. Chim.* 64, 47–51.

79. J. Bernasconi and H.J. Wiesmann 1976 Effective-medium theories for site-disordered resistance networks, *Phys. Rev. B* 13, 1131–1139.

80. S. Pal, S.R. Bisal, and S.P. Moulik 1992 Physicochemical studies on microemulsions: Test of theories of percolation, *J. Phys. Chem.* 96, 896–901.

81. S.P. Moulik and A.K. Rakshit 2006 Physicochemistry and applications of microemulsions, *J. Surf. Sci. Technol.* 22, 159–186.

82. S. Ajith, A.C. Jhon, and A.K. Rakshit 1994 Physicochemical studies of micro-emulsions pure, *Appl. Chem.* 66, 509–514; Z. Saidi, C. Matthew, J. Peyrelasse, and C. Boned 1990 Percolation and structural exponents for the viscosity of microemul-sions, *Phys. Rev. A* 42, 872–876; S. Ray, S.R. Bisal, and S.P. Moulik 1992 Studies on structure and dynamics of microemulsion II: Viscosity behavior of water-in-oil microemulsion, *J. Surf. Sci. Technol.* 8, 191–208.

83. A.C. John and A.K. Rakshit 1994 Phase behavior and properties of a microemulsion in the presence of NaCl, *Langmuir* 10, 2084–2087.

84. S. Ajith and A.K. Rakshit 1995 Studies of mixed surfactant microemulsion systems: Brij 35 with Tween 20 and sodium dodecyl sulfate, *J. Phys. Chem.* 99, 14778–14783.

85. B.K. Pal and S.P. Moulik 1991. Biological Microemulsions. III. The formation char-acteristics and transport properties of saffola–Aerosol OT–hexylamine–water system, *Ind. J. Biochem. Biophys.* 28, 174–184.

86. K. Mukherjee, D.C. Mukherjee, and S.P. Moulik 1997 Thermodynamics of micro-emulsion formation, *J. Colloid Interf. Sci.* 187, 327–333.

87. S.J. Chen, D.F. Evans, and B.W. Ninham 1984 Properties and structure of three component ionic microemulsions, *J. Phys. Chem.* 88, 1631–1634.

88. R.A. Mackay and R. Agarwal 1978 Conductivity measurements in nonionic micro-emulsions, *J. Colloid Interf. Sci.* 65, 225–231.

89. A. Katchalski and P.S. Curan 1967 *The Phenomenological Equations Relating Flows and Forces; Onsager's Law in Nonequilibrium Thermodynamics in Biophys-ics*, Harvard University Press, Cambridge, MA, Ch. 8, pp. 88–91.

90. A.L. Joshi and A.K. Rakshit 1997 Physicochemical studies of cyclohexane/CTAB/1-propanol/water microemulsion system in presence of PEG 400, *Indian J. Chem.* 36A, 38–44.

91. J.M. Tingey, J.L. Fulton, D.W. Matson, and R.D. Smith 1991 Micellar and bicontinu-ous microemulsions formed in both in near critical and supercritical propane with didocyldimethylammonium bromide and water, *J. Phys. Chem.* 95, 1445–1448.

92. S. Ray and S.P. Moulik 1995 Phase behavior, transport properties, and thermody-namics of water/AOT/alkanol microemulsion systems, *J. Colloid Interf. Sci.* 173, 28–98.

93. K. Maiti, D. Mitra, A.K. Panda, S.C. Bhattacharya, and S.P. Moulik 2007 Physico-chemical studies of octadecyltrimethylammonium bromide: A critical assessment of its solution behavior with reference to formation of micelles and microemulsion with *n*-butanol and *n*-heptane, *J. Phys. Chem. B* 111, 14175–14185.

94. M. Borkovec, H.F. Eicke, H. Hammerich, and B. Das Gupta 1988 Two percolation processes in microemulsions, *J. Phys. Chem.* 92, 206–211.

95. S.K. Hait, A. Sanyal, and S.P. Moulik 2002 Physicochemical studies on microemul-sions. 8. The effects of aromatic methoxy hydrotropes on droplet clustering and understanding of the dynamics of conductance percolation in water/oil microemul-sion systems, *J. Phys. Chem. B* 106, 12642–12650.

96. H. Mays 1997 Dynamics and energetics of droplet aggregation in percolating aot water-in-oil microemulsions, *J. Phys. Chem. B* 101, 10271–10280.

97. S. Ray, S.R. Bisal, and S.P. Moulik 1993 Structure and dynamics of microemul-sions. Part 1. Effect of additives on percolation of conductance and energetics of clustering in water–AOT–heptane microemulsions, *J. Chem. Soc. Faraday Trans.* 89, 3277–3282.

98. T.H. Wines and P. Somasundaran 2002, Effects of adsorbed block copolymer and comb-like amphiphilic polymers in solution on the electrical percolation and light scattering behavior of reverse microemulsions of heptane/water/AOT, *J. Colloid Interf. Sci.* 256, 183–189.

99. L. Garcia-Rio, J.R. Leis, J.C. Mejuto, M.E. Pena, and E. Iglesias 1994 Effects of additives on the internal dynamics and properties of water/AOT/isooctane microemulsions, *Langmuir* 10, 1676–1683.

100. J. Dasilva-Carvalhal, L. Garcia-Rio, D. Gormez-Diaz, J.C. Mejuto, and P. Rodriguez-Dafonte 2003 Influence of crown ethers on the electric percolation of AOT/isooctane/water (w/o) microemulsions, *Langmuir* 19, 5975–5983.

101. L. Garcia-Rio, P. Herves, J.C. Mejuto, J. Perez-Juste, and P. Rodriguez-Dafonte 2000 Effects of alkylamines on the percolation phenomena in water/AOT/isooctane microemulsions, *J. Colloid Interf. Sci.* 225, 259–264.

102. H.F. Eicke, M. Gauthier, and H. Hammerich 1993 A conductometric analysis of meso-gel formation by ABA block copolymers in w/o microemulsions, *J. Phys. II Fr.* 3, 255–258.

103. V. Papadimitriou, A. Xenakis, and P. Lianos 1993 Electric percolation of enzyme-containing microemulsions, *Langmuir* 9, 912–915.

104. J.P. Huruguen, M. Autheir, J.L. Greffe, and M.P. Pileni1991, Percolation process induced by solubilizing cytochrome *c* in reverse micelles, *Langmuir* 7, 243–249.

105. H. Yang, K. Erford, D.J. Kiserow, and L.B. McGown 2003 Effects of bile salts on percolation and size of AOT reversed micelles, *J. Colloid Interf. Sci.* 262, 531–535.

106. D. Schubel and G. Ilgenfritz 1997 Influence of polyethylene glycols on the percolation behavior of anionic and nonionic w/o microemulsions, *Langmuir* 13, 4246–4250.

107. D. Liu, L. Ma, H. Cheng, and Z. Zhao 1998 Investigation on the conductivity and microstructure of AOT/non-ionic surfactants/water/*n*-heptane mixed reverse micelles, *Colloids Surf. A* 135, 157–164.

108. R.K. Mitra and B.K. Pal 2005 Investigation on percolation in conductance of mixed reverse micelles, *Colloids Surf. A* 252, 243–259.

109. E. Acosta, E. Szekeres, D.A. Sabatini, and J.H. Harwell 2003 Net-average curvature model for solubilization and supersolubilization in surfactant microemulsions, *Langmuir* 19, 186–195.

3 Percolating Phenomenon in Microemulsions: Effect of External Entity

S. K. Mehta, Khushwinder Kaur, Gurpreet Kaur, and K. K. Bhasin

CONTENTS

3.1 INTRODUCTION

Mixtures of aqueous electrolytes, hydrocarbons, and amphiphilic compounds have been the subjects of extensive research, especially those systems forming amorphous isotropic solutions, called microemulsions. Several books and papers have treated this subject [1–5]. The term microemulsion was first introduced by Hoar and Schulman [5]. Microemulsions are thermodynamically stable, isotropic, transparent colloidal solutions of low viscosity, consisting of three components: a surfactant (amphiphile), a polar solvent (usually water), and a nonpolar solvent (oil) [1–7]. The surfactant monomers in these fluids reside at oil water interface and effectively lower the interfacial-free energy, resulting in the formation of optically clear, thermodynamically stable formulations. The innate formation of colloidal particles is typically up to nanometer scale; globular droplets each

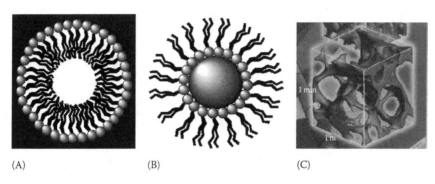

(A)   (B)   (C)

SCHEME 3.1   (A) oil in water (o/w), (B) water in oil (w/o), and (C) bicontinuous.

surrounded by monolayer of surfactant molecules and, thus, disperse in the bulk solvent. In microemulsion, solubilization of one phase into another is affected by the stubble balance of attractive and repulsive forces [8,9]. Depending upon the nature of the bulk solvent, microemulsions are classified as oil in water (o/w), water in oil (w/o), or reverse micelles and bicontinuous microemulsions (Scheme 3.1).

Ruckenstein and Chi [10] proposed the thermodynamic stability of the microemulsions by considering that the free energy of formation comprises of interfacial-free energy and interaction energy between droplets and entropy of dispersion. The interaction energy between droplets has been shown to be negligible and the free energy of formation can be zero or even negative if the interfacial tension is of the order of 10^{-2}–10^{-3} mN m^{-1}. The concept of hydrophilic–lipophilic balance temperature or phase inversion temperature at which maximum solubilization of oil in water and ultralow interfacial tensions are achieved has been also introduced [10,11].

The surfactant aggregates in reverse micelles are closely packed globules where the polar head group of the surfactant molecules occupies the interior of the aggregates whereas the hydrophobic tails extend into bulk apolar solvent, with water encapsulated in compartments. The water solubilization capacity is conventionally described in terms of a parameter [7,11–13], which is defined as molar ratio of water to surfactant, ω. It reflects the maximum amount of water that can be solubilized in the system before phase transition occurs. The nature of water in reverse micelles especially at low-water content is believed to be different from bulk water [14,15]. The highly structured yet heterogeneous water molecules in reverse micelles represent interesting models for water molecules present in biological systems such as membranes, which are more difficult to analyze experimentally. Therefore, they can also be viewed as compartmentalized liquids. The existence of distinct microenvironments provides the reverse micelles a particular ability to solubilize substances such as organochalcogens, aromatic heterocyclic compounds, polymers, drugs, etc. and modulate physiochemical behavior due to compartmentalization of reactants in different microenvironments. In fact, numerous applications [16–20] of microemulsions are based on compartmentalized state of minority solvent (the nanophase) and on its being dispersed in a medium of disparate polarity. The studies on reverse micellar organization and dynamics assume special significance in light of the fact that the general principles underlying their formation are common to other related assemblies such as micelles, bilayers, liposomes, and other biological membranes.

Under normal conditions, water in oil microemulsions represents very low-specific conductivity (ca. 10^{-9}–10^{-7} Ω^{-1} cm^{-1}). This conductivity is significantly greater than it would be if we consider the alkane, which constitutes the continuous medium and is the main component of the water in oil microemulsions ($\sim 10^{-14}$ Ω^{-1} cm^{-1}). This increase in the electrical conductivity of the microemulsions by comparison with that of the pure continuous medium is due to the fact that microemulsions are able to transport charges [21,22].

When we reach a certain volume of the disperse phase, the conductivity abruptly increases to give values of up to four orders of magnitude, which is greater than typical conductivity of water in oil microemulsions. This increase remains invariable after reaching the maximum value that is much higher than that for the microemulsion present before this transition occurs. Similar behavior is observed for the fixed composition of the microemulsion when either water or temperature or volume fraction is increased. This phenomenon is known as *electric percolation*, [21–32] the moment at which an abrupt transition occurs from poor electric conductor, system (10^{-7} Ω^{-1} cm^{-1}) to the system with fluid electric circulation (10^{-3} Ω^{-1} cm^{-1}). The mechanism proposed to explain the *electric percolation* phenomenon is based on the formation of channels exchanging matter between the dispersed water droplets and the continuous phase, as shown in Scheme 3.2.

As a consequence of ion transfer, it yields a sigmoidal $\sigma - \theta$, and $\sigma - \phi$ profile. The point of maximum gradient of the (d log σ/dθ, d logσ/dθ or d log σ/dω) profile corresponds to the transition of the percolation process and is designated as the threshold temperature (θ_c), threshold volume fraction (ϕ_c)or the threshold water content (ω_p) characteristic feature of a percolating system [21–32]. Moulik et al. [28] have proposed the sigmoidal Boltzmann equation (SBE) to determine the threshold characteristics of microemulsion systems. In conductance percolation, the equivalent equation can be written as

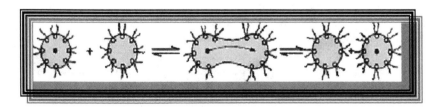

SCHEME 3.2   Proposed scheme for "fusion-mass transfer-fission."

$$\log \sigma = \log \sigma_f \left[1 + \left(\frac{\log \sigma_i - \log \sigma_f}{\log \sigma_f} \right) \{1 + \exp(\theta - \theta_c)/\Delta \theta\}^{-1} \right] \qquad (3.1)$$

where i, f, and c are the initial, final, and percolative stages. The composition of the system and other environmental conditions such as pressure and presence of additives control the threshold values.

3.2 PERCOLATION AND SCALING LAWS

The paper in 1978 by Lagues [33] was the first to interpret the dramatic increase in conductivity with droplet volume fraction for water in oil microemulsion in terms of percolation model and termed this physical situation as stirred percolation, referring to Brownian motion of the medium. This was, however, soon followed by several investigations. According to the most widely used theoretical model, which is based on the dynamic nature of microemulsions [33–38], there are two pseudo phases: one in which charge is transported by the diffusion of microemulsion globules and the other phase in which the change is conducted by the diffusion of the charge carrier itself inside the reverse micellar cluster. According to this theory, the conductivity along the oil dilution line can be reproduced in terms of two separate asymptotic power laws having different exponents below and above the percolation threshold.

$$\sigma = A(\theta_c - \theta)^{-s} \quad \text{where } \theta < \theta_c \text{ (below percolation)} \qquad (3.2)$$

$$\sigma = B(\theta - \theta_c)^{t} \quad \text{where } \theta_c < \theta \text{ (above percolation)} \qquad (3.3)$$

where A and B are free parameters and s and t are critical exponents. The slopes of the log σ versus $\log(\theta_c - \theta)$ plot for $\theta_c > \theta$ and the log σ versus $\log(\theta - \theta_c)$ for $\theta > \theta_c$ plot yield s and t parameters. The difference between the dynamic and static percolation is reflected in terms of deviation in the value of s and t from the predicted values. The critical exponent t generally ranges between 1.5 and 2, whereas the exponent s allows the assignment of the time-dependent percolation regime. Thus, $s > 1$ (generally around 1.3) identifies a dynamic percolation regime [33–38]. The static percolation (Scheme 3.3A) is related to the appearance of bicontinuous microemulsions, where a sharp increase in conductivity, due to both counter ions and, to lesser extent, surfactant ions, can be justified by a connected water path in the system. The conductivity transition is mainly caused due to the formation of a continuous connected disperse phase in the system. The dynamic percolation (Scheme 3.3B) is connected to rapid process of fusion–fission among the droplets. Transient water channels form when the surfactant interface breaks down during collisions or through the merging of droplets. In latter case, conductivity is mainly due to the motion of counter ions along the water channels.

In this chapter, we summarize some current experimental work, based on conductivity and spectroscopic studies, reporting the effects of interactions and

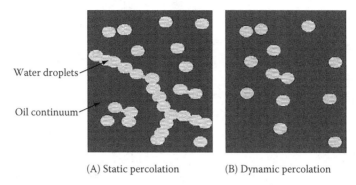

(A) Static percolation   (B) Dynamic percolation

SCHEME 3.3 Dynamics of droplet fusion.

dynamics on the percolation threshold and interface of microemulsions. We outline the effects of organic and inorganic derivatives on the percolation threshold and the modification of interface of the microemulsion system by the added external entities.

3.3 EFFECT OF EXTERNAL ENTITIES ON THE PERCOLATION PHENOMENON

3.3.1 ORGANOCHALCOGENIDES

The physical properties and dynamics of percolation of water/Sodium Bis (2-ethyl-hexyl) sulfosuccinate (AOT)/isooctane microemulsions are summarized as they are affected by the addition of water insoluble organochalcogenides [39] such as dipyridyl diselenide (Py_2Se_2), diphenyl diselenide (Ph_2Se_2), and diphenyl ditelluride (Ph_2Te_2). The organochalcogenides are bulky molecules insoluble in most of the solvents. Percolation threshold when investigated with increase in the concentration of different organochalcogens reveals sigmoidal behavior of log σ with increase in temperature. The threshold temperature, determined from the d log σ/dθ versus θ plot, has been depicted in Figure 3.1A. The conductivity of the systems decreases on addition of pyridyl and phenyl derivatives of Se except for 5 and 10 mM concentrations of Py_2Se_2. However, incorporation of Py increases the system conductance.

The value of θ_c varies in the order: Py < without additive < Py_2Se_2 < Ph_2Se_2 < Ph_2Te_2. Figure 3.1B shows comparison of all the additives at 30 mM. A comparison of Py- and Py_2Se_2-based systems with the microemulsion without any additive reveals that while Py favors percolation, Py_2Se_2 delays the process. Maitra et al. [40] have reported that in the percolative microemulsion the droplets retain their closed structure although infinite clusters are formed due to interdroplet interactions. The mutual contact and the bridging between the droplets are due to the presence of additives either at the interface (e.g., hydrotopes and bile salts) [28] or in the core of the droplets (e.g., poly(ethylene glycol)s) [41].

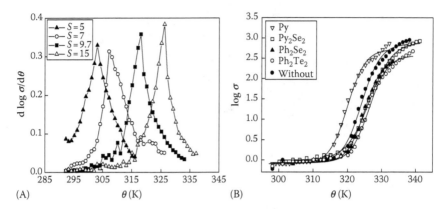

FIGURE 3.1 (A) Variation of d log $\sigma/d\theta$ with temperature to observe the percolation threshold (where S = [oil]/[AOT]). (B) Variation of specific conductance with temperature at 30 mM for additives (solid lines show SBE fitting).

In case of Py, the active N-end anchors the droplet surface while the aromatic ring offers the local hydrophobicity. This results in the straight bridging of the nanodroplets and the conduit formation is favored. Py_2Se_2 is a rigid nonplaner moiety in which one (Py–Se–)– unit is bent leaving the central horizontal plane. It contains an electronically active –(–Se–Se–)– along with the N-end. The molecule adheres to the droplet interface but the configuration results in the staggered bridging. The phenyl derivatives of selenium and tellurium have no such active sites except for –(–Se–Se–)– and –(–Te–Te–)–, respectively and remain in the dispersion medium, resisting the droplet fusion by enhancing the blockening effect and thus delaying the percolation threshold. Compared to Se, the compound of Te has been found to be more inert and susceptible to fast breakdown. This is due to the fact that comparatively Te is bulky and the bond length is longer in Te-compounds.

Analysis of the studied system with Equations 3.2 and 3.3 depicts the dynamic nature of percolation process. Figure 3.2 shows the plot of log σ versus $\log(\theta_c - \theta)$ with slope yielding the s and t parameters, respectively. For different sets of systems, i.e., with concentration variation of organochalcogens, ω variation for Py_2Se_2, and S variation of Ph_2Se_2 the value of exponent s lies in between 0.56–1.54, 0.30–1.09, and 0.91–1.58, respectively which are lower than the reported range (0.7–1.6) [33–38]. The value of t falls in the range 0.94–1.73, 1.05–1.67, and 0.92–1.86 for the said systems, which is also lower, compared to literature range (1.2–2.1) [33–38]. The value of both the s and t parameters has been lowered than the ideal predicted dynamic range.

The ΔG_{cl}^0 values calculated using the association model [42,43] are negative for all formulations, indicating the spontaneous nature of droplet clustering. The values of ΔH_{cl}^0 and ΔS_{cl}^0 are positive. The clustering phenomenon is accompanied by (1) the removal of oil barriers surrounding the dispersed droplets and (2) the

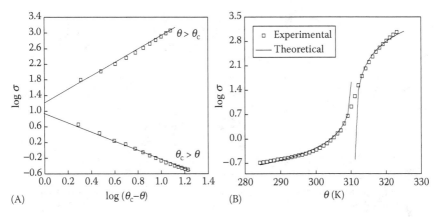

FIGURE 3.2 Variation of (A) $\log \sigma$ versus $\log(\theta_c - \theta)$ and (B) $\log \sigma$ versus θ for $Py_2Se_2 = 10\,mM$.

association of droplets. The first step is endothermic whereas the second is associated with the release of heat and thus exothermic. The individual magnitude of first overdoes the second and an overall endothermic effect dominates. Solvent disruption in the microenvironment of dispersed droplets compensates their organization during the clustering and hence account for positive ΔS_{cl}^0. In the presence of organochalcogens, the values of ΔH_{cl}^0 and ΔS_{cl}^0 are considerably low indicating an easily disturbable surroundings.

The fourier transform spectroscopy (FTIR) stretching of $-OH$ bond in the absence and in the presence of Py_2Se_2 and Ph_2Se_2 depicts a decrease in intensity of $-OH$ band with increase in ω value. The change in intensity becomes constant after a shoulder point (i.e., after $\omega = 16.85$). A shift in $-OH$ stretching has been observed at constant composition for organochalcogen compounds as: Ph_2Se_2 ($3448.2\,cm^{-1}$) < Py_2Se_2 ($3467.8\,cm^{-1}$) < Ph_2Te_2 ($3544.1\,cm^{-1}$), in comparison to a value of pure water $3400\,cm^{-1}$. The result indicates that organochalcogens interact with the core water. This confirms the inference drawn from conductivity results.

3.3.2 AROMATIC HETEROCYCLIC COMPOUNDS

The dynamics of water/AOT/isooctane microemulsions affected by the addition of aromatic heterocyclic compounds (Ar-Ht-C) [44], such as pyridine (Py), 2-aminopyridine (2-Ampy), 3-aminopyridine (3-Ampy), 2-amino-4-methylpyridine (2-Am-4-mpy), and 2-amino-6-methylpyridine (2-Am-6-mpy), reflect a significantly different physiochemical behavior as compared to the organochalcogenide molecules. The chosen compounds represent the relatively small organic molecules, which serve as starting materials to the synthetic Se/Te chemistry [45]. Apart from this, these compounds find wide industrial and biological applications [46]. Organochalcogenides, as depicted above, delay the percolation threshold as compared to the pure microemulsion whereas Ar-Ht-C has been found to favor the process. This is because organochalcogenides are large water insoluble molecules whereas Ar-Ht-C is relatively small water soluble compounds. The different solubilites and structures enforce these compounds to modify the interface in different ways, thus inducing different physiochemical behavior. The conductivity as a function of temperature and volume fraction for different concentrations (5, 15, 30, and 50 mM) depicts that the added Ar-Ht-C favors the percolation process. The value of θ_c or ϕ_c for all concentrations under study varies as 3-Ampy < 2-Ampy < py < 2-Am-6-mpy < 2-Am-4-mpy < without. All the Ar-Ht-C under study is water soluble entities and resides in the core of the reverse micelle. The active N-end anchors to the droplet surface and the NH_2 group enters into H-bonding with the water molecules present in the core of the microemulsion. The weak H-bonding results in softening of interface. This increases the feasibility of the formation of transient tube between the consecutive droplets hence favoring the percolation process.

The structural information on Ar-Ht-C confined within the AOT reverse micelles obtained by UV-vis absorbtion spectroscopy has been presented in Figure 3.3A. The shift in the band position with the change in Ar-Ht-C has been presented in Figure 3.3B. The Ar-Ht-C peak follows the order: 3-Ampy > 2-Ampy > 2-Am-6-mpy > 2-Am-4-mpy with corresponding values at 323, 312, 302, and 297 cm^{-1}. The above trend suggests that the presence of Ar-Ht-C moiety affects the bonding properties of the formulated microemulsion. A red shift of about 11 cm^{-1} observed with change in the position of $-NH_2$ from 2 to 3 positions indicates weakening of bonds between the water molecules by consecutive replacement of OH–O bonds of bound water by significantly weaker OH–N bonds.

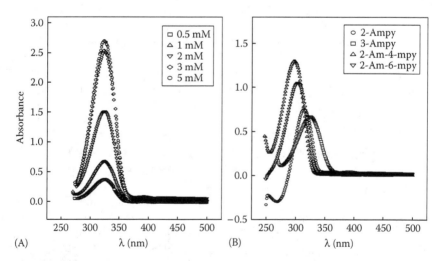

FIGURE 3.3 Variation of absorbance versus wavelength (A) at different concentrations of 3-Ampy (B) for different Ar-Ht-C. (From Mehta, S.K., Kaur, K.K., Sharma, S., and Bhasin, K.K., *Colloids Surf. A*, 298, 252, 2007.)

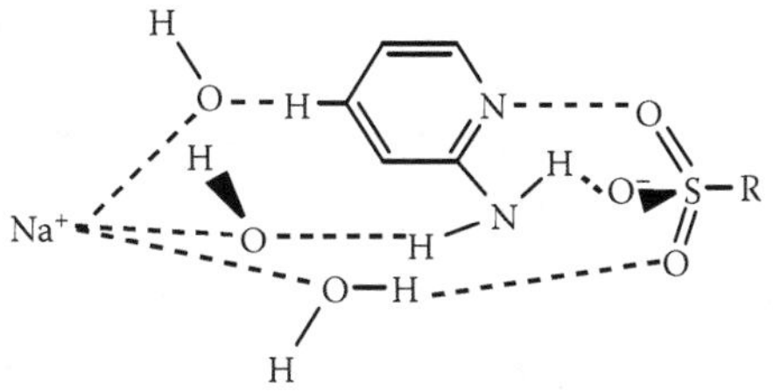

SCHEME 3.4 Proposed interaction between AOT and 2-Ampy. (From Mehta, S.K., Kaur, K.K., Sharma, S., and Bhasin, K.K., *Colloids Surf. A*, 298, 252, 2007.)

The further introduction of a hydrophobic moiety $-CH_3$ at 4 or 6 positions, residing towards the hydrophobic chain of AOT, shifts the peaks towards a lower wavelength of about $10\,cm^{-1}$ as compared to 2-Ampy. Methyl derivatives being slightly more bulky hinder the replacement of OH–O bonds with weaker OH–N bonds as a result of which the bonds are weakened to a slight lesser extent. The probable structural interaction has been presented in Scheme 3.4.

Application of scaling laws to the volume- and temperature-induced percolation process indicates the dynamic nature of percolation process. For volume-induced percolation, the value of s lies between 1.2 and 2.4 and the value of t is in the range of $1.2 - 2.1$, which is higher and in agreement with the reported range. The computed value of s for the temperature-induced percolation has been obtained between 1.0 and 2.1 and the value of t lies between 1.2 and 1.7, and is comparable to the reported range of 0.7–1.6 for s and 1.2–2.1 for t, respectively [33–38]. This further depicts that volume-induced percolation is more dynamic than temperature-induced percolation process.

The ΔG_{cl}^0 values estimated are negative for all formulations, indicating the spontaneous nature of droplet clustering. The values of ΔH_{cl}^0 and ΔS_{cl}^0 have been found to be positive accounting for a strong heat absorbing step in the percolation process. The values of ΔH_{cl}^0 and ΔS_{cl}^0 have been found to be considerably higher and follow the order: without > 2-Am-4-mpy > 2-Am-6-mpy > 2-Ampy > 3-Ampy. The low-positive enthalpy change in the presence of additives than in the absence indicates a relatively less organized, i.e., easily disruptable microsurrounding of the droplets. The results therefore indicate that the microenvironment of methylamine derivatives is more organized as compared to amine derivatives.

3.3.3 Acetyl-Modified Amino Acids

The temperature-induced percolation process is delayed by the incorporation of acetyl-modified amino acids (MAA) viz., *N*-acetyl-L-cysteine (NAC), *N*-acetyl-L-glycine (NAG), *N*-acetyl-L-aspartic acid (NAA) in water/AOT/isooctane microemulsion [47]. The value of θ_c varies as NAC > NAG > NAA > without. The observed behavior of MAA molecules is contrary to that found when other organic compounds like urea and thiourea are added to microemulsion systems [28]. This has been explained on the basis of binding of additives to the AOT head group, which induces greater disorder in the surfactant environment, thus promoting the interdroplet interactions. Bommarius et al. [48] have shown that the mechanism of solubilizate exchange involves the formation of short-lived dimers, which deform the surfactant layer followed by their separation. MAA molecules adhere to the interface and play some role in the interfacial region with regard to their capacity to H-bond with the water molecules in the microemulsion core. The system shows a dynamic behavior when analyzed with scaling laws, i.e., Equations 3.1 and 3.2. The value of s are in the range of 1.54–1.67, 1.40–1.95, and 1.20–1.76 and the value of t lies in the range of 1.61–2.00, 1.50–1.88, and 1.49–2.29 for NAG, NAA, and NAC, respectively representing the dynamic nature of the percolation process. The MAA when subjected to water-induced percolation process depicts a different physiochemical behavior. The values of ω_p follows the order NAG < NAC < NAA < without contrary to the temperature-induced percolation process. The curves in Figure 3.4 can be divided into two sections. The conductivity increases with the increase in water content at the initial stage. After reaching a maximum, it decreases until a critical ω_p value, followed by a sharp increase with the increase in water content, which suggests that an additional pathway for charge–carrier migration takes over.

The values of ΔH_{cl}^0 and ΔS_{cl}^0 have been found to be positive accounting for a strong heat-absorbing step in the percolation process. The considerably high values of ΔH_{cl}^0 and ΔS_{cl}^0 follow the order: without > NAA > NAG > NAC. The low-positive enthalpy change in the presence of MAA than in the absence indicates a relatively less organized, i.e., easily disruptable microsurrounding of the droplets. The results therefore indicate that the microenvironment of the system without MAA is more organized as compared to the presence of MAA.

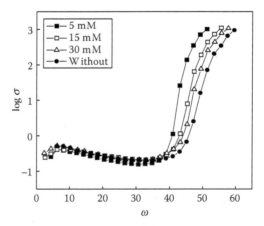

FIGURE 3.4   Variation of for NAG. (From Mehta, S.K., Kaur, K.K., Sharma, S., and Bhasin, K.K., J. *Colloid Interf. Sci.*, 314, 689, 2007.)

The FTIR stretching of –OH bond shows a blue shift of about 33.0, 32.8, and 22.2 cm⁻¹ with the incorporation of NAG, NAA, and NAC, respectively. This causes weakening of H-bonds among the water molecules with the incorporation of MAA suggesting that water–water H-bonds are being replaced by MAA–water H-bonds. The blue shift of about 1 cm⁻¹ for CO and SO_3^- appears to be modest when compared to the microemulsion without MAA. The peak assigned to COC (1210–1245 cm⁻¹) stretching shows a shift towards higher frequency of about 3.7, 2.8, and 2.5 cm⁻¹ for NAG, NAA, and NAC, respectively revealing interactions between the MAA and AOT head group. The OH, CO, and SO_3^- peaks follow the order NAA ~ NAC > NAG > without. The shift in peak positions of OH and C–O–C linkages indicates the interactions among AOT head group, water, and MAA molecules. The presence of specific interactions of MAA with the ester linkage of the AOT head group leads to formation of somewhat rigid interface, however, at the same time; the data reveal the presence of significant H-bonded interactions between the MAA and the water molecules.

3.3.4 DRUGS

Percolation phenomena have also been observed in the case of drug-incorporated nonionic microemulsion system. Electric conductivity has been measured as a function of water content ω for the oleic acid (oil), surfactant (Tween 80/cosurfactant(ethanol) mixture and variation of σ as a function of ω has been observed (Figure 3.5A). The behavior exhibits profile characteristic of percolative conductivity [33–38]. The conductivity is initially low in an oil–surfactant mixture but increases with increase in aqueous phase. The low conductivity below ω_p suggests that the reverse droplets are discrete (isolated droplets in a nonconducting oleic medium, forming w/o microemulsion) and have little interaction. When the water content is raised above ω_p, the value of σ increases linearly and steeply up

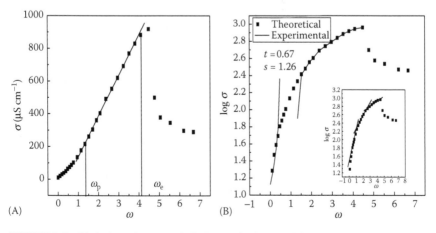

FIGURE 3.5   Variation of (A) σ and (B) log σ of microemulsion system with.

to kilobyte. The interaction between the aqueous domains becomes increasingly important and forms a network of conductive channel (bicontinuous microemulsion). With further increase in water content above ω_c, the σ shows a sharp decrease. The system becomes turbid that contributes to the dilution of o/w microemulsion where added water decreases the concentration of discrete oil droplets [49]. Figure 3.5B depicts the variation of log σ versus ω. The change in the slope of log σ (Figure 3.5B) has been interpreted, as a structural transition to o/w droplet from w/o [50], via bicontinuous phase. This transition takes place once the aqueous phase becomes continuous phase. Thus, the σ versus ω plots illustrate occurrence of three different structures (namely w/o, bicontinuous, o/w).

From the theoretical point of view, the post percolation stage shows a good agreement between the experimental and theoretical values of log σ versus ϕ_w for water-induced percolation as shown in Figure 3.5B. The values of 1.26 and 0.67 for s and t parameters indicate that a percolation phenomenon is static in nature. This attributes to the formation of continuous oil and water structures showing the presence of bicontinuous microemulsions.

The assimilation of different hydrophobic and hydrophilic drugs in colloidal assemblies having polar and apolar microdomains is extensively used as drug delivery systems [51]. The presence of drug moieties may or may not affect the microstructure of the microemulsion formulations. As in the case of incorporation of first line, anti-TB drugs (rifampicin, isoniazid, and pyrazinamide) in above-mentioned Tween-based microemulsion did not show much change in the microstructure. A comparison of pure and drug-loaded microemulsion systems (Figure 3.6) shows that drug incorporation does not affect the microstructure of the microemulsion; it only delays the gel formation and the percolating threshold, which is more significant for rifampicin (highly hydrophobic) and pyrazinamide (water soluble). Change is insignificant for isoniazid (highly water

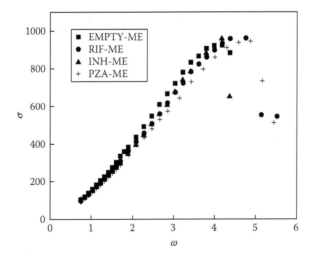

FIGURE 3.6 Comparison of conductivity of Tween 80 and drug-loaded microemulsion.

soluble), which is mainly present in the continuum region. The effect of drug is nearly negligible when the system is w/o but change is seen when transformation to o/w microemulsion has occurred on dilution. A value $-4.35\,\text{kJ mol}^{-1}$ of ΔG_{cl}^0 has been obtained for the pure system. On the other hand, ΔG_{cl}^0 is nearly same $(-5.17\,\text{kJ mol}^{-1})$ for the three drug samples incorporated in the media. The ΔG_{cl}^0 value for the pure and drug-loaded microemulsion supports the spontaneity of droplet clustering.

3.3.5 POLYMERS

The presence of polyethylene glycols such as ethylene glycol (EG), diethyleneglycol (DEG), triethyleneglycol (TEG), tetraeneglycol (TREG), and poly(ethylene glycol) (PEG) (400) shifts ϕ_c to lower-volume fraction in the order EG ~ PEG > DEG > TEG >TREG. The decrease of ϕ_c brought about by glycols is due to increase in the interdroplet interactions. This may be related to the decrease of interfacial rigidity. Binding of additive to the interfacial layer reduces the curvature of the surfactant, thereby favoring the opening of the AOT layers during interdroplet collisions. This association also results in greater disorder in the surfactant environment, which could help promote attractive interactions [52]. The reduction in the value of ϕ_c with the addition of bigger glycol molecules (mono to tetra) implies favorable percolation. This is due to the greater association with the surfactant film of bigger glycol but PEG, however, follows a different trend and has ϕ_c nearly equal to EG. This is because the with addition of PEG to the system, the ends of some parts of PEG get bound to AOT molecules at the interface while the rest of PEG molecules go into the water pool. As a result, the elastic energy of PEG molecules tends to reduce the size of the droplets and decrease the attractive interactions [53]. It has also been reported [52] that increasing concentration of PEG facilitates percolation. The high concentration of the polymer not only saturate the interface, but also alters the environment of a small fraction of head groups giving rise to significant changes, in the ease with which interdroplet channels can be opened for ion transport. However, system subjected to temperature percolation shows a decrease in the percolation temperature θ_c, the value of which is even lowered on moving from mono to polyethylene glycol 400. The mechanism is expected to proceed through the formation of short-lived dimmer. PEG acts as spacers between surfactant monolayer, resulting in an increase in effective area of the polar head group of the surfactant. This favors the formation of positive curvature region between the fusing droplets. PEG, thus, acts as a bridge and helps in channel formation. Applying percolation laws to the volume- and temperature-induced percolation process, the values of s and t computed are in the range of 0.2–0.7 and 0.7–1.0, respectively. However, for temperature-induced process, the value of s lies between 0.77 and 1.6 and t falls in the range of 0.61–1.77. The values of these critical exponents are, thus, near the lower limit of dynamic percolation.

The ΔG_{cl}^0 values at ϕ_c for all the microemulsions having glycols are negative and follow the order EG < DEG < TEG < TREG. It may be inferred that bigger the glycol molecule makes the clustering of the droplets easy. The replacement of EG (10%) with PEG (10%) also increases the value of ΔG_{cl}^0 implying the addition of polymer favoring clustering.

ACKNOWLEDGMENTS

Based on the research carried over several years, SKM is thankful to DST and CSIR for financial support and Department of Chemistry and Centre of Advanced Studies in Chemistry, Panjab University, Chandigarh, India for providing necessary facilities.

SYMBOLS AND TERMINOLOGIES

σ conductivity
ϕ_c critical volume fraction of the disperse phase
ΔH_{cl}^0 enthalpy of clustering
ΔS_{cl}^0 entropy of clustering
S oil to surfactant molar ratio
θ_c percolation threshold temperature
ΔG_{cl}^0 standard free energy of clustering
Θ temperature
ϕ volume fraction
ω water to surfactant molar ratio

REFERENCES

1. Scriven, L. E. (1977). *Equilibrium bicontinuous structure. In: Micellization, Solubilization and Microemulsions*, [Proc. Int. symp.] K. L. Mittal (Ed.), New York, Plenum Press, 2, pp. 877–893.
2. (a) Prince, L. M. (1974). In: *Emulsions and Emulsion Technology*, K. J. (Ed.), Part I, Chapter 3, New York, Marcel Dekker, pp. 125–177; (b) Prince, L. M. (1977). *Schulman's Microemulsions* In: *Microemulsions: Theory and Practice*, Prince, L. M. (Ed.), New York, Academic Press, pp. 1–20.
3. Friberg, S. (1977). *Microemulsions and Micellar solutions. In: Microemulsions: Theory and Practice*, Prince, L. M. (Ed.), New York, Academic Press, pp. 133–146.
4. Adamson, A. W. (1969). A model for micellar emulsions. *J Colloid Interf Sci* 25, 261–267;
5. (a) Hoar, T. P. and Schulman, J. H. (1943). Transparent water-in-oil dispersions: The oleopathic hydro-micelle. *Nature (London)* 152, 102–103; (b) Schulman, J. H. and Riley, D. P. (1948). X-ray investigation of the structure of transparent oil-water disperse systems. *I J Colloid Sci* 4, 383–405; (c) Schulman, J. H., Stoeckenius, W., and Prince, L. M. (1959). Mechanism of formation and structure of micro emulsions by electron microscopy. *J Phys Chem* 63, 1677–1680; (d) Schulman, J. H., Matalon, R., and Cohen, M. (1951). X-ray and optical properties of spherical and cylindrical aggregates in long chain hydrocarbon polyethylene oxide systems. *Discuss Faraday Soc* 11, 117–121; (e) Schulman, J. H. and Cockbain, E. G. (1940a). Molecular interaction at oil–water interfaces, Part I. *Trans Faraday Soc* 36, 651–661; (f) Schulman, J. H. and Cockbain, E. G.

(1940b). Molecular interaction at oil–water interfaces, Part II. *Trans Faraday Soc* 36, 661–668; (g) Alexender, A. E. and Schulman, J. H. (1940). Molecular interaction at oil–water interfaces, Part III. *Trans Faraday Soc* 36, 960–964; (h) Schulman, J. H. and Mc Roberts, T. S. (1946). On the structure of transparent water and oil dispersions (solubilized oil). *Trans Faraday Soc* 42B, 165–170; (i) Schulman, J. H. and Friend, J. P. (1949). Light scattering investigation of the structure of transparent oil–water disperse systems II. *J Colloid Sci* 4, 457–505.

6. (a) Kahlweit, M., Strey, R., and Haase, D. (1985). Phase behavior of multicomponent systems water–oil–amphiphile–electrolyte. *J Phys Chem* 89, 163–170; (b) Kahlweit, M. and Strey, R. (1985). Phase behavior of ternary systems of the type H_2O–oil–nonionic amphiphile (microemulsions). *Angew Chem (Engl Ed)* 24, 654–668; (c) Kahlweit, M., Strey, R., Haase, D., Kuneida, H., Schmeling, T., Faulhaber, B., Borkobec, M., Eicke, H. F., Busse, G., Eggers, F., Funck, T., Richmann, H., Magid, L., Soderman, S., Stilbs, P., Winkler, J., Dittrich, A., and Jahn, W. (1987). How to study microemulsions. *J Colloid Interface Sci* 118, 436–453.

7. (a) Shah, D. O. and Hamlin, R. M., Jr. (1971). Structure of water in microemulsions: Electrical, birefringence, and nuclear magnetic resonance studies. *Science* 171, 483–485; (b) Shah, D. O., Tamjeedi, A., Falco, J. W., and Walker, R. D., Jr. (1972). Interfacial instability and spontaneous formation of microemulsions. *AIChE J* 18, 1116–1120.

8. McBain, M. E. L. and Hutchison, E. (1955). *Solubilization and Related Phenomenon.* New York, Academic Press.

9. Shinoda, K. and Kunieda, H. (1973). Conditions to produce so-called microemulsions: Factors to increase mutual solubility of oil and water by solubilizer. *J Colloid Interf Sci* 42, 381–387.

10. Ruckenstein, E. and Chi, J. C. (1975). Stability of microemulsions. *J Chem Soc Faraday Trans* 71, 690–1707.

11. Shinoda, K. and Friberg, S. (1975). Microemulsions: Colloidal aspects. *Adv Colloid Interf Sci* 4, 281–300.

12. Fendler, J. H. (1982). *Membrane Mimtic Chemistry.* New York, John Wiley.

13. Luisi, P. L. and Straub, B. E. (1984). *Reverse Micelles: Amphiphillic Structures in Apolar Media.* New York, Plenum Press.

14. Keh, E. and Valeur, B. (1981). Investigation of water-containing inverted micelles by fluorescence polarization. Determination of size and internal fluidity. *J Colloid Interf Sci* 79, 465–478.

15. Wong, M., Thomas, J. K., and Graetzel, M. (1976). Fluorescence probing of inverted micelles. The state of solubilized water clusters in alkane/diisooctyl sulfosuccinate (aerosol OT) solution. *J Am Chem Soc* 98, 2391–2397.

16. Solans, C. and Kunieda, H. (1977). How to prepare Microemulsions: Temperature insensitive microemulsions. In: *Industrial Applications of Microemulsions.* New York, Marcel Dekker, pp. 21–46.

17. Towey, T. F., Khan-Lodhi, A., and Robinson, B. H. (1990). Kinetics and mechanism of formation of quantum-sized cadmium sulphide particles in water–aerosol-OT–oil microemulsions. *J Chem Soc Faraday Trans* 86, 3757–3762.

18. Atik, S. S. and Thomas, J. K. (1982). Photochemistry in polymerized microemulsion systems. *J Am Chem Soc* 104, 5868–5874.

19. Ranganathan, D., Ranganathan, S., Singh, G. P., and Patel, B. K. (1993). Demonstration of exclusive α-pep tidation at the micellar interface. *Tetrahedron Lett* 34, 525–528.

20. Adachi, M., Harada, M., Shioi, A., and Sato, Y. (1991). Extraction of amino acids to microemulsion. *J Phys Chem* 95, 7925–7931.

21. Eicke, H. F., Borkovec, M., and Gupta, B. D. (1989). Conductivity of water-in-oil microemulsions: A quantitative charge fluctuation model. *J Phys Chem* 93, 314–317.

22. Kallay, N. and Chittofrati, A. (1990). Conductivity of microemulsions: Refinement of charge fluctuation model. *J Phys Chem* 94, 4755–4756.

23. Giustini, M., Palazzo, G., Colafemmina, G., Monica, M. D., Marcello, G., Giomini, M., and Ceglie, A. (1996). Microstructure and dynamics of the water-in-oil CTAB/*n*-pentanol/*n*-hexane/water microemulsion: A spectroscopic and conductivity study. *J Phys Chem* 100, 3190–3198.

24. D'Aprano, A., D'Arrigo, G., Paparelli, A., Goffredi, M., and Liveri, V. T. (1993). Volumetric and transport properties of water/AOT/*n*-heptane microemulsions. *J Phys Chem* 97, 3614–3618.

25. Feldman, Y., Kozlovich, N., Nir, I., Garti, N., Archipov, V., Idiyatullin, Z., Zuev, Y., and Fedotov, V. (1996). Mechanism of transport of charge carriers in the sodium bis(2-ethylhexyl) sulfosuccinate–water–decane microemulsion near the percolation temperature threshold. *J Phys Chem* 100, 3745–3748.

26. Hamilton, R. T., Billman, J. F., and Kaler, E. W. (1990). Measurements of interdroplet attractions and the onset of percolation in water-in-oil microemulsions. *Langmuir* 6, 1696–1700.

27. Garcia-Rio, L., Herves, P., Leis, J. R., and Mujeto, J. C. (1997). Influence of crow ethers and macrocyclic kryptands upon the percolation phenomenon in AOT/isooctane/ H₂O microemulsions. *Langmuir* 13, 6083–6088.

28. (a) Hait, S. K., Moulik, S. P., Rodgers, M. P., Burke, S. E., and Palepu, R. (2001). Physiochemical studies on microemulsions. 7. Dynamics of percolation and energetics of clustering in water/AOT/isooctane and water/AOT/decane w/o microemulsions in the presence of hydrotopes (sodium salicylate, α-napthol, β-napthol, resorcinol, catechol, hydroquinone, pyrogallol and urea) and bile salt (sodium cholate). *J Phys Chem* 105, 7145–7154; (b) Hait, S. K., Sanyal, A., and Moulik, S. P. (2002). Physiochemical studies on microemulsions. 8. The effects of aromatic methoxy hydrotropes on droplet clustering and understanding of dynamics of conductance percolation in water/oil microemulsions. *J Phys Chem B* 106, 12642–12650.

29. Borkovec, M., Eicke, H. F., Hammerich, H., and Gupta, B. D. (1988). Two percolation processes in microemulsions. *J Phys Chem* 92, 206–211.

30. Papadimitriou, V., Xenakis, A., and Lianos, P. (1993). Electric percolation in enzyme-cointaining microemulsions. *Langmuir* 9, 912–915.

31. Nazario, L. M. M., Hatton, T. A., and Crespo, J. P. S. G. (1996). Nonionic cosurfactants in AOT reversed micelles: Effect on percolation, size and solubilization site. *Langmuir* 12, 6326–6335.

32. Huruguen, J. P., Authier, M., Greffe, J. L., and Pileni, M. P. (1991). Percolation process induced by solubilizing cytochrome *c* in reverse micelles. *Langmuir* 7, 243–249.

33. (a) Lagues, M. (1978). Study of structure and electrical conductivity in microemulsions: Evidence for percolation mechanism and phase inversion. *Phys (France) Lett* 39, L487–L491; (b) Lagues, M. (1979). Electrical conductivity of microemulsions: A case of stirred percolation. *J Phys Lett* 40, 331–333; (c) Lagues, M. and Sauterey, C. (1980). A structural description of microemulsions. Small-angle neutron scattering and electrical conductivity study. *J Phys Chem* 84, 1532–1535; (d) Lagues, M. and Sauterey, C. (1980). Percolation transition in water in oil microemulsions. Electrical conductivity measurements. *J Phys Chem* 84, 3503–3508.

34. (a) Bug, A. L. R., Safran, S. A., Grest, G. S., and Webman, I. (1985). Do interactions raise or lower a percolation threshold. *Phys Rev Lett* 55, 1896–1899; (b) Safran, S. A.,

Webman, I., and Grest, G. S. (1986). Percolation in interacting colloids. *Phys Rev A* 32, 506–511; (c) Grest, G. S., Webman, I., Safran, S. A., and Bug, A. L. R. (1986). Dynamic percolation in microemulsions. *Phys Rev A* 33, 2842–2845.

35. Boned, C., Peyrelasse, J., and Saidi, Z. (1993). Dynamic percolation of spheres in a continuum: The case of microemulsions. *Phys Rev E* 47, 5732–5737.

36. Cametti, C., Codastefno, P., Tartaglia, P., Chen, S., and Rouch, J. (1992). Electrical conductivity and percolation phenomena in water-in-oil microemulsions. *Phys Rev A* 45, R5358–R5361.

37. Straley, J. P. (1977). Critical exponents for the conductivity of random resistor lattices. *Phys Rev B* 15, 5733–5737.

38. Bhattacharya, S., Stokes, J. P., Kim, M. W., and Huang, J. S. (1985). Percolation in an oil-continuous microemulsion. *Phys Rev Lett* 55, 1884–1857.

39. Mehta, S. K., Sharma, S., and Bhasin, K. K. (2005). On the temperature percolation in w/o microemulsions in the presence of organic derivatives of chalcogens. *J Phys Chem B* 109, 9751–9759.

40. Maitra, A., Mathew, C., and Varshney, M. (1990). Closed and open structure aggregates in microemulsions and mechanism of percolative conduction. *J Phys Chem* 94, 5290–5292.

41. (a) Meier, W. (1996). Poly(oxyethylene) adsorption in water/oil microemulsions: A conductivity study. *Langmuir* 12, 1188–1192; (b) Meier, W. (1996). Structured polymer networks from o/w-microemulsions and liquid crystalline phases. *Langmuir* 12, 6341–6345.

42. Ray, S., Bisal, S. R., and Moulik, S. P. (1993). Structure and dynamics of microemulsions. Part 1—Effect of additives on percolation of conductance and energetics of clustering in water–AOT–heptane microemulsions. *J Chem Soc Faraday Trans* 89, 3277–3282.

43. Alexandridis, P., Holzwarth, J. F., and Hatton, T. A. (1993). Interfacial dynamics of water-in-oil microemulsion droplets: Determination of the bending modulus using iodine laser temperature jump. *Langmuir* 9, 2045–2052.

44. Mehta, S. K., Kaur, K. K., Sharma, S., and Bhasin, K. K. (2007). Incorporation of aromatic heterocyclic compounds in reverse micelles: A physiochemical and spectroscopic approach. *Colloids Surf A* 298, 252–261.

45. Craig, L. C. (1934). A study of the preparation of alpha-pyridyl halides from alpha-aminopyridine by the diazo reaction. *J Am Chem Soc* 56, 231–232.

46. Mugesh, G., Mont, W. W., and Sies, H. (2001). Chemistry of biologically important synthetic organoselenium compounds. *Chem Rev* 101, 2125–2180.

47. Mehta, S. K., Kaur, K. K., Sharma, S., and Bhasin, K. K. (2007). Behavior of acetyl modified amino acids in reverse micelles: A non-invasive and physiochemical approach. *J Colloid Interface Sci* 314, 689–698.

48. Bommarius, A. S., Holzwarth, J. F., Wang, D. I. C., and Hatton, T. A. (1990). Coalescence and solubilizate exchange in a cationic four-component reversed micellar system. *J Phys Chem* 94, 7232–7239.

49. Podlogar, F., Gašperlin, M., Tomšič, M., Jamnik, A., and Bešter-Rogač, M. (2004). Structural characterization of water–Tween 40®–Imwitor 308®–isopropyl myristate using different experimental methods. *Int J Pharm* 276, 115–128.

50. Lv, F. F., Zheng, L. Q., and Tung, C. -H. (2005). Phase behavior of the microemulsions and the stability of the chloramphenicol in the microemulsion-based ocular drug delivery system. *Int J Pharm* 301, 237–246.

51. Lawerence, M. J. and Rees, G. D. (2000). Microemulsion based media as novel drug delivery systems. *Adv Drug Deliv Rev* 45, 89–121.

52. Garcai-Rio, L., Leis, J. R., Mejuto, J. C., Pena, M. E., and Iglesias, E. (1994). Effects of additives on the internal dynamics and properties of water/AOT/isooctane microemulsions. *Langmuir* 10, 1676–1683.

53. Suarez, M. J. and Lang, J. (1995). Effect of addition of water-soluble polymers in water-in-oil microemulsions made with anionic and cationic surfactants. *J Phys Chem* 99, 4626–4631.

4 Influence of Polyethylene Glycols and Polyethylene Glycol Dimethyl Ethers upon the Internal Dynamics of Water in Oil Microemulsions

A. Cid, L. García-Río, D. Gómez-Díaz, and J. C. Mejuto

CONTENTS

4.1 INTRODUCTION

Microemulsions are stable, transparent solutions of water, oil, and surfactant, either with or without a cosurfactant. Microemulsions have been described as consisting of spherical droplets of a disperse phase separated from a continuous phase by a film of surfactant [1–4] and they are highly dynamic structures. The components organize themselves in time and space by means of different interactions or collisions, giving rise to coalescence and redispersion processes. Numerous studies have been carried out with the aim of determining the structure, dimensions, and internal dynamics of these systems. Among these studies we can cite those involving ultrasedimentation [5], different dispersion techniques [6], time-resolved fluorescence [7], and nuclear magnetic resonance [8]. As these systems provide both organic and aqueous environments, microemulsions can simultaneously dissolve both hydrophobic and hydrophilic compounds, with each compound distributed between water, organic solvent, and surfactant film in accordance with its physicochemical properties. Due to their microheterogeneous structure, microemulsions have found a growing number of scientific and technological applications: they afford control over the size of synthesized microparticles [9] and they have numerous applications in the fields of solubilization and extraction [10–13]. Microemulsions have also been used to simulate complex biological structures [14–19] (particularly in terms of the behavior of trapped water). In keeping with this ever-expanding range of applications, there is an increasing interest in studying the details of chemical [1], photochemical [20], and enzymocatalytic [1,21,22] processes in microemulsions. In particular, microemulsions—like phase transfer catalysis systems—are able to enhance reactions between nonhydrosoluble organic substrates and hydrosoluble reagents and, as a result, the kinetics of numerous reactions in microemulsions have been studied [2,23–31].

Electric conductivity measurements constitute a very useful technique for obtaining information about micellar interactions [32–43]. An AOT-based microemulsion at room temperature [44] presents a very low specific conductivity (ca. 10^{-9}–10^{-7} Ω^{-1} cm^{-1}). This conductivity, however, is significantly greater than it would be if we considered the alkane that constitutes the continuous medium and that is, without doubt, the main component of a microemulsion of water in oil (w/o microemulsions) ($<10^{-14}$ Ω^{-1} cm^{-1}). This increase in the conductivity of the microemulsions, by comparison with that which would be presented by the pure continuous medium, is due to the fact that the microemulsions are capable of transporting charges. When we reach a certain temperature, the conductivity is abruptly increased to give values up to four orders of magnitude greater than the typical conductivity of the w/o microemulsions. This increase reaches a maximum value, from which it remains invariable at a specific conductivity value, which is much higher than that which the microemulsions present before this transition occurs. This phenomenon is known as conductance percolation, and the moment at which there occurs an abrupt transition from a system which is poor as an electric conductor ($<10^{-7}$ Ω^{-1} cm^{-1}) to a system in which there is a fluid circulation of electric charge ($<10^{-3}$ Ω^{-1} cm^{-1}) is termed the percolation threshold.

The mechanism for conductance percolation consists of the formation of channels that allow the exchange of matter between the dispersed water droplets in the continuous phase. Therefore, it is necessary for there to be an effective collision between two water droplets of the microemulsion, causing the droplets to fuse together. Subsequently an exchange of matter between the water droplets takes place (allowing the charge conduction), which in turn brings about the separation of the droplets by means of a process of fission.

In the literature, there are numerous studies about the influence of different additives upon the electrical percolation of microemulsions [45–51]. Recently, the influence of crown ethers and aza crown ethers on the percolation temperature has been studied [45–48]. The observed behavior at low additive concentrations is an increase in the percolation temperature of the microemulsions, whereas at high concentrations there is a reduction in the percolation threshold. The additives are simply acyclic analogs of the crown ethers. These acyclic hosts generally exhibit less cation affinity than the corresponding one cyclic analogs, as a result of unfavorable enthalpic and entropic effects, but they may adopt wrapping conformations similar to those of the crown ethers in the presence of suitable metal cations. The extra flexibility of the polymers, however, also allows them to engage in multiple bridging and helical binding modes unknown for the crown ethers.

In the present work, the influence of polyethylene glycols and polyethylene glycol dimethyl ethers on the percolative phenomenon has been studied. The polyethylene glycols and polyethylene glycol dimethyl ethers were chosen on the basis of chain length (the number of polymeric units). On the basis of this criterion, four polyethylene glycols (ethylene glycol, diethylene glycol, triethylene glycol, and tetraethylene glycol) and polyethylene glycol dimethyl ethers (ethylene glycol dimethyl ether, diethylene glycol dimethyl ether, triethylene glycol dimethyl ether, and tetraethylene glycol dimethyl ether) were studied.

4.2 MATERIALS AND METHODS

The AOT was supplied by Aldrich (purity 98%). Given its highly hygroscopic nature, it was vacuum-dried and used without any further purification. The additives were supplied by Fluka, being the maximum purity commercially available. The microemulsions were prepared by weight. The composition of the microemulsions remained constant and equal to [AOT] = 0.5 M (referred to the total volume of the microemulsion) and $W = [H_2O]/[AOT] = 22.2$. The conductivity was measured using a Crison GPL 32 conductivimeter. Details about the experimental procedure are described elsewhere [48,49].

Traditionally, the percolation temperature (T_p) was obtained in terms of the sigmoidal Boltzmann equation (SBE) proposed by Moulik and coworkers [50]. Percolation temperatures obtained by both methods are compatible. In the following discussion we will use T_p values obtained from the SBE.

4.3 RESULTS AND DISCUSSION

We determined the percolation temperature of AOT/isooctane/water microemulsions ([AOT] = 0.5 M and $W = 22.2$) in the presence of different polyethylene glycols and polyethylene glycol dimethyl ethers concentrations. The obtained results are shown in Figures 4.1 and 4.2, respectively. In all cases, we have observed a decrease in the percolation temperature of the system as the polyethylene glycols and polyethylene glycol dimethyl ethers concentration increases. This behavior is significantly different from that observed with crown ethers, where we had observed [48,49] an increase in the percolation temperature of the system as the crown ether concentration increased, until a maximum value was

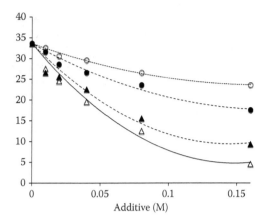

FIGURE 4.1 Influence of mono- (O), di- (●), tri- (▲), and tetraethylene glycol dimethyl ether (△) upon the electric percolation of W = 22.2 and [AOT] = 0.5 M AOT/isooctane/water microemulsions.

reached, from which the percolation temperature decreased. The effect exerted by polyethylene glycols and polyethylene glycol dimethyl ethers is similar to that of other small organic molecules [51], which is considered as a consequence of two different effects: (1) the inclusion into the surfactant film and (2) the replacement of water molecules from the interface.

In addition, as Moulik et al. [45] had proposed for the influence of hydrotropes on percolation phenomena, the bridging of water droplets by the polyethylene glycols and polyethylene glycol dimethyl ethers may be also responsible for the decrease in the percolation temperature. The incorporation of polyethylene glycols

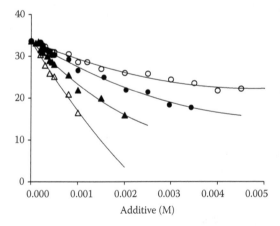

FIGURE 4.2 Influence of mono- (O), di- (●), tri- (▲), and tetraethylene glycol (△) upon the electric percolation of W = 22.2 and [AOT] = 0.5 M AOT/isooctane/water microemulsions.

and polyethylene glycol dimethyl ethers into the surfactant film will modify its geometry because of two different effects: (1) the additive acts as "spacers" within the surfactant monolayer, increasing the effective area of the polar head-groups of AOT. This phenomenon would imply a decrease in the curvature parameter of the surfactant, favoring the formation of positive curvature regions, and hence would favor fusion between droplets, speeding up the percolative process and (2) the association of additives to the surfactant film increases the degree of disorder in the interfacial region, which would imply a decrease in the rigidity of the AOT film and hence an increase of the deformability of the film. It is interesting to establish the role played by the organic molecules in the interfacial region in relation to their capacity to replace water molecules in the vicinity of the polar area of the surfactant film. Previous studies [52] suggest that moderate concentrations of different organic molecules open the interface and facilitate the penetrability of water, whereas high concentrations cause water replacement at the interface. In this sense, these results imply that molecules such as the ureas or thioureas [51] could play an important role in the solvation of the AOT head-groups and that it would justify in part their capacity to modify the percolation threshold of the microemulsions. Polyethylene glycols and polyethylene glycol dimethyl ethers would be able to present similar behavior with regard to their capacity for the water molecules within the solvation "sphere" of the AOT head-groups. As has been established, the fusion of a pair of droplets depends on their mutual contact [45,46]. The presence of an additive able to bridging the water droplets can assist such linkage, enhancing easier fusion, and lowering the percolation temperature. Moreover, the assistance of polyethylene glycols and polyethylene glycol dimethyl ethers in the formation of the channels as intermembrane bridges will favor the transport of Na^+ through the channel. The different behavior of crown ethers and polyethylene glycols and polyethylene glycol dimethyl ethers can be explained by taking into account that polyethylene glycols and polyethylene glycol dimethyl ethers exhibit less cation affinity than their cyclic analogs, as a result of unfavorable enthalpic and entropic effects. The higher stability of cation–crown ether complexes as compared with cation–polyethylene glycols and polyethylene glycol dimethyl ethers complexes can be explained in terms of complexation Gibbs free energy. The Gibbs energy [53] for the K^+–18-crown ether complex ($\Delta G_0 = -34.8$ kJ mol^{-1}) is much lower than the corresponding value for the K^+ equivalent polyethylene glycol dimethyl ether ($\Delta G_0 = -11.4$ kJ mol^{-1}). As we quote, the net effect of the crown ethers on the percolation temperature is a combination of two effects: (1) the complexation of metallic cations and their incorporation into the interface, which impedes the emergence of the percolative process and (2) their association with the surfactant film as organic compounds, increasing the head area of AOT and favoring the formation of positive curvatures and, consequently, facilitating the exchange of matter between droplets. The lower complexation ability of polyethylene glycols and polyethylene glycol dimethyl ethers implies that the quantity of Na^+ solubilized at the interface is much lower than in the case of crown ethers; hence the main effect should be the polyethylene glycols and polyethylene glycol dimethyl ethers association at the surfactant film. The additive influence on the conductance percolation can be related to the geometry of polyethylene glycols and polyethylene glycol dimethyl ethers. The percolation temperature decreases with increasing length of the additive. At a constant value of polyethylene glycols and polyethylene glycol dimethyl ethers concentration, δT_p (difference between the percolation temperature in the presence and in the absence of polyethylene glycols and polyethylene glycol dimethyl ethers) increases with increasing the number of oxygen atoms in the polyethylene glycols and polyethylene glycol dimethyl ethers. The distribution of the polyethylene glycols and polyethylene glycol dimethyl ethers through the microemulsion will be governed by their octanol–water partition coefficient (log P). We must note that all the polyethylene glycols and polyethylene glycol dimethyl ethers used in this work are very soluble in water irrespective of the pH. In fact the log P values are -0.5 ± 0.3, -0.8 ± 0.4, -1.2 ± 0.5, and -1.5 ± 0.5, respectively, for mono-, di-, tri-, and tetraethylene glycols and ethylene glycol dimethyl ethers. These values indicate that water solubility increases with increasing number of oxygen atoms of the glyme, so that the percentage of polyethylene glycols and polyethylene glycol dimethyl ethers at the aqueous microenvironment increases with increasing length of the polyethylene glycols and polyethylene glycol dimethyl ethers (analogous behavior is observed for the polyethylene glycols). From data in Figures 4.1 and 4.2, we observe a clear dependence between percolation temperature and polyethylene glycols and polyethylene glycol dimethyl ethers concentration. On increasing polyethylene glycols and polyethylene glycol dimethyl ethers concentration 16 times a decrease of $\delta T_p = 9°C$, $14°C$, $17°C$, and $21°C$ occurs for mono-, di-, tri-, and tetraethylene glycols and ethylene glycol dimethyl ethers, respectively. We can refer this decrease in percolation temperature to the number of oxygen atoms of the polyethylene glycols and polyethylene glycol dimethyl ethers, obtaining values of $(\delta T_p/O_x) = 4.5°C$, $4.7°C$, $4.3°C$, and $4.2°C$ per oxygen atom for mono-, di-, tri-, and tetraethylene glycols and ethylene glycol dimethyl ethers, respectively. The fact that $(\delta T_p/O_x)$ is independent of the nature of the polyethylene glycols and polyethylene glycol dimethyl ethers (and hence independent of its log P value) means that the main effect on percolation temperature is the number of monomer units interacting with AOT head-groups. The decrease of T_p value in the presence of polyethylene glycols and polyethylene glycol dimethyl ethers is opposite to the behavior observed for long-chain alcohols when they are added to the microemulsions as cosurfactants. Some authors [54–56], using dynamic light scattering, studied droplet size and interactions and concluded that the interactions decreased according to their structure. It is well known that changes in the interactions between droplets are correlated with changes in the value of the percolation temperature. An increase occurs in the value of the percolation temperature, and a decrease in the interaction between microdroplets correlates with this increase. These authors [54–56] made the distinction between short- and long-chain alcohols: the former increased interactions, while the later decreased them. A decrease in interaction was also noticed for other long-chain molecules such as Aracel (a nonionic long-chain surfactant) [55,56]. With regard to long-chain alcohols [57–59] when decanol is added to the system the resulting microemulsion has a higher percolation temperature, and the higher the concentration of alcohol the greater the effect on the percolation temperature. If the cosurfactant added is a poly(oxyethylene)alkyl ether the reverse behavior is observed [56]. The percolation temperature decreases and continues to fall as the concentration increases. The addition of oligoethylene glycols and polyethylene glycols in AOT microemulsions strongly affects the phase boundaries and the percolation behavior of the microemulsion. It is clear that there are significant differences between the inclusion of long-chain alcohols and long-chain alkylamines and the inclusion of polyethylene glycols and polyethylene glycol dimethyl ethers in the AOT film. The position of alcohols and amines inside the interface would probably be perpendicular to the water droplet surface, acting as spacers between the surfactant molecules. The polyethylene glycols and polyethylene glycol dimethyl ethers molecule would be located at the head-group region of the AOT film, replacing water molecules from the hydration sphere of the AOT head-group. This conclusion is supported by results obtained by Nazario et al. [60]. These authors have worked with alkyl polyoxyethylenes (C_iE_j) as cosurfactants. The C_iE_j surfactants have their polar head-groups within the water pools, and tend to produce more flexible films as they try to curve the interface away from the water pool. In this case, for polyethylene glycols and polyethylene glycol dimethyl ethers, the percolation begins at lower temperatures than for the AOT reverse micelles.

ACKNOWLEDGMENTS

Financial support from the Xunta de Galicia (PGIDT03-PXIC20905PN and PGIDIT04TMT209003PR) and the Ministerio de Ciencia y Tecnología (Project BQU2002-01184) is gratefully acknowledged.

SYMBOLS AND TERMINOLOGIES

AOT	bis-2-ethylhexylsulfosuccinate sodium
W	relation between volume of H_2O/alkane
PEG	polyethylene glycol
κ	specific conductivity
ΔG_0	Gibbs energy
δT_p	difference between the percolation temperature in presence and absence of polyethylene glycol dimethyl ethers

REFERENCES

1. P.L. Luisi and B.E. Straub, eds. *Reverse Micelles*, 1984, Plenum Press, New York.
2. M.P. Pileni, ed., *Structure and Reactivity in Reverse Micelles*, 1989, Elsevier, Amsterdam.
3. M. Zulauf and H.F. Eicke, Inverted micelles and microemulsions in the ternary system H_2O/aerosol-OT/isooctane as studied by photon correlation spectroscopy, 1979, *J. Phys. Chem.* 83, 480–486.
4. J. Eastoe, B.H. Robinson, D.C. Steytler, and D.T. Leeson, Structural studies of microemulsions stabilised by aerosol-OT, 1991, *Adv. Colloid Sci.* 36, 1–31.
5. H.F. Eicke and J. Rehak, Helv. On the formation of water/oil microemulsions, 1976, *Chim. Acta* 59, 2883–2891.
6. R. Zana, ed., *Surfactant Solutions. New Methods of Investigations*, 1987, Marcel Dekker, New York.
7. S.S. Atik and J.K. Thomas, Photochemical studies of an oleate oil in water, 1981, *J. Am. Chem. Soc.* 103, 3543–3551.
8. C. Chachaty, Applications of NMR methods to the physical chemistry of micellar solutions, 1987, *Prog. NMR Spectrom.* 19, 3543.
9. J.H. Fendler, Atomic and molecular clusters in membrane mimetic chemistry, 1987, *Chem. Rev.* 87, 877–899.
10. P. Mukerjee and A. Ray, Charge-transfer interactions and the polarity at the surface of micelles of long-chain pyridinium iodides, 1966, *J. Phys. Chem.* 70, 2144–2149.
11. K.L. Mittal, ed. *Micellization, Solubilization and Microemulsions*, 1977, 1, Plenum Press, New York.
12. J. Funasaki, Micellar effects on the kinetics and equilibrium of chemical reactions in salt solutions, 1979, *J. Phys. Chem.* 83, 1998–2003.
13. M.S. Fernández and P. Fromherz, Lipoid pH indicators as probes of electrical potential and polarity in micelles, 1977, *J. Phys. Chem.* 83, 1755–1760.
14. M. Wong, J.K. Thomas, and M. Gräzel, Fluorescence probing of inverted micelles. The state of solubilized water clusters in alkane/diisooctyl sulfosuccinate (aerosol OT) solution, 1976, *J. Am. Chem. Soc.* 98, 2391–2397.
15. G. Bakale, G. Beck, and J.K. Thomas, Electron capture in water pools of reversed micelles, 1981, *J. Phys. Chem.* 85, 1062–1064.
16. P.E. Zinsli, Inhomogeneous interior of aerosol OT microemulsions probed by fluorescence and polarization decay, 1979, *J. Phys. Chem.* 83, 3223–3231.
17. E. Keh and B. Valeur, Investigation of water-containing inverted micelles by fluorescence polarization. Determination of size and internal fluidity, 1981, *J. Colloid Interface Sci.* 79, 465–478.
18. J.K. Thomas, Radiation-induced reactions in organized assemblies, 1980, *Chem. Rev.* 80, 283–299.
19. W. Marcel, 1986, Proteins and peptides in water restricted environments, *Proteins: Struct. Funct. Genet.* 1, 4.
20. E.A. Lissi and D. Engel, Incorporation of *n*-alkanols in reverse micelles in the AOT/*n*-heptane/water system, 1992, *Langmuir* 8, 452–455.
21. R.M.D. Verhaert and R. Hilhorst, Enzymes in reverse micelles: 4. Theoretical analysis of one-substrate/one-product conversion and suggestions for efficient application, 1991, *Recl. Trav. Chim. Pays-Bas.* 110, 236–246.
22. Y.L. Khmelnitsky, I.N. Neverova, V.I. Polyakov, V.Y. Grinberg, A.V. Levashov, and K. Martinek, Kinetic theory of enzymatic reactions in reversed micellar systems. Application of the pseudophase approach for partitioning substrates, 1990, *Eur. J. Biochem.* 190, 155–159.
23. F.M. Menger, J.A. Donohue, and R.F. Williams, Catalysis in water pools, 1973, *J. Am. Chem. Soc.* 95, 286–288.

24. R. Da-Rocha-Pereira, D. Zanette, and F. Nome, Application of the pseudophase ion-exchange model to kinetics in microemulsions of anionic detergents, 1990, *J. Phys. Chem.* 94, 356–361.
25. L. García-Rio, J.C. Mejuto, and M. Pérez-Lorenzo, Ester aminolysis by morpholine in AOT-based water-in-oil microemulsions, 2006, *J. Colloid Interface Sci.* 301, 624–630.
26. L. García-Rio, J.C. Mejuto, and M. Pérez-Lorenzo, First evidence of simultaneous different kinetic behaviors at the interface and the continuous medium of w/o microemulsions, 2006, *J. Phys. Chem. B* 110, 812–819.
27. E. Fernández, L. García-Río, J.R. Leis, J.C. Mejuto, and M. Pérez-Lorenzo, Michael addition and ester aminolysis in w/o AOT-based microemulsions, 2005, *New J. Chem.* 29, 1594–1600.
28. L. García-Rio, J.C. Mejuto, and M. Pérez-Lorenzo, Modification of reactivity by changing microemulsion composition. Basic hydrolysis of nitrophenyl acetate in AOT/isooctane/water systems, 2004, *New J. Chem.* 28, 988–995.
29. L. García-Río, P. Hervés, J.C. Mejuto, and P. Rodríguez-Dafonte, Nitrosation reactions in water/AOT/xylene microemulsions, 2006, *Ind. Eng. Chem. Res.* 45, 600–606.
30. E. Fernández, L. García-Río, J.C. Mejuto, and M. Pérez-Lorenzo, Evidence for compartmentalization of reagents in w/o microemulsions, 2007, *Colloid. Surf. A* 295, 284–287.
31. L. García-Río, P. Hervés, J.C. Mejuto, J. Pérez-Juste, and P. Rodríguez-Dafonte, Comparative study of nitroso group transfer in colloidal aggregates: Micelles, vesicles and microemulsions, 2003, *New J. Chem.* 27, 372–380 and references therein.
32. S. Bhattacharya, J.P. Stokes, M.W. Kim, and J.S. Huang, Percolation in an oil-continuous microemulsion, 1985, *Phys. Rev. Lett.* 55, 1884–1887.
33. H.F. Eicke, M. Borkovec, and B. Das-Gupta, Conductivity of water-in-oil microemulsions: A quantitative charge fluctuation model, 1989, *J. Phys. Chem.* 93, 314–317.
34. N. Kallay and A. Chittofrati, Conductivity of microemulsions: Refinement of charge fluctuation model, 1990, *J. Phys. Chem.* 94, 4755–4756.
35. A. Maitra, C. Mathew, and M. Varshney, Closed and open structure aggregates in microemulsions and mechanism of percolative conduction, 1990, *J. Phys. Chem.* 94, 5290–5292.
36. L. Mukhopadhyay, P.K. Bhattacharya, and S.P. Moulik, Additive effects on the percolation of water/AOT/decane microemulsion with reference to the mechanism of conduction, 1990, *S. P. Colloid. Surf. A* 50, 295–308.
37. R.T. Hamilton, J.F. Billman, and E.W. Kaler, Measurements of interdroplet attractions and the onset of percolation in water-in-oil microemulsions, 1990, *Langmuir* 6, 1696–1700.
38. J.P. Hurugen, M. Authier, J.L. Greffe, and M.P. Pileni, Percolation process induced by solubilizing cytochrome c in reverse micelles, 1991, *Langmuir* 7, 243–249.
39. S. Paul, S. Bisal, and S.P. Moulik, Physicochemical studies on microemulsions: Test of the theories of percolation, 1992, *J. Phys. Chem.* 96, 896–901.
40. S. Ray, S.R. Bisal, and S.P. Moulik, Structure and dynamics of microemulsions. Part 1. Effect of additives on percolation of conductance and energetics of clustering in water-AOT-heptane microemulsions, 1993, *J. Chem. Soc. Faraday Trans.* 89, 3277–3282.
41. P. Alexandridis, J.F. Holzwarth, and T.A. Hatton, Thermodynamics of droplet clustering in percolating AOT water-in-oil microemulsions, 1995, *J. Phys. Chem.* 99, 8222–8232.
42. L. García-Río, J.R. Leis, J.L. López-Fontán, J.C. Mejuto, V. Mosquera, and P. Rodríguez-Dafonte, Mixed micelles of alkylamines and cetyltrimethylammonium chloride, 2005, *J. Colloid Interface Sci.* 289, 521–529.
43. L. García-Río, J.R. Leis, J.C. Mejuto, V. Mosquera, and P. Rodríguez-Dafonte, Stability of mixed micelles of cetylpyridinium chloride and linear primary alkylamines, 2007, *Colloid. Surf. A* 309, 216–223.

44. J. Eastoe, B.H. Robinson, D.C. Steytler, and D.T. Leeson, Structural studies of microemulsions stabilized by aerosol-OT, 1991, *Adv. Colloid Sci.* 36, 1–31.

45. S.P. Moulik, G.C. De, B.B. Bhowmik, and A.K. Panda, Physicochemical studies on microemulsions. 6. Phase behavior, dynamics of percolation, and energetics of droplet clustering in water/AOT/n-heptane system influenced by additives (sodium cholate and sodium salicylate), 1999, *J. Phys. Chem. B* 103, 7122–7129.

46. S.K. Hait, A. Sanyal, and S.P. Moulik, Physicochemical studies on microemulsions. 8. The effects of aromatic methoxy hydrotropes on droplet clustering and understanding of the dynamics of conductance percolation in water/oil microemulsion systems, 2002, *J. Phys. Chem. B* 106, 12642–12650.

47. S.K. Hait, S.P. Moulik, M.P. Rodgers, S.E. Burke, and R. Palepu, Physicochemical studies on microemulsions. 7. Dynamics of percolation and energetics of clustering in water/AOT/isooctane and water/AOT/decane w/o microemulsions in presence of hydrotopes (sodium salicylate, α-naphthol, β-naphthol, resorcinol, catechol, hydroquinone, pyrogallol and urea) and bile salt (sodium cholate), 2001, *J. Phys. Chem. B* 105, 7145–7154.

48. J. Dasilva-Carvalhal, L. García-Río, D. Gómez-Díaz, J.C. Mejuto, and P. Rodríguez Dafonte, Influence of crown ethers on the electric percolation of AOT/isooctane/water (w/o) microemulsions, 2003, *Langmuir* 19, 5975–5983.

49. J. Dasilva-Carvalhal, D. Fernández-Gándara, L. García-Río, and J.C. Mejuto, Influence of aza crown ethers on the electric percolation of AOT/isooctane/water (w/o) microemulsions, 2006, *J. Colloid Interface Sci.* 301, 637–643.

50. S.K. Hait, S.P. Moulik, and R. Palepu, Refined method of assessment of parameters of micellization of surfactants and percolation of w/o microemulsions, 2002, *Langmuir* 18, 2471–2476.

51. L. García-Río, J.R. Leis, J.C. Mejuto, M.E. Peña, and E. Iglesias, Effects of additives on the internal dynamics and properties of water/AOT/isooctane microemulsions, 1994, *Langmuir* 10, 1676–1683.

52. Y.S. Kang, H.J.D. McManus, and L. Kevan, Electron magnetic resonance studies on the photoionization of N-alkylphenothiazines in micellar solutions: Effect of urea on the radical photoyield, 1992, *J. Phys. Chem.* 96, 10049–10055.

53. J.W. Steed and J.L. Atwood, *Supramolecular Chemistry*, 2000, Wiley, Chichester.

54. L. Mukhopadhyay, P.K. Bhattacharya, and S.P. Moulik, Additive effects on the percolation of water/AOT/decane microemulsion with reference to the mechanism of conduction, 1990, *Colloid. Surf.* 50, 295–308.

55. M.J. Hou, M. Kim, and D.O. Shah, A light scattering study on the droplet size and interdroplet interaction in microemulsions of AOT–oil–water system, 1988, *Langmuir* 4, 398–404.

56. G. Gianmonna, F. Goffredi, V. Turco-Liveri, and G. Vassallo, Water structure in water/AOT/n-heptane microemulsions by FTIR spectroscopy, 1992, *J. Colloid Interface Sci.* 154, 411–415.

57. K.L. Mittal and B. Lindman, eds., *Surfactants in Solution*, 1984, 3, Plenum Press, New York.

58. L.M.M. Nazario, T.A. Hatton, and J.P.S.G. Crespo, Nonionic cosurfactants in bis(ethylhexyl) sodium sulfosuccinate reversed micelles: Effect on percolation, size, and solubilization site, 1996, *Langmuir* 12, 6326–6335.

59. S. Ray and S.P. Moulik, Phase behavior, transport properties, and thermodynamics of water/AOT/alkanol microemulsion systems, 1995, *J. Colloid Interface Sci.* 173, 28–33.

60. L.M.M. Nazario, J.P.S.G. Crespo, J.F. Holzwarth, and T.A. Hatton, Dynamics of AOT and AOT/nonionic cosurfactant microemulsions. An iodine-laser temperature jump study, 2000, *Langmuir* 16, 5892–5899.

5 Microemulsions with Mixed Nonionic Surfactants

Monzer Fanun

CONTENTS

5.1 INTRODUCTION

Microemulsions are spontaneously forming, thermodynamically stable, homogeneous low viscous, and optically isotropic solutions. These macroscopic homogeneous mixtures are heterogeneous on a nanometer scale [1–4]. Reverse or inverse microemulsions have nanoscopic water droplets dispersed in a pseudocontinuous phase of oil, and the low radius of curvature in this case has opposite sign to the radius of curvature in the oil-in-water (o/w) example. Due to their unique properties, microemulsions have been used in a variety of techno-logical applications, including environmental protection, nanoparticle forma-tion, personal care product formulations, drug delivery systems, and chemical reaction media [5–10]. In most industrial applications, the use of mixed surfac-tants system is preferred over the use of pure surfactant-based systems [11]. The determinations of the phase behavior of water/mixed surfactants/oil system quantitatively is of eminent significance for further investigation and possible applications. The study of the phase behavior of mixed nonionic surfactants in water and oil demonstrated that the amount of surfactants present at the water–oil interface determines the extent of water and oil mutual solubilization [12–16]. This amount of surfactants depends on factors like surfactant's chemical structure and hydrophilicity, the monomeric solubility of surfactants in oil and water, and the presence of additives. It was found that the water-in-oil (w/o), solubi-lization increases dramatically when nonionic surfactants, whose hydrophilic–lipophilic balances are far separated, are mixed [12–16]. At constant temperature, the hydrophilic–lipophilic balances of mixed surfactants in surfactants mono-layers inside the microemulsion are the same in the case where oil is fixed. It was also found that the total surfactant concentration needed to solubilize an equal amount of water and oil increases with the increase in the lipophilicity (molecular weight) of the oil [12–16]. Diverse and complex dynamics that are highly dependent on the system thermodynamics (i.e., composition and the resulting microstructure) and the dynamics of the constituent molecules are exhibited by microemulsions [1,4,17]. The structure of microemulsions can be idealized as a set of interfaces dividing polar and apolar domains. Mixtures of water, oil, and surfactant exhibit a rich variety of microstructures, ranging from spherical micelles, rodlike micelles, and bicontinuous microemulsions to ordered liquid crystalline phases. Depending on the composition of the system of oil, water, and surface-active agent(s), the microstructure of a microemul-sion may exist as w/o droplets, o/w droplets, or a bicontinuous structure [18–27]. Understanding of the microemulsion's properties is needed for any scientific or industrial application of these systems; thus much work has been done over the last decade in this particular area [28–35]. Various techniques have been employed for this characterization, including electrical conductiv-ity, dynamic viscosity, ultrasonic velocity, light scattering (dynamic and static), pulsed field gradient nuclear magnetic resonance (NMR) spectroscopy, small-angle scattering (SAS) methods, and others [36–51]. In this review paper, we report on the formulation and properties of newly formulated microemulsions of important potential applications including food, pharmacy, nanoparticles technology, and chemical and biochemical synthesis [52–57]. These systems are composed of water/sucrose laurate/ethoxylated mono-di-glyceride/oil. The chemical structures of the components used are presented in Figure 5.1.

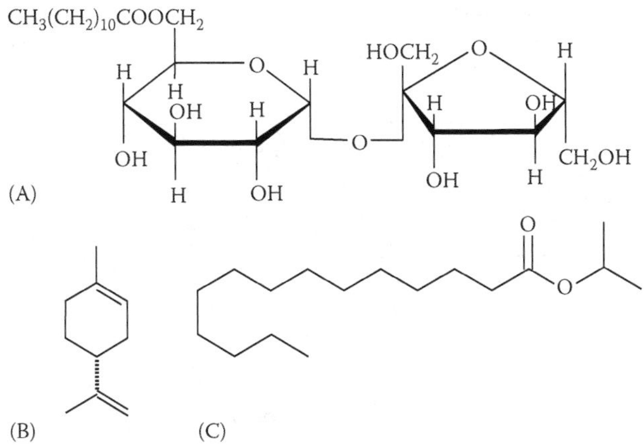

CH$_3$(CH$_2$)$_{10}$COOCH$_2$

(A)

(B)   (C)

FIGURE 5.1   Chemical structures of (A) sucrose laurate, (B) R (+)-LIM, and (C) IPM.

5.2   PHASE BEHAVIOR

A number of factors influence the water solubilization in mixed nonionic surfactants/oil mixtures that include the type of surfactants, surfactants mixing ratio, and the presence of additives such as alcohols or electrolytes and temperature [11]. The major applications of surfactants used additives to improve their properties [58,59]. Among these additives, alcohols hold a special place by being the most frequently used. The presence of alcohol influences the extent of the microemulsions regions and their internal structure. The roles of alcohol in microemulsions are to delay the occurrence of liquid crystalline phases, to increase the fluidity of the interfacial layer separating oil and water, to decrease the interfacial tension between the microemulsion phase and excess oil and water, and to increase the disorder in these interfacial layers as well as their dynamic character. The miscibility of water, oil, surfactant or mixed surfactants, and cosurfactant is a composition-dependent variable [12–16,60]. A number of authors [52–57,61,62] reported on the effect of polar oils such as triglycerides, middle- or long-chain alcohols or fatty acids, fatty acid ester on the water solubilization, and properties of microemulsions for different technological applications. These oils influence the surfactant layer curvature in aggregates or self-organized structures when solubilized. Other authors [52,55,56,63,64] studied cyclic hydrocarbons effect on microemulsions formation. The placement of the solubilized oil in the surfactant aggregates highly affects the change in surfactant layer curvature. If oil tends to penetrate in the surfactant palisade layer and locates near the interface of the water–lipophilic surfactant moiety [65,66], the curvature would be less positive or negative. The curvature being defined as positive when the surfactant film is convex toward water. On the other hand, if oil is solubilized deep inside the aggregates and tends to form an oil pool (swelling), the opposite change in surfactant layer curvature is often induced [67]. It is known that the penetration tendency of amphiphilic or polar oils and cyclic hydrocarbons is large. Cyclic hydrocarbons tend to penetrate in the surfactant layer and widen the effective cross-sectional area per surfactant. As a result, the surfactant layer curvature becomes less positive or negative. On the other hand, the surfactant layer curvature tends to be more positive upon addition of oils that do not penetrate in the surfactant layer. Significant advances in the understanding of phase and microstructural transitions of self-assembling fluids through different techniques have been reported [68–70].

5.2.1 Phase Behavior without Additives

The pseudoternary phase diagram study at 25°C of the water/sucrose laurate/ethoxylated mono-di-glyceride/oil systems for oils R(+)-limonene (R (+)-LIM) and isopropylmyristate (IPM) reveals the presence of an isotropic and low-viscosity area which is a microemulsion one-phase region (1ϕ), the remainder of the phase diagram represents the multiphase ($M\phi$) region based on visual identification. A small microemulsion phase region was observed (i.e., the total one-phase area (A_T) equals 5%) in the ternary system water/sucrose laurate/R (+)-LIM. No microemulsion regions were observed in the ternary systems water/sucrose laurate/IPM systems. Adding ethoxylated mono-di-glyceride to the ternary system enhances the formation of the one-phase microemulsion in the R (+)-LIM-based system and a microemulsion region is observed on the systems based on IPM. For ethoxylated mono-di-glyceride content in the mixture of (ethoxylated mono-di-glyceride + sucrose laurate) equals 25 wt%, the total one-phase microemulsion area (A_T) equals to 22% and 10% for R (+)-LIM, and IPM-based systems, respectively. At equal amounts of ethoxylated mono-di-glyceride and sucrose laurate in the surfactants mixture (i.e., surfactants mixing ratio (w/w) equals unity), the one-phase microemulsion region appears from the first drop of water added. This one-phase region extends overall the water contents range along the dilution line N60 in the system based on R (+)-LIM. The total one-phase area (A_T) equals 62% and 49% for the R (+)-LIM, and IPM-based systems, respectively. The phase diagrams are presented in Figure 5.2. The total one-phase area (A_T) decreases to the value of 47% in the case of R (+)-LIM-based system by increasing the ethoxylated mono-di-glyceride weight ratio in the mixed surfactants to 75 wt%, and increases to the value of 60% in the case of IPM-based system. In the ternary system water/ethoxylated mono-di-glyceride/oil, the total one-phase area (A_T) of the microemulsion phase region decreases to the value of 30%, and 52% in the R (+)-LIM, and IPM-based systems, respectively. Figure 5.3 represents the variation of the total one-phase area (A_T) as a function of ethoxylated mono-di-glyceride content in the mixed surfactants for the system investigated in this section. It seems therefore, as expected, that the mixture of surfactants enhances the surfactants partitioning at the interface, thus increasing the stability of the amphiphilic film. It is considered that surfactant monolayers at the interface of water and oil domains inside the microemulsions are directly related to the solubilization of water and oil. The monodisperse solubilities of sucrose laurate and ethoxylated mono-di-glyceride

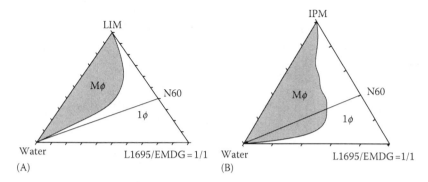

FIGURE 5.2 Pseudoternary phase diagrams of the water/sucrose laurate/ethoxylated mono-di-glyceride/oil systems at 25°C. The mixing ratio (w/w) of ethoxylated mono-di-glyceride/sucrose laurate equals unity. The oils were (A) R (+)-LIM and (B) IPM. The one-phase region is designated by 1φ, and the multiple phase regions are designated by Mφ.

in R (+)-LIM and IPM are very small. This means that the oil domains in the microemulsion phase are almost the same as bulk oil phase when solubilization is large. It can be assumed that the monomeric solubilities of sucrose laurate (S_{L1695}) and that of ethoxylated mono-di-glyceride (S_{EMDG}) in the water phase forming the microemulsions are similar to their respective critical micelle concentrations (CMC), which equal 3.4×10^{-4} and 1.1×10^{-5} M for sucrose laurate and ethoxylated mono-di-glyceride, respectively. Since surfactant

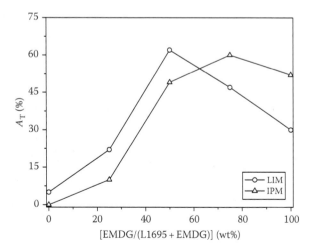

FIGURE 5.3 Variation of the total one-phase area (A_T) as a function of ethoxylated mono-di-glyceride/sucrose laurate mixing ratios (w/w) in water/sucrose laurate/ethoxylated mono-di-glyceride/oil systems at 25°C. The oils were R (+)-LIM and IPM. The lines serve as guides to the eyes.

molecules at the water–oil interface inside microemulsions are directly related to the solubilization, it is important to estimate the mixing fraction of each surfactant. The surfactant content at interface could be obtained by simple mass balance equations as follows:

$$C_{EMDG} = X^{min} X_{EMDG}^{min} - \frac{(1 - X^{min})S_{EMDG}}{2(1 - S_{L1695}) - S_{EMDG}} \tag{5.1}$$

and

$$C_{L1695} = X^{min}(1 - X_{EMDG}^{min}) - \frac{(1 - X^{min})S_{L1695}}{2(1 - S_{L1695}) - S_{EMDG}} \tag{5.2}$$

where

C_{L1695} and C_{EMDG} indicate the weight of sucrose laurate and ethoxylated mono-di-glyceride at the water–oil interface

X^{min} is the minimum weight fraction of mixed surfactants capable of solubilizing equal amounts of water and oil in the microemulsions

X^{min} is the minimum weight fraction of the lipophilic surfactant (in our case the ethoxylated mono-di-glyceride) corresponding to X^{min}

$C_{L1695} + C_{EMDG}$ is the weight fraction of total surfactants in surfactants monolayer at the water–oil interface inside the mixed surfactants microemulsion system and is directly related to the net maximum solubilizing power of the mixed surfactants

The value of $C_{L1695} + C_{EMDG}$ is much smaller than the C_{L1695} and C_{EMDG} in single surfactant-based system, which indicates that the mutual solubilization of water and oil increases due to the mixing of surfactants. At equal amounts of water and oil, the values of $C_{EMDG}/(C_{L1695} + C_{EMDG})$ depend on the surfactants mixing ratio. Table 5.1 shows the values of the mixing fractions of ethoxylated

TABLE 5.1

Mixing Fractions of Ethoxylated Mono-Di-Glyceride at the Water–Oil Interfaces in Water/Sucrose Laurate/Ethoxylated Mono-Di-Glyceride/Oil Microemulsions

IPM		R (+)-LIM		
S_{EMDG}^S	$C_{EMDG}/(C_{L1695} + C_{EMDG})$	S_{EMDG}^S	$C_{EMDG}/(C_{L1695} + C_{EMDG})$	Surfactants Mixing Ratio
		0.148	0.150	1/3
0.208	0.220	0.122	0.120	1/1
0.220	0.23	0.147	0.150	3/1

mono-di-glyceride surfactant at the water–oil interface for R (+)-LIM and IPM. From Table 5.1, we can see that a minimum value of $C_{EMDG}/(C_{L1695} + C_{EMDG})$ is obtained at surfactants mixing ratio equals unity which corresponds to maximum water solubilization as shown in Figure 5.3. Another way to determine the surfactants content at the interface of water–oil in the microemulsion systems is by calculating the mixing weight fraction of the surfactant at the interface using the equation:

$$X_{EMDG} = S^S_{EMDG} + \frac{S_{EMDG}S^S_{L1695} - S_{L1695}S^S_{EMDG}}{1 - S_{L1695} - S_{EMDG}} R_{ow}\left(\frac{1}{X} - 1\right) \qquad (5.3)$$

where

X_{EMDG} represents the weight fraction of ethoxylated mono-di-glyceride (the lipophilic surfactant) in the total mixed surfactants

S^S_{L1695} and S^S_{EMDG} represent the mixing weight fraction at the water–oil interface of sucrose laurate and ethoxylated mono-di-glyceride, respectively

R_{ow} is the weight fraction of o/w + oil

X is the weight fraction of mixed surfactants in the microemulsions

To do the calculations, we estimate that the water and oil are pure and do not dissolve in each other. By plotting X_{EMDG} versus $((1/X)-1)$, a straight line is obtained. S^S_{EMDG} is the intercept and should be equal to the value of $C_{EMDG}/(C_{L1695} + C_{EMDG})$. The obtained values of S^S_{EMDG} are in good agreement with the values obtained for $C_{EMDG}/(C_{L1695} + C_{EMDG})$ at equal amounts of water and oil in the microemulsions as shown in Table 5.1. In other words, it is assumed that the sucrose laurate molecules are present only at the surfactant layers inside the microemulsion phase. The ethoxylated mono-di-glyceride molecules are distributed between the microwater domains and the interface inside the microemulsion phase in a one-phase microemulsions. Kuneida et al. [5,12–16] reported on similar results obtained with mixtures of sucrose monolaurate and polyethylene glycol alkyl ether systems in the presence of heptane, decane, and hexadecane oils. The solubilization capability increases with the mixing of surfactants in particular when surfactants with different hydrophilic–lipophilic balances are mixed [12–16]. The monomeric solubility of lipophilic surfactant (ethoxylated mono-di-glyceride) in oil is low as was reported in our previous study [56] and its mixing with sucrose laurate enables us to obtain large solubilization capacity of water and oil. The difference in the behavior of the studied oils is explained in terms of oil lipophilicity and penetration in the surfactants palisade layer. Solubilization and placement of polar oils in aggregates or self-organized structures influences the surfactant layer curvature and the structure of the self-assembled systems. If the oils have a tendency to penetrate in the surfactant palisade layer and locate near the interface of the water-lipophilic surfactant moiety, the curvature would be less positive or negative [62]. IPM, which is a polar oil, has large penetration tendency in the surfactant palisade layer and this layer will be less positive or less convex toward water.

For this reason, the solubilization capacity of the microemulsions formulated with IPM is lower than that observed with R (+)-LIM. The oil behavior can also be governed by the chain compatibility between oil and surfactants. The change in the solubilization capacity behavior of the two oils when the mixing ratio (w/w) of ethoxylated mono-di-glyceride increases to 3/1 or in the quaternary systems water/ethoxylated mono-di-glyceride/oil could be attributed to the better chain compatibility between the mixed surfactants chains and the IPM chain length.

5.2.2 Phase Behavior with Additives

5.2.2.1 Alcohol as the Additive

The pseudoternary phase diagrams of the water/sucrose laurate/ethoxylated mono-di-glyceride/(oil + ethanol) systems for the oils R (+)-LIM and IPM at 25°C were studied. The mixing ratios, R, (w/w) of ethoxylated mono-di-glyceride/sucrose laurate were varied. The mixing ratio (w/w) of ethanol/oil was kept constant and equals unity. Figure 5.4 presents the pseudoternary phase diagrams at the mixing ratios, R, (w/w) of ethoxylated mono-di-glyceride/sucrose laurate equals unity for the two oils. The phase behavior indicates the presence of an isotropic and low-viscosity area that is a microemulsion one-phase region (1ϕ), the remainder of the phase diagram represents the multiphase ($M\phi$) region based on visual identification. The total one-phase area (A_T) of microemulsion phase region observed in the quaternary system water/sucrose laurate/oil + ethanol equals 35% and 10% for the systems based on R (+)-LIM and IPM, respectively.

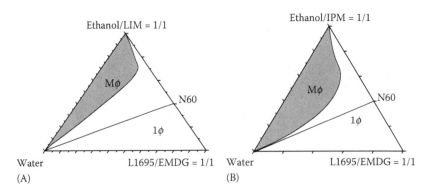

FIGURE 5.4 Pseudoternary phase diagrams of the water/sucrose laurate/ethoxylated mono-di-glyceride/oil + ethanol systems at 25°C. The mixing ratios (w/w) of ethanol/oil and that of ethoxylated mono-di-glyceride/sucrose laurate equal unity. The oils were (A) R (+)-LIM and (B) IPM. The one phase region is designated by 1ϕ, and the multiple phase regions are designated by $M\phi$.

Adding ethoxylated mono-di-glyceride to the quaternary system enhances the formation of the one-phase microemulsion in the two systems. For ethoxylated mono-di-glyceride/sucrose laurate, the mixing ratio equals 3/1, the total one-phase microemulsion area (A_T) equals to 60 and 57 for R (+)-LIM and IPM-based systems, respectively. At equal amounts of ethoxylated mono-di-glyceride and sucrose laurate in the surfactants mixture (i.e., surfactants mixing ratio (w/w) equals unity), the one-phase region extends overall the water contents range along the dilution line N60 in the system based on the two oils. The total one-phase area (A_T) equals 74 and 73 for the R (+)-LIM and IPM-based systems, respectively. The total one-phase area (A_T) decreases to the value of 66% and 70% in the case of R (+)-LIM and IPM-based systems, respectively. In the quaternary systems, water/ethoxylated mono-di-glyceride/oil + ethanol, the total one-phase area (A_T) of the microemulsion phase region decreases to the value of 45% and 65% in the R (+)-LIM and IPM-based systems, respectively. Figure 5.5 represents the variation of the total one-phase area (A_T) as a function of ethoxylated mono-di-glyceride content in the mixed surfactants for the systems investigated in this section. Using Equations 5.1 through 5.3 to evaluate the surfactant content and mixing weight fraction of the surfactant at the interface, we can see in Table 5.2 that the values of $C_{EMDG}/(C_{L1695} + C_{EMDG})$ are smaller than those determined in the absence of ethanol and a minimum value of $C_{EMDG}/(C_{L1695} + C_{EMDG})$ is obtained at surfactants mixing ratio equals unity which corresponds to maximum water solubilization as shown in Figure 5.5.

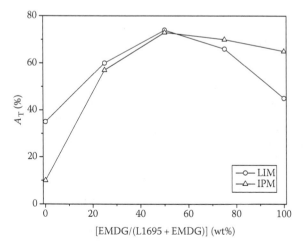

FIGURE 5.5 Variation of the total one-phase area (A_T) as a function of ethoxylated mono-di-glyceride/sucrose laurate mixing ratios (w/w) in water/sucrose laurate/ethoxylated mono-di-glyceride/oil + ethanol systems at 25°C. The ethanol/oil weight ratio equals unity. The oils were R (+)-LIM and IPM. The lines serve as guides to the eyes.

TABLE 5.2

Mixing Fractions of Ethoxylated Mono-Di-Glyceride at the Water–Oil Interfaces in Water/Sucrose Laurate/Ethoxylated Mono-Di-Glyceride/ Oil + Ethanol Microemulsions

	IPM		R (+)-LIM		
S_{EMDG}^s	$C_{EMDG}/(C_{L1695} + C_{EMDG})$	S_{EMDG}^s	$C_{EMDG}/(C_{L1695} + C_{EMDG})$	Surfactants Mixing Ratio
0.113	0.120	0.055	0.060	1/3
0.095	0.103	0.036	0.040	1/1
0.107	0.115	0.065	0.070	3/1

The reason for the selection of the ethanol/oil weight ratio equals unity becomes apparent when we consider the phase behavior of water/sucrose laurate/ ethoxylated mono-di-glyceride/oil + ethanol at different mixing ratios (w/w) of ethanol/oil varying from zero to three and a mixing ratio (w/w) of ethoxylated mono-di-glyceride/sucrose laurate equals unity. Figure 5.6 represents the variation of the total one-phase area (A_T) as a function of ethanol/oil mixing ratios (w/w) for the system investigated in this section. It is seen that the weight ratio of ethanol/ oil equal unity coincides more or less with the points where the curves in Figure 5.6 begin to level off. It is evidently superfluous to use higher ethanol/oil weight ratios to get the same degree of water solubilization. From a molecular point of

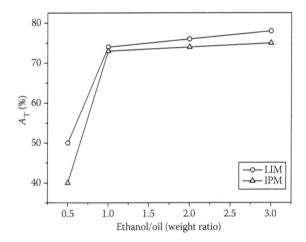

FIGURE 5.6 Variation of the total one-phase area (A_T) as a function of the ethanol/oil mixing ratios (w/w) in water/sucrose laurate/ethoxylated mono-di-glyceride/oil + ethanol systems at 25°C. The mixing ratio (w/w) of ethoxylated mono-di-glyceride/sucrose laurate equals unity. The oils were R (+)-LIM and IPM. The lines serve as guides to the eye.

view, the ascending branch of the plot reflects the gradual migration of alcohol molecules to the interface, thereby facilitating water solubilization within the microemulsion [40]. The addition of ethoxylated surfactant to sucrose esters and a short-chain alcohol (ethanol) to the oil phase increases the microemulsion area significantly compared to the systems free of ethanol. The change in phase behavior caused by the addition of mixed surfactants and ethanol interpreted in terms of film properties; that is, the flexibility of the surfactant film increased because the liquid crystal phase destabilized in favor of a microemulsion phase that even connects the aqueous with the oil phase. It can also be shown that

$$A_{CO} = \frac{X_{MS} \, a_{CO}^{MS} + X_{ethanol} \, a_{CO}^{ethanol}}{X_{MS} \, S_{MS} + X_{ethanol} \, S_{ethanol}} \tag{5.4}$$

Similarly

$$A_{CW} = \frac{X_{MS} \, a_{CW}^{MS} + X_{ethanol} \, a_{CW}^{ethanol}}{X_{MS} \, S_{MS} + X_{ethanol} \, S_{ethanol}} \tag{5.5}$$

where

A_{co} or A_{cw} terms designate the energies per molecule for the interactions of the mixed surfactants and the ethanol "a" (shown as superscripts on these terms) with oil (O) or water (W)

C denotes an amphiphile (i.e., a surfactant or an alcohol)

$X_{ethanol}$ is the molar fraction of the alcohol at the oil–water interface. a_{CO}^{MS}, $a_{CO}^{ethanol}$, a_{CW}^{MS}, and $a_{CW}^{ethanol}$ are, respectively, the interaction energy of mixed surfactants and the ethanol in the C-layer with the oil and water layers

Thus, since $X_{MS} + X_{ethanol} = 1$, the R-ratio can be written as

$$R = \frac{a_{CO}^{ethanol} - \left(a_{CO}^{ethanol} - a_{CO}^{MS}\right)X_{MS}}{a_{CW}^{ethanol} - \left(a_{CW}^{ethanol} - a_{CW}^{MS}\right)X_{MS}} \tag{5.6}$$

In the derivation of Equation 5.6, some simplifying and contestable assumptions have been made [50,51]. However, the main conclusions that may be inferred from this equation remain valid: R depends on the alcohol interfacial concentration via $X_{ethanol}$ and on the type of alcohol through $a_{CO}^{ethanol}$ and $a_{CW}^{ethanol}$. Clearly, $X_{ethanol}$ is related to the (total) ethanol concentration in solution water–ethanol, and to its chemical structure, as these parameters determine the partition of the alcohol between the aqueous, oleic, and interfacial regions. Increasing the added alcohol concentration does not entail enhanced water solubilization. The ratios $a_{CW}^{MS}/a_{CW}^{ethanol}$ and $S_{MS}/S_{ethanol}$ (where S_i denotes the molecular surface area occupied by the component i at the interface) determine whether water solubilization (and consequently A_T) is increased, is decreased, or remains constant with increasing $X_{ethanol}$ [71,72].

5.2.2.2 Effect of Added Salt

Figure 5.7 presents the pseudoternary phase diagrams of the systems water + NaCl/sucrose laurate/ethoxylated mono-di-glyceride/(oil + ethanol) for the oils R (+)-LIM, and IPM at 25°C. The mixing ratios (w/w) of sucrose laurate/ethoxylated mono-di-glyceride and that of ethanol/oil equal unity. The concentration of NaCl in water equals 0.01 M. In Figure 5.7A, the phase diagram of water + NaCl/ sucrose laurate/ethoxylated mono-di-glyceride/(R (+)-LIM + ethanol) is practically very similar as presented in the previous section. Figure 5.7B shows the phase diagrams of water + NaCl/sucrose laurate/ethoxylated mono-di-glyceride/ (IPM + ethanol), which is different from that presented in the previous section indicating the effect of adding salt on the phase behavior. Figure 5.8 presents the variation of the total one-phase region as a function of ethoxylated mono-di-glyceride content in the mixed surfactants for this system. From Figure 5.8, it could be shown that practically no effect of using 0.01 M NaCl aqueous solution instead of pure water in the case of R (+)-LIM-based system. A maximum of 5% decrease was observed in the total one-phase area in these systems. Similar independence of phase behavior in the presence of a small amount of electrolyte is reported in the literature [64,73,74]. In the case of IPM, addition of brine solution instead of water to the mixture of sucrose laurate/ethoxylated mono-di-glyceride/(IPM + ethanol) causes a decrease of about 13% in the one-phase region. Other authors [1–4] reported similar effect of adding salt to the microemulsions. The significant shrinkage of the microemulsion region upon replacing pure water with NaCl solution for the formulation of microemulsion in the case of IPM can be related to a decrease in the interactions among surfactant headgroups induced by a higher Na + binding at the interface.

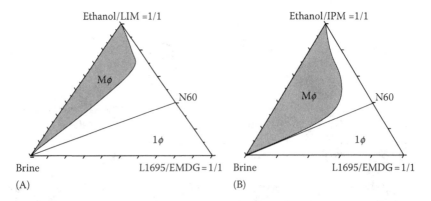

FIGURE 5.7 Pseudoternary phase diagrams of the brine/sucrose laurate/ethoxylated mono-di-glyceride/oil + ethanol systems at 25°C. The mixing ratios (w/w) of oil/ethanol and that of sucrose laurate/ethoxylated mono-di-glyceride equal unity. The concentration of NaCl in water is 0.01 M. The one phase region is designated by 1ϕ, and the multiple phase regions are designated by $M\phi$. (A) R (+)-LIM and (B) IPM.

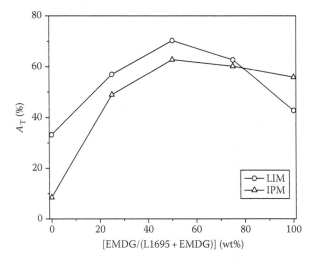

FIGURE 5.8   Variation of the total one-phase area (A_T) as a function of the of ethoxy-lated mono-di-glyceride/sucrose laurate mixing ratio (w/w) in brine/sucrose laurate/ethoxylated mono-di-glyceride/oil + ethanol systems at 25°C. The ethanol/oil mixing ratio (w/w) equals unity. The concentration of NaCl in water is 0.01 M. The oils were R (+)-LIM and IPM. The lines serve as guides to the eyes.

5.3   TRANSPORT PROPERTIES

The microstructure of (w/o) microemulsions is reasonably well described in terms of spherical droplets [1,4,17]. So far, with increasing water volume fraction, discrepancies with respect to the hard sphere model have been discussed in terms of strong interparticle interactions, microstructural transitions, and percolation behaviors, which are oil- and temperature-dependent [75–80]. Clusters of droplets which, in turn, generate water networks throughout the w/o phase are produced by the short-range attractive interactions. As a sequence, important changes of the transport properties such as electrical conductivity and dynamic viscosity occur. The phenomena have been described in terms of "percolation" [81,82]. According to the percolation model, the conductivity remains low up to a certain volume fraction (ϕ_c) of water at constant temperature, when the temperature reaches a value T_c at constant water volume fraction ϕ, or when the water-to-surfactant molar ratio increases. It must be emphasized that these conducting w/o droplets below ϕ_c are isolated from each other embedded in nonconducting continuum oil phase and hence contribute very little to the conductance. However, as the volume fraction of water reaches the percolation threshold ϕ_c, some of these conductive droplets begin to contact each other and form clusters which are sufficiently close to each other. The number of such clusters increases very rapidly above the percolation threshold ϕ_c, giving rise to the observed changes of properties, in particular to the increase of electrical conductivity. The electrical conductivity σ above

ϕ_c has been attributed to the transfer of counterions from one droplet to another through water channels opening between droplets during sticky collisions through transient merging of droplets [1]. The existence and position of this threshold depends on the interactions between droplets which control the duration of the collision and the degree of the interface overlapping, hence the probability of merging. Building up of conductivity needs attractive interactions and ϕ_c decreases when the strength of these interdroplet interactions increases as predicted by recent theoretical calculations [83]. A theoretical model of Safran et al. [83], which is based on the dynamical picture of percolation, has been utilized to analyze the conductivity results of the systems presented in the previous section. According to the theory

$$\sigma = A(\phi_c - \phi)^{-s} \quad \text{if } \phi < \phi_c \tag{5.7}$$

$$\sigma = B(\phi - \phi_c)^t \quad \text{if } \phi > \phi_c \tag{5.8}$$

The critical exponent t generally ranges between 1.5 and 2, whereas the exponent s allows assignment of the time-dependent percolation regime. Thus, $s < 1$ (generally around 0.6) identifies a "static percolation" regime, and $s > 1$ (generally around 1.3) identifies a "dynamic percolation" regime. The static percolation is related to the appearance of bicontinuous microemulsions [84]. The dynamic percolation [83,85] is related to rapid processes of fusion–fission among the droplets. Transient water channels form when the surfactant interface breaks down during collisions or through the merging of droplets. The interpretation of these interactions in terms of static or dynamic percolation is obviously strictly dependent on the timescale of the experimental technique. Experiments carried out within the whole microemulsion phase revealed a variety of interactions upon moving on water dilution line N60 in the water + NaCl/sucrose laurate/ethoxylated mono-di-glyceride/oleic phase systems. Close to the oil corner, the interactions can be described in terms of charge fluctuations among small w/o droplets. With increasing surfactant and water content, the lifetime of the contacts among the droplets increases up to the identification of a structural transition from closed water domains to a connected water network through a gradual variation of the interconnectivity. The comprehensive view of the physicochemical properties and microstructures of microemulsions requires the use of a variety of experimental methods [86–89].

Measurement of viscosity of microemulsions provides information on the flow properties of the systems [86–89]. The study of the viscosity of microemulsions as a tool for the assessment of fluids circulations within the droplets was performed in early stage of microemulsions investigation [76]. The study of Ktistis [90] confirmed that the change in the mass ratio of different components of the microemulsions produced a systematic change in the viscosity of the system. The viscosity of the microemulsions plays a major role in their intended applications as reported by Haβe and Keipert [91] in their study on the development and characterization of microemulsions for ocular applications. Shia et al. [92,93] affirmed that the spacing between molecules at microemulsions interfaces is extremely important in determining the physicochemical properties of such interfaces. Among these properties, surface viscosity is much affected by the chain length in mixed surfactants systems for technological applications. Constantinides and Scalart [94] studied the effect of changing oil chain length on the viscosity of microemulsions. They confirmed that the change from long chain to medium chain glycerides in microemulsions based on the same surfactants and the same composition affects highly the viscosity of the microemulsions. They reported that the viscosity increases as the chain length of the glyceride increases. In their study, Mehta and Bala [64] reported on a trend toward enhanced water-like character of the viscosity of the dispersed phase at high-volume fractions of water. Acharya et al. [95,96] studied the shear viscosities of microemulsions designed for pharmaceutical applications at different shear rates and temperatures and evaluated the activation parameters of viscous flow. Djordjecic et al. [73,97] demonstrated that the measurement of apparent viscosity of microemulsions used as drug delivery vehicles for an amphiphilic drug well described the microemulsions microstructure gradual changes. A percolation transition was predicted in pharmaceutically usable microemulsions systems prepared from water and IPM using mixed surfactants as reported by Podlogor et al. [98]. Mitra and Paul [99] revealed that the viscosity of microemulsions is affected by the mixing ratio (w/w) of mixed surfactants. They also observed that the presence of additives influences the viscosity of the microemulsions. They also evaluated the thermodynamic parameters of viscous flow. Rodriguez-Abreu et al. [100] reported on a study of the reheological behavior of water/sucrose hexadecanoate/cosurfactant/oil microemulsions systems. They found that the viscosity of the wormlike micellar solutions is affected by the alkyl chain length of the oils that tend to solubilize in the micellar core. Other authors [101–106] reported on the flow properties of microemulsions that include the consistency of flow, the flow behavior index. These studies revealed variable features: rise and decrease as well as a viscosity maximum have been observed depending on the composition, temperature, and other factors [53,107,108].

5.3.1 Electrical Conductivity (σ)

The electrical conductivity was measured on the (water + sodium chloride)/sucrose laurate/ethoxylated mono-di-glyceride/oil microemulsions for R (+)-LIM and IPM where the mixing ratio (w/w) of ethoxylated mono-di-glyceride/sucrose laurate equals unity along the dilution line N60. The concentration of sodium chloride in water is 0.01 M. Figure 5.9 displays the influence of water volume fraction on the electrical conductivity (σ). As the volume fraction of water increases, the electrical conductivity increases exponentially. The increase in the electrical conductivity as function of water volume fraction is due to the increase in the fraction of sodium chloride ions that are not enclosed in the core of the microemulsions. The high values of electrical conductivity at high-water volume fractions are explained by the fact that the sodium chloride ions are present in the external phase, which is the water. These results permit to distinguish

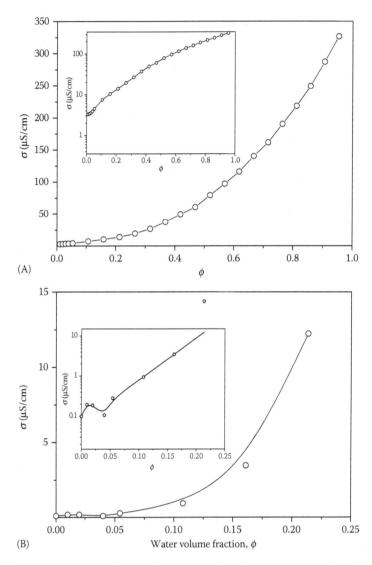

FIGURE 5.9 Variation of the electrical conductivity (σ) of (water + sodium chloride)/ sucrose laurate/ethoxylated mono-di-glyceride/oil system as function of water volume fraction along the dilution line N60 at 25°C. The mixing ratio (w/w) of ethoxylated mono-di-glyceride/sucrose laurate equals unity. Sodium chloride concentration in water is 0.01 M. The phase diagram is presented in Figure 5.7. (A) R (+)-LIM and (B) IPM. The lines serve as guides to the eye.

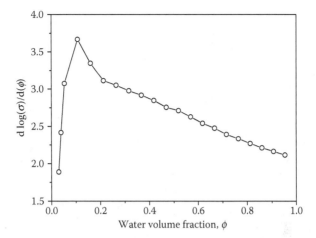

FIGURE 5.10 Plot of d log(σ)/d(ϕ) as function of water volume fraction for the system presented in Figure 5.9A. The lines serve as guides to the eye.

between w/o and o/w microemulsions. Similar results and interpretations were provided by a number of authors [54,55,57,87,88] who studied electrical conductivity variations as function of water volume fraction as a method of characterization of microemulsions interdroplet interactions. Figure 5.10 shows a plot of d(log σ)/dϕ as function of water volume fraction for the system presented in Figure 5.9A. The changes observed in the curve presented in Figure 5.10 have been attributed to the occurrence of a percolation transition. Φ_c, s, and t and prefactors A and B in the theoretical model of Safran et al. [83] were determined by numerical analysis with adjustment by the least squares method using simultaneously Equations 5.7 and 5.8. The computed s and t values are 0.16 and 1.0. The resulting ϕ_c values obtained in this manner are close to the values obtained by the numerical estimate of the maximum of d(log σ)/dϕ versus ϕ (Figure 5.10). There is a reasonable agreement between calculated (by Equations 5.7 and 5.8) and experimental values within the prescribed range of composition with a mean deviation of 5%. The above equations are valid only near ϕ_c and cannot be extrapolated to infinite dilution and unit concentration. In addition, these are not applicable at the immediate vicinity of ϕ_c, where there is a continuous variation within a narrow interval around the percolation threshold. The electrical conductivity was measured also on the (water + sodium chloride)/sucrose laurate/ethoxylated mono-di-glyceride/oil + ethanol along the dilution line N60 at 25°C (see Figure 5.7). The mixing ratio (w/w) of ethoxylated mono-di-glyceride/sucrose laurate and that of oil/ethanol equal unity. The concentration of sodium chloride in water is 0.01 M. All samples appeared transparent and isotropic. The influence of water volume fraction change on the electrical conductivity σ with respect to volume fraction of water was measured. The increase in the water volume fraction induces the increase in electrical conductivity exponentially in all systems. These changes have been attributed to the occurrence of a percolation transition [57,61]. In this percolation model, the conductivity remains low up to a certain volume fraction (ϕ_c) of water. It must be emphasized that these conducting w/o droplets below ϕ_c are isolated from each other embedded in nonconducting continuum oil phase and hence contribute very little to the conductance. However, as the volume fraction of water reaches the percolation thresholds (see Table 5.4), some of these conductive droplets begin to contact each other and form clusters which are sufficiently close to each other. The number of such clusters increases very rapidly above the percolation threshold, giving rise to the observed changes of properties, in particular to the increase of electrical conductivity. The water volume fraction ϕ above ϕ_c has been attributed to the transfer of counterions from one droplet to another through water channels opening between droplets during sticky collisions through transient merging of droplets [1]. Φ_c, s, and t and prefactors A and B were determined by numerical analysis with adjustment by the least squares method using simultaneously Equations 5.7 and 5.8. The computed s and t values are presented in Table 5.3. The resulting ϕ_c values obtained in this manner are close to the values obtained by the numerical estimate of the maximum of $d(\log\sigma)/d\phi$ versus ϕ (Figures 5.11 and 5.12). The calculated values of s and t indicate that the percolation process is static. The effect of temperature on electrical conductivity was also determined for all water volume fractions of microemulsion samples in the one-phase region for the microemulsions formulated using the three different oils. The energetics of conduction flow in microemulsion systems were calculated using an Arrhenius type equation [1,95,96]

$$\sigma = A\exp(\Delta E_{cond}/RT) \tag{5.9}$$

where
σ is the conductance
A is a preexponential constant
ΔE_{cond} is the activation energy of conduction

TABLE 5.3
Values of the Experimental and Theoretical Parameters of the Percolation Process in the Water + Sodium Chloride/ Sucrose Laurate/Ethoxylated Mono-Di-Glyceride/ Oil Microemulsions Studied at 25°C

Theoretical (ϕ_c)	s	t	Experimental (ϕ_c)	Oil-Based Microemulsion
0.05	0.55	1.5	0.04	R (+)-LIM
0.12	0.64	1.7	0.11	IPM

Note: Mixing ratio (w/w) of sucrose laurate/ethoxylated mono-di-glyceride equals unity. The concentration of NaCl in water is 0.01 M.

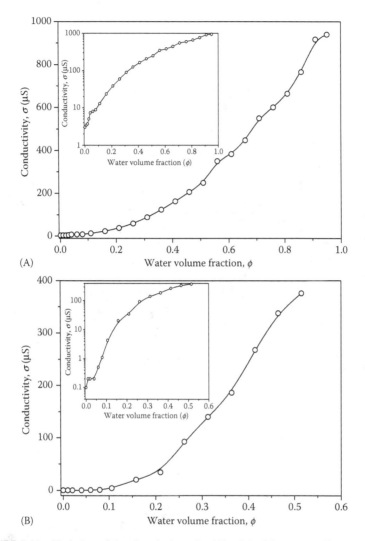

FIGURE 5.11 Variation of the electrical conductivity (σ) of (water + sodium chloride)/sucrose laurate/ethoxylated mono-di-glyceride/oil + ethanol system as function of water volume fraction along the dilution line N60 at 25°C. The mixing ratios (w/w) of ethanol/oil and ethoxylated mono-di-glyceride/sucrose laurate equal unity. Sodium chloride concentration in water is 0.01 M. The phase diagram is presented in Figure 5.7. The lines serve as guides to the eye. (A) R (+)-LIM and (B) IPM.

The plots between ln σ and $1/T$ were obtained for these systems which produced good linear fits (Figure 5.13). ΔE_{cond} values were calculated from the slopes and presented as function of water volume fraction for the microemulsions based on the two oils in Figure 5.14. Similar behaviors of ΔE_{cond} were reported [99, 108–110]. Five different regions can be identified in the activation energy of conduction

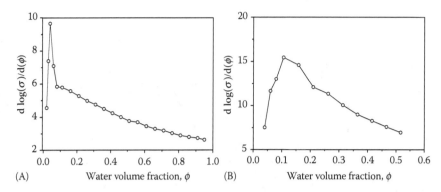

FIGURE 5.12 Plot of d log(σ)/d(ϕ) as function of water volume fraction for the system presented in Figure 5.11. The lines serve as guides to the eye. (A) R (+)-LIM and (B) IPM.

flow (ΔE_{cond}) as function of water volume fraction in the microemulsion system based on the R (+)-LIM at the dilution line N60 as presented in Figure 5.14A. At 0 to 0.04 water volume fractions (region I), the ΔE_{cond} decreases dramatically from 17.04 to 12.25 kJ/mol. The decrease in the ΔE_{cond} in the first region can be attributed to the increasing interactions between the w/o droplets. The hydroxyl and ethoxy groups of the surfactants strongly bind water. Adding water to the mixture

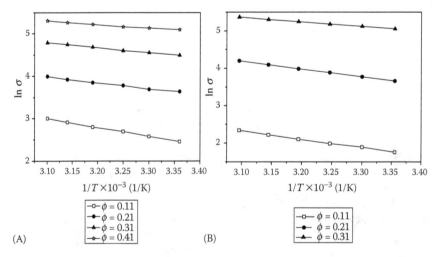

FIGURE 5.13 Semilog plot of electrical conductivity logσ as function of temperature of the system water + sodium chloride)/sucrose laurate/ethoxylated mono-di-glyceride/oil + ethanol system along the dilution line N60 at 25°C. The mixing ratios (w/w) of oil/ethanol and that of sucrose laurate/ethoxylated mono-di-glyceride equal unity. The concentration of NaCl in water is 0.01 M. The phase diagrams are presented in Figure 5.7. (A) R (+)-LIM and (B) IPM.

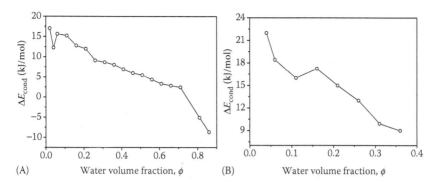

FIGURE 5.14 Plot of the activation of energy of conductive flow as function of water volume fraction of the (water + sodium chloride)/sucrose laurate/ethoxylated mono-di-glyceride/oil + ethanol systems along the dilution line N60 at 25°C. The mixing ratios (w/w) of oil/ethanol and that of sucrose laurate/ethoxylated mono-di-glyceride equal unity. The concentration of NaCl in water is 0.01 M. The phase diagrams are presented in Figure 5.7. (A) R (+)-LIM and (B) IPM.

of surfactants and oil induces the swelling mechanism. At 0.04 to 0.06 water volume fractions (region II), the ΔE_{cond} increases from 12.25 to 15.66 kJ/mol. This increase in the ΔE_{cond} in this region could be attributed to the reorganization of the clusters. At 0.06 to 0.26 water volume fractions (region III), the ΔE_{cond} decreases from 15.66 to 9.21 kJ/mol; the clusters form open channels and the transport of ions becomes easier and therefore the ΔE_{cond} decreases. In the fourth region (region IV) for water volume fractions from 0.26 to 0.71, a slow decrease in the ΔE_{cond} is observed. This fairly slow decrease in the ΔE_{cond} could be associated with the fact that the system has been transformed into a bicontinuous phase, and the interfacial area remains almost unchanged and the rate of transport of ions decreases slowly. For water volume fractions between 0.71 and 0.86 (region V), a sharp decrease is observed in the ΔE_{cond} indicating the transfer from bicontinuous to o/w structure. Three different regions are observed in the activation energy of conduction flow (ΔE_{cond}) as function of water volume fraction in the microemulsion system based on the IPM at the dilution line N60 as presented in Figure 5.14B. At 0 to 0.11 water volume fractions (region I), the ΔE_{cond} decreases dramatically from 22.00 to 16.00 kJ/mol. The decrease in the ΔE_{cond} in this region can be attributed to the increasing interactions between the w/o droplets. At 0.11 to 0.16 water volume fractions (region II), a slight increase from 16.00 to 17.23 kJ/mol is observed in the ΔE_{cond}. This increase in this region could be attributed to the reorganization of the clusters. At 0.16 to 0.36 water volume fractions (region III), ΔE_{cond} decreases from 17.23 to 9.00 kJ/mol indicating that the clusters form open channels and the transport of ions becomes easier and therefore the decrease in the ΔE_{cond} decreases. For water volume fractions above 0.36, the emulsification of w/o fails and phase separation is observed. The ΔE_{cond} curves as function of water volume fraction of the systems based on the two oils behave in the same manner but the

R (+)-LIM-based systems have the lower values of ΔE_{cond} compared to IPM. A second important observation is that the percolation threshold in the microemulsions based on R (+)-LIM is very low (ϕ_c = 0.04) while the IPM-based microemulsions, the percolation threshold is higher and equals 0.11. The third observation is the fact that the percolation thresholds are very low for nonionic surfactants-based systems. The first two observations indicate that the structure, shape, and molecular volume of oil play different roles in determining the percolation thresholds and hence ΔE_{cond} values. The effective carbon number of the oil, which is 7 and 14 for R (+)-LIM and IPM, respectively, plays the major role. The carbon number of the oil plays an important role in determining the interactions responsible for the determination of the electrical conductivities of these microemulsions. R (+)-LIM tends to be solubilized inside the palisade layer of the mixed surfactants aggregates at an initial stage of solubilization and the mixed surfactants layer curvature changes to be less positive. On the other hand, IPM is mainly solubilized deep inside of the mixed surfactants aggregate and makes an oil pool. Hence, with IPM, the mixed surfactants layer curvature tends to be more positive. The low values of percolation thresholds are attributed to the surfactants mixing effect and to the presence of alcohol in the system. These low values are a real expression of the synergetic effect of surfactants mixing.

The different regions in the ΔE_{cond} curves (Figure 5.14A) are an indication of the microstructure transition along the dilution line. The first region indicates the formation of w/o microstructure and the beginning of ion transfer. The second region indicates the transition from w/o microstructure to clusters with closed droplets and the rate of ion transfer becomes slower. The third region indicates the transition from clusters to open channels. The fourth region is observed only in the case of R (+)-LIM-based microemulsion which indicates the transition from percolated microemulsions to a developed bicontinuous structures. The fifth region indicates the transition from bicontinuous to o/w microstructures.

5.3.2   DYNAMIC VISCOSITY (η)

5.3.2.1   Viscosity without Additives

The formation of microemulsions is a dynamic self-organizing phenomenon where aggregating–disaggregating processes operate in conjunction. In their process dynamics, exchange of matters between different phases continuously occur resulting in an overall equilibrium. The flow of microemulsions under stress and the transport of molecules through them are also of potential importance in the study of microemulsion dynamics. The study of the dynamic viscosity of microemulsions can provide information on the intrinsic and derived processes in the microemulsion system, as well as furnish knowledge on the overall geometry of the particles of the dispersed phase [1,97,111,112]. Viscosity measurements evidenced the dependence of the size and shape of microemulsions droplets on the amount of solubilized water [113–116]. The understanding of structural consistencies in microemulsions has also been attempted from viscosity measurements

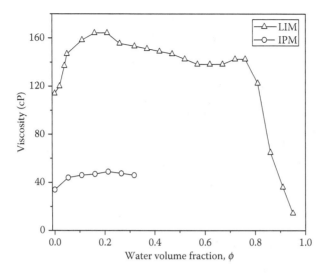

FIGURE 5.15 Variation of dynamic viscosity (η) of water/sucrose laurate/ethoxylated mono-di-glyceride/oil system as function of water volume fraction along the dilution line N60 at 25°C. The mixing ratio (w/w) of ethoxylated mono-di-glyceride/sucrose laurate equals unity. The lines serve as guides to the eye. The phase diagram is presented in Figure 5.2.

by others [117,118]. Figure 5.15 shows the variation in the dynamic viscosity as function of water volume fraction in the water/sucrose laurate/ethoxylated mono-di-glyceride/oil systems for R (+)-LIM and IPM along the dilution line N60, the mixing ratio of sucrose laurate/ethoxylated mono-di-glyceride equals unity (see phase diagram in Figure 5.2). The existence of two peaks in viscosity values with increasing water volume fraction implying the existence of at least three microstructural regions: one before, one after, and the other in between two peaks. The maxima are indicative of the transition from a mono- to a bicontinuous structure. The gradual increase of viscosity in the range of 0–0.2 water volume fraction suggests that the dispersed phase is present in the form of droplets in the microemulsion systems. These results are interpreted considering the clustering of water droplets with the mixed surfactants by the formation of a mixed monomolecular layer of the surfactants couple. For water volume fraction above 0.20, the viscosity of the system decreased and continues to decrease as the water volume fraction increases. At 0.7 water volume fraction, the viscosity increases again and reaches a second maximum at water volume fraction equals 0.75. The increase of dynamic viscosity for water volume fractions below 0.20 indicates attractive interaction and aggregation of droplets of water phase including molecular reorganization on the interface where the w/o microemulsions are present. The slow decrease in dynamic viscosity for water volume fractions between 0.20 and 0.70 indicates a transition from w/o microemulsions droplets to a bicontinuous structure. The increase in dynamic viscosity for water contents between 0.70 and 0.75 indicates a structural transition from bicontinuous structure to an o/w microemulsions microstructure. The sharp decrease in dynamic viscosity for water volume fractions above 0.80 indicates that the water, which is the least viscous component of the microemulsion system, becomes the outer phase and o/w microemulsions are formed. The percolation phenomenon in microemulsion essentially involves droplet association, i.e., clustering and fusion. It must, therefore, have a direct influence on the internal structure and hence viscosity. The study of viscosity confirmed the presence of two percolation processes for this microemulsion system. The structural inversion of the w/o microemulsion to the o/w type without any phase separation takes place in two stages. With increasing water volume fraction, the oil-continuous microemulsion transforms into the bicontinuous form at roughly 0.15 water volume fraction (water percolation threshold), and then at 0.75 water volume fraction the bicontinuous form transforms into the water continuous structure (oil percolation threshold). Other authors [64,97,111,112] reported about a similar behavior of dynamic viscosity of microemulsions based on nonionic surfactant in the presence of cosurfactants oil and water. Djordjevic et al. [97] demonstrated also similar effect of increasing water volume fraction on the dynamic viscosity based on nonionic surfactants in the presence of water and IPM microemulsion system. Like electrical conductivity, viscosity may also follow scaling type equations in the w/o microemulsion region [119–121].

$$\eta = A\left(\phi - \phi_c\right)^{-\mu} \quad \text{if } \phi > \phi_c \tag{5.10}$$

$$\eta = B\left(\phi_c - \phi\right)^{-s} \quad \text{if } \phi < \phi_c \tag{5.11}$$

where
ϕ is the volume fraction of water
ϕ_c is the percolation threshold
A and B are parameters
μ and s are scaling exponents

The slopes of the log η versus log($\phi_c - \phi$) plot for $\phi_c > \phi$ and the log η versus log($\phi - \phi_c$) for $\phi > \phi_c$ plot yield s and μ parameters. The average values of μ and s in Equations 5.10 and 5.11 are 1.04 and 0.18. These values are fairly near to the t and s values obtained for conductivity percolation. The estimated scaling parameters which are in good agreement with the experimental values, obtained with both electrical conductivity and dynamic viscosity, signify that these microemulsion systems show an interdependence of the viscosity–conductivity, especially at the stage of water percolation.

5.3.2.2 Viscosity with Alcohol as the Additive

Dynamic viscosities of water/sucrose laurate/ethoxylated mono-di-glyceride/oil + ethanol systems studied at 25°C are presented in Figure 5.16. The oils were R (+)-LIM

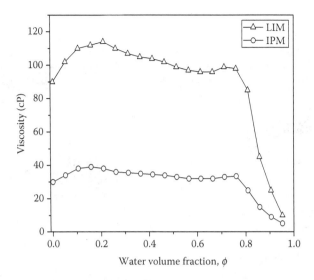

FIGURE 5.16 Variation of dynamic viscosity (η) of water/sucrose laurate/ethoxylated mono-di-glyceride/oil + ethanol system as function of water volume fraction along the dilution line N60 at 25°C. The mixing ratios (w/w) of ethoxylated mono-di-glyceride/ sucrose laurate and ethanol/oil equal unity. The lines serve as guides to the eye. The phase diagram is presented in Figure 5.4.

and IPM. The mixing ratios of ethanol/oil and ethoxylated mono-di-glyceride/ sucrose laurate both equal unity (phase diagrams are presented in Figure 5.4). The estimated μ as s values are presented in Table 5.4. These values differ from those obtained for the electrical conductivity due to the difference in behavior in the conductivity and the viscosity of the system. The viscosities of these mixed

TABLE 5.4

Values of the Experimental and Theoretical Parameters of the Percolation Process in the Water + Sodium Chloride/ Sucrose Laurate/Ethoxylated Mono-Di-Glyceride/ Oil + Ethanol Microemulsions Studied at 25°C

Theoretical (ϕ_c)	s	μ	Experimental (ϕ_c)	Oil-Based Microemulsion
0.17	0.18	1.04	0.18	R (+)-LIM
0.14	0.15	1.0	0.13	IPM

Note: Mixing ratios (w/w) of oil/ethanol and that of sucrose laurate/ethoxy- lated mono-di-glyceride equal unity. The concentration of NaCl in water is 0.01 M.

surfactant systems passed through a maximum, a minimum, and another maximum with increasing water content at 298 K as was previously reported [53,106,107]. This indicated that the w/o system was converted to the o/w system through a bicontinuous structure. Such a behavior was reported earlier for polyoxyethylene [95–97] oleyl alcohol (Brij-97)/butanol/dodecane/water; SDS/propanol/cyclohexane/water; and CTAB/1-propanol/cyclohexane/water systems [99,122]. O/w systems have been observed to have viscosity values lower than the w/o systems. Viscosity values for a particular weight percentage of water were increased with changing the oil type. Ajith and Rakshit [122] have also reported that for a mixed SDS + Brij-35/1-propanol/heptane/water system viscosity increased with increasing mole fraction of Brij-35 in the total surfactant content at a fixed microemulsion composition and temperature. The increase in viscosity in the w/o microemulsions is explained by an increase in dispersed droplet size and enhanced attractive interaction between the droplets. The maxima in viscosity for the studied systems occurred at different water volume fractions depending on the oil used in the formulation of the microemulsions. Yaghmur et al. [123] have also reported about structural transitions from w/o via bicontinuous phase to o/w for Winsor IV food grade microemulsions studied by pulsed gradient spin-echo NMR, conductivity, and viscosity measurements. The maximum in viscosity was obtained for systems containing equal weight percentages of oil and water at all temperatures and surfactant mixtures where bicontinuous structure prevailed with radius of curvature tending to infinity resulting in high viscosity. Newtonian behavior of the fluids was assumed since the systems were found to be low viscous under the studied conditions. The viscosities were also measured at different temperatures for samples along the N60 dilution line. Figure 5.17 represents the variation of dynamic viscosity values (MPa·s) of the water/sucrose laurate/ethoxylated mono-di-glyceride/R (+)-LIM + ethanol systems studied at different temperatures varying from 293 to 323 K to get the energetics of the viscous flow. Figure 5.17 depicts the viscosity behavior of the mixed surfactant systems at different temperatures with different volume fractions of water (ϕ). Increasing the water volume fraction in the microemulsion systems increases the viscosities. The viscosities of the samples decrease with the increase of temperature. The activation enthalpy ΔH_{vis} (which can be equated with the energy of activation for viscous flow, ΔE_{vis}) was obtained from the equation [1].

$$\eta = \left(Nh/V\right)\exp\left(\Delta H_{vis}/RT\right)\exp\left(-\Delta S_{vis}/R\right) \qquad (5.12)$$

where
  h is the Plank constant
  N is the Avogadro number
  V is the molar volume
  ΔS_{vis} is the entropy of activation for the viscous flow and other terms have their usual significance

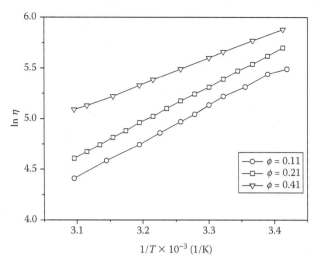

FIGURE 5.17   Plots of ln η on the systems water/sucrose laurate/ethoxylated mono-di-glyceride/R (+)-LIM + ethanol as function of water volume fraction along the dilution line N60 at different temperatures. The mixing ratios (w/w) of ethanol/oil and that of ethoxylated mono-di-glyceride/sucrose laurate equal unity. The phase diagram is presented in Figure 5.4A.

It follows from the equation that

$$\ln \eta = \left[\ln(Nh/V) - \Delta S_{vis}/R \right] + \Delta H_{vis}/RT \tag{5.13}$$

Assuming ΔS_{vis} be independent of temperature, a straight-line plot between ln η and T^{-1} is expected from the slope of which ΔH_{vis} can be calculated. Herein all the studied samples of this system produced good linear fit for ln η versus T^{-1} (Figure 5.17). The system water/sucrose laurate/ethoxylated mono-di-glyceride/IPM + ethanol also gave similar behaviors as presented in Figure 5.18. This observation was in contrast to the reports made by Acharya et al. [95] for microemulsion formulations using EO and other vegetable oils where non-Newtonian flow was observed and in accordance with Mitra and Paul [99]. The Gibbs free energy of activation (ΔG_{vis}) and the entropy of activation (ΔS_{vis}) were obtained from the relations

$$\Delta G_{vis} = RT \ln(\eta V/hN) \tag{5.14}$$

and

$$\Delta S_{vis} = (\Delta H_{vis} - \Delta G_{vis})/T \tag{5.15}$$

The effect of temperature on the activation parameters for the viscous flow for water/sucrose laurate/ethoxylated mono-di-glyceride/R (+)-LIM + ethanol systems

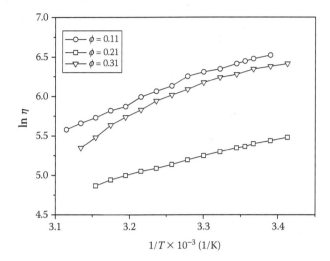

FIGURE 5.18 Plots of ln η on the systems water/sucrose laurate/ethoxylated mono-di-glyceride/IPM + ethanol as function of water volume fraction along the dilution line N60 at different temperatures. The mixing ratios (w/w) of ethanol/oil and that of ethoxylated mono-di-glyceride/sucrose laurate equal unity. The phase diagram is presented in Figure 5.4B.

studied is presented in Figure 5.19A. The ΔH_{vis} values of the viscous flow behave in the same manner as the viscosity rises and falls. These ΔH_{vis} values remain constant as function of temperature. The ΔG_{vis} values steadily decreased with increasing temperature for all the studied samples indicating thinning by the influence of temperature. ΔS_{vis} values increased with increasing temperatures suggesting

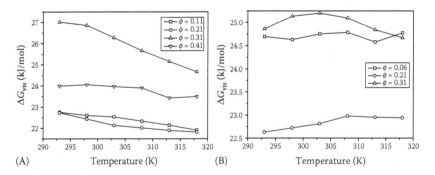

FIGURE 5.19 Plots of ΔG_{vis} as function of temperature for samples of different water volume fractions along the dilution line N60 on the systems water/sucrose laurate/ethoxylated mono-di-glyceride/oil + ethanol. The mixing ratios (w/w) of ethanol/oil and that of ethoxylated mono-di-glyceride/sucrose laurate equal unity. The phase diagram is presented in Figure 5.4. (A) R (+)-LIM and (B) IPM.

ordering or organization of the flowing fluid under the used shear rate. The ΔH_{vis} values the viscous flow for water/sucrose laurate/ethoxylated mono-di-glyceride/IPM + ethanol systems studied behave in the same manner as the viscosity rises and falls. These ΔH_{vis} values remain constant as function of temperature. The ΔG_{vis} values (Figure 5.19) increased and decreased with increasing temperature for the studied samples indicating thickening and thinning by the influence of temperature. ΔS_{vis} values decreased or increased with increasing temperatures suggesting ordering or disordering of the flowing fluid under the used shear rate. The positive ΔS_{vis} values decreased with increasing water content in the mixed surfactant for all the compositions, indicating dispersing of bulky polar head of ethoxylated mono-di-glyceride at the interface. The enthalpy–entropy compensation temperature is an important parameter of microemulsion systems. The linearly correlated ΔS_{vis} and ΔH_{vis} (with a correlation coefficient of 0.999) for various studied systems at different compositions (mentioned earlier) yielded a compensation temperature of around 265 and 349 K for the system based on R (+)-LIM and IPM-based systems, respectively (see Figure 5.20). It refers to an overall temperature that fits the H_{vis}, and S_{vis}, data into a linear correlation. The value of the compensation temperature of the viscous flow is around the average of the experimental temperatures, in the case of the IPM-based system. In the case of R (+)-LIM-based system, the compensation temperature is much lower than the average experimental temperature. Earlier Acharya et al. [95,96] obtained compensation temperatures of 298 and 312 K against the average experimental temperatures of 294 and 308 K for viscous flow of Brij-30/ethanol/EO/water and Brij-52/ethanol or isopropanol/coconut oil/water systems, respectively. Such isokinetic effects are very often observed in the dynamics of chemical processes [95,96]. The observed compensation temperature for the studied systems revealed interplay of similar internal physical processes.

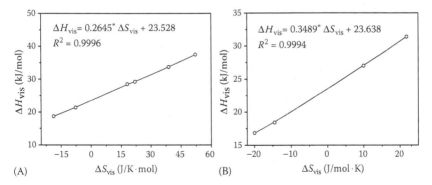

FIGURE 5.20 The ΔH_{vis} and ΔS_{vis} compensation plots for samples of different water volume fractions along the dilution line N60 on the systems water/sucrose laurate/ethoxylated mono-di-glyceride/oil + ethanol. The mixing ratios (w/w) of ethanol/oil and that of ethoxylated mono-di-glyceride/sucrose laurate equal unity. The phase diagrams are presented in Figure 5.4. (A) R (+)-LIM and (B) IPM.

Other authors [100] reported on the effect of adding ethoxylated surfactants to the sucrose mono fatty acid ester micelles in water. The shape of the micelles changes from spherical or very short rod micelles to long wormlike micelles due to the reduction of the average section area of each surfactant molecule at the interface. This change in the micelles shape causes the viscosity to increase. When oil is solubilized in the micellar aggregate, it induces a change in the micellar shape depending on the oil type and structure, which is expected to be reflected in the change in viscosity. The viscosity values obtained for the microemulsion samples based on R (+)-LIM were lower compared to the viscosities obtained for IPM-based microemulsions. It is clear here that another factor beside the molecular volume of the oil plays a role in determining the viscosities of the microemulsions. This factor is the oil effective carbon number that varies from 7 to 14 for R (+)-LIM and IPM, respectively. The carbon number of the oil plays an important role in determining the interactions responsible for the determination of the viscosities of these microemulsions. R (+)-LIM is a small hydrocarbon molecule while IPM is saturated long chain oil. R (+)-LIM tends to be solubilized inside the palisade layer of the mixed surfactants aggregates at an initial stage of solubilization and the mixed surfactants layer curvature changes to be less positive. On the other hand, IPM is mainly solubilized deep inside of the mixed surfactants aggregate and makes an oil pool. Hence with IPM the mixed surfactants layer curvature tends to be more positive. When water is solubilized in the microemulsions, different parts of the mixed surfactants are solubilized in the water phase. The solubility of ethoxylated mono-di-glyceride is higher in water and in the two oils compared to sucrose laurate and hence the mixing fraction of ethoxylated mono-di-glyceride in the total surfactants would be different from the original mixing ratio. The water volume fraction where the maximum in viscosity observed in the microemulsion systems based in the two oils vary depending on the oil. The lower-water volume fraction was observed in the IPM-based microemulsions followed by R (+)-LIM-based microemulsions.

5.4　MICROSTRUCTURE PROPERTIES

Microemulsion's typical size enables the use of SAS, which includes the small-angle x-ray scattering (SAXS) and neutron scattering (SANS) techniques to investigate their structure and size. These techniques are ideal for quantitatively investigating structure and interactions in micellar solutions because the length scales probed include both the size of the aggregate and inter-aggregate distance.

5.4.1　Microstructure without Additives

SAXS scattering probes the pertinent colloidal length scales of 1–100 nm and therefore is the method of choice for determining the size of colloidal particles. The scattering intensity depends on the different scattering length densities of the particles and the solvent. For SAXS, the scattering length density is propor-tional to the electron density, which is a linear function of the number region of aggregates composed of surfactant molecules [18–30]. However, at the time of these experiments, SAXS data were not measured on absolute scale and so all data are presented in arbitrary units. Only the polar headgroups and water regions are visible with SAXS experiments due to the fact that their electron densities are higher than the electron density of the surrounding oil. In this section, we used the SAXS technique to investigate the water/sucrose laurate/ethoxylated mono-di-glyceride/ oil microemulsion systems as a function of the water volume fraction (ϕ) along the dilution line N60 at 25°C. The oils were R (+)-LIM and IPM. The mixing ratio (w/w) of ethoxylated mono-di-glyceride/sucrose laurate equals unity (see phase diagram Figure 5.2). Characteristic profiles of the systems are presented in Figure 5.21 as an example of what happens in these systems. From Figure 5.21, we observe that in each case, the scattering profile exhibits a single intensity maximum at $q \neq 0$, followed by a high-angle tail. With increasing water volume fraction, the position of the maximum moves to a lower angle. For given water volume fraction, the peak width varies inversely with the weight ratio of mixed surfactants to oil. According to the Teubner–Strey equation (Equation 5.16) [124], we were able to derive from values of the periodicity, d, correlation length, ξ, and the amphiphilicity factor, f_a, as described in the experimental section (Equations 5.17 through 5.20).

$$I(q) = (1/a_2 + c_1 q^2 + c_2 q^4) + b \qquad (5.16)$$

with the constants a_2, c_1, c_2, and b obtained by using the Levenburg–Marquardt procedure [125]. Such a functional form is simple and convenient for the fitting of spectra. Equation 5.2 corresponds to a real space correlation function of the form

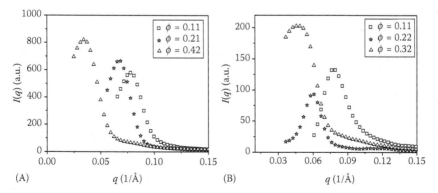

FIGURE 5.21 SAXS curves for samples whose compositions lie along the N60 dilution line on the water/sucrose laurate/ethoxylated mono-di-glyceride/oil system at 25°C. The mixing ratios (w/w) of sucrose laurate/ethoxylated mono-di-glyceride equal unity. The phase diagram is presented in Figure 5.2. (A) R (+)-LIM and (B) IPM.

$$\gamma(r) = (\sin kr/kr)e^{-r/\xi} \tag{5.17}$$

The correlation function describes a structure with periodicity $d = (2\pi/k)$ damped as a function of correlation length ξ. This formalism also predicts the surface to volume ratio, but because this ratio is inversely related to the correlation length and therefore must go to zero for a perfectly ordered system; calculated values are frequently found to be too low [126]. d and ξ are related to the constants in Equation 5.16 by [124]:

$$d = [(1/2)\,((a_2/c_2))^{1/2} - (c_1/4c_2)]^{-1/2} \tag{5.18}$$

$$\xi = [(1/2)((a_2/c_2))^{1/2} + (c_1/4c_2)]^{-1/2} \tag{5.19}$$

A third parameter, which can also be defined, is f_a, the amphiphilicity factor [124,127–129], which relates to the behavior of the correlation function $\gamma(r)$ and reflects the ability of the surfactant to impose order on the microemulsion:

$$f_a = \left(c_1/[4a_2c_2]^{1/2}\right) \tag{5.20}$$

The dependence of the d parameter on the volume fraction of water, ϕ was plotted in Figure 5.22. In Figure 5.22, we find that the periodicity, d, increases linearly with the increase in the water volume fraction for the systems based on the two

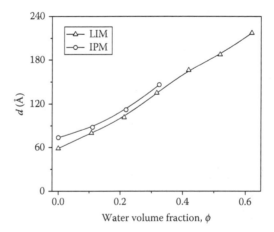

FIGURE 5.22 Microemulsions periodicity, d, for the systems water/sucrose laurate/ethoxylated mono-di-glyceride/oil at 25°C. The mixing ratios (w/w) of ethoxylated mono-di-glyceride/sucrose laurate equal unity. The oils were R (+)-LIM and IPM. The lines serve as guides to the eye. The phase diagrams are presented in Figure 5.2.

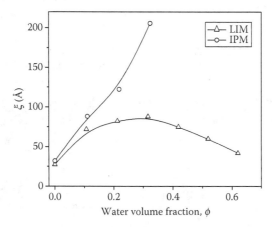

FIGURE 5.23 Microemulsions correlation length, ξ, for the systems water/sucrose laurate/ethoxylated mono-di-glyceride/oil at 25°C. The mixing ratios (w/w) of ethoxylated mono-di-glyceride/sucrose laurate equal unity. The oils were R (+)-LIM and IPM. The lines serve as guides to the eye. The phase diagrams are presented in Figure 5.2.

oils. In the case of IPM-based microemulsions, the periodicity values are the highest compared to R (+)-LIM. Plots of d versus ϕ can probe the dimensionality of swelling along the dilution lines. The linear relationship between the periodicity, d, and the water volume fraction indicates monodimensional swelling of the microemulsions droplets. Equation 5.19 was used to determine the values of the correlation length, ξ for the systems presented in Figure 5.21. Figure 5.23 presents the variation of the correlation length, ξ, as function of water volume fraction ϕ. Initially, the growth of ξ parallels that of d, with the former being smaller than the latter. In the R (+)-LIM-based system, ξ reaches a maximum at ϕ equals 0.3 then decreases while in the system based on IPM ξ increases in a monotonic fashion over the whole range of water dilution until phase separation. The higher values were obtained in the system based on IPM followed by the R (+)-LIM-based systems. The microemulsion system with the higher ξ values is the more ordered. The behavior of the correlation length, ξ, as function of water volume fraction ϕ can be explained as follows: when the water is the dispersed phase, increasing the water volume fraction increases the size of the scattering units and the correlation length,ξ, whereas when water is in the bulk, increasing the water volume fraction dilutes the scattering units and ξ decreases. Equation 5.20 was used to calculate the amphiphilicity factor, f_a, in the systems based on the two oils (see Tables 5.5 and 5.6). In all cases, f_a is with values negative. The lower values of f_a were obtained in the case of the system based on IPM which lies between -0.98 and -0.86, consistent with oscillatory behavior of the correlation function $\gamma(r)$ and the appearance of a well-defined scattering peak at $q \neq 0$. Within this range, for more negative f_a values, the microemulsions are more ordered. As we mentioned

TABLE 5.5

Values of the Amphiphilicity Factor (f_a) for the Water/Sucrose Laurate/Ethoxylated Mono-Di-Glyceride/R (+)-LIM Microemulsion System as a Function of the Water Volume Fraction (ϕ) along the Dilution Line N60

Amphiphilicity Factor (f_a)	Water Volume Fraction (ϕ)
−0.79	0
−0.94	0.11
−0.94	0.21
−0.84	0.32
−0.72	0.42
−0.72	0.52
−0.72	0.62

Note: Mixing ratio (w/w) of ethoxylated mono-di-glyceride/ sucrose laurate equals unity. Values of f_a calculated from equation 5.20 from data obtained at 25°C.

previously in the case of ξ values, it is clear that the microemulsions formed with IPM are more ordered compared to R (+)-LIM-based microemulsions. It seems that the effective carbon number of the oil plays the major role in determining the degree of order of the microemulsions formed. The effective carbon numbers of the oils are 7 and 14 for R (+)-LIM and IPM, respectively. It is the better chain length compatibility between the IPM and the mixed surfactants tails that improves the order in the microemulsions based on this oil.

TABLE 5.6

Values of the Amphiphilicity Factor (f_a) for the Water/ Sucrose Laurate/Ethoxylated Mono-Di-Glyceride/IPM Microemulsion System as a Function of the Water Volume Fraction (ϕ) along the Dilution Line N60

Amphiphilicity Factor (f_a)	Water Volume Fraction (ϕ)
−0.86	0.00
−0.96	0.11
−0.96	0.22
−0.98	0.32

Note: Mixing ratio (w/w) of ethoxylated mono-di-glyceride/sucrose laurate equals unity. Values of f_a calculated from Equation 5.20 from data obtained at 25°C.

5.4.2 Microstructure with Alcohol as the Additive

In this section, we used the SAXS technique to investigate the water/sucrose laurate/ethoxylated mono-di-glyceride/oil + ethanol microemulsion systems as a function of the water volume fraction (ϕ) along the dilution line N60. The oils were R (+)-LIM and IPM. The mixing ratios (w/w) of ethoxylated mono-di-glyceride/sucrose laurate and that of ethanol/oil equal unity (see phase diagram Figure 5.4). Characteristic profiles of the systems are presented in Figure 5.24 as an example of what happens in these systems. From Figure 5.24, we observe that, in each case, the scattering profile exhibits a single intensity maximum at $q \neq 0$, followed by a high-angle tail. With increasing water volume fraction, the position of the maximum moves to a lower angle. For given water volume fraction, the peak width varies inversely with the weight ratio of mixed surfactants to oil. By fitting all the scattering curves (Figure 5.24) to the Teubner–Strey equation [124] (Equation 5.16), we were able to derive from values of the periodicity, d, correlation length, ξ, and the amphiphilicity factor, f_a, as described in the experimental section (Equations 5.18 through 5.20). The dependence of the d parameter on the volume fraction of water, ϕ, was plotted in Figure 5.25. In Figure 5.25, we find that the periodicity, d, increases linearly with the increase in the water volume fraction for the systems based on the two oils. In the case of IPM-based microemulsions, the periodicity values are the highest compared to R (+)-LIM. Plots of d versus ϕ can probe the dimensionality of swelling along the dilution lines. The linear relationship between the periodicity, d, and the water volume fraction indicates monodimensional swelling of the microemulsions droplets. Equation 5.19 was used to determine the values of the correlation length, ξ, for the systems presented in Figure 5.24. Figure 5.26 presents the variation of the correlation length, ξ, as function of water volume

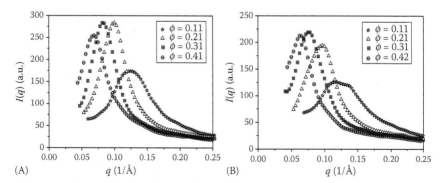

FIGURE 5.24 SAXS curves for samples whose compositions lie along the N60 dilution line on the water/sucrose laurate/ethoxylated mono-di-glyceride/oil + ethanol system at 25°C. The mixing ratios (w/w) of ethanol/oil and that of sucrose laurate/ethoxylated mono-di-glyceride equal unity. The phase diagram is presented in Figure 5.4. (A) R (+)-LIM and (B) IPM.

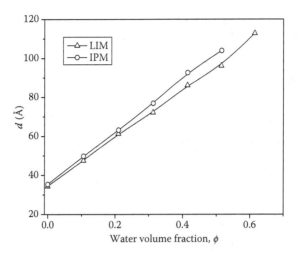

FIGURE 5.25 Microemulsions periodicity, d, for the systems water/sucrose laurate/ ethoxylated mono-di-glyceride/oil + ethanol at 25°C. The mixing ratios (w/w) of ethanol/ oil and that of ethoxylated mono-di-glyceride/sucrose laurate equal unity. The oils were R (+)-LIM and IPM. The lines serve as guides to the eye. The phase diagrams are presented in Figure 5.4.

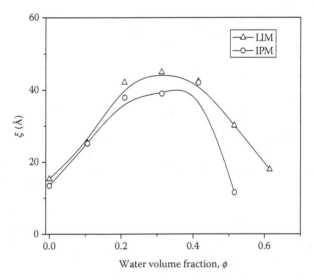

FIGURE 5.26 Microemulsions correlation length, ξ, for the systems water/sucrose laurate/ethoxylated mono-di-glyceride/oil + ethanol at 25°C. The mixing ratios (w/w) of ethanol/oil and that of ethoxylated mono-di-glyceride/sucrose laurate equal unity. The oils were R (+)-LIM and IPM. The lines serve as guides to the eye. The phase diagrams are presented in Figure 5.4.

fraction ϕ. Initially, the growth of ξ parallels that of d, with the former being smaller than the latter. ξ reaches a maximum at ϕ equals 0.3 in the system based on R (+)-LIM and 0.4 in the case of IPM while d increases in a monotonic fashion over the whole range of water dilution. The increase and decrease of the correlation length, ξ, as function of water volume fraction ϕ can be explained as reported in the previous section. Equation 5.20 was used to calculate the amphiphilicity factor, f_a, (see Tables 5.7 and 5.8). The higher values were obtained in the system based on R (+)-LIM followed by the IPM-based one. The microemulsion system with the higher ξ values is the more ordered. The behavior of the correlation length, ξ, as function of water volume fraction ϕ is explained as follows: when the water is the dispersed phase, increasing the water volume fraction increases the size of the scattering units and the correlation length, ξ, whereas when water is in the bulk, increasing the water volume fraction dilutes the scattering units and ξ decreases. Equation 5.16 was used to calculate the amphiphilicity factor, f_a, in the systems based on the two oils (see Tables 5.7 and 5.8). In all cases, f_a is with negative values. The lower values of f_a were obtained in the case of the system based on R (+)-LIM, which lies between −0.92 and −0.67, consistent with oscillatory behavior of the correlation function $\gamma(r)$ and the appearance of a well-defined scattering peak at $q \neq 0$. Within this range, for more negative f_a values, the microemulsions are more ordered. As we mentioned previously in the case of ξ values, it is clear that the microemulsions formed with IPM are less ordered compared to R (+)-LIM-based microemulsions. In these systems, the role of ethanol is to tune or adjust the phase

---

TABLE 5.7

Values of the Amphiphilicity Factor (f_a) for the Water/ Sucrose Laurate/Ethoxylated Mono-Di-Glyceride/ R (+)-LIM + Ethanol Microemulsion System as a Function of the Water Volume Fraction (ϕ) along the Dilution Line N60

Amphiphilicity Factor (f_a)	Water Volume Fraction (ϕ)
−0.81	0
−0.86	0.11
−0.92	0.21
−0.90	0.31
−0.84	0.41
−0.70	0.51
−0.67	0.61

Note: Mixing ratios (w/w) of ethanol/oil and that of sucrose laurate/ethoxy-lated mono-di-glyceride equal unity. Values of f_a calculated from Equation 5.20 from data obtained at 25°C.

TABLE 5.8

Values of the Amphiphilicity Factor (f_a) for the Water/Sucrose Laurate/Ethoxylated Mono-Di-Glyceride/IPM + Ethanol Microemulsion System as a Function of the Water Volume Fraction (ϕ) along the Dilution Line N60

Amphiphilicity Factor (f_a)	Water Volume Fraction (ϕ)
−0.78	0
−0.84	0.11
−0.90	0.21
−0.88	0.31
−0.81	0.41
−0.60	0.52

Note: Mixing ratios (w/w) of ethanol/oil and that of sucrose laurate/ethoxylated mono-di-glyceride equal unity. Values of f_a calculated from Equation 5.20 from data obtained at 25°C.

behavior to bring the one-phase microemulsion region into the experimental window of composition and temperature. Addition of ethanol to nonionic microemulsions will generally increase the solubility of the surfactant in the aqueous phase and enhance water solubilization. The chain length of the oils controls the penetration of oil and ethanol into the surfactant films and thus influences the distribution of the components between different domains. Oil penetration is dependent on chain length of the oil and surfactant. Alcohols are more associated with the interfaces when oil penetration is sterically hindered. Ethanol redistributes mainly between water and the interface-headgroup region of the surfactant. Part of it replaces surfactant molecules in the micelles, which increases the available interface and results in a higher number of micelles with shrinking size. Upon adding ethanol to the systems, the effective volume fractions of the dispersed phase remain nearly constant, while the interaction radii of the micelles decrease. Ethanol reduces the size of the micelles, which is in good agreement with the shrinking interaction radii derived from the structure factor. Consequently, the number of particles per unit volume is lowered, but these fewer particles have a stronger scattering power. At the same time the effective volume fraction of interaction stays approximately the same in the presence of ethanol, which, in combination, implies that the number of micelles per unit volume and, therefore, also the available interface must be increased. Consequently, ethanol reduces the surfactant aggregation number and replaces surfactant molecules in the interface [28].

5.5 DIFFUSION PROPERTIES

Studies of microemulsions using NMR technique have demonstrated a considerable structural variability including o/w and w/o droplet type structures as well as bicontinuous structures depending on composition. Low- and equal-diffusion coefficient of the surfactant and the dispersed medium characterize the two extremes o/w and w/o. The solvent molecules in these extremes have a diffusion coefficient close to what is found in the neat liquid, reduced only by obstruction from the dispersed particles. High-diffusion coefficients for all the constituents characterize the bicontinuous structures [130–136]. Dynamic light scattering (DLS) studies of microemulsions provide information about their dynamic behavior by measurements of the intensity auto-autocorrelation functions [35–42]. DLS measurements describe the collective dynamical process, whose physical origin is the local dynamics (Brownian fluctuations and shape fluctuations) of the trapped droplets. In the case of diluted monodisperse systems, these functions can be described by single exponentials and from the observed relaxation time the apparent diffusion coefficient of the particles can be obtained [43–48]. Depending on the temperature, a lot of different microstructures can be formed in a microemulsion. DLS is more sensitive to structural changes than other scattering techniques and does not perturb the equilibrium microstructures [49,50].

5.5.1 DIFFUSION WITHOUT ADDITIVES

We used the diffusion coefficient obtained by NMR analysis to characterize the microstructure changes in the microemulsions system water/sucrose laurate/ethoxylated mono-di-glyceride/oil along the dilution line N60. The oils were R (+)-LIM and IPM, the surfactants mixing ratio (w/w) equals unity (see phase diagram in Figure 5.2). To evaluate the self-diffusion data in terms of microstructure, the calculation of the relative diffusion coefficient, D/D_0, of the different components of the microemulsion especially oil and water is needed [133]. Relative diffusion coefficients were obtained by dividing water (D^{water}) and oil (D^{oil}) diffusion coefficients in the microemulsion by the diffusion coefficient of water in the pure water phase (D_0^{water}) and oil in the neat phase (D_0^{oil}). It is well documented [130–136] that if the D/D_0 values of water and oil differ by more than 1 order of magnitude, discrete particles of the slowly diffusing solvent are implied, whereas if the D/D_0 values of water and oil are of the same order of magnitude, a bicontinuous structure is suggested. Figure 5.27 shows the relative diffusion coefficients of water and R (+)-LIM as a function of the water content (wt %). One can clearly see that, at the two extremes of aqueous-phase concentrations (up to 0.20 and above 0.70 water volume fractions), the D^{water}/D_0^{water} values are easily interpreted, while the in-between regions are somewhat more difficult to explain since gradual changes take place. As Figure 5.27 indicates that microemulsions containing up to 0.20 water volume fraction have a confined water molecules microstructure, since the relative diffusion coefficients of water and R (+)-LIM differ by more than 1 order of magnitude. Microemulsions containing 0.20–0.70 water have a microstructure

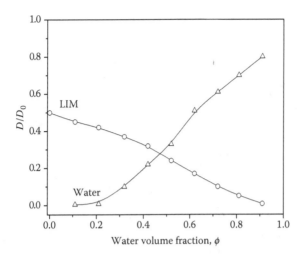

FIGURE 5.27 Relative diffusion coefficients of water (Δ) and R (+)-LIM (O) as function of water volume fractions for samples along the dilution line N60 of water/sucrose laurate/ethoxylated mono-di-glyceride/R (+)-LIM system at 25°C. The mixing ratio (w/w) of ethoxylated mono-di-glyceride/sucrose laurate equals unity. The lines serve as guides to the eye. The phase diagram is presented in Figure 5.2.

where neither the water nor the oil can be distinguished as confined phases, as the diffusion coefficients of water and R (+)-LIM are of the same order of magnitude. Relative diffusion coefficient results for water volume fractions above 0.70 indicate that the oil phase is the confined one since the relative diffusion coefficients of water and R (+)-LIM differ by more than 1 order of magnitude. The diffusion coefficients of sucrose laurate and ethoxylated mono-di-glyceride at low-water contents are very low. The values of their diffusion coefficients at 0.11 water volume fraction are 0.013×10^{-5} and $0.011 \times 10^{-5}\,cm^2/s$ for sucrose laurate and ethoxylated mono-di-glyceride, respectively. For water volume fraction equals 0.91, the diffusion coefficient of sucrose laurate increases and equals $0.063 \times 10^{-5}\,cm^2/s$. On the other hand, the diffusion coefficient of ethoxylated mono-di-glyceride equals the same value as for 0.10 water volume fraction (i.e., $0.011 \times 10^{-5}\,cm^2/s$). These results suggest that the ethoxylated mono-di-glyceride is bordered by the more mobile molecules of sucrose laurate at the oil–water interface. NMR results provide a clear picture of structural transitions along the N60 dilution line studied. As in the dynamic viscosity results, three different regions are observed. The first region indicates the presence of w/o droplets microstructure. The second region indicates the transition from w/o microstructure to a bicontinuous microemulsion. In the third region, a transition from a bicontinuous microemulsion to an o/w microstructure occurs and a discrete o/w microstructure can be present. The behavior of the IPM-based system before the emulsification failure (Figure 5.28) is similar to the R (+)-L/M-based system with the fact that the relative diffusion

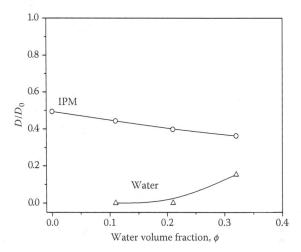

FIGURE 5.28   Relative diffusion coefficients of water (Δ) and IPM (O) as function of water volume fractions for samples along the dilution line N60 of water/sucrose laurate/ethoxylated mono-di-glyceride/IPM system at 25°C. The mixing ratio (w/w) of ethoxylated mono-di-glyceride/sucrose laurate equals unity. The lines serve as guides to the eye. The phase diagram is presented in Figure 5.2.

co-efficients of the IPM and water are lower than those observed for R (+)-LIM and water. This behavior indicates that the molecular volume of the oil plays a role in determining the diffusion properties of the system.

5.5.2 DIFFUSION WITH ALCOHOL AS THE ADDITIVE

Figure 5.29 shows the relative diffusion coefficients of water and R (+)-LIM in water/sucrose laurate/ethoxylated mono-di-glyceride/(R (+)-LIM + ethanol) microemulsion system where the mixing ratios (w/w) of sucrose laurate/ethoxylated mono-di-glyceride and that of ethanol/R (+)-LIM equal unity as a function of the water content (wt%) along the dilution line N60 (see phase diagram Figure 5.4). It is clear that the diffusion coefficient behavior of R (+)-LIM and water at the two extremes of aqueous-phase concentrations (below 0.20 and above 0.70 water volume fraction should be interpreted in the same way as in the previous section. As Figure 5.29 indicates, microemulsions containing below 0.20 water volume fractions have a discrete w/o microstructure, since the relative diffusion coefficients of water and R (+)-LIM differ by more than 1 order of magnitude. For water volume fractions between 0.20 and 0.70, a bicontinuous microstructure is envisaged, as the diffusion coefficients of water and R (+)-LIM are of the same order of magnitude. Increasing the water volume faction to above 0.70 induces the formation of discrete o/w microstructure. From the self diffusion-NMR results, it is clear that there are microstructure transitions along the dilution line N60

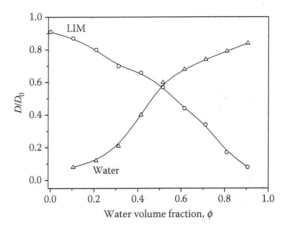

FIGURE 5.29 Relative diffusion coefficients of water (Δ) and R (+)-LIM (O) in micro-emulsions system water/sucrose laurate/ethoxylated mono-di-glyceride/(R (+)-LIM + ethanol) as function of water volume fractions along the dilution line N60 at 25°C. The mixing ratios (w/w) of ethanol/R (+)-LIM and sucrose laurate/ethoxylated mono-di-glyceride equal unity. The lines serve as guides to the eye. The phase diagram is presented in Figure 5.4A.

investigated. Figure 5.30 presents the relative diffusion coefficients of IPM and water in the water/sucrose laurate/ethoxylated mono-di-glyceride/(IPM + ethanol) where the mixing ratios (w/w) of sucrose laurate/ethoxylated mono-di-glyceride and ethanol/IPM equal unity. One can clearly see that the general diffusion coefficient behaviors of IPM and water are not very different from those of

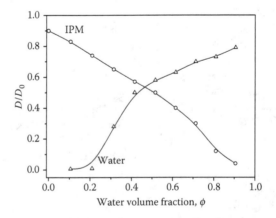

FIGURE 5.30 Relative diffusion coefficients of water (Δ) and IPM (O) in microemulsions system water/sucrose laurate/ethoxylated mono-di-glyceride/(IPM + ethanol) as function of water volume fractions along the dilution line N60 at 25°C. The mixing ratios (w/w) of ethanol/R (+)-LIM and sucrose laurate/ethoxylated mono-di-glyceride equal unity. The lines serve as guides to the eye. The phase diagram is presented in Figure 5.4B.

R (+)-LIM and water. It can also be seen that, at the two extremes of water volume fractions (below 0.30 and above 0.70), the D^{water}/D_0^{water} values are easily interpreted, while the in-between regions are somewhat more difficult to explain since gradual changes take place. The structural changes in the presence of IPM are more pronounced than those in the presence of R (+)-LIM. As Figure 5.30 indicates, microemulsions containing below 0.30 water volume fractions, and solubilizing IPM, have a discrete w/o microstructure, since the relative diffusion coefficients of water and R (+)-LIM differ by more than 1 order of magnitude. Microemulsions solubilizing IPM and containing 0.30–0.70 water volume fractions have a bicontinuous microstructure, as the diffusion coefficients of water and IPM are of the same order of magnitude. Increasing the water volume fraction to above 0.70 induces the formation of discrete o/w microstructure, as the relative diffusion coefficients of water and IPM differ by more than 1 order of magnitude. From self-diffusion NMR results, it is clear that penetration of oil affects the diffusion properties of the different components of the microemulsions. The different regions in the diffusion coefficients curves (Figures 5.29 and 5.30) are an indication of the microstructure transition along the dilution line. While the general behavior of the diffusion coefficients is the same for microemulsions based on R (+)-LIM and IPM, the transition point from one microstructure to another is different. Figure 5.30 indicates that solubilization of IPM influences the transition from w/o to bicontinuous microstructure and furthers to o/w microstructure. In R (+)-LIM microemulsions, the formation of bicontinuous microstructure occurs when the microemulsion contains 0.20–0.70 water volume fractions (Figure 5.29), whereas in a microemulsion containing IPM, bicontinuous microstructure starts at high-water volume fractions (i.e., 0.30) and continues up to water volume fractions of 0.30 to 0.70 (Figure 5.30). It seems that IPM disturbs both the flexibility of the micelle and the spontaneous curvature. As a result, the interface changes into a flatter curvature (bicontinuous) at an early stage of water concentration, more so in the presence of R (+)-LIM than in IPM.

The observed diffusion coefficient of the ethoxylated mono-di-glyceride (D^{EMDG}) which is the one that has higher-molecular volume between the two mixed surfactants used sucrose laurate and ethoxylated mono-di-glyceride is related to the self-diffusion coefficients of the aggregated and the free molecules, D_{mic}^{EMDG} and, D_{free}^{EMDG}, respectively, by

$$D^{EMDG} = p_{mic}D_{mic}^{EMDG} + (1 - p_{mic})D_{free}^{EMDG} \qquad (5.21)$$

where D^{EMDG} denotes the ethoxylated mono-di-glyceride experimental self-diffusion coefficient, D_{mic}^{EMDG} and D_{free}^{EMDG} represent the translational self-diffusion coefficient of micellized ethoxylated mono-di-glyceride surfactant, respectively.

The measured values of D_{mic}^{EMDG} is 8×10^{-11} and $7 \times 10^{-11} \, m^2/s$ in the case of R (+)-LIM and IPM-based systems, respectively. The diffusion coefficient of free ethoxylated mono-di-glyceride D_{free}^{EMDG} is $7 \times 10^{-10} \, m^2/s$. p_{mic} is the fraction of micellized surfactant which equals 1-CMC/c, where c is the overall mixed surfactant concentration. For the high-mixed surfactants content of the present study, the contribution from free surfactant molecules is negligible because the CMC of ethoxylated mono-di-glyceride is very low (1.12×10^{-5} M at 25°C). Hence, a decreasing diffusion coefficient with increasing water concentration reflects micellar growth [50,51,136–141]. The picture of the microstructures that emerges from the data over the whole water content range studied is that at low-water contents values there are probably spherical droplets of w/o. As the water content increases, the w/o microstructure is no longer optimal and a flattening of the interfacial film is approached. Both water and oil diffuses rapidly indicate an average equilibrium bicontinuous arrangement where neither oil nor water is confined into closed domains. The characteristics of microstructures are in accordance with the electrical conductivity and dynamic viscosity results presented in the previous part.

DLS is a fast method for obtaining information on the size of the particles in relatively simple and dilute systems, but in the case of the complex solutions investigated here it would not be possible to qualitatively interpret the respective results without the additional information from static experiments. So DLS and SAS methods are complementary techniques and the results support each other. Nevertheless, to extract quantitative information on dynamic measurements of such complex mixtures and to fully understand the results, we would need much more a priori information on these systems. The DLS technique was used to investigate the water/sucrose laurate/ethoxylated mono-di-glyceride/oil + ethanol microemulsion systems as a function of temperature at water volume fractions equals 0.91 along the dilution line N60. The oils were R (+)-LIM and IPM. The mixing ratios of ethoxylated mono-di-glyceride/sucrose laurate and ethanol/oil equal unity (Figure 5.4). The variations of the hydrodynamic radius (R_H) as function of temperature in the water-rich region are presented in Figure 5.31. The values of the hydrodynamic radius in the system based on IPM are higher than those observed in the R (+)-LIM-based systems, a cause of the higher-effective chain length of IPM. R_H is seen to increase with the temperature increase indicating that micelles grow in size as temperature increases and may be a change occurred on the shape in the micelles, that is, sphere to rod or sphere to disk transitions. The microemulsion droplets deform by thermal fluctuations. The droplets may undergo attractive interactions that lead to aggregation between the droplets. Two mechanisms have been used to explain the origin of the attractive interactions: interpenetration of the tails of the surfactants molecules residing on different droplets or fusion of droplets which lower the curvature energy. Similar results of the behavior of the hydrodynamic radius as function of temperature were reported in our previous studies [142,143].

5.6 IMAGING PROPERTIES

The water/sucrose laurate/ethoxylated mono-di-glyceride/R (+)-LIM microemulsion samples were imaged by the cryogenic transmission electron microscopy (Cryo-TEM). For low-water contents, the concentrated microemulsions cause beam damage and clear images were difficult to obtain [144–146]. The micrographs of microemulsions samples at high-water volume fractions (0.70, 0.80, and 0.90) show spheroidal-swollen micelles as shown in Figure 5.32. In these

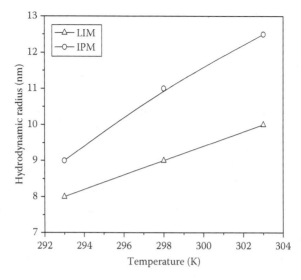

FIGURE 5.31 Hydrodynamic radius (R_H) for the water/sucrose laurate/ethoxylated mono-di-glyceride/oil + ethanol microemulsion system as a function of temperature for a water volume fractions equals 0.90. The oils were R (+)-LIM and IPM. The mixing ratios (w/w) oil/ethanol and that of sucrose laurate/ethoxylated mono-di-glyceride equal unity. Values of R_H are calculated from Equation 5.7. The lines serve as guides to the eye. The phase diagrams are presented in Figure 5.4.

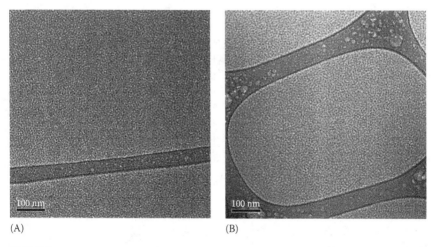

(A)   (B)

FIGURE 5.32 Cryo-TEM images of vitrified microemulsion samples of water/ sucrose laurate/ethoxylated mono-di-glyceride/R (+)-LIM system at (A) 0.81 and (B) 0.91 water volume fractions. The mixing ratio (w/w) of ethoxylated mono-di-glyceride/sucrose laurate equals unity. The phase diagram is presented in Figure 5.2. Bar = 100 nm.

preparations, suspended particles are pushed to the edge of the holes of the holey carbon film, giving rise to particle crowding and in essence changing the local concentration. In Figure 5.32, away from the edge, in a thin area of the specimen, one sees individual micelles. The specimen in this study is extremely sensitive to the electron beam. To overcome electron beam damage, these micrographs were taken with minimal electron exposure [144–146]. Cryo-TEM provides an important measure of the dimension of the microemulsion system.

5.7 CONCLUSION

The properties of microemulsion with mixed nonionic surfactants and different oils were reviewed. The water solubilization capacity study revealed that mixed surfactants improve the water solubilization capacity in the microemulsions compared to the microemulsion systems based on single surfactants. The molar ratios of mixed surfactants play an important role in determining the maximum water solubilization. The chemical structure of the oil affects its penetration in the surfactants palisade layer and determines the extent of water solubilization. Adding ethanol to the system improves the water solubilization capacity of the microemulsions and makes the system more organized. The study of the transport properties indicated that percolation thresholds are observed, which vary depending on the oil used in the microemulsion formulation. Low values of percolation thresholds observed in these systems are a clear expression of the synergistic effect of the surfactants mixing. The activation energy of conduction flow has a minimum value at the percolation threshold. The Gibbs free energy of viscous flow increased or decreased with increasing temperature indicating thickening and thinning by the influence of temperature. The observed enthalpy–entropy compensation temperature for the studied systems revealed interplay of similar internal physical processes during microemulsions flow. Microstructure and diffusion properties studied by SAXS, NMR, and DLS revealed a monodimensional water swelling in the microemulsions droplets. Diffusion coefficients confirm that the system undergoes a continuous structural transition of microemulsions from w/o via bicontinuous phase to o/w. The difference in estimated parameter values for the two oils suggests that the R (+)-LIM molecules are present in the palisade layer of the mixed surfactants aggregates. On the other hand, IPM molecules are not solubilized in the palisade layer. The carbon number of the oil plays a dominant role in determining the conduction and viscous flow parameters.

SYMBOLS AND TERMINOLOGIES

A_T	total one-phase microemulsion area
C	weight of surfactant at the water–oil interface
CMC	critical micelle concentration
Cryo-TEM	cryogenic transmission electron microscopy
Δ	diffusion time (time between the two gradients in the pulse sequence)

D	self-diffusion coefficient measured by NMR
δ	gradient pulse length
d	periodicity or characteristic length
ΔE_{cond}	activation energy of conduction
ΔH_{vis}	activation enthalpy of viscous flow
ΔG_{vis}	Gibbs free energy of activation of viscous flow
ΔS_{vis}	entropy of activation for the viscous flow
D_H	hydrodynamic diameter
DLS	dynamic light scattering
D_0	self-diffusion coefficient of neat materials measured by NMR
D_z	translational diffusion coefficients measured by DLS
D/D_0	relative diffusion coefficient
EMDG	ethoxylated mono-di-glyceride
EtOH	ethanol
I	measured signal intensity
I_0	signal intensity for $G = 0$
ϕ	water volume fraction
f_a	amphiphilicity factor
G	z-gradient strength
ϕ_c	percolation threshold
γ	gyro magnetic ratio for the ^1H nucleus
$\gamma(r)$	correlation function
HLB	hydrophilic–lipophilic balance
H	viscosity
$I(q)$	intensity
IPM	isopropylmyristate
k_B	Boltzmann's constant
L1695	sucrose laurate
LIM	R (+)-limonene
$M\phi$	multiphase region
NaCl	sodium chloride
NMR	nuclear magnetic resonance
1ϕ	one-phase region
o/w	oil in water
PGSE	pulsed gradient spin echo NMR
p_{mic}	fraction of micellized surfactant
q	scattering vector = $(4\pi/\lambda)\sin\theta$, where λ is the x-ray wavelength and 2θ is the scattering angle
R_H	hydrodynamic radius
R_{ow}	weight fraction of oil in water + oil
S	monomeric solubility
σ	electrical conductivity
SAS	small-angle scattering
SANS	small-angle neutron scattering

SAXS	small-angle x-ray scattering
SE	sucrose esters
w/o	water in oil
w/w	weight per weight
ξ	correlation length
X^{min}	minimum weight fraction of mixed surfactants capable of solubilizing equal amounts of water and oil

REFERENCES

1. Moulik, S.P. and Paul, B.K. 1998 Structure, dynamics and transport properties of microemulsions, *Adv. Colloid Interf. Sci.* 78:99–195.
2. Wormuth, K., Lade, O., Lade, M., and Schomacker, R. Microemulsions. In: *Handbook of Applied Surface and Colloid Chemistry.* Krister H. (Ed.) John Wiley & Sons, West Sussex, England, 2001, pp. 55–74.
3. Kumar, P. and Mital, K.L. *Handbook of Microemulsion Science and Technology,* Marcel Dekker, New York, 1999.
4. Sjoblom, J., Lindberg, R., and Friberg, S.E. 1996 Microemulsions-phase equilibria characterization, structure, applications and chemical reactions, *Adv. Colloid Interf. Sci.* 95:125–287.
5. Solans, C. and Kunieda, H. (Eds.) *Industrial Applications of Microemulsions. Surfactant Science Series* Vol. 66. Marcel Dekker, New York, 1997.
6. Brusseau, M.L., Sabatini, D.A., Gierke, J.S., and Annable, M.D. (Eds.) *Innovative Subsurface Remediation, Field Testing of Physical, Chemical, and Characterization Technology; ACS Symposium Series 725,* American Chemical Society, Washington, DC, 1999.
7. Sabatini, D.A. and Knox, R.C. (Eds.) *Transport and Remediation of Subsurface Contaminants, Colloidal, Interfacial and Surfactant Phenomena; ACS Symposium Series 491,* American Chemical Society, Washington, DC, 1992.
8. Texter, J. (Ed.) *Reactions and Synthesis in Surfactant Systems. Surfactant Science Series,* Vol. 100. Marcel Dekker, New York, 2001.
9. Lawrence, M.J. and Rees, G.D. 2000 Microemulsion-based media as novel drug delivery systems, *Adv. Drug Deliv. Rev.* 45:89–121.
10. Terjarla, S. 1999 Microemulsions: An overview and pharmaceutical applications, *Crit. Rev. Ther. Drug Carrier Syst.* 16:461–521.
11. Ogino, K. and Abe, M. (Eds.) *Mixed Surfactant Systems, Surfactant Science Series 46,* Marcel Dekker Inc., New York, 1992.
12. Kunieda, H., Ushio, N., Nakano, A., and Miura, M. 1993 Three phase behavior in a mixed sucrose alkanoate and polyethyleneglycol alkyl ether system, *J. Colloid Interf. Sci.* 159:37–44.
13. Kunieda, H. and Yamagata, M. 1993 Mixing of nonionic surfactants at water–oil interface in microemulsions, *Langmuir* 9:3345–3351.
14. Kunieda, H., Nakano, A., and Akimura, M. 1995 The effect of mixing of surfactants on solubilization in a microemulsion system, *J. Colloid Interf. Sci.* 170:78–84.
15. Kunieda, H., Nakano, A., and Pes, M.A. 1995 Effect of oil on the solubilization in microemulsion systems including nonionic surfactant mixture, *Langmuir* 11:3302–3306.

16. Pes M. A., Aramaki, K., Nakamura, N., and Kunieda, H. 1996 Temperature-insensitive microemulsions in a sucrose monolalkanoate system, *J. Colloid Interf. Sci.* 178:666–672.

17. Tondre, C. Dynamic processes in microemulsions. In: *Dynamics of Surfactant Self-Assemblies: Micelles, Microemulsions, Vesicles and Lyotropic Phases.* Raoul, Z. (Ed.) Taylor & Francis, Boca Raton, FL, 2005, pp. 233–298.

18. Tomšič, M., Bešter-Rogač, M., Jamnik, A., Kunz, W., Touraud, D., Bergmann, A., and Glatter, O. 2006 Ternary systems of nonionic surfactant Brij 35, water and various simple alcohols: Structural investigations by small-angle X-ray scattering and dynamic light scattering, *J. Colloid Interf. Sci.* 294:194–211.

19. Brunner-Popela, J., Mittelbach, R., Strey, R., Schubert, K.-V., Kaler, E.W., and Glatter, O. 1999 Small-angle scattering of interacting particles. III. D2O-C12E5 mixtures and microemulsions with *n*-octane, *J. Chem. Phys.* 110:10623–10632.

20. Glatter, O., Strey, R., Schubert, K.-V., and Kaler, E.W. 1996 Small angle scattering applied to microemulsions. *Ber. Bunsen/Phys. Chem. Chem. Phy.* 100:323–335.

21. Strey, R., Glatter, O., Schubert, K.-V., and Kaler, E.W. 1996 Small-angle neutron scattering of $D_2O-C_{12}E_5$ mixtures and microemulsions with *n*-octane: Direct analysis by Fourier transformation, *J. Chem. Phys.* 105:1175–1188.

22. Reimer, J., Soderman, O., Sottmann, T., Kluge, K., and Strey, R. 2003 Microstructure of alkyl glucoside microemulsions: Control of curvature by interfacial composition, *Langmuir* 19:10692–10702.

23. Bumajdad, A., Eastoe, J., Griffiths, P., Steytler, D.C., Heenan, R.K., Jian, R., Lu, R.J., and Timmins, P. 1999 Interfacial compositions and phase structures in mixed surfactant microemulsions, *Langmuir* 15:5271–5278.

24. von Corswant, C., Engstrom, S., and Olle Soderman, O. 1997 Microemulsions based on soybean phosphatidylcholine and triglycerides. Phase behavior and microstructure, *Langmuir* 13:5061–5070.

25. Bumajdad, A., Eastoe, J., Nave, S., Steytler, D.C., Heenan, R.K., and Grillo, I. 2003 Compositions of mixed surfactant layers in microemulsions determined by small-angle neutron scattering, *Langmuir* 19:2560–2567.

26. Silas, J.A. and Kaler, E.W. 2003 Effect of multiple scattering on SANS spectra from bicontinuous microemulsions, *J. Colloid Interf. Sci.* 257:291–298.

27. Yaghmur, A., De Campo, L., Sagalowicz, L., Leser, M.E., and Glatter, O. 2005 Emulsified microemulsions and oil-containing liquid crystalline phases, *Langmuir* 21:569–577.

28. De Campo, L., Yaghmur, A., Garti, N., Leser, M.E., Folmer, B., and Glatter, O. 2004 Five-component food-grade microemulsions: Structural characterization by SANS, *J. Colloid Interf. Sci.* 274:251–267.

29. Yaghmur, A., De Campo, L., Aserin, A., Garti, N., and Glatter, O. 2004 Structural characterization of five-component food grade oil-in-water nonionic microemulsions, *Phys. Chem. Chem. Phys.* 6:1524–1533.

30. Glatter, O., Orthaber, D., Stradner, A., Scherf, G., Fanun, M., Garti, N., Clément, V., and Leser, M.E. 2001 Sugar-ester nonionic microemulsion: Structural characterization, *J. Colloid Interf. Sci.* 241:215–225.

31. Papadimitriou, V., Sotiroudis, T.G., and Xenakis, A. 2007 Olive oil microemulsions: Enzymatic activities and structural characteristics, *Langmuir* 23:2071–2077.

32. Salazar-Alvarez, G., Björkman, E., Lopes, C., Eriksson, A., Svensson, S., and Muhammed, M. 2007 Synthesis and nonlinear light scattering of microemulsions and nanoparticle suspensions, *J. Nanoparticle Res.* 9:647–652.

33. Silva, E.J., Zaniquelli, M.E.D., and Loh, W. 2007 Light-scattering investigation on microemulsion formation in mixtures of diesel oil (or hydrocarbons) + ethanol + additives, *Energy Fuels* 21:222–226.
34. Tomšič, M., Podlogar, F., Gašperlin, M., Bešter-Rogač, M., and Jamnik, A. 2006 Water-Tween 40®/Imwitor 308®-isopropyl myristate microemulsions as delivery systems for ketoprofen: Small-angle X-ray scattering study, *Int. J. Pharm.* 327:170–177.
35. Holderer, O., Frielinghaus, H., Monkenbusch, M., Allgaier, J., Richter, D., and Farago, B. 2007 Hydrodynamic effects in bicontinuous microemulsions measured by inelastic neutron scattering, *Eur. Phys. J. E* 22:157–161.
36. Goddeeris, C., Cuppo, F., Reynaers, H., Bouwman, W.G., and Van Den Mooter, G. 2006 Light scattering measurements on microemulsions: Estimation of droplet sizes, *Int. J. Pharm.* 312:187–195.
37. Shukla, A. and Neubert, R.H.H. 2006 Diffusion behavior of pharmaceutical o/w microemulsions studied by dynamic light scattering, *Colloid Polym. Sci.* 284:568–573.
38. Shukla, A. and Neubert, R.H.H. 2005 Investigation of w/o microemulsion droplets by contrast variation light scattering, *Pramana J.Phys.* 65:1097–1108.
39. Shukla, A., Graener, H., and Neubert, R.H.H. 2004 Observation of two diffusive relaxation modes in microemulsions by dynamic light scattering, *Langmuir* 20:8526–8530.
40. Shukla, A., Kiselev, M.A., Hoell, A., and Neubert, R.H.H. 2004 Characterization of nanoparticles of lidocaine in w/o microemulsions using small-angle neutron scattering and dynamic light scattering, *Pramana J. Phys.* 63:291–295.
41. De Geyer, A., Molle, B., Lartigue, C., Guillermo, A., and Farago, B. 2004 Dynamics of caged microemulsion droplets: A neutron spin echo and dynamic light scattering study, *Physica B: Condens. Mat.* 350:200–203.
42. Shukla, A., Krause, A., and Neubert, R.H.H. 2003 Microemulsions as colloidal vehicle systems for dermal drug delivery, Part IV: Investigation of microemulsion systems based on a eutectic mixture of lidocaine and prilocaine as the colloidal phase by dynamic light scattering. *J. Pharm. Pharmacol.* 55:741–748.
43. Shukla, A., Janich, M., Jahn, K., and Neubert, R.H.H. 2003 Microemulsions for dermal drug delivery studied by dynamic light scattering: Effect of interparticle interactions in oil-in-water microemulsions, *J. Pharm. Sci.* 92:730–738.
44. Malcolmson, C., Barlow, D.J., and Lawrence, M.J. 2002 Light-scattering studies of testosterone enanthate containing soybean oil/C18:1E10/water oil-in-water microemulsions, *J. Pharm. Sci.* 91:2317–2331.
45. Mihailescu, M., Monkenbusch, M., Allgaier, J., Frielinghaus, H., Richter, D., Jakobs, B., and Sottmann, T. 2002 Neutron scattering study on the structure and dynamics of oriented lamellar phase Microemulsions, *Phys. Rev. E* 66:041504/1–041504/13.
46. Lee Jr., C.T., Psathas, P.A., Ziegler, K.J., Johnston, K.P., Dai, H.J., Cochran, H.D., Melnichenko, Y.B., and Wignall, G.D. 2000 Formation of water-in-carbon dioxide microemulsions with a cationic surfactant: A small-angle neutron scattering study, *J. Phys. Chem. B* 104:11094–11102.
47. Choi, S.M., Chen, S.H., Sottmann, T., and Strey, R. 1997 Measurement of interfacial curvatures in microemulsions using small-angle neutron scattering, *Physica B: Condens. Mat.* 241–243:976–978.
48. Gradzielski, M., Langevin, D., Sottmann, T., and Strey, R. 1996 Small angle neutron scattering near the wetting transition: Discrimination of microemulsions from weakly structured mixtures, *J. Chem. Phys.* 104:3782–3787.

49. Hellweg, T., Brulet, A., and Thomas Sottmann, T. 2000 Dynamics in an oil-continuous droplet microemulsion as seen by quasielastic scattering techniques, *Phys. Chem. Chem. Phys.* 2:5168–5174.

50. Hellweg, T. and Regine von Klitzing, R. 2000 Evidence for polymer-like structures in the single phase region of a dodecane/$C_{12}E_5$/water microemulsion: A dynamic light scattering study, *Physica A* 283:349–358.

51. Olsson, U., Shinoda, K., and Lindman, B. 1986 Change of the structure of microemulsions with the hydrophile–lipophile balance of nonionic surfactant as revealed by NMR self-diffusion studies, *J. Phys. Chem.* 90:4083–4088.

52. Fanun, M. 2007 Propylene glycol and ethoxylated surfactant effects on the phase behavior of water/sucrose stearate/oil systems, *J. Disper. Sci. Technol.* 28:1244–1253.

53. Fanun, M. 2007 Conductivity, viscosity, NMR and diclofenac solubilization capacity studies of mixed nonionic surfactants microemulsions, *J. Mol. Liq.* 135:5–13.

54. Fanun, M. 2007 Structure probing of water/mixed nonionic surfactants/caprylic capric triglyceride, *J. Mol. Liq.* 133:22–27.

55. Fanun, M. and Salah Al-Diyn, W. 2007 Structural transitions in the system water/mixed nonionic surfactants/R (+)-limonene studied by electrical conductivity and self-diffusion-NMR, *J. Disper. Sci. Technol.* 28:165–174.

56. Fanun, M. and Salah Al-Diyn, W. 2006 Temperature effect on the phase behavior of the systems water/sucrose laurate/ethoxylated-mono-di-glyceride/oil, *J. Disper. Sci. Technol.* 27:1119–1127.

57. Fanun, M. and Salah Al-Diyn, W. 2006 Electrical conductivity and self diffusion NMR studies of the system: Water/sucrose laurate/ethoxylated mono-di-glyceride/isopropylmyristate systems, *Colloids Surf., A* 277:83–89.

58. Kahlweit, M., Busse, G., and Faulhaber, B. 1995 Preparing microemulsions with alkyl monoglucosides and the role of *n*-alkanols, *Langmuir* 11:3382–3387.

59. Alany, R.G., Rades, T., Agatonovic-Kustin, S., Davies, N.M., and Tucker, I.G. 2000 Effects of alcohols and diols on the phase behaviour of quaternary systems, *Int. J. Pharm.* 196:141–145.

60. Sato, T., Acharya, D.P., Kaneko, M., Aramaki, K., Singh, Y., Ishitobi, M., and Kunieda, H. 2006 Oil-induced structural change of wormlike micelles in sugar surfactant system, *J. Disp. Sci. Technol.* 27:611–616.

61. Paul, B.K. and Mitra, R.K. 2005 Water solubilization capacity of mixed reverse micelles: Effect of surfactant component, the nature of the oil, and electrolyte concentration, *J. Colloid Interf. Sci.* 288:261–279.

62. Kunieda, H., Horii, M., Koyama, M., and Sakamoto, K. 2001 Solubilization of polar oils in surfactant self-organized structures, *J. Colloid Interf. Sci.* 236:78–84.

63. Szekeres, E., Acosta, E., Sabatini, D.A., and Harwell, J.H. 2006 Modeling solubilization of oil mixtures in anionic microemulsions: II. Mixtures of polar and nonpolar oils, *J. Colloid Interf. Sci.* 294:222–233.

64. Mehta, S.K. and Bala, K. 2000 Tween based microemulsions: A percolation study, *Fluid Phase Equilibr.* 172:197–209.

65. Kunieda, H., Ozawa, K., and Huang, K.-L. 1998 Effect of oil on the surfactant molecular curvatures in liquid crystals, *J. Phys. Chem. B* 102:831–838.

66. Aramaki, K. and Kunieda, H. 1999 Solubilization of oil in a mixed cationic liquid crystal, *Colloid Polym. Sci.* 277:34–40.

67. Kunieda, H., Umizu, G., and Aramaki, K. 2000 Effect of mixing oils on the hexagonal liquid crystalline structures, *J. Phys. Chem. B* 104:2005–2011.

68. Raoul, Z. (Ed.) *Dynamics of Surfactant Self-Assemblies: Micelles, Microemulsions, Vesicles and Lyotropic Phases*, Taylor & Francis, Boca Raton, FL, 2005.

69. Hellweg, T. 2002 Phase structures of microemulsions, *Curr. Opin. Colloid Interf. Sci.* 7:50–56.

70. Larson, R.G. *The Structure and Rheology of Complex Fluids*, Oxford University Press, New York, 1999.
71. Ezrahi, S., Tuval, E., Aserin, A., and Garti, N. 2005 The effect of structural variation of alcohols on water solubilization in nonionic microemulsions 1. From linear to branched amphiphiles-general considerations, *J. Colloid Interf. Sci.* 291:263–272.
72. Ezrahi, S., Tuval, E., Aserin, A., and Garti, N. 2005 The effect of structural variation of alcohols on water solubilization in nonionic microemulsions 2, Branched alcohols as solubilization modifiers: results and interpretation, *J. Colloid Interf. Sci.* 291:273–281.
73. Djordjevic, L., Primorac, M., Stupar, M., and Krajisnik, D. 2005 In vitro release of diclofenac diethylamine from caprylocaproyl macrogloglycerides based microemulsions, *Int. J. Pharm.* 296:73–79.
74. Yaghmur, A., Aserin, A., Antalek, B., and Garti, N. 2003 Microstructure considerations of new five-component Winsor IV food-grade microemulsions studied by pulsed gradient spin–Echo NMR, conductivity, and viscosity, *Langmuir* 19:1063–1068.
75. Matsumoto, S.P. and Sherman, P. 1969 The viscosity of microemulsions, *J. Colloid Interf. Sci.* 30:525–536.
76. Ray, S., Bisal, S., and Moulik, S. 1992 Studies on structure and dynamics of microemulsions II: Viscosity behavior of water-in-oil microemulsions, *J. Surface Sci. Technol.* 8:191–208.
77. Berghenholtz, J., Romagnoli, A., and Wagner, N. 1995 Viscosity, microstructure, and interparticle potential of $AOT/H_2O/n$-decane inverse microemulsions, *Langmuir* 11:1559–1570.
78. Roe, J.E., Ramanan, D.D., Hornak, J.P., and Kotlarchyk, M. 1996 Application of dense microemulsions to magnetic resonance imaging, *Physica A* 231:359–367.
79. Feldman, Y., Kozlovich, N., Nir, I., and Garti, N. 1995 Dielectric relaxation in sodium bis(2-ethylhexyl) sulfosuccinate–water–decane microemulsions near the percolation temperature threshold, *Phys. Rev. E* 51:478–491 and references therein.
80. Cametti, C., Codastefano, P., Tartaglia, P., Chen, S.H., and Rouch, J. 1992 Electrical conductivity and percolation phenomena in water-in-oil microemulsions, *J. Phys. Rev. A* 45:R5358–R5361 and references therein.
81. Feldman, Y., Kozlovich, N., Nir, I., Garti, N., Archpov, V., Idiyatullin, Z., Zuev, Y., and Fedotov, V. 1996 Mechanism of transport of charge carriers in the sodium bis(2-ethylhexyl) sulfosuccinate–water–decane microemulsion near the percolation temperature threshold, *J. Phys. Chem.* 100:3745–3748.
82. Ponton, A., Bose, T.K., and Delbos, G. 1991 Dielectric study of percolation in an oil–continuous microemulsion, *J. Chem. Phys.* 94:6879–6886.
83. Safran, S.A., Grest G.S., Bug A., and Webman, I. Percolation in interacting microemulsions. In *Microemulsion Systems*. Rosano, H. and Clausse, M. (Eds.) Marcel Dekker Inc., New York, 1987, pp. 235–245.
84. De Gennes, P.G. and Taupin, C. 1990 Microemulsions and the flexibility of oil/water interfaces, *J. Phys. Chem.* 94:8407–8413.
85. Grest, G., Webman, I., Safran, S., and Bug, A. 1986 Dynamic percolation in microemulsions, *Phys. Rev. A* 33:2842–2845.
86. Fanun, M., Wachtel, E., Antalek, B., Aserin, A., and Garti, N. 2001 A study of the microstructure of four-component sucrose ester microemulsions by SAXS and NMR, *Colloids Surf., A* 180:173–186.
87. Kahlweit, M., Strey, R., Haase, D., Kunieda, H., Schmeling, T., Faulhaber, B., Borkovec, M., Eicke, H.F., Busse, G., Eggers, F., Funck, T., Richmann, H., Magid, L., Soderman, O., Stilbs, P., Winkler, J., Dittrich, A., and Jahn, W. 1987 How to study microemulsions, *J. Colloid Interf. Sci.* 118:436–453.

88. Kahlweit, M., Busse, G., and Winkler, J. 1993 Electric conductivity in microemulsions, *J. Chem. Phys.* 99:5605–5614.
89. Soderman, O. and Nyden, M. 1999 NMR in microemulsions. NMR translational diffusion studies of a model microemulsion, *Colloids Surf., A* 158:273–280.
90. Ktistis, G. 1990 A viscosity study on oil-in-water microemulsions, *Int. J. Pharm.* 61:213–218.
91. Haβe, A. and Keipert, S. 1997 Development and characterization of microemulsions for ocular application, *Euro. J. Pharm. Biopharm.* 43:179–183.
92. Shiao, S.Y., Patist, A., Free, M.L., Chhabra, V., Huibers, P.D.H., Gregory, A., Patel, S., and Shah, D.O. 1997 The importance of sub-angstrom distances in mixed surfactant systems for technological processes, *Colloids Surf., A* 128:197–208.
93. Shiao, S.Y., Chhabra, V., Patist, A., Free, M.L., Huibers, P.D.H., Gregory, A., Patel, S., and Shah, D.O. 1998 Chain length compatibility effects in mixed surfactant systems for technological applications, *Adv. Colloid Interf. Sci.* 74:1–29.
94. Constantinides, P.P. and Scalart, J.P. 1997 Formulation and physical characterization of water-in-oil microemulsions containing long- versus medium-chain glycerides, *Int. J. Pharm.* 158:57–68.
95. Acharya, A., Sanyal, S.K., and Moulik, S.P. 2001 Physicochemical investigations on microemulsification of eucalyptol and water in presence of polyoxyethylene (4) lauryl ether (Brij-30) and ethanol, *Int. J. Pharm.* 229:213–226.
96. Acharya, A., Moulik, S.P., and Sanyal, S.K. 2002 Physicochemical investigations of microemulsification of coconut oil and water using polyoxyethylene2-cetyl ether (Brij-52) and isopropanol or ethanol, *J. Colloid Interf. Sci.* 245:163–170.
97. Djordjevic, L., Primorac, M., Stupar, M., and Krajisnik, D. 2004 Characterization of caprylocaproyl macrogolglycerides based microemulsion drug delivery vehicles for an amphiphilic drug, *Int. J. Pharm.* 271:11–19.
98. Podlogar, F., Gasperlin, M., Tomsic, M., Jamnik, A., and Bester Rogac, M. 2004 Structural characterisation of water–Tween 40®/Imwitor 308®–isopropyl myristate microemulsions using different experimental methods, *Int. J. Pharm.* 276:115–128.
99. Mitra, R.K. and Paul, B.K. 2005 Physicochemical investigations of microemulsification of eucalyptus oil and water using mixed surfactants (AOT + Brij-35) and butanol, *J. Colloid Interf. Sci.* 283:565–577.
100. Rodriguez-Abreu, C., Aramaki, K., Tanaka, Y., Arturo Lopez-Quintela, M., Ishtobi, M., and Kunieda, H. 2005 Wormlike micelles and microemulsions in aqueous mixtures of sucrose esters and nonionic cosurfactants, *J. Colloid Interf. Sci.* 291:560–569.
101. Mahji, P.R. and Moulik, S.P. 1999 Physicochemical studies on biological macro- and microemulsions VI: Mixing behaviours of eucalyptus oil, water and polyoxyethylene sorbitan monolaurate (Tween-20) assisted by *n*-butanol and cinnamic alcohol, *J. Disp. Sci. Technol.* 20:1407–1427.
102. Snabre, P. and Porte, G. 1990 Viscosity of the L3 phase in amphiphilic systems, *Europhys. Lett.* 13:641–645.
103. Chen, C.-M. and Warr, G.G. 1992 Rheology of ternary microemulsions, *J. Phys. Chem.* 96:9492–9497.
104. Warr, G.G. 1995 Shear and elongational rheology of ternary microemulsions, *Colloids Surf., A* 103:273–279.
105. Gradzielski, M., Valiente, M., Hoffman, H., and Egelhaaf, S. 1998 Structural changes in the isotropic phase of the ternary surfactant system: Tetradecyldimethylamine oxide/benzyl alcohol/water, *J. Colloid Interf. Sci.* 205:149–160.
106. Montalvo, G., Rodenas, E., and Valienete, M. 2000 Effects of cetylpyridinium chloride on phase and rheological behavior of the diluted $C_{12}E_4$/benzyl alcohol/water system, *J. Colloid Interf. Sci.* 227:171–175.

107. Fanun, M. Phase behavior, structure evolution and diclofenac solubilization studies on mixed nonionic surfactants microemulsions. In: *Colloid and Surface Research Trends.* Fong, P.A. (Ed.), Nova Science Publisher, New York, 2007, pp. 107–146.

108. Paul, B.K. and Mitra, R.K. 2006 Conductivity of reverse micellar systems of water/ AOT + Brij-56 or Brij-58/IPM and their percolation under varied concentrations of amphiphiles and different additives, *Colloids Surf., A* 273:129–140.

109. Mitra, R.K. and Paul, B.K. 2005 Investigation on percolation in conductance of mixed reverse micelles, *Colloids Surf., A* 252:243–259.

110. Mitra, R.K. and Paul, B.K. 2005 Effect of NaCl and temperature on the water solubilization behavior of AOT/nonionics mixed reverse micellar systems stabilized in IPM oil, *Colloids Surf., A* 255:165–180.

111. Mehta, S.K., Dewan, R.K., and Kiran Bala. 1994 Percolation phenomenon and the study of conductivity, viscosity, and ultrasonic velocity in microemulsions, *Phys. Rev. E* 50:4759–4762.

112. Ezrahi, S., Aserin, A., and Garti, N. 1998 Structural evolution along water dilution lines in nonionic systems, *J. Colloid Interf. Sci.* 202:222–224.

113. Ekwall, P., Mandell, L., and Frontell, K. 1970 Some observations on binary and ternary aerosol OT systems, *J. Colloid Interf. Sci.* 33:215–235.

114. Matsumoto, S. and Sherman, P. 1969 The viscosity of microemulsions, *J. Colloid Interf. Sci.* 30:525–536.

115. Ajith, S. and Rakshit, A.K. 1995 Effect of NaCl on a nonionic surfactant microemulsion system, *Langmuir* 11:1122–1126.

116. Rakshit, A.K. and Ajith, S. 1995 Studies of mixed surfactant microemulsion systems: Brij 35 with Tween 20 and sodium dodecyl sulfate, *J. Phys. Chem.* 99:14778–14783.

117. Maitra, A., Mathew, C., and Varshney, M. 1990 Closed and open structure aggregates in microemulsions and mechanism of percolative conduction, *J. Phys. Chem.* 94:5290–5292.

118. MacKay, R.A. and Agarwal, R. 1978 Conductivity measurements in nonionic microemulsions, *J. Colloid Interf. Sci.* 65:225–231.

119. Peyrelasse, J. and Boned, C. 1990 Conductivity, dielectric relaxation, and viscosity of ternary microemulsions: The role of the experimental path and the point of view of percolation theory, *Phys. Rev. A* 41:938–953.

120. Saidi, Z., Matthew, C., Peyrelasse, J., and Boned, C. 1990 Percolation and critical exponents for the viscosity of microemulsions, *Phys. Rev. A* 42:872–876.

121. Saidi, Z., Boned, C., and Peyrelasse, J., Viscosity-percolation behavior of waterless Microemulsions: a curious temperature effect, In: *Trends in Colloid and Interface Science VI* (C. Helm, M. Losche, H. Mohwaldt, Eds.) Springer, The Netherlands, 1992, pp. 301–316.

122. Ajith, S., Jhon, A.C., and Rakshit, A.K. 1994 Physicochemical studies of microemulsions, *Pure Appl. Chem.* 66:509–514.

123. Yaghmur, A., Aserin, A., Antalek, B., and Garti, N. 2003 Microstructure considerations of new five-component Winsor IV food-grade microemulsions studied by pulsed gradient Spin–Echo NMR, conductivity, and viscosity, *Langmuir* 19:1063–1068.

124. Teubner, M. and Strey, R. 1987 Origin of the scattering peak in microemulsions, *J. Chem. Phys.* 87:3195–3200.

125. Teukolsky, S.A., Vetterling, W.T., and Flannery, B.P. *Numerical Recipes in C: The Art of Scientific Computing,* 2nd ed., Cambridge University Press, New York, 1992, pp. 683–688.

126. Billman, J.F. and Kaler, E.W. 1991 Structure and phase behavior in four-component nonionic microemulsions, *Langmuir* 7:1609–1617.

127. Schubert, K.-V. and Strey, R. 1991 Small-angle neutron scattering from microemulsions near the disorder line in water/formamide–octane-C_iE_j systems, *J. Chem. Phys.* 95:8532–8445.
128. Schubert, K.-V., Strey, R., Kline, S.R., and Kaler, E.W. 1994 Small angle neutron scattering near Lifshitz lines: Transition from weakly structured mixtures to microemulsions, *J. Chem. Phys.* 101:5343–5355.
129. Gradzielski, M., Langevin, D., Sottmann, T., and Strey, R. 1996 Small angle neutron scattering near the wetting transition: Discrimination of microemulsions from weakly structured mixtures, *J. Chem. Phys.* 104:3782–3787.
130. Nilsson, P.-G. and Lindman, B. 1983 Water self-diffusion in nonionic surfactant solution, Hydration and obstruction effects, *J. Phys. Chem.* 87:4756–4761.
131. Lindman, B., Shinoda, K., Jonstromer, M., and Shinohara, A. 1988 Change of organized solution (microemulsion) structure with small changes in surfactant composition as revealed by NMR self-diffusion studies, *J. Phys. Chem.* 92:4702–4706.
132. Bastogne, F., Nagy, B.J., and David, C. 1999 Quaternary '*N*-alkylaldonamide-brine-decane-alcohol' systems. Part II: microstructure of the one-phase microemulsion by NMR spectroscopy, *Colloids Surf., A* 148:245–257.
133. El-Seoud, O. 1997 Use of NMR to probe the structure of water at interfaces of organized assemblies, *J. Mol. Liq.* 72:85–103.
134. Olsson, U., Nagai, K., and Wennerstrom, H. 1988 Microemulsions with nonionic surfactants, 1. diffusion process of oil molecules, *J. Phys. Chem.* 92:6675–6679.
135. Soderman, O. and Nyden, M. 1999 NMR in microemulsions, NMR translational diffusion studies of a model microemulsion, *Colloids Surf., A* 158:273–280.
136. Ko, C.J., Ko, Y.J., Kim, D.M., and Park, H.J. 2003 Solution properties and PGSE-NMR self-diffusion study of $C_{18:1}E_{10}$/oil/water system, *Colloids Surf., A* 216:55–63.
137. Ceglie, A., Das, K.P., and Lindman, B. 1987 Effect of oil on the microscopic structure in four-component cosurfactant microemulsions, *J. Colloid Interf. Sci.* 115:115–120.
138. Reimer, J., Soderman, O., Sottmann, T., Kluge, K., and Strey, R. 2003 Microstructure of alkyl glucoside microemulsions: Control of curvature by interfacial composition, *Langmuir* 19:10692–10702.
139. Stubenrauch, C. and Findenegg, G.H. 1998 Microemulsions supported by octyl monoglucoside and geraniol. 2. An NMR self-diffusion study of the microstructure, *Langmuir* 14:6005–6012.
140. Fukuda, K., Soderman, O., Lindman, B., and Shinoda, K. 1993 Microemulsions formed by alkyl polyglucosides and an alkyl glycerol ether, *Langmuir* 9:2921–2925.
141. Nilsson, P.-G., Wennerstrom, H., and Lindman, B. 1983 Structure of micellar solutions of nonionic surfactants. Nuclear magnetic resonance self-diffusion and proton relaxation studies of poly (ethylene oxide) alkyl ethers, *J. Phys. Chem.* 87:1377–1385.
142. Koppel, D.E. 1972 Analysis of macro molecular polydispersity in intensity correlation spectroscopy: The method of cumulants, *J. Chem. Phys.* 57:4814–4820.
143. Provencher, S.W. 1979 Inverse problems in polymer characterisation: direct analysis of polydispersity with photon correlation spectroscopy, *Macromol. Chem.* 180:201–209.
144. Talmon, Y. 1996 Transmission electron microscopy of complex fluids: The state of art, *Ber. Bunsenges. Phy. Chem.* 100:364–372.
145. Regev, O., Ezrahi, S., Aserin, A., Garti, N., Wachtel, E., Kaler, E., Khan, A., and Talmon, Y. 1996 A study of the microstructure of a four-component nonionic microemulsion by Cryo-TEM, NMR, SAXS, and SANS, *Langmuir* 12:668–674.
146. Frederik, P.M. and Sommerdijk, N. 2005 Spatial and temporal resolution in cryo-electron microscopy—A scope for nano-chemistry, *Curr. Opin. Colloid interf. Sci.* 10:245–249.

6 Simple Alcohols and Their Role in the Structure and Interactions of Microemulsion Systems

Matija Tomšič and Andrej Jamnik

CONTENTS

6.1 INTRODUCTION

Microemulsion systems with their inner structure in the colloidal domain have been the subject of many theoretical and experimental studies due to their very broad applicability [1–4]. The name "microemulsion" was first introduced by Hoar and Schulman in 1943 as the name for a clear or transparent system obtained by titration of a milky white emulsion with a medium-chain length alcohol (e.g., 1-pentanol or 1-hexanol) [5]. A more general definition of the term "microemulsion" was given later by Danielsson and Lindman, who described it as a "system, composed of water, oil and an amphiphilic component, being an optically isotropic and thermodynamically stable liquid solution" [6].

Molecules of an amphiphilic surfactant in a microemulsion system specifically arrange to form an interfacial area between the water component and the oil component, which are otherwise immiscible. Various cosurfactants can be added to the system in order to improve the oil-solubilizing efficiency of the surfactant. Most frequently simple alcohols are used for this purpose and, therefore, the phenomena associated with their incorporation in microemulsion systems have drawn a lot of attention. Their specific effects on the molecular organization in surfactant systems as a function of the molecular hydrocarbon chain length are thoroughly described for the case of different ionic-surfactant systems around the critical micelle concentration (CMC) in the review article of Zana [7].

In the present work we review the effects of alcohols on microemulsion structure and illustrate them with the example of nonionic surfactant Brij 35 ($C_{12}E_{23}$; dodecyl-poly[ethylene oxide-23] ether) in alcohol and water systems, at moderate surfactant concentrations well above its CMC [8]. This investigation was based on the generalized indirect Fourier transformation (GIFT) [9–14] of the small-angle x-ray scattering (SAXS) data, providing the results on the structure and interactions in these ternary systems. The GIFT method represents a generalization of the indirect Fourier transformation (IFT) method [15–17] to concentrated systems with interacting scattering particles. It separates the scattering intensity into the contribution arising from the interparticle correlations, which is described by a liquid-type repulsive interaction model, and the scattering contribution arising from the particle form, which is model-free and as such yields model-free results on the particle geometry. In spite of the fact that these specific systems based on the surfactant Brij 35 do not contain a real oil component, their phase and structural behavior resemble those of microemulsions; therefore, they are usually referred to as microemulsions. Moreover, they have also been recognized useful in similar applications to microemulsions, e.g., biocatalytic synthesis of aldehydes [18,19].

As pure simple alcohols themselves show some interesting structure on the molecular level, findings about molecular organization in pure alcohols are first outlined in Section 6.2. Similarly, alcohol/water binary systems also show an interesting propensity to structural organizational phenomena that are delineated in Section 6.3, which is divided into Section 6.3.1 concerned with the short-chain and Section 6.3.2 with the long-chain alcohol/water mixtures. In Section 6.4, we exploit the structural SAXS study of Brij 35/alcohol/water ternary systems [8] in order to present, review, and comment on various effects that simple alcohols can have on the structure and interactions in microemulsion systems. This section is divided into Section 6.4.1 dealing with the effects of short-chain alcohols behaving as cosolvents, Section 6.4.2 describing the cosurfactant behavior of medium-chain alcohols, and Section 6.4.3 devoted to the long-chain alcohols with properties similar to real oil components. Finally, the conclusions and perspectives are given in the last part, Section 6.5.

6.2 MOLECULAR ORGANIZATION IN PURE LIQUID ALCOHOLS

As primary alcohols exhibit a rather peculiar behavior recent decades have witnessed intense theoretical and experimental research on the thermodynamic and structural properties of these simple liquids [20–51]. However, investigating their structural and dynamic behavior in the bulk turned out to be far from trivial. A great deal of work has also been performed on investigating their crystal structure [52–57].

Early x-ray diffraction studies on primary liquid alcohols [20–22] indicated that these pure simple liquids show some interesting structural features that are reflected in two distinct small-angle scattering peaks. These investigations are important for the fact that they belong to the pioneering work on x-ray diffraction of liquids. In Figure 6.1 we show the diffraction curves for the primary alcohols from methanol to 1-undecanol that Stewart and Morrow measured already in 1927 [21]. The earliest x-ray data were interpreted solely on the basis of the Bragg law. A considerable effort was made later to interpret the x-ray and also neutron diffraction data of simple alcohols by various rather involved Fourier transform-based methods [23–30]. The requirement for separation of the intra- and intermolecular contributions to the diffraction patterns is a common problem of these methods, which consequently yield information on the intra- and intermolecular structure via various correlation structure factors related to the corresponding real-space radial distribution functions. Nevertheless, such separation is usually not a straightforward task, and, therefore, the development of contrast variation in the neutron diffraction method [27–30] and the development of various, more or less realistic intermolecular interaction potentials for liquid alcohols were rather helpful in this respect [31–38].

A great deal of attention was paid to H-bonding in primary alcohols. In a thorough study of pure liquid ethanol Narten and Habenschuss showed already in 1980s that the H-bonding distance in pure liquid ethanol is approximately 2.8 Å [25]. This value was later confirmed by numerous studies on simple liquid alcohols that utilized molecular dynamics (MD) and Monte Carlo (MC) investigations, yielding distances of intermolecular O–O correlations from 2.70 to 2.81 Å [29–34,40]. In Figure 6.2 we show the oxygen–oxygen radial distribution function $g_{O-O}(r)$ for pure primary liquid alcohols from ethanol to 1-hexanol. The H-bonding distances correspond to the positions of the first correlation peak in the $g_{O-O}(r)$ function.

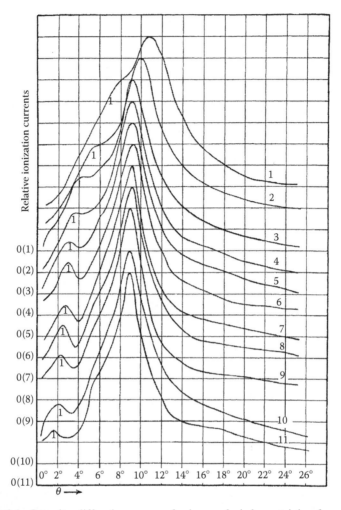

FIGURE 6.1 Intensity–diffraction curves of primary alcohols containing from 1 to 11 carbon atoms. (Reprinted from Stewart, G.W. and Morrow, R.M., *Phys. Rev.*, 30, 323, 1927. With permission.)

As it turns out, the strong tendency of alcohol molecules to H-bond represents the prevailing force for the rather well-organized microheterogeneous structure in the bulk alcohols [26,27,40–45]. Namely, as is nowadays well accepted, the alcohol molecules sequentially H-bond into larger flexible linear aggregates (winding chains), although the reported average number of alcohol molecules that form such aggregates varies strongly for a specific alcohol (tetramers, pentamers, hexamers). Nevertheless, some studies suggest that the structure of liquid alcohols is dominated by cyclic aggregates and/or lasso structures of various sizes [47–50]. It has been reported [49] that the thermodynamic and spectroscopic experimental data can only be fitted on the basis of such structures.

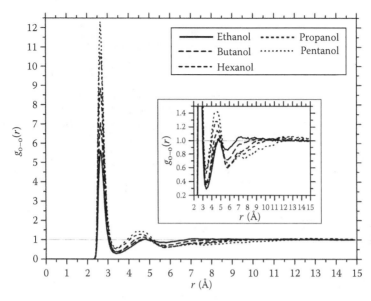

FIGURE 6.2 Oxygen–oxygen pair correlation function $g_{O-O}(r)$ of pure primary liquid alcohols from ethanol to 1-hexanol at 25°C. Inset: Magnification of the second correlation peaks and the succeeding minima. (Reprinted from Tomšič, M., Jamnik, A., Fritz-Popovski, G., Glatter, O., and Vlček, L., *J. Phys. Chem. B*, 111, 1738, 2007. With permission.)

In Figure 6.3 we show the results of configurational-bias MC simulations based on the transferable potential for the phase equilibria—united atom (TraPPE-UA) force field [38] for alcohols from ethanol to 1-hexanol [40]. These results confirm the presence of flexible linear aggregates, complexly branched aggregates, and aggregates closed in cyclic or lasso structures, in all of the studied alcohols. In some configurations also percolated (endless) aggregates were found. These structures were divided into two main types, i.e., linear and cyclic, and were summed yielding the results presented in Figure 6.3a and b, respectively. These results suggest that it is not trivial to state the average number of alcohol molecules forming an aggregate due to the broad distribution of these numbers. Further, they also indicate the presence of a minor fraction of cyclic aggregates. A picturesque representation of the molecular aggregates in pure 1-butanol is given in the insets of Figure 6.3, where the hydrocarbon parts of the 1-butanol molecules are hidden and only the –OH groups are shown for better visibility.

The primary liquid alcohols show two distinct scattering peaks in the SAXS regime up to $q = 2.5\text{Å}^{-1}$ (scattering vector $q = 4\pi/\lambda \cdot \sin[\vartheta/2]$, with ϑ being the scattering angle and λ the wavelength of the radiation). In Figure 6.4 we present the experimental SAXS curve of liquid 1-butanol together with the results of the so-called theoretical contrast matching experiment. The latter relates the organization of atoms (butanol molecules) in the MC box with the details of the overall scattering curve and consequently reveals the origin of the two scattering peaks [40].

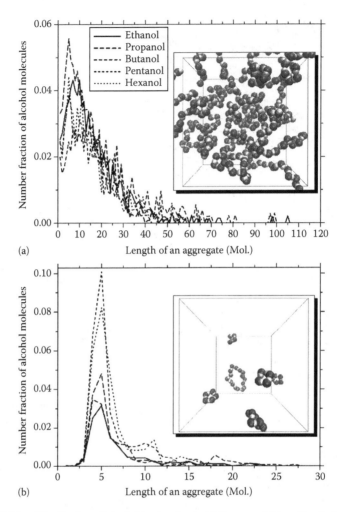

FIGURE 6.3 Distribution of alcohol molecules in (a) linear and (b) cyclic alcohol aggregates in bulk liquid primary alcohols from ethanol to 1-hexanol at 25°C. (Reprinted from Tomšič, M., Jamnik, A., Fritz-Popovski, G., Glatter, O., and Vlček, L., *J. Phys. Chem. B*, 111, 1738, 2007.) Insets: Schematic examples of liquid 1-butanol—only the atoms of the –OH groups (–OH skeletons) are shown. In the bottom inset only the –OH groups of the 1-butanol molecules that are H-bonded into closed cyclic aggregates are shown. Insets were created using VMD (http://www.ks.uiuc.edu/Research/vmd/; [148]).

In such a computational contrast matching experiment one calculates the scattering contribution of a specific atom type in the MC configuration and compares it to the overall scattering curve. In order to be able to perform such an experiment a special procedure for calculating the scattering intensities from the MC results has been recently developed [39,40].

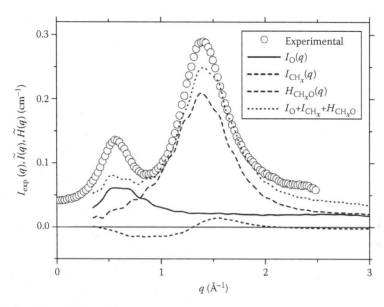

FIGURE 6.4   A "theoretical contrast matching experiment" on the example of liquid 1-butanol at 25°C. Summation of the theoretically smeared calculated partial scattering contributions to the scattering of 1-butanol $I_{CH_x}(q)$, $I_O(q)$, and $H_{CH_xO}(q)$ results in a curve that already shows all the important features of the experimentally obtained SAXS curve—the partial structure factors $H_{OH}(q)$, $H_{CH_xH}(q)$, and $H_{HH}(q)$ are missing for complete representation of the scattering contributions. (Reprinted from Tomšič, M., Jamnik, A., Fritz-Popovski, G., Glatter, O., and Vlček, L., *J. Phys. Chem. B*, 111, 1738, 2007. With permission.)

One can clearly see in Figure 6.4 that the outer alcohol scattering peak relates to the scattering contribution solely of the atoms of the hydrocarbon part of the alcohol molecules, whereas the inner alcohol scattering peak relates to the scattering contribution solely of the oxygen atoms of alcohol molecules. Namely, the $I_{CH_x}(q)$ contribution has a single pronounced peak at $q \approx 1.41$ Å^{-1}, which is practically the position of the outer alcohol scattering peak in the 1-butanol scattering curve. According to Bragg law, the correlations corresponding to a distance of approximately 4.5 Å are mainly expressed at this q value. These correlations match up to the average correlations between the CH_x pseudo-atoms in the side chains of the −OH skeletons of the alcohol aggregates. Similarly, the scattering contribution solely of the oxygen atoms $I_o(q)$ in the MC configurations of 1-butanol has a single pronounced peak expressed at $q \approx 0.59$ Å^{-1} indicating a distance of approximately 10.6 Å. This relatively long distance corresponds to the average inter-chain distance of the adjacent −OH skeletons, or better, to the average correlations between the oxygen atoms in the adjacent −OH skeletons [40]. Interestingly, one can notice that all the important features of the experimental scattering curve of 1-butanol can be obtained simply by summing up the $I_{CH_x}(q)$, $I_o(q)$, and $H_{CH_xO}(q)$ scattering contributions.

In summary, the above results indicate that all the studied alcohols showed practically the same structural features. The outer alcohol scattering peaks appeared at practically the same positions on the q scale (with the exception of ethanol [40]), indicating qualitatively similar packing of the hydrocarbon parts of the molecules in space for various alcohols, but the innermost alcohol scattering peaks shifted their position proportionally to the number of carbon atoms in the alcohol molecule. The latter fact corresponds to the increasing interskeleton distances due to the increase in the hydrocarbon part of the alcohol molecules.

6.3 PROPERTIES OF ALCOHOL/WATER BINARY SYSTEMS

Plain alcohol/water binary systems already with rather complex behavior represent the first evolutionary step toward even more complex ternary (microemulsion) systems. Though the latter is the primary interest of this book, we first review the work on binary mixtures, because understanding their behavior is crucial for a perception of the situation in even more complex systems that evolve from them.

A great deal of work with different experimental and theoretical techniques has been performed in the past on characterizing liquid binary mixtures of simple alcohols with water [46,58–78]. Water mixtures with a short-chain alcohol up to 1-propanol [50,58–81], which mix with water over the whole concentration regime, have been studied the most extensively. Nevertheless, one can also find some similar studies with long-chain alcohols that are practically insoluble in water [46,76,77,82–86]. The motivation for such abundant work lies in the anomalous behavior of such mixtures. Namely, when simple alcohols are mixed with water the entropy of the system seems to increase far less than would be expected for an ideal solution of randomly mixed molecules [75]—a phenomenon that is also clearly expressed in the anomalous behavior of some other measurable properties. Furthermore, one can find two completely different concepts to explain these anomalous properties existing in the literature, which makes these binary mixtures even more intriguing nowadays.

6.3.1 SHORT-CHAIN ALCOHOL/WATER MIXTURES

In the microemulsion systems the primary alcohols are frequently considered as cosurfactants, which are usually weakly amphiphilic molecules that help the amphiphilic surfactants to reduce the surface tension of the interface between the immiscible components of the system. In this way they usually enhance and emphasize the internal structure of the system at the colloidal level. Remarkably, the short-chain alcohols, which are sufficiently soluble in water, themselves show surfactant-like behavior in plain binary water mixtures. As was shown by Kahlweit et al. [87], this specific behavior can be observed from the break in the curves of surface tension versus molar fraction of alcohol in water. Similar breaks were observed by Zana et al. [7] in the curves of fluorescence intensity versus molar fraction of alcohol, where changes in the environment polarity are sensed by the pyrene fluorescence probe. Interestingly, with increasing the length of the alcohol molecule, in both cases even 1-butanol was not soluble enough in water to still show such a break. These breaks were interpreted as an indication of the molecular aggregation of the alcohol.

Furthermore, an extensive study of the dynamic and structural properties of water/alcohol systems by Onori and Santucci [73] based on adiabatic compressibility and surface tension measurements, infrared and near-infrared absorption spectra, and the dielectric relaxation method showed that two characteristic limiting molar fraction values can be designated for a specific alcohol. The adiabatic compressibility is a quantity that refers to the volume unit of solution irrespective of the number of molecules; therefore, the "excess" quantity $(\beta_S \cdot \bar{V}_m)^E/x_2$, which is expected to be zero for a two phase system, is used to rationalize such compressibility data [73]. The behavior of the excess quantity $(\beta_S \cdot \bar{V}_m)^E/x_2$ obtained from the adiabatic compressibility β_S and the mean molar volume $\bar{V}_m$ of an ethanol/water mixture is republished in Figure 6.5 and clearly shows the two characteristic molar fraction limits x_2^a and x_2^b. Similar results were also obtained for the other alcohols studied (methanol, 2-propanol, and t-butanol) and a trend of decreasing x_2^a and increasing x_2^b was observed with enlargement of the hydrophobic part of the alcohol molecule. As one can see in Figure 6.5, this excess quantity is practically constant below x_2^a and afterwards increases steeply. This implies that the hydration scheme, or better, the association state of the alcohol molecules, changes qualitatively at around x_2^a as more alcohol is added to the solution. The observed behavior suggests that the alcohol molecules are essentially monomolecularly dispersed and surrounded

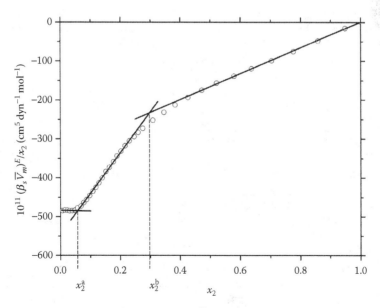

FIGURE 6.5 Plot of the excess quantity $(\beta_S \cdot \bar{V}_m)^E/x_2$ for ethanol/water mixtures at 25°C. (Reprinted from Onori, G. and Santucci, A., *J. Mol. Liq.*, 69, 161, 1996. With permission.)

by water molecules below x_2^a, but above this concentration the alcohol molecules start to aggregate and an interplay of the hydration and aggregation of the alcohol molecules dictates the structure of the system in this regime.

The concept that molecular organization in alcohol/water mixtures at low alcohol concentrations is dominated by the low entropy tetrahedral-like structure of water, sometimes also designated as clathrate-like structures, ice-like structures, "icebergs" or "water cages" [73,77,80,83,88], is well known nowadays. It explains how the behavior of alcohol molecules at low concentrations is governed by the phenomenon of hydrophobic hydration [75,89] that structurally enhances the H-bond network of water around the hydrophobic moiety. Accordingly, the situation in Figure 6.5 can be explained in terms of alcohol association, which starts to take place after reaching the x_2^a value, where already almost all the water present is engaged in forming the hydration structures. From this point on, the alcohol association changes the hydration scheme, leading to less compressible structures and consequently to gradual diminution of the monitored excess quantity $(\beta_S \cdot \bar{V}_m)^E/x_2$. Further, with increasing alcohol concentration the interactions between water and nonpolar alkyl groups gradually decrease, eventually leading to another breaking point at x_2^b beyond which the compressibility changes are exclusively based on the interactions between water and alcohol hydroxyl groups. The latter is confirmed by the infrared spectra, indicating that at these high alcohol concentrations there is practically no contact between the alcohol nonpolar alkyl groups and water—the hydrophobic part of the alcohol molecule practically only "sees" the other alcohol molecules [73].

Similarly, based on density and sound velocity measurements, Parke and Birch [88] reported apparent specific volume and apparent specific isentropic compressibility curves for the ethanol/water mixture. They observed a shallow minimum in the apparent specific isentropic compressibility curves at $x_2 \approx 0.05$ and an increase afterwards, which generally conforms to the situation presented in Figure 6.5.

Sato et al. [67–70], utilizing the time domain reflectometry, studied dynamic processes in alcohol/water mixtures over the whole concentration regime and obtained complex dielectric spectra. A three-process model turned out to be the most appropriate for evaluation of these data. It resolved the dominant low-frequency process with a relaxation time τ_1 ranging from 8.3 to 165 ps for the ethanol/water system in the regime $0 \leq x_2 \leq 1$. This process was assigned to the cooperative dynamics of the H-bond system and represents the lifetime of the locally stable configuration in the energy landscape of the liquid. Two additional processes with relaxation times $\tau_2 \approx 10$ ps and $\tau_3 \approx 1 - 2$ ps had to be used to reproduce the high-frequency part of the spectrum. According to the results on pure liquids [90–92], these two processes correspond to the motion of singly H-bonded alcohol monomers at the end of the alcohol aggregates and to the flipping motion of free OH, respectively. Furthermore, these results imply that at higher molar ratios of ethanol the water molecules tend to insert themselves into the structure of winding H-bonded alcohol aggregates, resulting in a decrease of the aggregate lengths and correspondingly an increase in the number of singly H-bonded alcohol monomers, the latter phenomenon enhancing the contribution of the τ_2 mode. Interestingly enough, at alcohol concentrations below $x_2 = 0.3$ only a two-process model is sufficient to describe the data, because the τ_1 process rapidly merges with the τ_2 process. Interestingly, this value coincides with the position x_2^b in Figure 6.5, above which water was explained as behaving as though it would interact practically exclusively with the alcohol –OH skeletons. Obviously, below this value where the two processes merge, the number of alcohol aggregates rapidly increases and consequently their length decreases, which is in good agreement with the qualitative change in the hydration scheme revealed by x_2^b in Figure 6.5.

The excess activation free energy ΔG^E, enthalpy ΔH^E, and entropy ΔS^E, and their partial molar quantities for alcohols ΔG_A^E, ΔH_A^E, and ΔS_A^E were also calculated for the dominating processes with τ_1 in methanol, ethanol, and 1-propanol water mixtures [67–69] and are depicted in Figure 6.6. These thermodynamic quantities were calculated according to the Eyring transition state theory [93]. Based on the curves in Figure 6.6 characteristic molar fraction values (0.30, 0.18, and 0.14 for methanol, ethanol, and 1-propanol water mixtures, respectively) were also reported above and below which the behavior of the partial molar excess activation quantities

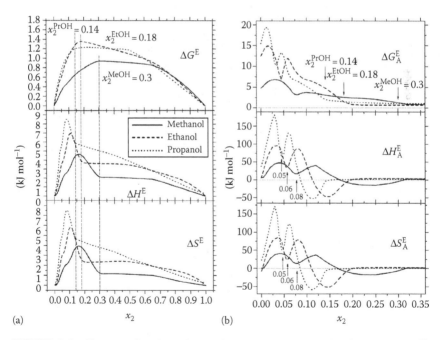

FIGURE 6.6 Concentration dependence of the (a) excess activation free energy ΔG^E, enthalpy ΔH^E, and entropy ΔS^E, and (b) excess activation free energy of alcohol ΔG_A^E, enthalpy ΔH_A^E, and entropy ΔS_A^E for methanol/water, ethanol/water, and 1-propanol/water mixtures at 25°C for the processes with the relaxation times τ_1. (Reprinted from Sato, T., Chiba, A., and Nozaki, R., *J. Chem. Phys.*, 113, 9748, 2000. With permission.)

are rather different [68]. Beyond these characteristic values the partial molar excess activation quantities are nearly zero, meaning that in terms of thermodynamics the alcohol molecules find themselves in a very similar environment as in pure liquid alcohols. Compared to the results presented in Figure 6.5 for the ethanol/water mixture, this characteristic molar fraction value is practically between the x_2^a and x_2^b values, meaning that at this point there is already a considerable aggregation of the alcohol molecules present. This makes perfect sense also in terms of the behavior of the thermodynamic properties shown in Figure 6.6, because the local nature of the aggregates is certainly approaching a situation similar to that in pure liquid alcohols as the aggregates grow and at some point become prevailing structure of these systems. However, below this characteristic molar fraction value the two maxima can be seen in the ΔH_A^E and $T\Delta S_A^E$ curves in Figure 6.6b. They can be clearly attributed to the hydrophobic hydration of two different kinds of saturated hydration structures in this concentration regime and the transition between the two, namely, to the tetrahedral-like structure of water around the alcohol molecules at very low alcohol concentrations and to nonclathrate shells with large cavities accompanied by molecular aggregation of alcohol molecules due to hydrophobic interactions when the alcohol concentration increases. It is worth noting that the position of the minima between the two peaks in ΔH_A^E and $T\Delta S_A^E$ plots of the ethanol/water mixture in Figure 6.6b ($x_2 \approx 0.06$) coincides well with the x_2^a value from Figure 6.5. The x_2^a value represents the breaking point between the low-concentration region with a tetrahedral-like structure of water (icebergs) and the region of its collapse due to the alcohol molecular association. This behavior and the transition point also exactly explain the appearance of the two peaks in the ΔH_A^E and $T\Delta S_A^E$ curves and the position of the minimum in-between.

To summarize, according to the results reviewed above, it seems that the structure of short-alcohol/water mixture systems is governed by H-bonded alcohol aggregates at high concentrations of the alcohol, but at high water concentrations the structure is dictated by the tetrahedral-like structure of water. At compositions in-between a more or less continuous transition between the two is observed with two characteristic breaking points; the collapse of the icebergs and the predomination of the alcohol-governed structure. On the basis of these results one could also easily imagine that the dominant mechanism by which short-chain alcohols influence micellization is through their effect on the structure of the solvent (water).

However, eventually we have to point out that in the last decade a number of investigations appeared in the literature claiming that the water structure in the vicinity of the hydrophobic solute is not really considerably perturbed [80,81,94,95] and calling for a revision of the icebergs concept for hydrophobic hydration. A study questioning this concept was made by Bowron et al. in 1998 [94,95] on water-rich t-butanol/water mixtures and pure t-butanol. They reported neutron diffraction results evaluated by the empirical potential structure refinement method (EPSR). Already at very low alcohol concentrations they could find direct solute–solute contacts. Furthermore, in 2002 Dixit et al. [80] published an EPSR neutron diffraction study on molecular segregation in a highly concentrated methanol/water mixture

($x_2 = 0.7$ or mass fraction $w_2 \approx 86\,\text{wt}\%$). As they explicitly state, the local structure of water in a concentrated methanol/water mixture is surprisingly close to its counterpart in pure water. The number of hydrogen bonds per molecule in solution was found to be not significantly different from that in the pure liquids before mixing. Although they could not calculate the entropy of mixing directly from their data, they did imply that the negative excess entropy observed in these systems arises from incomplete mixing at the molecular level, rather than from water restructuring. Based on these results for concentrated methanol/water mixture they further make a general claim that the polar interaction of water with the alcohol hydroxyl group is likely to be a far more potent influence on the thermodynamic properties of alcohol–water mixtures than any water restructuring induced by the hydrophobic methyl groups. With this claim they challenge the well-accepted explanation of excess enthalpies and entropies of solution in terms of iceberg formation in the water which surrounds a hydrophobic entity in aqueous solutions [80]. Shortly after this paper they also published a study with similar conclusions on the structure of dilute aqueous methanol ($x_2 = 0.05$ or mass fraction $w_2 \approx 12\,\text{wt}\%$) [81], the latter schematically depicted in Figure 6.7. Namely, as they explicitly state, the response of water molecules to the presence of the methanol solute molecules was not an ordering of the hydration shell of the nonpolar methanol group, as would be expected according to the hydrophobic hydration concept, but rather the compression of the second order water shell, with the sharpening of the second water–water correlation peak in the $g_{O-O}(r)$ function. The latter reduction of the freedom of water molecules could contribute to the deficient increase of the entropy compared to the ideal mixture observed on blending alcohol with water [75,96]. Here we would like to specifically point out that according to Figure 6.5 for alcohol/water systems around the x_2^a and x_2^b values, qualitative changes in the hydration/aggregation scheme are implied. Since icebergs are anyway to be expected at low alcohol concentrations below x_2^a, this second study by Dixit et al. [81] in the low concentration regime in principle provides more justification to question this concept.

However, a recent MD study of the ethanol/water mixture over the whole concentration regime by Zhang and Yang [97] shows that the water–water correlation is enhanced in ethanol/water mixtures compared to that in pure water, while the ethanol–ethanol hydrogen bonding structure is gradually broken as the water concentration increases.

To conclude, Koga et al. [83] recently published a thermodynamic study on the question of the existence of icebergs. Their motivation lay in the fact that the iceberg concept appeared to be in turmoil, but they believed that it could be sorted out by clarifying the conditions under which its consideration is justified. Namely, though not clearly stated, the original idea of iceberg formation was meant to be applicable to the infinite dilution of aqueous solutions of small nonelectrolytes [89]. Based on their results they explicitly explain that the formation of icebergs is actually only a local enhancement with a concomitant reduction of hydrogen-bond probability away from the solute. Namely, as the solute concentration increases, the probability of H-bonding of the bulk water away from the solute is reduced

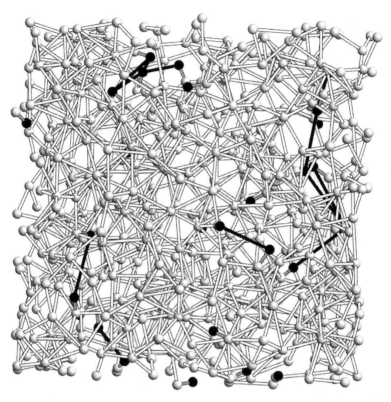

FIGURE 6.7 Simulated molecular box. Water and methanol oxygens are shown as white, but the methanol carbon atoms as black. Black bonds indicate methanol carbons in the first coordination shell, $r \leq 5.8$ Å, gray bonds join a carbon and oxygen atom of a methanol molecule, and white bonds join all oxygen atoms. Note that the snapshot shows just one face of the cubic simulation box. Thus, some water molecules are seen to be surrounded by only water molecules in all directions. However, as one moves away from such a water molecule, even if there are relatively few methanol molecules within the first O_w–O_w coordination shell (i.e., within 3.5 Å), there is a high probability of finding a methanol molecule in the second coordination shell, which peaks around 4.5 Å. Since a methanol molecule can be anywhere up to approximately 5 Å from the second-shell water molecule for the latter to be in the hydration shell of the former, it is concluded that most of the water molecules are within the first hydration shell of at least one methanol molecule. (Reprinted from Dixit, S., Soper, A.K., Finney, J.L., and Crain, J., *Europhys. Lett.*, 59, 377, 2002. With permission.)

progressively. When this probability is reduced to the bond percolation threshold of hexagonal ice connectivity, H-bond percolation is lost and a qualitatively different mixing scheme sets in. At this point two kinds of clusters appear in the solution; solute molecules form clusters of its own kind in the solute-rich region. Thus, "iceberg formation" is basically correct within a narrow range in the water-rich region for small nonelectrolyte solutes. To avoid further confusion, references made to the iceberg concept in the recent literature should be carefully clarified in terms of the concentration range and the size of the solute in question. Nevertheless, unfortunately still no direct structural evidence has been obtained for the existence of icebergs; therefore, such desirable structural investigations must be aimed at these narrow dilute concentration regimes where they are actually expected to exist.

6.3.2 Long-Chain Alcohol/Water Mixtures

In the literature [46,76,77,82–86] one can find some similar studies with long-chain alcohols, which no longer mix with water in all proportions. These alcohols are practically insoluble in water and vice versa, so only highly water-rich and alcohol-rich one-phase systems exist. 1-Butanol is the first in a series of alcohols bearing such properties. With increasing length of the nonpolar hydrocarbon chain, the regimes of mixing narrow considerably. Such long-chain alcohol/water systems were investigated mainly in the alcohol-rich regime.

The 1-octanol/water binary system seems to be the most interesting one in this series, because it has been accepted as a good reference system for studies of the partition coefficients of solutes [7,82,98–105]. Namely, 1-octanol has also been widely used as a mimetic surrogate for the much more complicated lipid molecules that comprise biological membranes, although it cannot form typical lipid structures such as bilayers or micelles. Accordingly, the partition coefficients between 1-octanol and water are widely used to predict biological activity and pharmacokinetic properties such as bioavailability, transport, and elimination. The great pharmaceutical interest in this system also motivated some more basic studies on structural phenomena in the 1-octanol/water mixture in the one-phase regime. Interestingly, the saturation concentration of water in 1-octanol is 2.30 M (ca. 5 wt%), but of 1-octanol in water only 0.0045 M (ca. 0.06 wt%).

Sassi et al. [84] studied the 1-octanol/water system spectroscopically utilizing the Raman, depolarized-Rayleigh, and Rayleigh–Brillouin methods. They reported that the 1-octanol liquid structure is essentially unchanged after the addition of water and that the microinhomogeneity of the liquid system can be considered to be established. This means that "pockets" containing high-density water on a scale smaller than mesoscopic are assumed to be present in the system, representing the phenomena of "molecular confinement." On the basis of MD simulations Best et al. [82] similarly found that the water molecules preferentially coordinate to other water molecules rather than to the hydroxyl groups of the alcohol in the 1-octanol/water system. In contrast the MD simulation study of MacCallum and Tieleman [46] suggests a rather drastic increase of the alcohol cluster dimensions in the proximity of water molecules. Nearly four times the number of molecules is expected in an average alcohol aggregate of the 1-octanol/water system in comparison to nonaqueous 1-octanol. The water molecules are expected to fit into the "tight" spaces near the polar regions of the aggregates that would be inaccessible to the 1-octanol molecules and due to H-bonding capability of water allows for

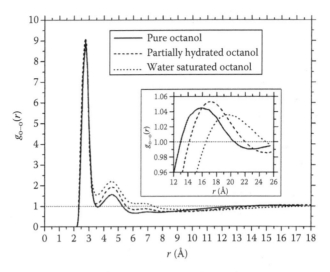

FIGURE 6.8 Oxygen–oxygen pair correlation function $g_{O-O}(r)$ of pure and hydrated 1-octanol. Inset: Magnification of part of the curves. (Reprinted from MacCallum, J.L. and Tieleman, D.P., *J. Am. Chem. Soc.*, 124, 15085, 2002. With permission.)

more branching of the aggregates. The growth of the polar regions with increasing water concentration can be seen from the increase of the second maximum of the oxygen–oxygen pair correlation functions for the 1-octanol/water system shown in Figure 6.8 [46]. The polar regions are also expected to become more spherical. However, a mass spectrometric analysis of the 1-butanol/water system performed by Wakisaka et al. [85] showed that in the alcohol-rich 1-butanol/water system 1-butanol molecules exist largely as self-association clusters, and that the monomeric water molecules are saturated in solution. In other words, water molecules can stay in the 1-butanol-rich phase by promoting 1-butanol self-association, which counteracts the destabilization resulting from contact between 1-butanol and water molecules.

Recently we conducted a short experimental SAXS study on the 1-pentanol/water system [40]. The results depicted in Figure 6.9 show that the structure of 1-pentanol indeed does not change appreciably after the addition of water, although slight changes in scattering were observed with the very sensitive SAXS method. The structure of pure liquid alcohols has been discussed already in the Section 6.2, where also the origin of the appearance of the two alcohol scattering peaks was explained. Interestingly, the movement of the inner scattering peak in Figure 6.9 is directly proportional to the changes in water concentration, and also the peak height (intensity) increases linearly in the same manner [40]. Furthermore, during movement the inner scattering peaks also slightly broaden. All these changes were interpreted as a sign of gradual uniform microscopic changes in some characteristic average distances (dimensions) inside the alcohol structure as

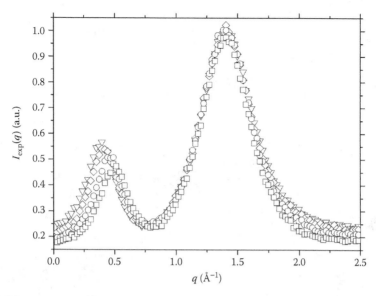

FIGURE 6.9 Experimental SAXS curves of 1-pentanol/water system with 0, 4, 8, and ca. 11 (saturated sample) wt% of water. To facilitate comparison the curves are normalized to the second peak height. (Reprinted from Tomšič, M., Jamnik, A., Fritz-Popovski, G., Glatter, O., and Vlček, L., *J. Phys. Chem. B*, 111, 1738, 2007. With permission.)

water is gradually introduced into the system. Since the inner scattering peak was ascribed to the correlations between adjacent −OH skeletons, the water molecules most probably distribute near the latter, broaden them slightly, and consequently also slightly increase the distances between them, leading to the observed shifts in the inner scattering peak position. Furthermore, the intensity of the inner peak increases, because water molecules increase the scattering contribution of −OH skeletons. In parallel this peak broadens, because the distribution of interskeleton distances becomes slightly broader with increasing hydration. In other words the −OH skeletons slightly swell in a plane perpendicular to the skeleton axis due to the presence of water molecules. Nevertheless, this incorporation of water into 1-pentanol obviously does not affect the intra- and intermolecular CH_x–CH_x correlations of the alcohol hydrocarbon tails, meaning that the alcohol structure stays qualitatively the same.

To conclude, it seems there are two main leads regarding the distribution of water in alcohol-rich alcohol/water mixtures in the literature. The first defends the hypothesis of water-rich microinhomogeneities in the mixture structure in terms of small "water-pockets," and the second speaks in favor of a more uniform distribution of water over the system—the hydration of the hydrophilic −OH chains and the enhancement of alcohol self-association. The inconsistencies in some of the findings of these studies indicate that structural investigations of primary alcohol/water binary mixtures are far from a trivial task.

6.4 ROLE OF SIMPLE ALCOHOLS IN MICROEMULSIONS: A REVIEW BASED ON THE EXAMPLE OF BRIJ 35/WATER/ALCOHOL TERNARY SYSTEMS

The behavior of liquid primary alcohols in various surfactant systems is of course not universally the same. It can depend on the type of surfactant and/or of the other components in the microemulsion system [7,8,18,19,106–132]; the latter fact justifies and increases the importance of the previous sections, which reveal the properties of pure alcohols and alcohol/water systems. The effects of alcohols also strongly depend on their partition between the water phase, oil phase, and interface surfactant film. Alcohol partition behavior in ionic surfactant systems was thoroughly reviewed by Zana [7]. In the following we focus more on their different roles and the variety of effects on the structure and intermolecular interactions in microemulsion systems. These phenomena will be presented for the example of a ternary system composed of the nonionic surfactant Brij 35 at moderate concentrations, water and one of the simple alcohols from ethanol to 1-decanol, and will be supported with an additional literature review [7,109–119].

In Figure 6.10 we show the phase diagrams of the studied Brij 35/alcohol/ water ternary systems [133], where one can follow the intriguing evolution of the microscopically homogeneous microemulsion region (dark zone) with increasing alcohol hydrocarbon chain length. These phase diagrams are given in wt% and, therefore, all concentrations in the following text are expressed in these units. The studied compositions in the one-phase regime are represented by the 5% Brij 35 lines starting on the water-rich side and going in the direction of increasing concentration of an individual alcohol; the corresponding group of samples is addressed as the 5% Brij 35 individual alcohol series [8]. It should be stressed that the 5% Brij 35/water binary system is actually considered as the reference system for all of the ternary systems presented in the following sections, and, therefore, all of the data are interpreted in relation to this binary system [106].

Since the corresponding SAXS data were evaluated by the GIFT method [9–17], a brief description of this technique is given in the following. To obtain a deeper insight into the method, the reader is encouraged to follow the original references. The GIFT technique is based on a representation of the scattering intensity $I(q)$ as a product of the so-called form factor $P(q)$ and the structure factor $S(q)$. The structure factor describes the interparticle interactions in reciprocal (q) space, whereas the form factor represents the scattering of a free particle, which can be written as the Fourier transformation of the so-called pair-distance distribution function $p(r)$. The latter function expresses the geometry of the scattering particles in real space [134,135]. On inspecting the course of this function essential information on the particle geometry can be obtained. From the abscissa value where the $p(r)$ function vanishes, the particle's maximal dimension can be estimated, and further, from the functional form (course) of this function, the shape and inner structure of the scattering particles can be deduced [136]. Information on interparticle interactions in the studied system is

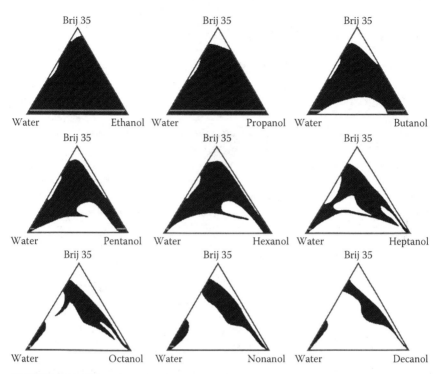

FIGURE 6.10   Evolution of ternary phase diagrams of the Brij 35/alcohol/water system with change in the alcohol from ethanol to 1-decanol at 25°C. Gray lines indicate the composition lines of the studied samples with 5% of Brij 35. (Reproduced from Touraud, D., Contribution à l'étude de microemulsions utilisables comme milieux réactionnels (thesis). Compiègne: Université de Technologie de Compiègne, 1991. With permission.)

obtained via the resulting structure factor data. A polydisperse system of hard spheres is usually chosen as a model for the interparticle correlations in microemulsions of nonionic surfactants, and the Percus–Yevick approximation [9] is used as a method for the calculation of $S(q)$. The averaged structure factor $S_{ave}(q)$ represents the average taken over the weighted contributions of partial structure factors for individual monodisperse systems. It is described by three parameters: the volume fraction ϕ, the interaction radius R, and the polydispersity μ, which is defined as the ratio between the width of the Gaussian size distribution σ and the size R at its maximum value, $\mu = \sigma/R$ [11]. However, in the case of particles that strongly deviate from monodisperse spheres this model is no longer exact and these three parameters do not have a correct, if any, physical meaning. Nevertheless, $P(q)$ and $p(r)$, respectively, can be determined successfully and one can still deduce the particle geometry [135,137–139]. Furthermore, the electron density contrast profile $\Delta\rho(r)$, which reveals the inner structure of the spherical scattering particles, can be obtained from the $p(r)$ function by a convolution square root operation utilizing the DECON technique [140–142].

Nevertheless, one must be somewhat cautious when applying these methods for the analysis and interpretation of the SAXS data. Firstly, their use is limited only to particulate systems containing more or less defined scattering particles and secondly, they are also limited by the actual quality of the model used for the description of the interparticle interactions. As shown later in the text, some problems occurred at higher alcohol concentrations, where the value of the polydispersity parameter in the GIFT evaluations ran too high to get reasonable results, and also, most probably for the same reasons, the fits of the DECON program, which is based on the monodisperse spherical model, were no longer satisfactory. Correspondingly, the DECON procedure can yield some numerical artifacts at very high values of r (sudden jumps in the electron density) that should not be taken as a reflection of the actual structure in the micelles, but rather as a consequence of the discrepancy between the actual situation and the DECON model—discrepancies from the spherical shape are of course the most strongly expressed at higher distances, r. Nevertheless, if such malfunctions of the methods are recognized, they can be used to provide useful information on the studied system.

6.4.1 "STRUCTURE BREAKING" EFFECT OF SHORT-CHAIN ALCOHOLS OR ALCOHOLS AS COSOLVENTS

As it turns out, alcohols up to 1-propanol can generally be considered under this title. In Figure 6.11a the experimental SAXS curves of the 5% Brij 35/ethanol/water ternary system at various ethanol concentrations are displayed [8]. Already these raw SAXS data (symbols) show that ethanol at low concentrations (up to 20%) does not have a noticeable effect on the geometry of scattering structures in comparison to the situation in the reference 5% Brij 35/water binary system (gray line). Namely, position of the peak remains practically constant with increasing alcohol concentration, as the different courses of scattering intensity may be ascribed to the lowering of the scattering contrast. This lowering is a direct consequence of the complete miscibility of ethanol and water, the electron density of ethanol being $\rho_{ethanol} = 267$ e$^-$/nm^3 and of water $\rho_{water} = 334$ e$^-$/nm^3. Such behavior is explicitly confirmed by the results of the GIFT evaluation of these SAXS spectra shown in Figure 6.11b and c, and in addition, by the DECON electron density profiles depicted in Figure 6.11d.

The course of the $p(r)$ functions in Figure 6.11b indicates inhomogeneous, more or less spherical structures, with a difference in the core and shell scattering contrast [11]. The maximal dimension of these micellar structures in the lower alcohol concentration range is approximately 10 nm, as estimated from the abscissa value where the $p(r)$ function vanishes. Further, the radius of the hydrophobic core of these core-shell type micelles is around 1.7 nm, obtained from the value where the electron density profile in Figure 6.11d changes sign. The geometry of these particles at low ethanol concentrations is practically the same as in the reference binary system without added alcohol.

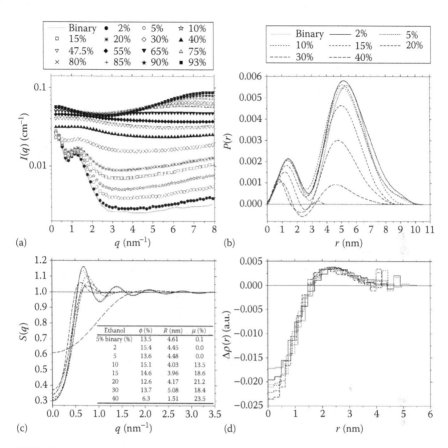

FIGURE 6.11 SAXS results for 5% Brij 35/ethanol/water ternary systems (black) with the result of the reference binary 5% Brij 35/water binary system (gray) at 25°C. (a) Experimental SAXS spectra on an absolute scale, (b) the GIFT results of some curves from Figure 6.11a: $p(r)$ functions and (c) averaged structure factor $S_{ave}(q)$. Inset: Parameters of the modelled $S_{ave}(q)$—volume fraction ϕ, interaction radius R, and polydispersity μ. (d) Electron density contrast profiles $\Delta\rho(r)$. The legend indicates the wt % of ethanol in the system. (Reprinted from Tomšič, M., Bešter-Rogač, M., Jamnik, A., Kunz, W., Touraud, D., Bergmann, A., and Glatter, O., *J. Colloid Interf. Sci.*, 294, 194, 2006. With permission.)

Upon increasing the concentration of ethanol, the latter begins noticeably to affect the micellar structures in the sense of decreasing their sizes. The most pronounced change in the SAXS spectra occurs at a concentration between 30% and 40% of ethanol, as also seen from the $p(r)$ functions in Figure 6.11b. These functions show only a slight decrease of micellar dimensions up to 30% of ethanol, whereas upon further increase in ethanol concentration up to 40% the maximal dimension decreases considerably. From the structure factor parameters shown in the inset of Figure 6.11c, one can see that the volume fraction ϕ remains nearly constant with increasing ethanol concentration and at higher concentrations even shows tendency to decrease slightly. The sudden drop of the volume fraction at 40% of ethanol indicates a pronounced effect of micellar breakdown. The value of the polydispersity parameter μ also increases considerably at higher ethanol concentrations. A high degree of polydispersity is a general feature of microemulsion systems, but in this case it points in the direction of the destabilization of micelles and in favor of the so-called micelle breaking effect. This effect has been observed previously in small-angle neutron scattering (SANS) measurements on similar systems [19].

Short-chain alcohols are actually well known to increase the critical micellar concentration of surfactants [7,79,110–114]. Usually this is attributed solely to the effect of alcohol on water, assigning significant importance to the situation in alcohol/water systems. Namely, their presence increases the hydrophobicity of the system and correspondingly increases the surfactant solubility; thus the short-chain alcohols are usually addressed as good cosolvents. Some studies actually report short-chain alcohols to have the properties of structure breaking agents also at higher surfactant concentrations [7,8,19,113,114]—decreasing the size of the micelles or preventing some characteristic structures forming in a specific regime.

Eventually, we can add further solid evidence for the structure breaking effect of short-chain alcohols to the example of the 5% Brij 35/alcohol binary systems with different alcohols. The 5% Brij 35/ethanol binary system is actually the limiting case of the 5% Brij 35 ethanol series shown in Figure 6.11. This proves that at least at the extreme right hand region of the 5% Brij 35 line in the surfactant/ethanol/water ternary phase diagram (see Figure 6.10), a complete breakdown of micelles is achieved [106]. Namely, in Figure 6.12a the SAXS curves of 5% Brij 35/alcohol systems with different alcohols are shown. In the regime of higher q values the SAXS spectra stay practically unchanged in comparison to those of pure alcohols (see dashed gray lines in Figure 6.12a). The latter spectra are, therefore, considered as background (solvent) scattering, which justifies their subtraction from those of Brij 35/alcohol binary systems. The resulting scattering functions refer solely to the scattering of the dispersed Brij 35 structures and are shown in the inset of Figure 6.12b. These curves exhibit a gradual increase of intensities on approaching the lower q values, but in the case of 1-butanol and higher alcohols we observe a steep upturn of intensity at very low angles, indicating the presence of an additional population of very large objects in these samples. The latter were found to be the cause of the phase separation phenomena [106]. Nevertheless, these SAXS data were evaluated with the "cutoff" IFT technique [106] resulting in the $p(r)$ functions depicted in Figure 6.12b. The latter seem to be very similar in all samples and represent the smaller population of scattering particles with maximal dimensions around 2.7 nm (with some side wings up to 5 nm). This population was also clearly evidenced by dynamic light scattering of the 5% Brij 35/ethanol sample [106]. Furthermore, in the case of the ethanol and 1-propanol samples the aggregation numbers of Brij 35 could be calculated [143] and yielded 0.91 and 1.02 for ethanol and 1-propanol, respectively. Inside the

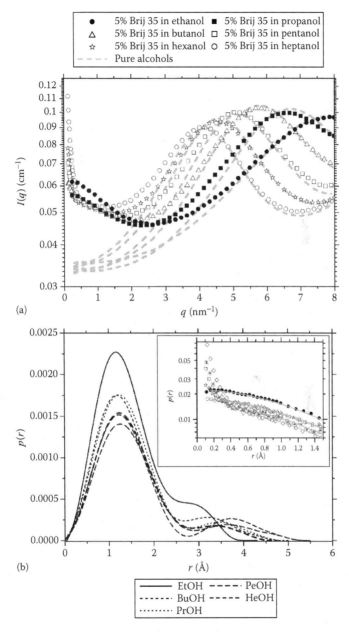

FIGURE 6.12 (a) Experimental SAXS curves of 5% Brij 35/alcohol binary systems (symbols) and corresponding pure alcohols (dashed gray lines) at 25°C. (b) Corresponding $p(r)$ functions obtained with the cutoff IFT method. Inset: Fits of the cutoff IFT evaluation technique (gray curves) to the SAXS data (symbols). (Reproduced from Tomšič, M., Bešter-Rogač, M., Jamnik, A., Kunz, W., Touraud, D., Bergmann, A., and Glatter, O., *J. Phys. Chem. B*, 108, 7021, 2004. With permission.)

estimated error of the method this indicates the presence of monomers in both samples, or in other words, a complete breakdown of the micelles, i.e., molecular solutions of Brij 35 molecules in pure alcohols.

Similarly, we show in Figure 6.13 the SAXS results for the 5% Brij 35/ 1-propanol series. This series shows practically the same qualitative features as already explained in the case of ethanol, the only difference being the fact that 1-propanol seems to be even more efficient in its structure breaking effect. The drop of micellar size from 10 to ca. 7 nm is achieved already by the substitution of 20%

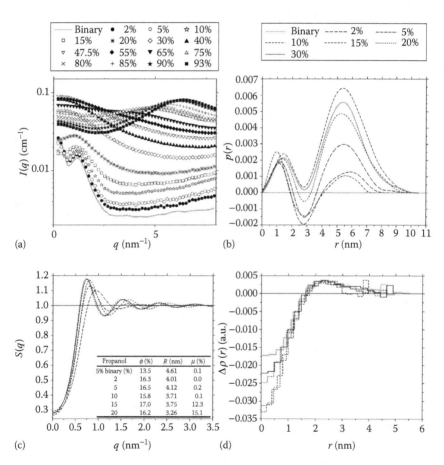

FIGURE 6.13   SAXS results for 5% Brij 35/propanol/water ternary systems (black) with the result of the reference binary 5% Brij 35/water binary system (gray) at 25°C. (a) Experimental SAXS spectra on an absolute scale, (b) the GIFT results of some curves from Figure 6.11a: $p(r)$ functions and (c) averaged structure factor $S_{ave}(q)$. Inset: Parameters of the modelled $S_{ave}(q)$—volume fraction ϕ, interaction radius R, and polydispersity μ. (d) Electron density contrast profiles $\Delta\rho(r)$. The legend indicates the wt % of propanol in the system. (Reprinted from Tomšič, M., Bešter-Rogač, M., Jamnik, A., Kunz, W., Touraud, D., Bergmann, A., and Glatter, O., *J. Colloid Interf. Sci.*, 294, 194, 2006. With permission.)

of water with 1-propanol, whereas correspondingly the core radius does not change considerably (see Figure 6.13d). The additional $-CH_2$ unit in the 1-propanol molecule weakens the polarity of 1-propanol compared to ethanol. As it is still completely miscible with water, 1-propanol behaves as an even better cosolvent.

However, another interesting feature can be observed from the results of the 5% Brij 35 1-propanol series. The electron density profiles depicted in Figure 6.13d show the increasing negative contrast of the micellar core with increasing 1-propanol concentration. On first sight this looks surprising, because one expects that the entry of 1-propanol into the micellar core would decrease its negative contrast due to the slightly higher electron density of propanol in respect to the C_{12} tail of the surfactant ($\rho_{propanol} = 273$ e$^-$/nm^3, $\rho_{dodecane} = 260$ e$^-$/nm^3). However, binary mixtures of primary alcohols and a hydrophobic solvent (octane) have positive excess volumes in the region of lower alcohol concentrations [144]; therefore, the density of such mixtures is lower than it would be if the mixtures were ideal. If one anticipates that alcohols also act similarly in the hydrophobic surroundings of the hydrophobic cores of the micelles, this would mean a lower electron density of the core and correspondingly higher negative scattering contrast. As mentioned above, such an increase of the scattering contrast of the micellar core was actually observed more markedly in the 1-propanol samples and, as is shown in Figures 6.13d, 6.14d, and 6.15d, it becomes even more pronounced for the higher alcohols that show a much higher tendency to enter the micelles. Nevertheless, the obtained trend of increasing micellar core contrast with raising the higher alcohol concentration can be at least qualitatively explained in this way, and is also very interesting from the interaction point of view. However, one should also be aware that the DECON routine is the most uncertain at very low values of r due to the lack of scattering information for this regime.

6.4.2 Medium-Chain Alcohols as Cosurfactants Forming an Interfacial Film

1-Butanol is the first alcohol in the studied series that is only partially miscible with water. Consequently, it shows a much stronger tendency to incorporate itself in the micelles, which is confirmed in Figure 6.14 showing the SAXS results for the 5% Brij 35/1-butanol/water ternary system. As seen already from the phase diagrams in Figure 6.10, systems with 1-butanol and higher alcohols show much more complex phase behavior than ethanol and 1-propanol ternary mixtures. At 5% Brij 35 concentration one is able to increase the concentration of 1-butanol only to about 10% in order not to leave the monophasic region of the phase diagram. At higher concentrations of 1-butanol, which obviously already behaves more as a cosurfactant here, the amount of Brij 35 is obviously no longer sufficient to incorporate all the 1-butanol molecules into the micelles, giving rise to phase separation.

The trends of the $p(r)$ functions, interparticle interaction data, and electron density profiles $\Delta\rho(r)$ shown in Figure 6.14 are also slightly changed compared to those for the ethanol and 1-propanol series. At lower concentrations of 1-butanol

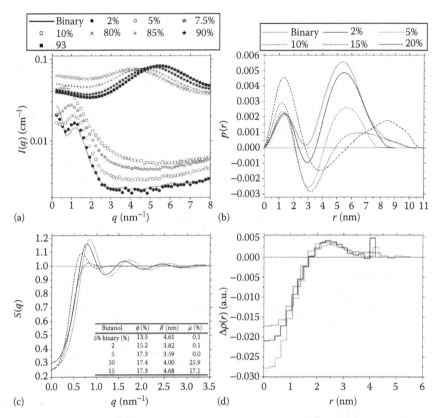

FIGURE 6.14 SAXS results for 5% Brij 35/butanol/water ternary systems (black) with the result of the reference binary 5% Brij 35/water binary system (gray) at 25°C. (a) Experimental SAXS spectra on an absolute scale, (b) the GIFT results of some curves from Figure 6.11a: $p(r)$ functions, and (c) averaged structure factor $S_{ave}(q)$. Inset: Parameters of the modelled $S_{ave}(q)$—volume fraction ϕ, interaction radius R, and polydispersity μ. (d) Electron density contrast profiles $\Delta\rho(r)$. The legend indicates the wt% of butanol in the system. (Reprinted from Tomšič, M., Bešter-Rogač, M., Jamnik, A., Kunz, W., Touraud, D., Bergmann, A., and Glatter, O., *J. Colloid Interf. Sci.*, 294, 194, 2006. With permission.)

the micellar core dimensions slightly increase, whereas the micelles as a whole slightly shrink in the beginning in comparison to the reference binary system (gray line in Figure 6.14b), but at 1-butanol concentrations above 5% the overall micellar size starts to increase. The structure factor polydispersity values follow this trend. Polydispersity is very low for samples with q 1-butanol concentration below 7.5%, indicating more or less spherical scattering particles, whereas at higher concentrations it increases markedly pointing to the elongation of the scattering particles [8]. Namely, it is a well-known fact that for systems with elliptical (elongated) particles GIFT yields increased polydispersity parameter values,

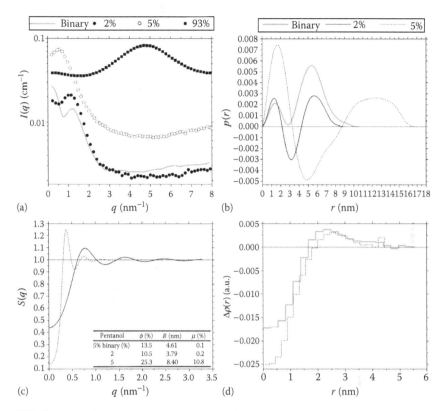

FIGURE 6.15 SAXS results for 5% Brij 35/pentanol/water ternary systems (black) with the result of the reference binary 5% Brij 35/water binary system (gray) at 25°C. (a) Experimental SAXS spectra on an absolute scale, (b) the GIFT results of some curves from Figure 6.11a: $p(r)$ functions, and (c) averaged structure factor $S_{ave}(q)$. Inset: Parameters of the modelled $S_{ave}(q)$—volume fraction ϕ, interaction radius R, and polydispersity μ. (d) Electron density contrast profiles $\Delta\rho(r)$. The legend indicates the wt% of pentanol in the system. (Reprinted from Tomšič, M., Bešter-Rogač, M., Jamnik, A., Kunz, W., Touraud, D., Bergmann, A., and Glatter, O., *J. Colloid Interf. Sci.*, 294, 194, 2006. With permission.)

which arise from the adjustment for the discrepancy between the actual symmetry of the system and the spherical symmetry of the structure factor model [145]. Nevertheless, in parallel according to the $p(r)$ functions in Figure 6.14b, the micellar dimensions start to grow and demonstrate the elongation of the micelles at higher 1-butanol concentrations. It is interesting that the volume fraction parameter ϕ stays rather constant for the studied samples although one might expect it to increase too. However, it is known that at higher deviations from spherical symmetry, where the model used for description of the interparticle interactions worsens, none of the structure factor parameters has any real physical meaning. In any case, the evaluation of the 1-butanol SAXS data was found to be the most challenging in this study. It is worth noting that 1-butanol has already been found previously to be the member of the alcohol series that shows highly complex behavior in microemulsion systems [7,114].

The evaluation of the particle electron density profiles in the 5% Brij 35 1-butanol series was successful only up to 5% of 1-butanol in the sample, which additionally implies elongation of the scattering particles at higher 1-butanol concentrations. The electron density profiles in Figure 6.14d indicate a much more pronounced increase of the negative core scattering contrast than in the previous series, confirming enhanced incorporation of 1-butanol into the micelles. A similar tendency of 1-butanol (and also of 1-octanol) has been reported on the basis of electron paramagnetic resonance and proton matrix electron nuclear double resonance studies of the SDS micellar system [146]. These studies clearly showed that the depth of the alcohol penetration into the micelle is directly dependent on the hydrophobicity of the alcohol itself. Namely, both 1-butanol and 1-octanol were found in the hydrophobic micellar core, though the transfer into the core was found to be much deeper for 1-octanol. These findings are also in excellent agreement with the structural results on systems with higher alcohols presented later in the text.

The monophasic domains in Figure 6.10 are even narrower in the case of the Brij 35/1-pentanol/water system. As seen from the SAXS results for this system shown in Figure 6.15, only two samples from the water-rich side and one sample from the alcohol-rich side were prepared. Nevertheless, these results are once again very similar to the previously presented series, with the only difference that the observed trends are again much more pronounced for the higher alcohol. Compared to 1-butanol, the incorporation of the alcohol into the micelles is even stronger for 1-pentanol, which is shown via an earlier and stronger increase of the core radius, an earlier initial decrease of the overall micellar size (9 nm at 2% of 1-pentanol) with its subsequent stronger increase (16.5 nm at 5% of 1-pentanol), and an earlier increase of the core scattering contrast. As the trends of the structure factor parameters are rather similar, practically the same conclusions can be drawn for 1-pentanol.

The only exceptional point is the huge increase of the volume fraction parameter from approximately 10% to approximately 25% just on increasing the 1-pentanol concentration by 3 mass %. It is highly unlikely that this drastic increase would reflect the actual situation in the system. However, it occurs in parallel with a great increase of the polydispersity parameter, indicating a distinct elongation of the initially spherical scattering particles. When going from spherical to highly nonspherical scattering structures, the model describing the interparticle interactions becomes worse. As a consequence, the values of all the structure factor parameters no longer have a clear physical meaning. Therefore, a drastic (nonphysical) change in the volume fraction parameter strongly implies a considerable elongation of the scattering particles in this sample. Furthermore, the sudden very pronounced increase of the $p(r)$ functions with the increase of the 1-pentanol concentration from 2% to 5%, and its adopted functional form, suggests that the particles most probably elongate into oblate ellipsoids in this case [147].

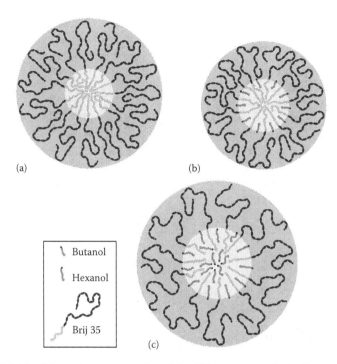

(a)   (b)

Butanol

Hexanol

Brij 35

(c)

FIGURE 6.16 Schematic representation of the different types of micellar structures in (a) Brij 35/water binary system, (b) Brij 35/water /1-butanol ternary system, and (c) Brij 35/ water /1-hexanol ternary system. The scheme represents only the structural trends and no detailed information about aggregation should be drawn from it. (Reprinted from Tomšič, M., Bešter-Rogač, M., Jamnik, A., Kunz, W., Touraud, D., Bergmann, A., and Glatter, O., *J. Colloid Interf. Sci.*, 294, 194, 2006. With permission.)

To summarize and explain the observed trends we use the simple sketch depicted in Figure 6.16. In Figure 6.16a, a spherical inhomogeneous core-shell micelle of the binary Brij 35/water system is represented. Once the added alcohol is of a sufficiently nonpolar nature it starts to behave as a cosurfactant and incorporates itself into such a micelle. Consequently, in Figure 6.16b the still spherical micelle of the ternary Brij 35/1-butanol/water system is shown. Due to their already rather nonpolar nature, the molecules of 1-butanol can partly insert themselves into the micellar core and cause a slight increase in its size in comparison to that of the binary system sketched in Figure 6.16a. This is achieved by a slight stretching of the C_{12} hydrocarbon tails of Brij 35 molecules caused by the 1-butanol molecules organizing themselves in between these tails. In parallel the surfactant aggregation number decreases slightly [7], and, therefore, the poly(oxy-ethylene) parts of the Brij 35 molecules can lie closer to the core, resulting in a thinner hydrophilic micellar shell and causing the overall size of the micelle to decrease initially. With further increase of the medium-chain alcohol concentration the micelles in Figure 6.16b can only elongate either in one or in two directions. Namely, if the alcohol is not hydrophobic enough to fully enter the space in the close vicinity of the core center as shown in Figure 6.16c, the micellar core dimensions are limited at least in one direction to twice the length of the fully stretched C_{12} hydrocarbon chain of the Brij 35 molecule. With increasing medium-chain alcohol concentration (1-butanol or 1-pentanol), therefore, at some point we observe elongation of the micelles. Of course one can imagine that with increasing length of the hydrocarbon tail of the alcohol molecules they should eventually be able to fully enter the micellar core and behave similarly to the hydrophobic oil molecules sketched in Figure 6.16c. In the latter case the core is allowed to grow in all directions and the spherical symmetry can be maintained longer. In parallel the hydrophilic shell becomes more and more hydrated since the poly(oxy-ethylene) parts of the Brij 35 molecules are looser and looser.

At this point we would just like to remind the reader of the results on alcohol-rich samples that were presented in Figures 6.11a through 6.15a. According to the broad and pronounced alcohol interaction scattering peak observed in the SAXS curves at moderate q values, the structure in this alcohol-rich regime is obviously directed by the structure of the alcohol itself. The situation is very reminiscent of the results for alcohol/water systems presented in Section 6.3, only that additional dispersed surfactant particles are also present in these systems contributing to the scattering peak at very low scattering angles. The latter fact emphasizes the importance of the corresponding binary system results in understanding the behavior of more complex ternary systems.

6.4.3 LONG-CHAIN ALCOHOLS AS "OIL-PHASES" ENTERING THE MICELLAR CORE

In this subsection we will show that the structure sketched in Figure 6.16c can also be observed in the SAXS results of 5% Brij 35/alcohol/water ternary systems, proving that all three entitled types of behavior in Section 6.4 can be confirmed structurally. For 1-hexanol and higher alcohols only one sample with the composition 5% Brij 35/2% alcohol/water was prepared from the very narrow water-rich monophasic regions in Figure 6.10. All these SAXS data are shown in Figure 6.17 together with some curves for short-chain alcohols already shown previously in order to collect examples giving evidence of the structural situations depicted in Figure 6.16. However, we have to stress that the scattering curves in Figure 6.17 should be individually compared to the reference 5% Brij 35/water binary system, rather than compared with each other, due to the different alcohol molar fractions in these samples. Only in this way it is possible directly to address the influence of a specific alcohol on the microemulsion structure. On going to higher alcohols, their hydrophobic nature becomes increasingly important and the interplay between their behavior as cosurfactant and the real oil phase starts to direct the structural situation in such ternary systems. In Figure 6.17 we again see that ethanol and 1-propanol as good cosolvents practically do not change the situation at these low alcohol and moderate surfactant concentrations in respect to the reference

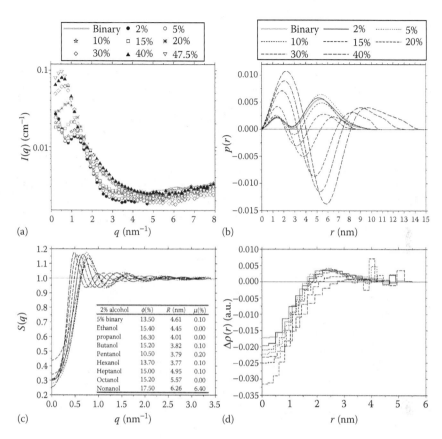

FIGURE 6.17 SAXS results for 5% Brij 35/2% alcohol/water ternary systems with alcohols from ethanol to 1-nonanol (black) and the result of the reference binary 5% Brij 35/water binary system (gray) at 25°C. (a) Experimental SAXS spectra on an absolute scale, (b) GIFT results of some curves from Figure 6.11a: $p(r)$ functions and (c) averaged structure factor $S_{ave}(q)$. Inset: Parameters of the modelled $S_{ave}(q)$—volume fraction ϕ, interaction radius R, and polydispersity μ. (d) Electron density contrast profiles $\Delta\rho(r)$. (Reprinted from Tomšič, M., Bešter-Rogač, M., Jamnik, A., Kunz, W., Touraud, D., Bergmann, A., and Glatter, O., *J. Colloid Interf. Sci.*, 294, 194, 2006. With permission.)

binary system. Both 1-butanol and 1-pentanol at these concentrations show a prevailing cosurfactant-type behavior, decreasing the overall size of the micelles and concomitantly increasing the radius of the core. However, 1-hexanol at this concentration already reverses the trend to decreased overall micellar size, because it shows micelles of comparable size as in the reference binary system, but with much bigger core radii. Obviously in 1-hexanol the two types of behavior, namely as cosurfactant/real oil phase, are practically balanced and as the alcohol chain length increases further the alcohols increasingly behave as real oil phases. 1-heptanol, 1-octanol, and 1-nonanol, therefore, increase the core even more with a concomitant large increase of the overall micellar dimension. We have to stress that according to the very low polydispersity parameters obtained in practically all cases (with the exception of a 1-nonanol sample that was later found to be on the limit of phase separation) the scattering particles were spherical. With increasing alcohol hydrocarbon tail the positive contrast of the hydrophobic micellar shell was decreasing, which can be observed in the electron density profiles in Figure 6.17d. This originates in the strong hydration of the hydrophobic shell. We also tried to investigate the 1-decanol sample in this series, but inferred from its pronounced turbidity, we found it to be a two-phase system already at this concentration.

All these results agree well with the findings on the cosurfactant behavior of primary alcohols in the literature indicating that the alcohols make the interfacial films of these surfactant/water systems increasingly hydrophobic [109,110, 118,119] until at some point they are correspondingly forced to phase separate (clouding).

6.5 CONCLUSIONS AND PERSPECTIVES

We started this review with a brief insight into the structure of the simplest, but still rather complex, systems of pure liquid primary alcohols. It is well accepted that alcohol molecules sequentially H-bond into chains comprising large branched winding structures and even some closed cyclic structures. Such organization creates some inhomogeneities of electron density at an intermolecular level that bring about two distinct characteristic scattering peaks in the SAXS spectra. We were able to successfully link the details expressed in the SAXS curves with the MC simulation structural information on the molecular level and indicate the origin of the two characteristic alcohol scattering peaks. These structural insights give a much clearer picture of the situation in binary and also ternary systems containing primary alcohols.

There is certainly still some ambiguity present in the recent literature regarding the actual structure in alcohol/water binary systems. While some recent investigations claim that the iceberg concept should be thoroughly revised since there are no direct structural proofs that they actually exist [80], some recent papers can be found claiming that icebergs do exist [83], the latter arguing that the right question to be answered is up to which concentration in the water-rich region of the alcohol/water mixture and up to which size of the alcohol (solute) molecules can we find such structures. Of course, there is a mass of literature with thermodynamic, spectroscopic, and similar results speaking in favor of their existence, but nevertheless, further effort and probably also scientific method development still needs to be made to find solid structural proof for their actual existence.

Based on the example of SAXS results for ternary systems composed of the nonionic surfactant Brij 35, water and various primary alcohols from ethanol to 1-decanol, we reviewed the various structural and interactional situations in such microemulsion systems. Three specific types of alcohol behavior depending exclusively on their individual chemical nature were revealed; alcohols behaving as cosolvents inducing the so-called structure breaking effect, as cosurfactants enhancing the structure of the media by increasing the hydrophobicity of the surfactant interfacial layer, and as oil-phases able to occupy the central space of the hydrophobic micellar core. Of course, there is no sharp boundary between the two successive types of behavior (cosolvent/cosurfactant and cosurfactant/oil-phase); the systems rather are organized according to the interplay of these specific trends, depending strongly on the hydrophobicity of the alcohol molecules. In the limiting cases of very short and very long alcohol chains, respectively, with the corresponding predominating hydrophilic and hydrophobic nature of the alcohols, they behave strictly as cosolvents or as real oil phases. In the case of medium-chain length the individual characters may partly be fused, giving rise to the complex structural organization of the alcohols in these systems.

Pure simple alcohols, alcohol/water binary systems, and ternary or even quaternary microemulsion systems containing alcohols are all very complex in terms of molecular organization. As these systems have been extensively studied using numerous experimental and theoretical methods, it seems that the majority of questions concerning their structural, interactional, and dynamic features have nowadays been resolved. However, we can be certain that with the evolution of new scientific techniques these systems will still provoke many more detailed investigations and probably open a lot of interesting scientific questions.

ACKNOWLEDGMENTS

Some of the material republished here is based on research of the authors carried out over several years and supported by several sources: the Slovenian Research Agency and the Federal Ministry for Education, Science, and Culture of Austria. We thank all the coworkers involved in individual studies related to the subjects reviewed here, especially Professor Otto Glatter, Dr. Gerhard Fritz, Dr. Lukáš Vlček, Professor Werner Kunz, and Dr. Alexander Bergmann.

SYMBOLS AND TERMINOLOGIES

r	distance
$g(r)$	interparticle radial distribution function
q	scattering vector
λ	wavelength of radiation
ϑ	scattering angle
$I(q)$	scattering intensity
$I_\alpha(q)$	contribution of atoms of type α to the total scattering intensity (alpha is O or CH_X)
$H_{\alpha\beta}(q)$	scattering contribution of cross-term-correlations between atoms of type α and β (alpha and beta are O, H, or CH_X)
β_S	adiabatic compressibility
$\bar{V}_m$	mean molar volume
x	molar fraction
τ_1	relaxation time
ΔG^E	excess activation free energy

ΔH^E excess activation enthalpy
ΔS^E excess activation entropy
ΔG_A^E partial molar excess activation free energy of component A
ΔH_A^E partial molar excess enthalpy for alcohol of component A
ΔS_A^E partial molar excess entropy for alcohol of component A
w mass fraction
$P(q)$ form factor
$S(q)$ structure factor
$p(r)$ pair-distance distribution function
ϕ volume fraction
R interaction radius
μ polydispersity parameter
σ width of the Gaussian size distribution
ρ electron density
$\Delta\rho(r)$ scattering contrast

REFERENCES

1. Paul, B. K. and Moulik, S. P. (1997). Microemulsions: An overview. *J. Disper. Sci. Technol.*, 18, 301–367.
2. Tenjarla, S. (1999). Microemulsions: An overview and pharmaceutical applications. *Crit. Rev. Ther. Drug.*, 16, 461–521.
3. Kumar, P. and Mittal, K. L. (1999). *Handbook of Microemulsion Science and Technology.* New York: Marcel Dekker.
4. Solans, C. and Kunieda, H. (1997). *Industrial Applications of Microemulsions.* New York: Marcel Dekker.
5. Hoar, T. P. and Schulman, J. H. (1943). Transparent water-in-oil dispersions: The oleopathic hydro-micelle. *Nature*, 152, 102–103.
6. Danielsson, I. and Lindman, B. (1981). The definition of micro-emulsion. *Colloid. Surface*, 3, 391–392.
7. Zana, R. (1995). Aqueous surfactant-alcohol systems: A review. *Adv. Colloid Interf. Sci.*, 57, 1–64.
8. Tomšič, M., Bešter-Rogač, M., Jamnik, A., Kunz, W., Touraud, D., Bergmann, A., and Glatter, O. (2006). Ternary systems of nonionic surfactant Brij 35, water and various simple alcohols: Structural investigations by small-angle x-ray scattering and dynamic light scattering. *J. Colloid Interf. Sci.*, 294, 194–211.
9. Glatter, O. (1983). Data treatment. In O. Glatter and O. Kratky (Eds.), *Small Angle X-Ray Scattering* (pp. 119–165). London: Academic Press Inc. London Ltd.
10. Brunner-Popela, J. and Glatter, O. (1997). Small-angle scattering of interacting particles. 1. Basic principles of a global evaluation technique. *J. Appl. Crystallogr.*, 30, 431–442.
11. Weyerich, B., Brunner-Popela, J., and Glatter, O. (1999). Recently Schmidt-Rohr reported that at least five different approaches are possible to calculate scattering intensities. *J. Appl. Crystallogr.*, 32, 197–209.
12. Brunner-Popela, J. and Glatter, O. (1999). Small-angle scattering of interacting particles. III. D2O-C12E5 mixtures and microemulsions with *n*-octane. *J. Chem. Phys.*, 110, 10623–10632.
13. Bergmann, A., Fritz, G., and Glatter, O. (2000). Solving the generalized indirect Fourier transformation (GIFT) by Boltzmann simplex simulated annealing (BSSA). *J. Appl. Crystallogr.*, 33, 1212–1216.

14. Fritz, G. and Glatter, O. (2006). Structure and interaction in dense colloidal systems: Evaluation of scattering data by the generalized indirect Fourier transformation method. *J. Phys-Condens. Mat.*, 18, S2403–S2419.

15. Glatter, O. (1980). Computation of distance distribution functions and scattering functions of models for small angle scattering experiments. *Acta Phys. Austriaca*, 52, 243–256.

16. Glatter, O. (1977). Data evaluation in small angle scattering: Calculation of the radial electron density distribution by means of indirect Fourier transformation. *Acta Phys. Austriaca*, 47, 83–102.

17. Glatter, O. (1977). A new method for the evaluation of small-angle scattering data. *J. Appl. Crystallogr.*, 10, 415–421.

18. Schirmer, C., Liu, Y., Touraud, D., Meziani, A., Pulvin, S., and Kunz, W. (2002). Horse liver alcohol dehydrogenase as a probe for nanostructuring effects of alcohols in water/nonionic surfactant system. *J. Phys. Chem. B*, 106, 7414–7421.

19. Meziani, A., Touraud, D., Zradba, A., Pulvin, S., Pezron, I., Clausse, M., and Kunz, W. (1997). Comparison of enzymatic activity and nanostructures in water/ethanol/Brij 35 and water/1-pentanol/Brij 35 systems. *J. Phys. Chem. B*, 101, 3620–3625.

20. Keesom, W. H. and Smedt, J. (1923). *Proc. Roy. Soc. Amsterdam*, 26, 112.

21. Stewart, G. W. and Morrow, R. M. (1927). X-Ray diffraction in liquids: Primary normal alcohols. *Phys. Rev.*, 30, 323–244.

22. Stewart, G. W. and Skinner, E. W. (1928). X-Ray diffraction in liquids: A comparison of certain primary normal alcohols and their isomers. *Phys. Rev.*, 31, 1–9.

23. Zachariasen, W. H. (1935). The liquid "structure" of methyl alcohol. *J. Chem. Phys.*, 3, 158–160.

24. Harvey, G. G. (1938). Fourier analysis of liquid methyl alcohol. *J. Chem. Phys.*, 6, 111–114.

25. Narten, A. H. and Habenschuss, A. (1984). Hydrogen bonding in liquid methanol and ethanol determined by x-ray diffraction. *J. Chem. Phys.*, 80, 3387–3391.

26. Akiyama, I., Ogawa, M., Takase, K., Takamuku, T., Yamaguchi, T., and Othori, N. (2004). Liquid structure of 1-propanol by molecular dynamics simulations and x-ray scattering. *J. Solution Chem.*, 33, 797–809.

27. Benmore, C. J. and Loh, Y. L. (2000). The structure of liquid ethanol: A neutron diffraction and molecular dynamics study. *J. Chem. Phys.*, 112, 5877–5883.

28. Montague, D. G., Gibson, I. P., and Dore, J. C. (1982). Structural studies of liquid alcohols by neutron-diffraction. 2. Deuterated ethyl-alcohol C2D5OD. *Mol. Phys.*, 47, 1405–1416.

29. Weitkamp, T., Neufeind, J., Fischer, H. E., and Zeidler, M. D. (2000). Hydrogen bonding in liquid methanol at ambient conditions and at high pressure. *Mol. Phys.*, 98, 125–134.

30. Adya, A. K., Bianchi, L., and Wormald, C. J. (2000). The structure of liquid methanol by H/D substitution technique of neutron diffraction. *J. Chem. Phys.*, 112, 4231–4241.

31. Jorgensen, W. L. (1981). Quantum and statistical mechanical studies of liquids.11. Transferable intermolecular potential functions—Application to liquid methanol including internal-rotation. *J. Am. Chem. Soc.*, 103, 341–345.

32. Jorgensen, W. L. (1986). Optimized intermolecular potential functions for liquid alcohols. *J. Phys. Chem.*, 90, 1276–1284.

33. Gao, J., Habibollazadeh, D., and Shao, L. (1995). A polarizable intermolecular potential function for simulation of liquid alcohols. *J. Phys. Chem.*, 99, 16460–16467.

34. Jorgensen, W. L., Maxwell, D. S., and Tirado-Rives, J. (1996). Development and testing of the OPLS all-atom force field on conformational energetics and properties of organic liquids. *J. Am. Chem. Soc.*, 118, 11225–11236.

35. Haughney, M., Ferrario, M., and McDonald, I. R. (1986). Pair interactions and hydrogen-bond networks in models of liquid methanol. *Mol. Phys.*, 58, 849–853.

36. Haughney, M., Ferrario, M., and McDonald, I. R. (1987). Molecular-dynamics simulation of liquid methanol. *J. Phys. Chem.*, 91, 4934–4940.

37. Van Leeuwen, M. E. (1996). Prediction of the vapour–liquid coexistence curve of alkanols by molecular simulation. *Mol. Phys.*, 87, 87.

38. Chen, B., Potoff, J. J., and Siepmann, J. I. (2001). Monte Carlo calculations for alcohols and their mixtures with alkanes. Transferable potentials for phase equilibria. 5. United-atom description of primary, secondary, and tertiary alcohols. *J. Phys. Chem. B*, 105, 3093–3104.

39. Tomšič, M., Fritz-Popovski, G., Vlček, L., and Jamnik, A. (2007). Calculating small-angle x-ray scattering intensities from Monte Carlo results: Exploring different approaches on the example of primary alcohols. *Acta Chim. Slov.*, 54, 484–491.

40. Tomšič, M., Jamnik, A., Fritz-Popovski, G., Glatter, O., and Vlček, L. (2007). Structural properties of pure simple alcohols from ethanol, propanol, butanol, pentanol, to hexanol: Comparing Monte Carlo simulations with experimental SAXS data. *J. Phys. Chem. B*, 111, 1738–1751.

41. Vahvaselka, S. K., Serimaa, R., and Torkkeli, M. (1995). Determination of liquid structures of the primary alcohols methanol, ethanol, 1-propanol, 1-butanol and 1-octanol by x-ray scattering. *J. Appl. Crystallogr.*, 28, 189–195.

42. Guàrdia, E., Martí, J., Padró, J. A., Saiz, L., and Komolkin, A. V. (2002). Dynamics in hydrogen bonded liquids: Water and alcohols. *J. Mol. Liq.*, 96–97, 3–17.

43. Padró, J. A., Saiz, L., and Guàrdia, E. (1997). Hydrogen bonding in liquid alcohols: A computer simulation study. *J. Mol. Struct.*, 416, 243–248.

44. Saiz, L., Padró, J. A., and Guàrdia, E. (1999). Dynamics and hydrogen bonding in liquid ethanol. *Mol. Phys.*, 97, 897–905.

45. Saiz, L., Padró, J. A., and Guàrdia, E. (1997). Structure and dynamics of liquid ethanol. *J. Phys. Chem. B*, 101, 78–86.

46. MacCallum, J. L. and Tieleman, D. P. (2002). Structures of neat and hydrated 1-octanol from computer simulations. *J. Am. Chem. Soc.*, 124, 15085–15093.

47. Sarkar, S. and Joarder, R. N. (1993). Molecular clusters and correlations in liquid methaanol at room-temperature. *J. Chem. Phys.*, 99, 2033–2039.

48. Sarkar, S. and Joarder, R. N. (1994). Molecular clusters in liquid ethanol at room-temperature. *J. Chem. Phys.*, 100, 5118–5122.

49. Ludwig, R. (2005). The structure of liquid methanol. *Chem. Phys. Chem.*, 6, 1369–1375.

50. Allison, S. K., Fox, J. P., Hargreaves, R., and Bates, S. P. (2005). Clustering and microimmiscibility in alcohol–water mixtures: Evidence from molecular-dynamics simulations. *Phys. Rev. B*, 71, 5.

51. Ko, J. H. and Kojima, S. (2001). Brillouin-scattering study of primary alcohols by using an angular dispersion-type Fabry-Perot interferometer. *J. Korean Phys. Soc.*, 39, 702–707.

52. Tauer, K. J. and Lipscomb, W. N. (1952). On the crystal structures, residual entropy and dielectric anomaly of methanol. *Acta Crystallogr.*, 5, 606–612.

53. Jönsson, P.-G. (1976). Hydrogen bond studies. CXIII. The crystal structure of ethanol at 87 K. *Acta Crystallogr. B*, 32, 232.

54. Weng, S. X., Decker, J., Torrie, B. H., and Anderson, A. (1988). Raman and far infrared spectra of solid ethanol C_2H_5OH, C_2H_5OD and C_2D_5OD. *Indian J. Pure Ap. Phys.*, 26, 76–80.

55. Torrie, B. H., Weng, S. X., and Powell, B. M. (1998). Structure of the alpha-phase of solid methanol. *Mol. Phys.*, 67, 575–581.

56. Mooij, W. T. M., Eijck, B. P., and Kroon, J. (1999). Transferable ab initio intermolecular potentials. 2. Validation and application to crystal structure prediction. *J. Phys. Chem. A*, 103, 9883–9890.
57. Taylor, R. and Macrae, C. F. (2001). Rules governing the crystal packing of mono- and dialcohols. *Acta Crystallogr. B*, 57, 815–827.
58. Hayashi, H., Nishikawa, K., and Iijima, T. (1990). Small-angle x-ray scattering study of fluctuations in 1-propanol water and 2-propanol water-systems. *J. Phys. Chem.*, 94, 8338–8338.
59. Nishikawa, K. and Iijima, T. (1993). Small-angle x-ray scattering study of fluctuations in ethanol and water mixtures. *J. Phys. Chem.*, 97, 10824–10828.
60. Matsumoto, M., Nishi, N., Furusawa, T., Saita, M., Takamuku, T., Yamagami, M., and Yamaguchi, T. (1995). Structure of clusters in ethanol–water binary-solutions studied by mass-spectroscopy and x-ray diffraction *Bull. Chem. Soc. Jpn.*, 68, 1775–1783.
61. Takamuku, T., Yamaguchi, T., Asato, M., Matsumoto, M., and Nishi, N. (2000). Structure of clusters in methanol–water binary solutions studied by mass spectrometry and x-ray diffraction. *Z. Naturforsch.*, 55a, 513–525.
62. Takamuku, T., Maruyama, H., Watanabe, K., and Yamaguchi, T. (2004). Structure of 1-propanol–water mixtures investigated by large-angle x-ray scattering technique. *J. Solution Chem.*, 33, 641–660.
63. Takaizumi, K. (2005). A curious phenomenon in the freezing–thawing process of aqueous ethanol solution. *J. Solution Chem.*, 34, 597–612.
64. Gonzalez-Salgado, D. and Nezbeda, I. (2006). Excess properties of aqueous mixtures of methanol: Simulation versus experiment. *Fluid Phase Equil.*, 240, 161–166.
65. Zhu, T., Chen, G. Q., Yu, R. P., Liu, Y., and Ni, X. W. (2006). Experiment study of alcohol–water mixtures absorption and fluorescence spectra induced by UV-light spectroscopy and spectral analysis, 26, 291–294.
66. Misawa, M. (2002). Mesoscale structure and fractal nature of 1-propanol aqueous solution: A reverse Monte Carlo analysis of small angle neutron scattering intensity. *J. Chem. Phys.*, 116, 8463–8468.
67. Sato, T., Chiba, A., and Nozaki, R. (2002). Composition-dependent dynamical structures of monohydric alcohol–water mixtures studied by microwave dielectric analysis. *J. Mol. Liq.*, 96–97, 327–339.
68. Sato, T., Chiba, A., and Nozaki, R. (2000). Composition-dependent dynamical structures of 1-propanol–water mixtures determined by dynamical dielectric properties *J. Chem. Phys.*, 113, 9748–9758.
69. Sato, T., Chiba, A., and Nozaki, R. (1999). Dynamical aspects of mixing schemes in ethanol–water mixtures in terms of the excess partial molar activation free energy, enthalpy, and entropy of the dielectric relaxation process. *J. Chem. Phys.*, 110, 2508–2521.
70. Sato, T. and Buchner, R. (2004). Dielectric relaxation processes in ethanol/water mixtures *J. Phys. Chem. A*, 108, 5007–5015.
71. Misawa, M., Sato, T., Onozuka, A., Maruyama, K., Mori, K., Suzuki, S., and Otomo, T. (2007). A visualized analysis of small-angle neutron scattering intensity: Concentration fluctuation in alcohol–water mixtures. *J. Appl. Crystallogr.*, 40, S93–S96.
72. Takamuku, T., Saisho, K., Nozawa, S., and Yamaguchi, T. (2005). X-ray diffraction studies on methanol–water, ethanol–water, and 2-propanol–water mixtures at low temperatures. *J. Mol. Liq.*, 119, 133–146.
73. Onori, G. and Santucci, A. (1996). Dynamical and structural properties of water/alcohol mixtures. *J. Mol. Liq.*, 69, 161–182.

74. Misawa, M., Inamura, Y., Hosaka, D., and Yamamuro, O. (2006). Hydration of alcohol clusters in 1-propanol–water mixture studied by quasielastic neutron scattering and an interpretation of anomalous excess partial molar volume *J. Chem. Phys.*, 125, 6.

75. Franks, F. and Ives, D. J. G. (1966). The structural properties of alcohol–water mixtures. *Quart. Rev. Chem. Soc.*, 20, 1–44.

76. Liltorp, K., Westh, P., and Koga, Y. (2005). Thermodynamic properties of water in the water-poor region of binary water plus alcohol mixtures. *Can. J. Chem. Revue Canadienne de Chimie*, 83, 420–429.

77. Yano, Y. F. (2005). Correlation between surface and bulk structures of alcohol–water mixtures *J. Colloid Interf. Sci.*, 384, 255–259.

78. Wensink, E. J. W., Hoffmann, A. C., and van Maaren, P. J. (2003). Dynamic properties of water/alcohol mixtures studied by computer simulation. *J. Chem. Phys.*, 119, 7308–7317.

79. Guo, J. H., Luo, Y., Augustsson, A., Kashtanov, S., Rubensson, J. E., Shuh, D. K., Agren, H., and Nordgren, J. (2003). Molecular structure of alcohol–water mixtures. *Phys. Rev. Let.*, 91, 4.

80. Dixit, S., Crain, J., Poon, W. C. K., Finney, J. L., and Soper, A. K. (2002). Molecular segregation observed in a concentrated alcohol–water solution. *Nature*, 416, 829–832.

81. Dixit, S., Soper, A. K., Finney, J. L., and Crain, J. (2002). Water structure and solute association in dilute aqueous methanol. *Europhys. Lett.*, 59, 377–383.

82. Best, S. A., Merz, K. M., and Reynolds, C. H. (1999). Free energy perturbation study of octanol/water partition coefficients: Comparison with continuum GB/SA calculations. *J. Phys. Chem. B*, 103, 714–726.

83. Koga, Y., Nishikawa, K., and Westh, P. (2004). "Icebergs" or no "Icebergs" in aqueous alcohols?: Composition-dependent mixing schemes. *J. Phys. Chem. A*, 108, 3873–3877

84. Sassi, P., Paolantoni, M., Cataliotti, R. S., Palombo, F., and Morresi, A. (2004). Water/alcohol mixtures: A spectroscopic study of the water-saturated 1-octanol solution. *J. Phys. Chem. B*, 108, 19557–19565.

85. Wakisaka, A., Mochizuki, S., and Kobara, H. (2004). Cluster formation of 1-butanol–water mixture leading to phase separation. *J. Solution Chem.*, 33, 721–732.

86. Bowron, D. T. and Moreno, S. D. (2003). Structural correlations of water molecules in a concentrated alcohol solution. *J. Phys-Condens. Mat.*, 15, S121–S127.

87. Kahlweit, M., Busse, G., and Jen, J. (1991). Effects of alcohols on the phase-behaviour of microemulsions. *J. Phys. Chem.*, 95, 5588.

88. Parke, S. A. and Birch, G. G. (1999). Solution properties of ethanol in water. *Food Chem.*, 67, 241–246.

89. Frank, H. S. and Evans, M. W. (1945). Free volume and entropy in condensed systems III. Entropy in binary liquid mixtures; partial molal entropy in dilute solutions; structure and thermodynamics in aqueous electrolytes. *J. Chem. Phys.*, 13, 507–532.

90. Barthel, J., Bachhuber, K., Buchner, R., and Hetzenauer, H. (1990). Dielectric spectra of some common solvents in the microwave region—Water and lower alcohols. *Chem. Phys. Lett.*, 165, 369–373.

91. Barthel, J. and Buchner, R. (1991). High-frequency permittivity and its use in the investigation of solution properties. *Pure Appl. Chem.*, 63, 1473–1482.

92. Buchner, R., Barthel, J., and Stauber, J. (1999). The dielectric relaxation of water between 0 degrees C and 35 degrees C. *Chem. Phys. Lett.*, 306, 57–63.

93. Hill, N. E, Vaughan, W. E., Price, A. H., and Davies, M. (1969). *Dielectric Properties and Molecular Behaviour*, London: Reinhold.

94. Bowron, D. T., Finney, J. L., and Soper, A. K. (1998). Structural investigation of solute–solute interactions in aqueous solutions of tertiary butanol. *J. Phys. Chem. B*, 102, 3551–3563.

95. Bowron, D. T., Soper, A. K., and Finney, J. L. (2001). Temperature dependence of the structure of a 0.06 mole fraction tertiary butanol–water solution. *J. Chem. Phys.*, 114, 6203–6219.

96. Franks, F. and Desnoyers, J. E. (1985). Alcohol–water mixtures revisited. In F. Franks (Ed.), *Water Science Reviews* (pp. 171–232). Cambridge: Cambridge University Press.

97. Zhang, C. and Yang, X. (2005). Molecular dynamics simulation of ethanol/water mixtures for structure and diffusion properties. *Fluid Phase Equil.*, 231, 1–10.

98. Leo, A., Hansch, C., and Elkins, D. (1971). Partition coefficients and their use. *Chem. Rev.*, 71, 525–616.

99. Bodor, N. and Huang, M. J. (1992). An extended version of a novel method for the estimation of partition-coefficients. *J. Pharm. Sci.*, 81, 272–281.

100. Sangster, J. (1997). *Octanol–Water Partition Coefficients: Fundamentals and Physical Chemistry*. Chichester: Wiley.

101. Carrupt, P. Testa, B., and Gaillard, P. (1997). Computational approaches to lipophilicity: Methods and applications. In K.B. Lipkowitz and D. Boyd (Eds.), *Reviews in Computational Chemistry* (pp. 241–345). New York: Wiley-VCH.

102. Hansch, C. and Leo, A. (1995). *Exploring QSAR Fundamentals and Applications in Cchemistry and Biology*. Washington: American Chemical Society.

103. Nasal, A., Sznitowska, M., Bucinski, A., and Kaliszan, R. (1995). Hydrophobicity parameter from high-performance liquid-chromatography on an immobilized artificial membrane column and its relationship. *J. Chrom. A*, 692, 83–89.

104. Martinez, M. N. and Amidon, G. L. (2002). A mechanistic approach to understanding the factors affecting drug absorption: A review of fundamentals. *J. Clin. Pharmacol.*, 42, 620–643.

105. Smith, R. N., Hansch, C., and Ames, M. M. (1975). Selection of a reference partitioning system for drug design work. *J. Pharm. Sci.*, 64, 599–606.

106. Tomšič, M., Bešter-Rogač, M., Jamnik, A., Kunz, W., Touraud, D., Bergmann, A., and Glatter, O. (2004). Nonionic surfactant Brij 35 in water and in various simple alcohols: Structural investigations by small-angle x-ray scattering and dynamic light scattering. *J. Phys. Chem. B*, 108, 7021–7032.

107. Preu, H., Schirmer, C., Tomsic, M., Bester-Rogac, M., Jamnik, A., Belloni, L., and Kunz, W. (2003). Light, neutron, x-ray scattering, and conductivity measurements on aqueous dodecyltrimethylammonium bromide/1-hexanol solutions. *J. Phys. Chem. B*, 107, 13862–13870.

108. Preu, H., Zradba, A., Rast, S., Kunz, W., Hardy, E. H., and Zeidler, M. D. (1999). Small angle neutron scattering of D_2O–Brij 35 and D_2O–alcohol–Brij 35 solutions and their modelling using the Percus–Yevick integral equation. *Phys. Chem. Chem. Phys.*, 1, 3321–3329.

109. Caponetti, E., Lizzio, A., Triolo, R., Griffith, W. L., and Johnson, J. S. (1992). Alcohol partition in a water-in-oil microemulsion from small-angle neutron-scattering. *Langmuir*, 8, 1554–1562.

110. Strey, R. and Jonströmer, M. (1992). Role of medium-chain alcohols in interfacial films of nonionic microemulsions. *J. Phys. Chem.*, 96, 4537–4542.

111. Nishikido, N., Moroi, Y., Uehara, H., and Matuura, R. (1974). Effect of alcohols on micelle formation of nonionic surfactants in aqueous-solutions. *Bull. Chem. Soc. Jpn.*, 47, 2634–2638.

112. Griffiths, P. C., Hirst, N., Paul, A., King, S. M., Heenan, R. K., and Farley, R. (2004). Effect of ethanol on the interaction between poly(vinylpyrrolidone) and sodium dodecyl sulfate. *Langmuir*, 20, 6904–6913.

113. Caponetti, E., Chilura Martino, D., Floriano, M. A., and Triolo, R. (1997). Localization of *n*-alcohols and structural effects in aqueous solutions of sodium dodecyl sulfate. *Langmuir*, 13, 3277–3283.

114. Førland, G. M., Samseth, J., Gjerde, M. I., Høiland, H., Jensen, A. Ø., and Mortensen, K. (1998). Influence of alcohol on the behavior of sodium dodecylsulfate micelles. *J. Colloid Interf. Sci.*, 203, 328–334.

115. Moya, S. E. and Schulz, P. C. (1999). The aggregation of the sodium dodecyl sulfate *n*-octanol water system at low concentration. *Colloid Polym. Sci.*, 277, 735–742.

116. Cinelli, S., Onori, G., and Santucci, A. (1999). Effect of 1-alcohols on micelle formation and protein folding. *Colloid. Surface. A*, 160, 3–8.

117. Bockstahl, F. and Duplâtre, G. (2001). Effect of 1-pentanol on size and shape of sodium dodecyl sulfate micelles as studied by positron annihilation lifetime spectroscopy. *J. Phys. Chem. A*, 105, 13–18.

118. van Bommel, A. and Palepu, R. M. (2004). *n*-Alkanol induced clouding of Brij (R) 56 and the energetics of the process. *Colloid. Surface. A*, 233, 109–115.

119. Bharatiya, B., Guo, C., Ma, J. H., Hassan, P. A., and Bahadur, P. (2005). Aggregation and clouding behavior of aqueous solution of EO-PO block copolymer in presence of *n*-alkanols. *Eur. Polym. J.*, 43, 1883–1891.

120. Regev, O., Ezrahi, S., Aserin, A., Garti, N., Wachtel, E., Kaler, E. W., Khan, A., and Talmon, Y. (1996). A study of the microstructure of a four-component nonionic microemulsion by cryo-TEM, NMR, SAXS, and SANS. *Langmuir*, 12, 668–674.

121. Ezrahi, S., Wachtel, E., Aserin, A., and Garti, N. (1997). Structural polymorphism in a four-component nonionic microemulsion. *J. Colloid Interf. Sci.*, 191, 277–290.

122. Li, X. F. and Kunieda, H. (2003). Catanionic surfactants: Microemulsion formation and solubilization. *Curr. Opin. Colloid Interf. Sci.*, 8, 327–336.

123. Yaghmur, A., Aserin, A., and Garti, N. (2002). Phase behavior of microemulsions based on food-grade nonionic surfactants: Effect of polyols and short-chairs alcohols. *Colloid. Surface. A*, 209, 71–81.

124. Compere, A. L., Griffith, W. L., Johnson, J. S., Caponetti, E., Chillura-Martino, D., and Triolo, R. (1997). Alcohol partition in a water-in-oil microemulsion: Small-angle neutron-scattering contrast measurements. *J. Phys. Chem. B*, 101, 7139–7146.

125. Patist, A., Axelberd, T., and Shah, D. O. (1998). Effect of long chain alcohols on micellar relaxation time and foaming properties of sodium dodecyl sulfate solutions. *J. Colloid Interf. Sci.*, 208, 259–265.

126. Gonzalez-Perez, A., Czapkiewicz, J., Del Castillo, J. L., and Rodriguez, J. R. (2003). Micellar behavior of tetradecyldimethylbenzylammonium chloride in water–alcohol mixtures. *J. Colloid Interf. Sci.*, 262, 525–530.

127. Gonzalez-Perez, A., Galan, J. J., and Rodriguez, J. R. (2003). The solubilization of alcohols in micellar solutions—Estimation of thermal parameters. *J. Therm. Anal. Calorim.*, 72, 471–479.

128. Gunaseelan, K. and Ismail, K. (2003). Estimation of micellization parameters of sodium dodecyl sulfate in water + 1-butanol using the mixed electrolyte model for molar conductance. *J. Colloid Interf. Sci.*, 258, 110–115.

129. Thimons, K. L., Brazdil, L. C., Harrison, D., and Fisch, M. R. (1997). Effects of pentanol isomers on the growth of SDS micelles in 0.5 M NaCl. *J. Phys. Chem. B*, 101, 11087–11091.

130. Palazzo, G., Lopez, F., Giustini, M., Colafemmina, G., and Ceglie, A. (2003). Role of the cosurfactant in the CTAB/water/n-pentanol/n-hexane water-in-oil microemulsion. 1. Pentanol effect on the microstructure. *J. Phys. Chem. B*, 107, 1924–1931.
131. Ezrahi, S., Tuval, E., Aserin, A., and Garti, N. (2005). The effect of structural variation of alcohols on water solubilization in nonionic microemulsions. 1. From linear to branched amphiphiles—General considerations. *J. Colloid Interf. Sci.*, 291, 263–272.
132. Gonzalez-Perez, A., Galan, J. J., and Rodriguez, J. R. (2004). Solubilization of butanol in dodecyldimethylethylammonium bromide micellar solutions. *Fluid Phase Equil.*, 224, 7–11.
133. Touraud, D. (1991). Contribution à l'étude de microemulsions utilisables comme milieux réactionnels (thesis). Compiègne: Université de Technologie de Compiègne.
134. Glatter, O. (1983). Interpretation. In O. Glatter and O. Kratky (Eds.), *Small Angle X-Ray Scattering* (pp. 167–196). London: Academic Press Inc. London Ltd.
135. Glatter, O. (2002). The inverse scattering problem in small angle x-ray scattering. In P. Lindner and T. Zemb (Eds.), *Neutron, X-Rays and Light: Scattering Methods Applied to Soft Condensed Matter*. Amsterdam: Elsevier.
136. Hansen, J. P. and McDonald, I. R. (1990). *The Theory of Simple Liquids*. London: Academic Press.
137. Glatter, O. (1981). Convolution square root of band-limited symmetrical functions and its application to small-angle scattering data. *J. Appl. Crystallogr.*, 14, 101–108.
138. Glatter, O. and Hainisch, B. (1984). Improvements in real-space deconvolution of small-angle scattering data. *J. Appl. Crystallogr.*, 17, 435–441.
139. Mittelbach, R. and Glatter, O. (1998). Direct structure analysis of small-angle scattering data from polydisperse colloidal particles. *J. Appl. Crystallogr.*, 31, 600–608.
140. Pusey, P. N. and Tough, R. J. A. (1985). *Dynamic Light Scattering. Applications of Photon Correlation Spectroscopy*. New York: Plenum.
141. Langowsky, J. and Brian, R. (1991). *Neutron, X-Ray and Light Scattering*. Amsterdam: North Holland.
142. Schnablegger, H. and Glatter, O. (1991). Optical sizing of small colloidal particles—An optimized regularization technique. *Appl. Opt.*, 30, 4889–4896.
143. Orthaber, D., Bergmann, A., and Glatter, O. (2000). SAXS experiments on absolute scale with Kratky systems using water as a secondary standard. *J. Appl. Crystallogr.*, 33, 218–225.
144. Garcia-Garabal, S., Segade, L., Cabeza, O., Franjo, C., Jimenez, E., Pineiro, M. M., and Andrade, M. I. P. (2003). Density, surface tension and refractive index of octane + 1-alkanol mixtures at T = 298.15 K. *J. Chem. Eng. Data*, 48, 1251–1255.
145. Mittelbach, P. and Porod, G. (1965). Kolloid. *Z. Z. Polym.*, 202, 40–49.
146. McManus, H. J. D., Kang, Y. S., and Kevan, L. (1993). Electron-paramagnetic-resonance and proton matrix electron-nuclear double-resonance studies of N,N,N',N'-tetramethylbenzidine photoionization in sodium dodecyl-sulfate micelles—Structural effects of added alcohols. *J. Chem. Soc. Faraday Trans.*, 89, 4085–4089.
147. Fritz, G. and Bergmann, A. (2004). Interpretation of small-angle scattering data of inhomogeneous ellipsoids. *J. Appl. Crystallogr.*, 37, 815–822.
148. Humphrey, W., Dalke, A., and Schulten, K. (1996). VMD Visual molecular dynamics. *J. Mol. Graphic*, 14, 33–38.

7 Formation and Characterization of Emulsified Microemulsions

Anan Yaghmur, Liliana de Campo, and Otto Glatter

CONTENTS

7.1 INTRODUCTION

In the literature, a surge of investigations have been devoted to the understanding of structural aspects of various biologically relevant self-assembled nanostructures [1–7]. This high interest has been stimulated by an increased body of evidence for their crucial roles in various biological processes [3,6,7]. In addition, intensive recent studies have focused on understanding the physicochemical properties of self-assembled nanostructures mimicking biological systems [8–10]. Among these systems, emulsions with internal well-ordered nanostructures set the stage for new potential applications in various areas of food and pharmaceutical technologies as well as other purposes [11–14]. For instance, these nanosystems are promising means for formulating novel functional foods, cosmetics, and drug nanocarriers [12,13].

Recently, a great deal of research has been carried out with regard to the physical and the interfacial properties of monolinolein (MLO)-based nanostructured emulsions [14–21]. This family of colloidal particles (cubosomes, hexosomes, micellar cubosomes, and emulsified microemulsion [EME]) has been denoted as Internally Self-Assembled Particles or "Somes" (ISASOMES) [14]. Our major goal has been to focus on understanding the various parameters that control the symmetry of the confined nanostructures in these dispersions and to learn how to significantly tune their interfacial film to more negative spontaneous curvatures. This can induce a transformation from inverted types of bicontinuous cubic phases (cubosomes) or hexagonal structures (hexosomes) to water-in-oil (W/O) microemulsion systems (EMEs). We were aware of the fact that the characterization of emulsions with internal W/O microemulsion has for many years been considered a difficult task. Thus, to prove that it was possible to form and characterize such nanostructured aqueous dispersions, it was important to find appropriate experimental methods for investigating the emulsion after the dispersion procedure as well as to face this scientific challenge by carrying out detailed systematic investigations on both dispersed and nondispersed bulk phases.

A microemulsion is constituted by a single isotropic and thermodynamically stable nanostructured solution of two immiscible components, i.e., an apolar solvent (oil) and a polar one (usually water), stabilized by a third component, a surfactant or a surfactant/cosurfactant mixture [22]. Historically, the concept of microemulsions was first introduced by Hoar and Schulman in 1943 [23]. In this chapter, we summarize our recent findings on how to confine this nanoscaled nonviscous structure in kinetically dispersed emulsion particles with sizes of a few hundred nanometers.

To form stable MLO-based EMEs, the following important requirements need to be fulfilled: (1) the system should be biphasic (Winsor-II-type system), i.e., the inverted type microemulsion systems should coexist with excess water, (2) the presence of an efficient stabilizer covering the outer surface of the kinetically stabilized droplets containing the internal W/O microemulsion system is needed, and (3) this internal microemulsion structure should be preserved after the dispersing procedure. It should here be pointed out that microemulsions are, at a certain temperature, known to be sensitive systems with respect to any changes in their composition that may lead to the loss of their thermodynamic stability. Such a loss would in turn cause the formation of normal emulsions or, in a worst case scenario, a phase separation into oleic- and aqueous-rich phases.

7.2 EMULSIFICATION OF A REVERSE MICELLAR SYSTEM (SO-CALLED L$_2$ PHASE)

This section deals with the emulsification of the oil-free MLO/water system. Our aim was to briefly describe the formation of emulsified inverse micellar system (L$_2$) at high temperatures.

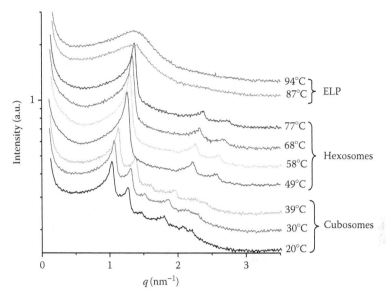

FIGURE 7.1   Scattering curves of an MLO-based dispersion at temperatures between 20°C and 94°C.

In the presence of a suitable stabilizer such as Pluronic F127, monoglycerides can be dispersed in excess water, resulting in stable submicron-sized cubosome particles [24–26]. In a similar study on an MLO/water system [15], cubosomes containing a bicontinuous cubic phase of symmetry Pn3m were formed.

A typical small angle x-ray scattering (SAXS) pattern for a dispersed MLO cubic Pn3m phase at 20°C can be seen in Figure 7.1. Increasing the temperature induces a structural transition in the interior of these dispersed particles from the cubic Pn3m-phase to the H_2 phase (hexosomes). At a certain temperature, the H_2 phase melts into a fluid isotropic, so-called inverse micellar solution (L_2 phase). Thus, this figure presents clear evidence for the formation of an emulsified L_2 phase upon heating.

It is important to note the following points that summarize our findings with regard to the oil-free systems:

1. Phase transitions in the interior of the dispersed droplets are reversible. For instance, when cooling the MLO-based aqueous dispersion from 94°C to 20°C, the molten interior of the droplets (L_2 phase) recrystallizes to an H_2-phase after which it transforms into a cubic Pn3m phase. The internal structures of the dispersed particles are thereby identical for a given temperature, irrespective of their thermal history. This means that the temperature can be used to unequivocally tune the internal structures of the particles, which is thus proof that the internal structures of the dispersed particles are in thermodynamic equilibrium.

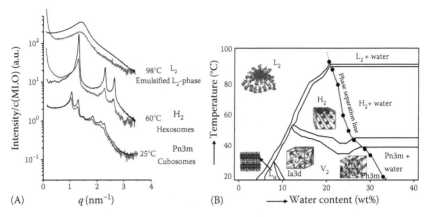

FIGURE 7.2 (A) Comparison at various temperatures of the scattering curves of an MLO–water bulk phase with excess water and an MLO-based dispersion. (B) Temperature-dependent phase diagram of MLO–water system.

2. Self-assembled nanostructure of the nondispersed MLO phase in equilibrium with excess water is practically identical to the nanostructure in the dispersed particles at each investigated temperature, as shown in Figure 7.2A.

The second point has very interesting consequences: first of all, neither the confined geometry imposed on these phases (the limited size of the particles) nor the polymer used for their stabilization (F127) considerably affected the internal nanostructure. Moreover, since the phase diagram of MLO–water system showed that the bulk phase was able to solubilize considerably more water at low temperatures (Figure 7.2B), the same concept was valid also for the dispersed particles. Similarly to the behavior of the binary fully hydrated nondispersed bulk phase coexisting with excess water, the internal nanostructure was in equilibrium with the surrounding excess water phase. Thus, the particles expelled water upon heating and reabsorbed water upon cooling in a reversible way. We termed this temperature-dependent behavior the "breathing mode." This mode is identical and equally valid to that observed for the oil-loaded ternary systems discussed in greater detail in the following sections.

7.3   EMULSIFICATION OF W/O MICROEMULSIONS

The previous section dealt with the formation of an emulsified L_2 phase at high temperature. As a next step, the main intention was to load these nanostructured droplets with oil in an attempt to lower the phase transition temperatures and form EMEs at room temperature. To do so, it must be ensured that the bulk-concentrated W/O microemulsion system is preserved after the applied dispersing procedure thus preventing a case in which the oil and/or water are expelled from the internal nanostructure.

In our study, the tetradecane (TC)-loaded MLO-based microemulsion was considered a Winsor-II-type system. This is a W/O microemulsion consisting of the ternary water/MLO/TC system that coexists with excess water (a biphasic sample denoted $\bar{2}$). The well-known effective emulsifier Pluronic F127 was used to stabilize the EMEs. A schematic description for the formation of these MLO-based EMEs is shown in Figure 7.3.

We investigated the impact of TC solubilization on the internal nanostructure of MLO-based aqueous dispersions stabilized by the polymer F127. Figure 7.4 shows the SAXS scattering curves at 25°C obtained from dispersions in which the α ratio value, defined as [(mass of oil)/(mass of MLO) × 100], was in the range of 0–60.

In the absence of oil ($\alpha = 0$), the diffraction pattern was indexed in accordance with a cubic Pn3m phase, denoted Q^{224}, (cubosomes). For the TC-loaded dispersion with $\alpha = 28$, the scattering curve of the dispersion displayed three peaks in the characteristic ratio for an H_2 phase (hexosomes). Intriguingly, a further increase of the TC concentration in the dispersion ($\alpha = 40$) gave rise to a scattering curve displaying more than seven peaks. These peaks were in the characteristic ratio for

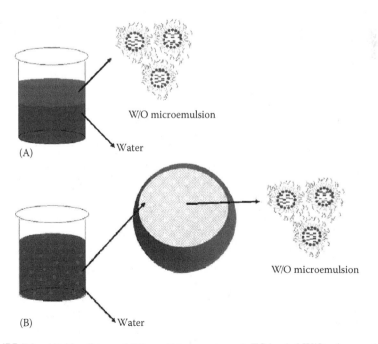

FIGURE 7.3   (A) Nondispersed Winsor-II-type system: A TC-loaded W/O microemulsion system coexisting with excess water. (B) Formation of F127-stabilized TC-loaded EME.

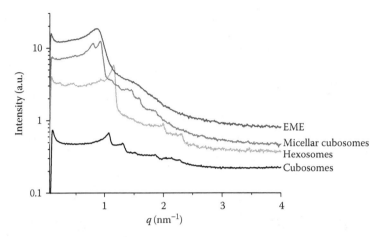

FIGURE 7.4 Effect of the amount of solubilized TC on the scattering curves of MLO-based emulsified systems at 25°C. In these dispersions, the α ratio [((mass of oil)/(mass of MLO)) $\times$ 100] varied from 0 (cubosomes), to 28 (hexosomes), 40 (micellar cubosomes), and 60 (EME).

a discontinuous micellar cubic structure of the type Fd3m, which is denoted Q^{227} (micellar cubosomes, emulsified micellar cubic phase). The mean lattice parameter, a, of the cubosomes, hexosomes, and micellar cubosomes were 8.6, 6.3, and 21.7 nm, respectively. Our results showed that the structural transformation from hexosomes to EME was not direct, but rather took place in a certain TC concentration range in which micellar cubosomes with the internal space group Fd3m (also denoted by the symbol "I_2") were formed. At higher TC concentrations ($\alpha \geq 60$), the scattering curve showed only one broad peak. This is typical for a concentrated microemulsion phase. In short, we found for the internal nanostructure that increasing the TC concentration led to structural transitions in the order V_2 (Pn3m) $\rightarrow$ $H_2 \rightarrow I_2$ (Fd3m) $\rightarrow$ W/O microemulsion.

To better understand the dispersed W/O systems, it was important to compare the internal structure of the dispersions to that of the corresponding bulk (nondispersed) W/O microemulsion based on the ternary MLO/TC/water system. In this context, it should be pointed out that the dispersed internal structures, as well as the structures of the nondispersed bulk samples, depended not only on the TC-content (α value), but also on the amount of solubilized water in the respective phase.

Figure 7.5 shows as an example of how the structures of a TC/MLO-system with $\alpha = 110$ change with increasing water content: at such high-TC contents, only an inverse microemulsion system could be formed. With an increasing water content in the single phase regions, the observed peaks moved to lower angles, a behavior corresponding to larger structures ("swelling" with water). Moreover, the scattering curves displayed a shoulder in addition to the first broad peak. This can be explained by the fact that the solubilized water increased the order in the inverse microemulsion system, which led to the appearance of a side maximum in

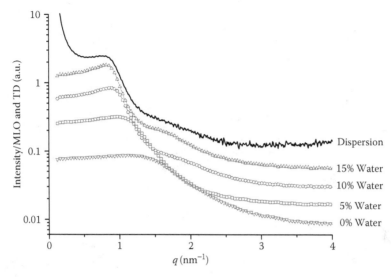

FIGURE 7.5 SAXS scattering curves comparing the MLO-based dispersion (black lines) with nondispersed bulk samples with varying water content (open symbols) at 25°C. Both dispersed and nondispersed W/O microemulsion systems had a constant α ratio of 110. The intensities were first normalized by the respective MLO and TC concentration, and then shifted by a constant arbitrary factor for the sake of visibility.

the scattering curves. This side maximum was directly related to the main maximum and therefore did not indicate the presence of any additional phase in the system. At 15% water, the water solubilization capacity of this sample was exceeded, and the W/O system came into equilibrium with excess water. Above this fully hydrated condition, no further change was observed concerning the peak positions and in the corresponding structure parameters (the d spacing for the L_2 phase, known also in literature as the *characteristic distance*). Figure 7.6 shows another example of an EME with an internal nanostructure identical to that of the corresponding nondispersed W/O microemulsion. This means that it was, in principle, possible to form EMEs without concomitant loss of the integrity of the W/O microemulsion.

It was evident from the very good agreement of the scattering curves of these EMEs with those of the corresponding nondispersed bulk samples coexisting with excess water (typical examples in Figures 7.5 and 7.6) that the structures were practically the same. This also implies that the samples had identical water contents. Moreover, the d-spacing value indicated that the TC/MLO mixing ratio was maintained (the oil was not expelled from the internal nanostructure). We thus believe that TC did not form normal emulsion droplets in addition to the EME, despite the high-energy input of the dispersing procedure, and furthermore, the polymer used to stabilize the particles did not disturb the internal structure. Owing to the identical nanostructure in both the dispersed and nondispersed phases, it

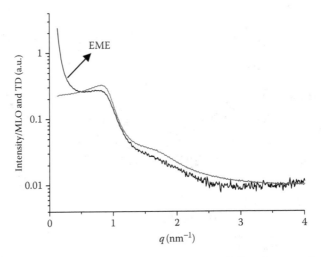

FIGURE 7.6 Effect of solubilized TC content at α ratio of 75 on the scattering curves of the MLO-based EME as well as on that of the nondispersed bulk W/O sample with excess water. The investigation was carried out at 25°C. The intensities were normalized by the respective MLO and TC concentrations. F127 was used to stabilize the dispersed particles.

was possible to calculate the water content in the dispersed droplets from the maximum water solubilization capacity of the bulk samples (fully hydrated systems). Without oil ($\alpha = 0$), the water content was determined to be approximately 32 wt% [15]. For instance, in the TC-loaded EMEs and their corresponding fully hydrated nondispersed samples with $\alpha = 75$ and 110, the water content decreased to less than 20 and 15 wt%, respectively.

In addition to the SAXS analysis, the submicron-sized dispersed EME particles were investigated by cryo-TEM, as shown in Figure 7.7. For the particles containing a significant amount of solublized TC ($\alpha = 75$), the observations showed structures that clearly differed from those previously obtained for the hexosomes and cubosomes [14,15,24–26]. No long range order was observed and the projected shape of the particles was almost circular, which is usually not the case for hexosomes and cubosomes. Rather, the outer shape of the dispersed droplets displayed a similarity to that found in O/W or W/O emulsion systems. However, in contrast to regular oil (or water) droplets in ordinary emulsion systems, the brightness inside the particles was not uniform but showed an internal arrangement. The fast Fourier transforms (FFTs) of the EME particle also revealed the presence of an internal structure. For this EME dispersion, the characteristic distance determined by SAXS was approximately 6.7 nm while the position of the diffuse brightness peak in the FFT (top insert in Figure 7.7) agreed fairly well with that value and indicated a characteristic distance of approximately 7.5 nm.

We also found that the average hydrodynamic radius of these particles, which was determined by dynamic light scattering, was in the range of 120–200 nm.

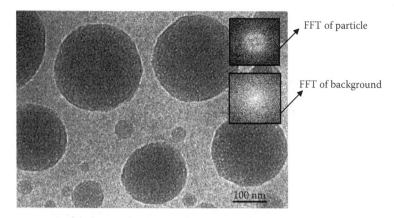

FFT of particle

FFT of background

FIGURE 7.7   Cryo-TEM image for EME particles with $\alpha = 75$. The top insert is an FFT of a particle revealing a characteristic distance of approximately 7.5 nm. The bottom insert is an FFT of the background revealing no characteristic distance.

In summary, SAXS and cryo-TEM investigations proved the possibility to disperse W/O microemulsion systems into particles, while preserving their nanostructure. In other words, the internal confined microemulsified water droplets were thermodynamically stable and not lost to the external emulsion (the continuous phase plus F127).

7.4   STABILITY OF THE INTERNAL W/O MICROEMULSIONS

Figure 7.8 shows the stability with respect to aging of a TC-loaded EME sample. Intriguingly, there was no change in the internal nanostructure of this dispersion after 4 months of storage at room temperature. The presented data show that the oil molecules were incorporated in the internal nanostructure of the mesophase particles, and no experimental evidence of a leakage with time was found: the solublized oil did not move from the internal W/O nanostructure to the continuous aqueous phase. In other words, there was no indication of any change in the internal nanostructure that could lead to phase separation. Thus, the formation of TC-rich normal O/W emulsion droplets in addition to an MLO-rich nanostructured aqueous phase could not be observed. This supports our hypothesis that the internal structure of the dispersed particles formed only through self-assembly principles.

7.5   REVERSIBLE WATER EXCHANGE IN THE ENTRAPPED W/O INTERIOR

Figure 7.9A shows scattering curves as functions of temperature for an MLO-based aqueous dispersion with an α value of 110, as well as its corresponding nondispersed samples during heating in the temperature range of 25°C–76°C.

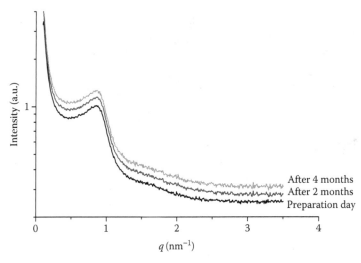

FIGURE 7.8 Stability with respect to aging: Scattering curves of MLO-based EME with α ratio of 75 are shown. The measurements were carried out at 25°C: after preparation (black line), after 2 months, and after 4 months. The curves have been shifted by a constant arbitrary factor for the sake of visibility. F127 was used to stabilize the dispersed particles.

As demonstrated in this figure, the formation of EMEs at room temperature was promoted in the presence of such high-TC contents. Increasing the temperature at constant oil content (and in the absence of oil as discussed in Ref. [15]) was expected to induce the dehydration of the MLO headgroup and simultaneously an increase in the kink states in the lipid acyl chains and thereby also their effective volume [14–16]. As a consequence, this influence enhanced the negative spontaneous curvature. Increasing the temperature caused shifts in the single broad peak of the W/O microemulsion to higher q values and reduced the characteristic distance.

At each investigated temperature, our results showed that the structure of the entrapped W/O interior was identical to that of the nondispersed W/O sample coexisting with excess water. This confirms our previous results [15] on oil-free nanostructured aqueous dispersions and demonstrates that the internal W/O structure of the dispersed particles at each temperature corresponds to an equilibrium phase.

For a detailed elucidation of the water exchange inside–outside the particles' internal W/O nanostructures, occurring during heating–cooling processes, a SAXS investigation was carried out on the aqueous EME dispersion. The scattering curves of this dispersion, presented in Figure 7.9B as a function of temperature (from 25°C to 76°C), reveal the structural reversibility of the confined nanostructures. In this dispersion, the broad peak displayed temperature-dependent shifts. It should be pointed out that at each temperature of investigation, the scattering curves were

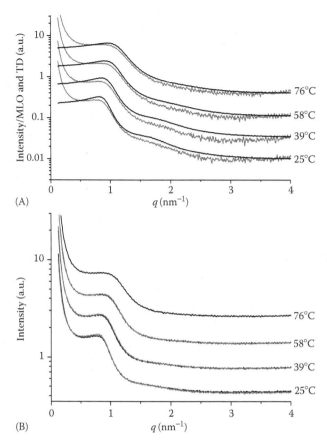

FIGURE 7.9 (A) Temperature dependence of scattering curves for the MLO-based EME dispersion (thin lines) in comparison with those from the nondispersed bulk W/O sample (thick lines). The effect of the temperature in the range of 25°C–76°C was determined for a system with an α value of 110. (B) Reversibility: the same dispersion was analyzed after first being heated to 25°C, 39°C, 58°C, and 76°C (thin lines), and then cooled back down to 58°C, 39°C, and 25°C (thick lines). It is important to note that the peak position of EME during cooling coincided with that obtained during heating and it is therefore difficult to distinguish between the two curves at 25°C, 39°C, and 58°C. The curves have been shifted by a constant arbitrary factor for the sake of visibility.

identical, which indicates that the internal W/O nanostructure, at a specific temperature, was independent of the thermal history. In other words, either heating or cooling to the required temperature led to the same structure. This fact was clear evidence that the formed W/O nanostructures in the kinetically stabilized particles were thermodynamic equilibrium structures similar to those of the dispersed and nondispersed oil-free systems (Figure 7.2B). As a consequence of heating, the water solubilization capacity of the nondispersed phases decreased upon increasing the temperature. The same behavior also occurred in the interior of the droplets inducing a water swelling–deswelling behavior during the heating/ cooling cycles (denoted as "breathing mode" [14,15]). This result reveals that there was a reversible exchange of water inside–outside the confined internal particle structures during the cooling and heating cycles.

An additional point is related to the temperature-induced direct and indirect transitions of hexosomes to EMEs. In our recent study [14], an example for direct hexosome-EME transitions with increasing temperature is presented. However, an indirect transition sometimes occurs via micellar cubosomes at certain amounts of solubilized oil. A detailed description for these structural transitions is given in Ref. [19].

7.6   MODULATION OF THE INTERNAL W/O MICROEMULSION SYSTEM

The internal W/O structure of the oil-loaded dispersed particles can be modulated by varying the lipid composition [18]. This can be achieved by, for instance, replacing a portion of the MLO by diglycerol monooleate (DGMO) or soybean phosphatidylcholine [18]. This means that these surfactants have a counter effect on the internal confined structure of the particles as opposed to that of the oil. In a recent study, we found that DGMO tuned back the aqueous dispersions from hexosomes to cubosomes at low-solubilized oil contents. However, an amount of surfactant higher than that of the oil was required to overcome the effect of loaded oil, which indicates that the solubilized oil had a stronger impact on the internal structure than DGMO. At a high content of solubilized TC content (at $\alpha = 75$ as presented in Figure 7.10), replacing a relatively large amount of MLO by DGMO

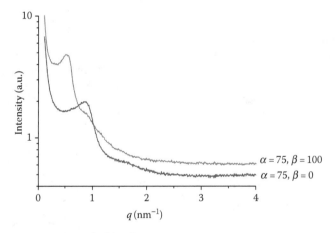

FIGURE 7.10   Effect of the DGMO content on the scattering curves of TC-loaded aqueous MLO-based dispersions at 25°C (the α ratio was 75). The scattering curves of the DGMO-free dispersions ($\beta = 0$) were compared to those containing DGMO ($\beta = 100$). The intensities have been shifted by a constant arbitrary factor for the sake of visibility.

($\beta = 100$, this term is defined as [mass of DGMO]/[mass of MLO] × 100) caused a significant shift of the peaks in the scattering curves to lower angles, corresponding to larger structures.

Our results revealed that DGMO was not efficient enough to tune back the internal structure from a W/O microemulsion to a hexagonal or a cubic phase. However, the increase in the lattice parameter due to the addition of DGMO was remarkable. Here, we demonstrate that the addition of DGMO significantly affected the internal structure of the TC-solubilized dispersions thus favoring the formation of W/O microemulsions with a large amount of solubilized water. DGMO functionalized the confined W/O nanostructure by altering its properties and increasing the size of its hydrophilic domains, thereby also strongly increasing the water solubilization capacity. This is of utmost interest for various applications. In particular, it is attractive for enhancing the solubilization of hydrophilic active molecules in internal W/O nanostructures.

7.7 FOOD-GRADE EMULSIFIED W/O MICROEMULSION SYSTEMS

In an attempt to study the ability to stabilize food-grade MLO-based aqueous dispersions, our first intention was to replace the solubilized oil (TC) with other oils such as triglycerides, and subsequently form EMEs comprising only materials acceptable in the food industry. Although our first aqueous dispersions were stabilized by F127, we were also able to form EMEs with various food-grade emulsifiers (data not shown). We found that it was possible to form food-grade EMEs, in which TC was replaced by triglycerides such as triolein, $R(+)$-limonene (the major constituent of citrus essential oils), and natural antioxidants.

7.8 SELF-ASSEMBLY AS THE DRIVING FORCE FOR THE FORMATION OF ISASOMES

Our findings have demonstrated that the process behind the formation of confined nanoscaled tunable hierarchical structures in kinetically stabilized particles is driven by the principles of self-assembly. In the absence of this driving force—a situation easily simulated by studying MLO-free dispersions—our results reveal the formation of "structureless" stable normal TC-in-water emulsion (Figure 7.11A). However, as shown above, well-defined nanostructures were formed in the presence of MLO, owing to the strong self-assembly interactions of MLO with TC/water mixtures.

On this basis, there is no evidence of a reorganization of the lipid (and lipid/oil) molecules during preparation of the dispersions. Thus, these constituents should behave as pseudo-components at all investigated temperatures: the solubilized oil (and water) was maintained in the nanostructure during the dispersing procedure. In other words, our results showed no indication for any significant change in the composition at the interfacial film. Furthermore, there was no implication for

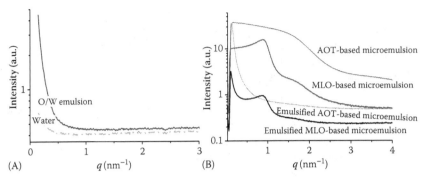

FIGURE 7.11   (A) Structureless normal emulsion of TC-in-water. The dispersion was prepared at 25°C in the absence of MLO. The SAXS patterns of this normal emulsion as well as that of water are presented. (B) A comparison of SAXS scattering curves of the MLO-based fully hydrated nondispersed phase and its corresponding dispersed sample with those of an AOT-based fully hydrated nondispersed W/O microemulsion system and its dispersion at 25°C. Both W/O microemulsion systems had constant α ratio of 75. F127 was used to stabilize the dispersed particles.

redistribution of the molecules between the dispersed phase (TC) and the continuous phase when a high-energy input was applied in the presence of the polymeric stabilizer F127. It should be pointed out that the reorganization of the lipid (and oil) molecules could be induced by replacing the primary surfactant MLO by any surfactant that would not remain bound to oil and water during the dispersing procedure. For instance, we found that the Winsor-II-type system based on a TC-loaded W/O microemulsion system of the well-known sodium bis(2-ethylhexyl) sulfosuccinate (AOT) lost its stability when it was emulsified in the presence of F127 (see Figure 7.11B).

7.9   SUMMARY

The present chapter describes our recent studies dealing with the formation of various TC-loaded emulsions with hierarchical internal nanostructures. In particular, we focused on studying the possible formation of emulsions confining W/O microemulsions. Our SAXS and cryo-TEM results revealed that such nanostructured emulsions could be formed either by solubilizing oil at room temperature or by increasing the temperature at a fixed oil concentration. In this context, solubilizing an oil such as TC in the dispersed MLO-based particles induced, at room temperature, the internal transformation from a bicontinuous cubic (Pn3m) phase via H_2 to a discontinuous micellar cubic (Fd3m) phase and ultimately to an L_2 phase (W/O microemulsion).

Figure 7.12 summarizes the transitions that occurred when TC was added to dispersed and nondispersed systems of MLO/water. In these emulsified particles, the confined W/O nanostructures were reversible structures, i.e., they existed in

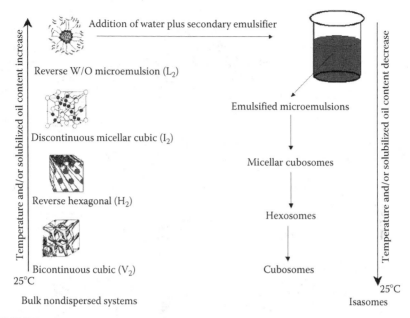

FIGURE 7.12   Formation of ISASOMES from monoglyceride-based mesophases.

thermodynamic equilibrium with the surrounding aqueous phase, and depended only on the actual temperature and oil content. Although a transition from bicontinuous cubic (Pn3m) via hexagonal H_2 to L_2 (inverse micellar solution) phases was observed upon increasing the temperature in the absence of TC in the MLO–water dispersions [15], a different transition scheme occurred upon addition of TC to the MLO–water dispersion at constant temperature.

Our study provides the basis for understanding the structural transitions that take place within monoglyceride-based aqueous dispersions after loading hydrophobic active molecules. It was also the first reported evidence in literature on the possible formation of EMEs at room temperature. These nanostructured emulsions are important as nanoparticulate carriers for enhancing the solubilization capacity of active guest molecules. The formation of such effective nanocarriers is of great interest for various pharmaceutical, food, and cosmetic applications.

SYMBOLS AND TERMINOLOGIES

AOT	sodium bis (2-ethylhexyl) sulfosuccinate
DGMO	diglycerol monooleate
EME	emulsified microemulsion
ELP	emulsified L_2 phase (inverse micellar solution)
F127	the Pluronic polymer PEO_{99}-PPO_{67}-PEO_{99}
Fd3m	discontinuous micellar cubic phase, it is also called I_2

FFT   fast Fourier transform
H_2 inverted type hexagonal liquid crystalline phase
ISASOMES   internally self-assembled particles or "Somes"
L_α lamellar phase
L_2 reverse micellar system (water-in-oil micelles)
SAXS   small angle x-ray scattering
MLO   monolinolein
TC   tetradecane
V_2 inverted type bicontinuous cubic phase
W/O microemulsion water-in-oil microemulsion system
Winsor-II-type system water-in-oil microemulsion coexists with excess water

REFERENCES

1. Hyde, S., Andersson, S., Larsson, K., Blum, Z., Landh, T., Lidin, S., and Ninham, B. W. 1997 *The Language of Shape: The Role of Curvature in Condensed Matter: Physics, Chemistry, Biology*, Amsterdam, Elsevier book series.

2. Couvreur, P. and Vauthier, C. 2006 Nanotechnology: Intelligent design to treat complex disease. *Pharm. Res.*, 23, 1417–1450.

3. Luzzati, V. 1997 Biological significance of lipid polymorphism: The cubic phases. *Curr. Opin. Struct. Biol.*, 7, 661–668.

4. Lindblom, G. and Rilfors, L. 1989 Cubic phases and isotropic structures formed by membrane lipids—possible biological relevance. *Biochim. Biophys. Acta*, 988, 221–256.

5. de Kruijff, B. 1997 Lipids beyond the bilayer. *Nature*, 386, 129–130.

6. Almsherqi, Z. A., Kohlwein, S. D., and Deng, Y. 2006 Cubic membranes: A legend beyond the flatland of cell membrane organization. *J. Cell Biol.*, 173, 839–844.

7. Patton, J. S. and Carrey, M. C. 1979 Watching fat digestion, *Science*, 204, 145–148.

8. Siegel, D. P. 1999 The modified stalk mechanism of lamellar/inverted phase transitions and its implications for membrane fusion. *Biophys. J.*, 76, 291–313.

9. Yaghmur, A., Laggner, P., Zhang, S., and Rappolt, M. 2007 Tuning curvature and stability of monoolein bilayers by short surfactant-like designer peptides. *PLoS ONE*, 2, e479.

10. Ellens, H., Siegel, D. P., Alford, D., Yeagle, P. L., Boni, L., Lis, L. J., Quinn, P. J., and Bentz, J. 1989 Membrane fusion and inverted phases. *Biochemistry*, 28, 3692–3703.

11. Mezzenga, R., Schurtenberger, P., Burbidge, A., and Michel, M. 2005 Understanding foods as soft materials. *Nat. Mat.*, 4, 729–740.

12. Leser, M. E., Sagalowicz, L., Michel, M., and Watzke, H. J. 2006 Self-assembly of polar food lipids. *Adv. Colloid Interf. Sci.*, 123–126, 125–136.

13. Boyd, B. J. 2005 Controlled release from cubic liquid-crystalline particles. In P. T. Spicer, and M. L. Lynch (Eds.), *Bicontinuous Liquid Crystals* (vol. 127, pp. 285–304). New York, Marcel Dekker.

14. Yaghmur, A., de Campo, L., Sagalowicz, L., Leser, M. E., and Glatter, O. 2005 Emulsified microemulsions and oil-containing liquid crystalline phases. *Langmuir*, 21, 569–577.

15. de Campo, L., Yaghmur, A., Sagalowicz, L., Leser, M. E., Watzke, H., and Glatter, O. 2004 Reversible phase transitions in emulsified nanostructured lipid systems. *Langmuir*, 20, 5254–5261.

16. Yaghmur, A., de Campo, L., Salentinig, S., Sagalowicz, L., Leser, M. E., and Glatter, O. 2006 Oil-loaded monolinolein-based particles with confined inverse discontinuous cubic structure (*Fd3m*). *Langmuir*, 22, 517–521.

17. Sagalowicz, L., Michel, M., Adrian, M., Frossard, P., Rouvet, M., Watzke, H. J., Yaghmur, A., de Campo, L., Glatter, O., and Leser, M. E. 2006 Crystallography of dispersed liquid crystalline phases studied by cryo-transmission electron microscopy. *J. Microsc.*, 221, 110–121.

18. Yaghmur, A., de Campo, L., Sagalowicz, L., Leser, M. E., and Glatter, O. 2006 Control of the internal structure of MLO-based isasomes by the addition of diglycerol monooleate and soybean phosphatidylcholine. *Langmuir*, 22, 9919–9927.

19. Guillot, S., Moitzi, C., Salentinig, S., Sagalowicz, L., Leser, M. E., and Glatter, O. 2006 Direct and indirect thermal transitions from hexosomes to emulsified microemulsions in oil-loaded monoglyceride-based particles. *Colloids Surf. A*, 291, 78–84.

20. Salonen, A., Guillot, S., and Glatter, O. 2007 Determination of water content in internally self-assembled monoglyceride-based dispersions from the bulk phase. *Langmuir*, 23, 9151–9154.

21. Moitzi, C., Guillot, S., Fritz, G., Salentinig, S., and Glatter, O. 2007 Phase reorganization in self-assembled systems through interparticle material transfer. *Adv. Mat.*, 19, 1352–1358.

22. Strey, R. 1994 Microemulsion microstructure and interfacial curvature. *Colloid Polym. Sci.*, 272, 1005–1019.

23. Hoar, T. P. and Schulman, J. H. 1943 Transparent water in oil dispersions: The oleopathic hydromicelles. *Nature*, 152, 102–103.

24. Larsson, K. 2000 Aqueous dispersions of cubic lipid–water phases. *Curr. Opin. Colloid Interf. Sci.*, 4, 64–69.

25. Gustafsson, J., Ljusberg-Wahren, H., Almgren, M., and Larsson, K. 1996 Cubic lipid–water phase dispersed into submicron particles. *Langmuir*, 12, 4611–4613.

26. Spicer, P. T. 2004 Cubosomes: Bicontinuous liquid crystalline nanoparticles. In J. A. Schwarz, C. Contescu, and K. Putyera, (Eds.), *Dekker Encyclopaedia of Nanoscience and Nanotechnology* (pp. 881–892). New York, Marcel Dekker.

8 Dynamics of Solvent and Rotational Relaxation of RTILs in RTILs-Containing Microemulsions

Debabrata Seth and Nilmoni Sarkar

CONTENTS

8.1 INTRODUCTION: DEFINITION OF THE RTILs- MICROEMULSIONS SYSTEMS AND BACKGROUND OF THIS REVIEW CHAPTER

Room temperature ionic liquids (RTILs), a class of neoteric solvent, have been extensively used as environment friendly "green substitutes" for toxic, hazardous, flammable, and volatile organic solvents [1–11]. Ray and Rakshit first reported preparation of RTILs in 1911 [12]. They prepared the nitrite salts of ethylamine, dimethylamine, and trimethylamine, although these salts are spontaneously decompose on standing. A few years latter Walden reported the first useful RTILs in 1914 [13]. He reported the physical properties of ethylammonium nitrate (EAN), which has a melting point of 12°C. Cations and anions of RTILs are generally organic and inorganic in nature, respectively. In RTILs coulombic interactions between ions are prominent in contrast to dipolar or multipolar interaction in volatile organic compounds (VOCs). By changing substituents in the cation or in the anion components various RTILs can be prepared for different purposes. By fine-tuning the cations and anions certain bulk properties like viscosity, conductivity, and density of RTILs can be changed. The key feature of RTILs is the nonvolatile nature and RTILs do not contribute VOCs in the global atmosphere. So it is better to use RTILs rather than VOCs for industrial use. Most of the RTILs have some common properties such as negligible vapor pressure, broad liquidous temperature range (−96°C to ~300°C), high conductivity, wide electrochemical windows, and ability to solvate various organic or inorganic species.

It is widely believed that RTILs exert no effective vapor pressure; it was found that RTILs were not always green [14]; and $[PF_6]^-$ ion-containing RTILs is very unstable to moisture and hydrolyzed to produce volatile, harmful, and corrosive HF, POF_3, etc. Recently, Baker and Baker [15] showed that anions of RTILs are hydrolyzed. Moreover, it was shown that RTILs could be distilled under reduced pressure and at high temperature [16–18]. The nonflammability of RTILs is often recommended as a safety advantage of using RTILs over VOCs. It was recently showed that many RTILs are not safe near fire, i.e., they are combustible due to their positive heat formation and oxygen content [19]. Many limitations and disadvantages of RTILs were found day by day, but RTILs are technologically advance solvents and can be designed to fit for a specific application.

Several photophysical and ultrafast spectroscopic studies were undertaken in these RTILs [20–69]. Aki et al. [20] determined the polarity of the imidazolium- and pyridinium-based RTILs using UV-Vis absorption and fluorescence spectroscopy. Muldoon et al. [21] determined the polarity of the RTILs using solvatochromic probes. Carmichael et al. [22] determined the polarity of several neat 1-alkyl-3-methylimidazolium-based RTILs using the solvatochromic dye Nile red. Using Prodan, pyrene, 1-pyrenecarboxaldehyde, Reichardt's betain dye, and Rhodamine 6G as the solvatochromic probes various bulk properties and polarity of various RTILs–cosolvents mixtures were determined [23–26]. Recently, one excellent review article was published by Reichardt to determine the polarity of RTILs by means of solvatochromic betaine dyes [27]. Femtosecond optical Kerr effect measurement was used to study the low frequency vibrational motions [28,29]. The photoisomerization reactions [30] and intramolecular excimer formation kinetics [31] have been studied in RTILs. There are also a few reports available on time-dependent solvation in neat RTILs [32–55]. The excitation wavelength-dependent fluorescence behavior of some dipolar molecules in RTILs was investigated by Samanta et al., which attributed to the unusual red edge effect in RTILs [56,57]. Recently, Hu and Margulis [58–60] predicted the heterogeneity in RTILs and also observed the red edge effect in RTILs using molecular dynamics simulation. They showed that the structures of imidazolium-based RTILs with longer alkyl chains are reminiscent of reverse micelles [58–60]. Recently, Paul et al. [61,62] reported the various optical properties of different neat RTILs using absorption and fluorescence spectroscopy. Recently, ultrafast dynamical studies and vibrational relaxation in RTILs and RTILs-containing reverse micelles were investigated [63,64]. Recently, Shirota and Castner prepared novel RTILs with trimethylsilylmethyl substituted imidazolium cation. The viscosity of these RTILs is lower than other alkyl substitute imidazolium cation-containing RTILs [65].

Water-in-oil microemulsions are used as an elegant model for biological membranes [70]. AOT (dioctylsulfosuccinate, sodium salt) is the most commonly used surfactant to form microemulsions. In presence of nonpolar organic solvent, surfactant molecules aggregate to form reverse micelles. The size of reverse micelle depends upon w value (w = [water or polar solvent]/[surfactant]). Other surfactants such as cetyltrimethyl-ammonium bromide (CTAB), sodium dodecyl sulfate (SDS), and triton X-100 (TX-100) form microemulsions in alkane solvent in the presence of medium chain length primary alcohols such as 1-pentanol, 1-hexanol, etc. Riter et al. [71,72] characterized various types of microemulsions using AOT as a surfactant and other polar solvents like acetonitrile, methanol, formamide, ethylene glycol and N, N-dimethyl formamide (DMF) as cosolvents. The polarities of RTILs are very close to short chain alcohols [20–22,27]. Therefore, RTILs may be used as green substitutes for common solvents to form micelles and microemulsions. Recently, micelles were prepared in different RTILs [73–79]. It was found that the CMC of the surfactant in RTILs is much higher than that in pure water. In pure water the CMC of Brij-35 is 0.06 mM, and it becomes ~120 mM in 1-butyl-3-methylimidazolium hexafluorophosphate ([Bmim][PF$_6$]) [73]. It is shown recently that in presence of RTILs some surfactants form microemulsions [80–89]. Several group prepared and characterized the RTILs-containing microemulsions [80–89] using phase behavior, cyclic voltammetry, UV-Vis, Fourier transform IR, ^{1}H NMR spectroscopy. The size of these RTILs-containing microemulsions was determined by freeze-fracture electron microscopy (FFEM), dynamic light scattering (DLS) and small-angle neutron scattering (SANS) measurements. The size of these RTILs-containing microemulsions is much larger than the microemulsions containing other polar solvents. Several groups prepared and characterized RTILs-containing ternary microemulsions, which excludes the use of nonpolar VOCs [85–89]. Recently, Liu et al. [89] showed formation of reverse micelle in carbon dioxide with RTILs domains. These kind of green microemulsions have potential application in different fields, such as preparation of nanoparticle and chemical synthesis [90,91]. The [Bmim][BF$_4$]/TX-100/cyclohexane microemulsions have been characterized by phase behavior; conductivity measurement, DLS and FFEM [80]. Recently, Gao et al. [85] prepared and characterized [Bmim][PF$_6$]/TX-100/ water-containing ternary microemulsions. They recognized three types of micro-structures: water-in-[Bmim][PF$_6$], [Bmim][PF$_6$]-in-water, and bicontinuous in the microemulsions. In another work Gao et al. [86] prepared and characterized [Bmim][PF$_6$]/tween 20/water-containing microemulsion. The role of added water on the microstructure of RTILs-containing microemulsions was studied [92,93]. It was shown that the addition of small amount of water stabilized microemulsions and increased the amount of solubilized RTILs in the microemulsions.

We have studied solvent and rotational relaxation of RTILs in water-in-[Bmim][PF$_6$], [Bmim][PF$_6$]-in-water microstructure in the microemulsions and in [Bmim][BF$_4$]/TX-100/cyclohexane microemulsions using Coumarin 153 (C-153) and Coumarin 151 (C-151) as probes [94–97]. Adhikari et al. [98] investigated solvation dynamics in 1-pentyl-3-methyl-imidazolium tetraflouroborate ([pmim] [BF$_4$])/TX-100/benzene microemulsion using femtosecond up-conversion technique. Gao et al. [85] showed that with the addition of [Bmim][PF$_6$] in TX-100/ water mixture the size of the droplet increases similar to other microemulsions. DLS measurements showed that the structure of the microemulsions is spherical and hydrodynamic diameter of [Bmim][PF$_6$]/TX-100/water microemulsions increases from 8.3 to 18.9 nm as [Bmim][PF$_6$]-to-surfactant molar ratio (R) increases from 0.17 to 0.41 (see Table 8.1). In water-in-[Bmim][PF$_6$] microstructure of the microemulsions with the gradual addition of water in [Bmim][PF$_6$]/ surfactant mixture, micelles are swollen by water and consequently the size of the microemulsions also increases at a fixed value of [Bmim][PF$_6$]-to-surfactant weight ratio (I). A typical size distribution graph from the DLS experiment is shown in Figure 8.1. The micrographs of [Bmim][BF$_4$]/TX-100/cyclohexane microemulsions obtained from FFEM techniques are shown in Figure 8.2. The composition of water, [Bmim][PF$_6$], and surfactant used to prepare different microemulsions and hydrodynamic diameters of microemulsions are listed in

TABLE 8.1

Composition and Hydrodynamic Diameter of [Bmim][PF$_6$]/TX-100/Water ([Bmim][PF$_6$]-in-Water) Microemulsions (R = [Bmim][PF$_6$]-to-TX-100 Molar Ratio)

System Number	Water (wt%)	TX-100 (wt%)	[Bmim] [PF$_6$] (wt%)	R	D_h (nm)
1	83.9	15.0	1.1	0.17	8.3
2	83.3	15.1	1.6	0.24	12.6
3	81.5	15.7	2.8	0.41	18.9

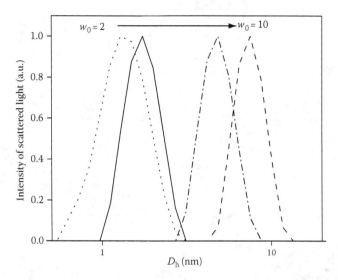

FIGURE 8.1 The size distribution graph for water-in-[Bmim][PF$_6$] microemulsions at different w_0 values with [Bmim][PF$_6$]-to-TX-100 weight ratio 1.5 at (i) $w_0 = 2$ (dotted line), (ii) $w_0 = 5$ (solid line), (iii) $w_0 = 8$ (dash dot line), and (iv) $w_0 = 10$ (dash line). All curves of the size distribution function are normalized to the same peak height. (From Seth, D., Chakraborty, A., Setua, P., and Sarkar, N., *J. Chem. Phys.*, 126, 224512/1, 2007. With permission.)

Table 8.1. In RTILs-containing microemulsions the solvent relaxation time is retarded compared to that in RTILs. Although this retardation is small compared to several 1000-fold retardation of solvent relaxation time of water in water-containing microemulsions compared to neat water [99–102].

The difference between [Bmim][PF$_6$]-in-water and water-in-[Bmim][PF$_6$] microemulsions is as follows: the previous system is oil-in-water and the latter is water-in-oil microemulsion. A schematic difference between these two microemulsions is shown in Scheme 8.1.

8.2 EXPERIMENTAL METHODS

C-153 and C-151 (laser grade, exciton) are used as received. 1-Butyl-3-methylimidazolium hexafluorophosphate ([Bmim][PF$_6$]) and 1-butyl-3-methyl-imidazolium tetrafluoroborate ([Bmim][BF$_4$]) are obtained from Acros chemicals and Fluka chemicals (98% purity) and purified using the literature procedure [46]. Cyclohexane (spectroscopic grade, Spectrochem, India) was used as received. The RTILs were dried in vacuum for ~24 h at 70°C–80°C before use. TX-100 and tween 20 were purchased from Aldrich and used as received. Triple distilled water was used to prepare all solutions. The structures of C-153, C-151, [Bmim][PF$_6$], [Bmim][BF$_4$], TX-100, and tween 20 are shown in Scheme 8.2. The weight fraction of the TX-100 in the [Bmim][PF$_6$]-in-water microemulsions is ~0.15, and the RTILs/TX-100 molar ratios (R) are 0.17, 0.24, and 0.41, respectively (Table 8.1). In the case of

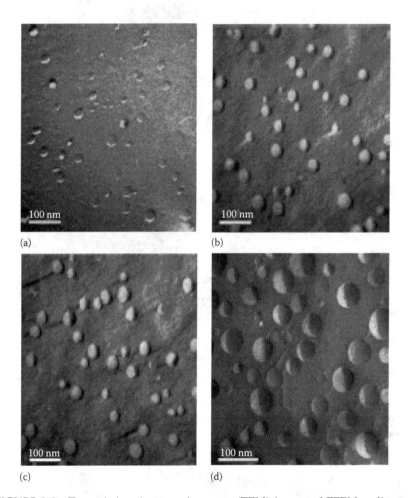

(a)   (b)

(c)   (d)

FIGURE 8.2 Transmission electron microscope (TEM) images of FFEM replicas of [Bmim][BF$_4$]/TX-100/cyclohexane microemulsions (weight fraction of TX-100 is 0.45): (a) $R = 0.2$, (b) $R = 0.5$, (c) $R = 1.0$, and (d) $R = 1.5$. (From Gao, H., Li, J., Han, B., Chen, W., Zhang, J., Zhang, R., and Yan, D., *Phys. Chem. Chem. Phys.*, 6, 2914, 2004. With permission.)

water-in-[Bmim][PF$_6$] microemulsions all experiments were conducted at [Bmim] [PF$_6$]-to-TX-100 weight ratio of 1.5 at different w_0 values, from $w_0 = 2$ to $w_0 = 10$, where $w_0 = $ [water]/[TX-100]. All experiments were carried out at temperature 298 K. In the case of [Bmim][PF$_6$]/tween 20/water microemulsions the solution was prepared at three different Is ($I = $ [Bmim][PF$_6$]-to-tween 20 weight ratio) value with $I = 0.05$, 0.18, and 0.25, respectively [86] (Table 8.2) at temperature 303 K. [Bmim][BF$_4$]/TX-100/cyclohexane microemulsions were prepared using literature procedure [80]. The weight fraction of the TX-100 in the microemulsions is 0.45 and the RTIL-to-TX-100 molar ratios (w) are 0.2, 1, and 1.5 in solvent and rotational relaxation studies. The temperature was kept at 308 K.

OH
C-151
C-153
Water
RTIL
TX-100
(a)
Water pool
RTIL
TX-100
(b)

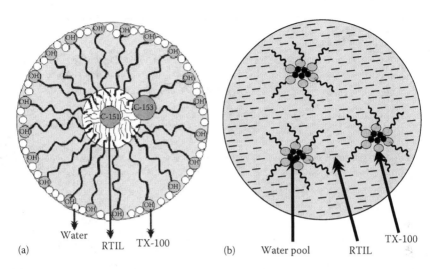

SCHEME 8.1 (a) $[Bmim][PF_6]$-in-water microemulsions. (From Seth, D., Chakraborty, A., Setua, P., and Sarkar, N., *Langmuir*, 22, 7768, 2006. With permission.) (b) Water-in-$[Bmim][PF_6]$ microemulsions. (From Seth, D., Chakraborty, A., Setua, P., and Sarkar, N., *J. Chem. Phys.*, 126, 224512/1, 2007. With permission.)

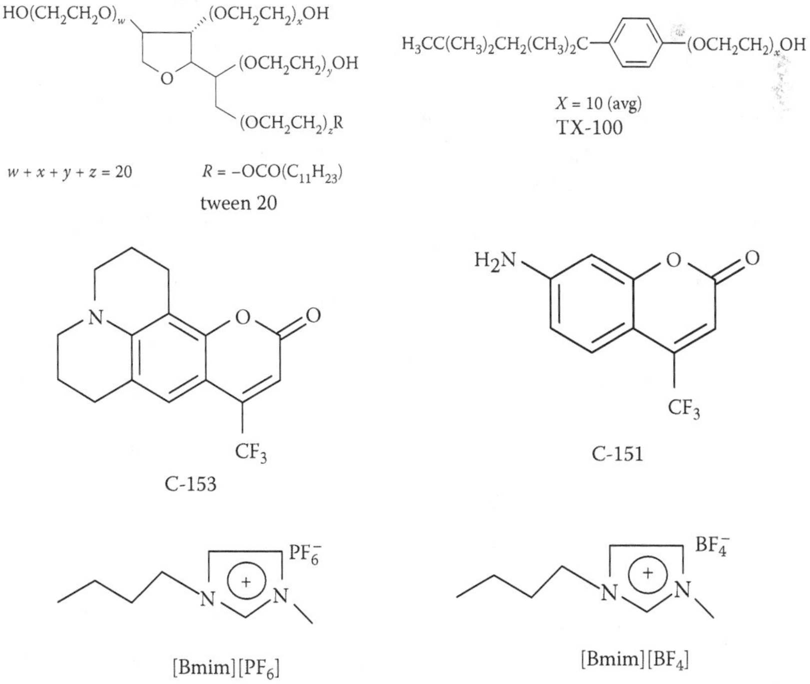

HO(CH2CH2O)w
(OCH2CH2)xOH
(OCH2CH2)yOH
(OCH2CH2)zR
w + x + y + z = 20
R = –OCO(C11H23)
tween 20
H3CC(CH3)2CH2(CH3)2C
(OCH2CH2)xOH
X = 10 (avg)
TX-100
CF3
C-153
H2N
CF3
C-151
PF6−
[Bmim][PF6]
BF4−
[Bmim][BF4]

SCHEME 8.2 Structures of surfactants, Coumarin dyes, and RTILs.

TABLE 8.2

Composition of the Tween 20/[Bmim][PF$_6$]/Water-Containing Microemulsions at Different *I* Value (*I* = [Bmim][PF$_6$]-to-Tween 20 Weight Ratio, *R* = [Bmim][PF$_6$]-to-Tween 20 Molar Ratio)

System Number	*I* Value	Tween 20 (wt%)	Water (wt%)	[Bmim][PF$_6$] (wt%)	*R*
1	0.05	14	85.3	0.7	0.216
2	0.18	11	87	2	0.785
3	0.25	10	87.5	2.5	1.08

The absorption and fluorescence spectra are collected using Shimadzu (model no: UV1601) spectrophotometer and a Spex Fluorolog-3 (model no: FL3-11) spectrofluorimeter. The fluorescence spectra are corrected for spectral sensitivity of the instrument. For steady-state experiments, all the samples are excited at 410 nm. The experimental setup for picosecond time correlated single photon counting (TCSPC) is as follows. Briefly, the samples are excited at 408 nm using picosecond diode laser (IBH, nanoled) and the signals are collected at magic angles (54.7°) using a Hamamatsu MCP PMT (3809U). The instrument response function of our setup is ~90 ps. The same setup is used for anisotropy measurements. For the anisotropy decays, we used a motorized polarizer in the emission side. The emission intensities at parallel ($I_{\parallel}$) and perpendicular ($I_{\perp}$) polarizations were collected alternatively until a certain peak difference between parallel ($I_{\parallel}$) and perpendicular ($I_{\perp}$) decay was reached. The peak difference depended on the tail matching of the parallel ($I_{\parallel}$) and perpendicular ($I_{\perp}$) decays. The analysis of the data is done using IBH DAS 6 decay analysis software. The same software is also used to analyze the anisotropy data. The best fit of the fluorescence decays was judged by the χ^2 value 1–1.2. A Neslab RTE-7 Thermostat was used to maintain constant temperature. For viscosity measurement we used an advanced rheometer (TA instrument, AR 1000) at 298 K. For DLS measurements we used Nano ZS of Malvern instrument employing a 4 mW He–Ne laser ($\lambda = 632.8$ nm) and equipped with a thermostated sample chamber. All experiments were carried out at 173° scattering angle at 298 K.

8.3 RESULTS

8.3.1 STEADY-STATE ABSORPTION AND EMISSION SPECTRA

C-153 in water shows absorption peak at 434 nm. In an ~15 wt% TX-100 solution in water the absorption peak has been blue shifted to 416 nm. With the addition of [Bmim][PF$_6$] to this solution the absorption peak of C-153 has been red shifted to 427 nm. With further addition of [Bmim][PF$_6$], i.e., with increasing [Bmim][PF$_6$]/TX-100 molar ratio (*R*) the absorption peak of C-153 remains unchanged to

TABLE 8.3

Hydrodynamic Diameter (D_h) of [Bmim] [PF$_6$]/TX-100/Water (Water-in-[Bmim] [PF$_6$]) Microemulsions at Different w_0 Values with [Bmim][PF$_6$]-to-TX-100 Weight Ratio 1.5

System	D_h (nm)
$w_0 = 2$	1.5
$w_0 = 5$	1.9
$w_0 = 8$	5.0
$w_0 = 10$	7.6

427 nm (Table 8.4). In the case of C-151, the absorption peak maximum in water is at 364 nm and in an ~15 wt% TX-100 solution in water the absorption peak of C-151 is red shifted to 374 nm. With the addition of [Bmim][PF$_6$] to this solution the absorption peak is further red shifted to 382 nm. With a further increase in R the absorption peak maximum of C-151 remains unchanged at 380 nm. Thus, from the absorption spectra it is clear that the microenvironment of C-153 and C-151 in [Bmim][PF$_6$]-in-water microemulsions is completely different from that of a bulk water solvent or a TX-100/water binary system. The representative absorption spectra and peak positions of C-153 and C-151 in microemulsions are shown in Figure 8.3 and Table 8.4, respectively.

TABLE 8.4

Absorption and Emission Maxima, Fluorescence Lifetime of C-153 and C-151 in [Bmim][PF$_6$]/TX-100/Water ([Bmim][PF$_6$]-in-Water) Microemulsions and TX-100-Water Mixture, and Bulk Viscosity of the Medium

System	λ_{max}^{Abs} (nm)	λ_{max}^{Flu} (nm)	$<\tau>$ (ns)	Viscosity (cP)
C-153 system 1	427	528	4.52	5.09
C-153 system 2	427	528	4.50	5.27
C-153 system 3	427	528	4.52	5.87
C-151 system 1	382	484	5.00	
C-151 system 2	380	484	4.90	
C-151 system 3	380	484	4.60	
C-153 in 15 wt% TX-100/water mixture	416	528	4.52	
C-151 in 15 wt% TX-100/water mixture	374	484	4.95	

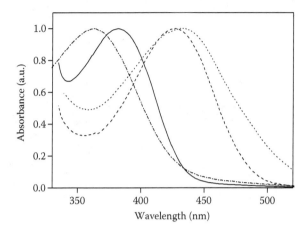

FIGURE 8.3 Absorption spectra of C-151 in [Bmim][PF$_6$]/TX-100/water ([Bmim][PF$_6$]-in-water) microemulsions at $R = 0.17$ (solid line) and in water (dashed-dotted line), and C-153 in [Bmim][PF$_6$]-in-water microemulsions at $R = 0.17$ (dashed line) and in water (dotted line). (From Seth, D., Chakraborty, A., Setua, P., and Sarkar, N., *Langmuir*, 22, 7768, 2006. With permission.)

The representative emission spectra of C-153 and C-151 in [Bmim][PF$_6$]-in-water microemulsions are shown in Figure 8.4a. The emission peaks in microemulsions are given in Table 8.4. The blue shift in the emission spectra compared to that of bulk solvent water confirms that both probes reside in a region in the microemulsions, which is less polar than water. With a gradual increase in R the

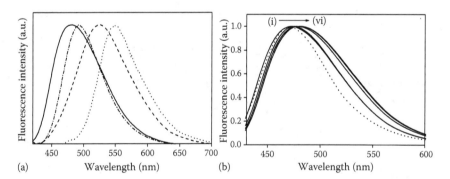

FIGURE 8.4 (a) Normalized emission spectra of C-151 in [Bmim][PF$_6$]/TX-100/water ([Bmim][PF$_6$]-in-water) microemulsions at $R = 0.41$ (solid line) and in water (dashed-dotted line), C-153 in [Bmim][PF$_6$]-in-water microemulsions at $R = 0.41$ (dashed line) and in water (dotted line). (From Seth, D., Chakraborty, A., Setua, P., and Sarkar, N., *Langmuir*, 22, 7768, 2006. With permission.) (b) Normalized emission spectra of C-151 in water-in-[Bmim][PF$_6$] microemulsions at (i) neat [Bmim][PF$_6$] (dotted line), (ii) $w_0 = 0$, (iii) $w_0 = 2$, (iv) $w_0 = 5$, (v) $w_0 = 8$, and (vi) $w_0 = 10$. (From Seth, D., Chakraborty, A., Setua, P., and Sarkar, N., *J. Chem. Phys.*, 126, 224512/1, 2007. With permission.)

emission maxima of both probes remain the same compared to those of the initial composition (see Table 8.4). With a gradual increase in R the micelles are swollen by [Bmim][PF$_6$] as a result the hydrodynamic diameter (D_h) increases, but the emission maxima of both probes do not change with loading of [Bmim][PF$_6$] in the microemulsions (see Tables 8.1 and 8.4). Therefore, from steady-state data we can conclude that both probes reside in the surface of the microemulsions. The observed emission peak of C-153 in [Bmim][PF$_6$] is at 532 nm and the emission peak of C-153 in microemulsions is at 528 nm, i.e., C-153 resides in a region which is less polar than [Bmim][PF$_6$], probably in the interface of TX-100 and [Bmim][PF$_6$] in the microemulsions. We have observed the emission maxima of C-151 remain fixed at 484 nm at three different compositions of microemulsions.

In tween 20/water binary mixture the absorption peak is blue shifted to 423 nm compared to water. With the addition of small amounts of [Bmim][PF$_6$] ($I = 0.05$) the absorption peak is further blue shifted to 415 nm, which clearly shows the change in microenvironment of C-153 molecules and also shows structural change of the medium from micelle to microemulsion. With further the addition of [Bmim][PF$_6$] the absorption peak is red shifted to 423 nm. Therefore, from absorption spectra we have observed a clear change in microenvironment around probe molecules. The emission peak of C-153 in water is at 549 nm. The emission peak of C-153 in tween 20/water binary mixture is at 531 nm. With the addition of [Bmim][PF$_6$] we have observed only a small change in the emission maximum (Table 8.5).

In water-in-[Bmim][PF$_6$] microemulsions at $w_0 = 2$, C-153 shows absorption peak at 426 nm. With an increase in w_0 value the peak maxima of absorption gradually increase. Whereas, C-151 in water shows absorption peak at 364 nm. In microemulsions the absorption peak of C-151 appeared at 370 nm. With an increase in w_0 the absorption peak of C-151 remains unchanged. The representative absorption maxima of C-153 and C-151 in microemulsions are tabulated in Table 8.6.

With an increase in w_0 value from 2 to 10 the emission peak of C-153 has been gradually red shifted from 523 to 529 nm. Thus with gradual loading of water

TABLE 8.5

Absorption and Emission Maxima of C-153 in Tween 20/[Bmim] [PF$_6$]/Water-Containing Microemulsions at Different I Values (I = [Bmim][PF$_6$]-to-Tween 20 Weight Ratio) and in Tween 20/Water Mixture

System	λ_{max}^{Abs} (nm)	λ_{max}^{Flu} (nm)
C-153 in microemulsions at $I = 0.05$	415	531
C-153 in microemulsions at $I = 0.18$	423	533
C-153 in microemulsions at $I = 0.25$	423	533
C-153 in tween 20/water binary mixture	423	531

TABLE 8.6

Absorption and Emission Maxima of C-153 and C-151 in Water-in-[Bmim] [PF$_6$] Microemulsions at Different w_0 Values and Bulk Viscosity

System	λ_{max}^{Abs} (nm)	λ_{max}^{em} (nm)	Viscosity (cP)
C-153 at $w_0 = 0$	425	518	357.5
C-153 at $w_0 = 2$	426	523	233.2
C-153 at $w_0 = 5$	427	527	160.0
C-153 at $w_0 = 8$	428	528	121.0
C-153 at $w_0 = 10$	428	529	107.3
C-151 at $w_0 = 0$	370	472	
C-151 at $w_0 = 2$	370	475	
C-151 at $w_0 = 5$	370	477	
C-151 at $w_0 = 8$	370	480	
C-151 at $w_0 = 10$	370	481	
C-151 in [Bmim][PF$_6$]	358	470	160
C-153 in [Bmim][PF$_6$]	426	532	

molecule, polarity inside the core of the microemulsions increases. The emission peaks of C-153 in neat [Bmim][PF$_6$] and water are, respectively, at 532 and 549 nm. The polarity sensed by C-153 inside the core of the microemulsions is less compared to pure water and neat [Bmim][PF$_6$]. In the case of C-151 we have also observed similar trend in the emission peak position. The emission peaks of C-153 and C-151 with an increase in w_0 are tabulated in Table 8.6. The representative emission spectra of C-151 in water-in-[Bmim][PF$_6$] microemulsions are shown in Figure 8.4b. The maximum solubility of water in [Bmim][PF$_6$] is 0.07 mol/L or 1.26 g/L [103,104]. In our experiment total concentration of water at $w_0 = 2$ is 28.33 g/L or 1.57 mol/L, which is 22 times higher than the maximum solubility limit of water in [Bmim][PF$_6$]. With an increase in w_0 from 2 to 10 the amount of water increases gradually. So it is not possible to report the absorption and emission maximum of the probes in the water/[Bmim][PF$_6$] mixture in the absence of surfactant.

C-153 in cyclohexane shows two strong absorption peak, one at 394 nm and another at 377 nm though the peak at 377 nm has maximum absorbance. After the addition of TX-100 and [Bmim][BF$_4$] one strong peak at ~410 nm is observed. The representative emission peaks of C-153 in [Bmim][BF$_4$]/TX-100/cyclohexane microemulsions are listed in Table 8.7. In cyclohexane C-153 shows a strong emission peak at 452 nm. With the addition of TX-100 in cyclohexane the emission spectrum is red shifted to 505 nm. Due to the addition of [Bmim][BF$_4$] the emission peak of C-153 has been gradually red shifted to ~516 nm at $w = 1.5$. The reported emission peak of C-153 in neat [Bmim][BF$_4$] is ~537 nm [32]. Thus it clearly indicates that microenvironment surrounding the probe molecules in microemulsions is not like neat [Bmim][BF$_4$].

TABLE 8.7
Emission Maxima of C-153 in [Bmim][BF$_4$]/
TX-100/Cyclohexane Microemulsions
at Different w_0 Values

W	λ_{em}^{max} (nm)
0	505
0.2	508
1	515
1.5	516

8.3.2 TIME-RESOLVED STUDIES

8.3.2.1 Solvation Dynamics

Solvation refers to reorientation of polar solvent molecule around a dipole instantaneously created in a polar solvent. Solvation dynamics can be measured through the time depended shifting of spectrum of the solvated probe molecules. The solvation time (τ_s) is defined as the time taken for the solvent molecules to go from randomly state to the fully solvated state (Scheme 8.3). The solvated species

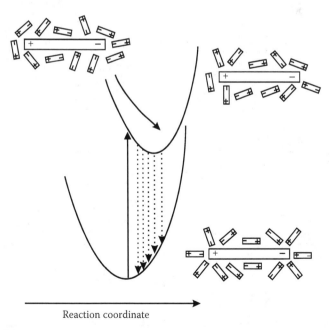

Reaction coordinate

SCHEME 8.3 Creation and solvation of a dipole in a polar solvent (sticks denote solvent dipoles).

lowers the energy of the system, thus red shift in the emission spectra is observed. This phenomenon is known as time-dependent Stokes' shift (TDSS).

We have observed a dynamic Stokes' shift in the emission spectra of C-153 and C-151 in the microemulsions at different R or w_0 values. The fluorescence decay of both probes is markedly dependent on the emission wavelength. At short wavelength, a fast decay is observed. At the red edge of the emission spectra the decay profile consists of a clear rise followed by the usual decay (which is shown in Figure 8.5). The time-resolved emission spectra (TRES) has been constructed following the procedure of Maroncelli and Fleming [105]. Each time-resolved emission spectrum was fitted by lognormal function to extract the peak frequencies. These peak frequencies were then used to construct the decay of solvation correlation function ($C(t)$), which is defined as

$$C(t) = \frac{v(t) - v(\infty)}{v(0) - v(\infty)} \tag{8.1}$$

where $v(0)$, $v(t)$, and $v(\infty)$ are the peak frequencies at time zero, t, and infinity, respectively.

The representative TRES of C-153 and C-151 are shown in Figure 8.6. The decay of $C(t)$ was fitted by biexponential, triexponential, and stretched exponential functions. The biexponential model is superior to other models.

We have also fitted $C(t)$ by stretch exponential function:

$$C(t) = \exp\left(-\left(\frac{t}{t_0}\right)^{\beta}\right) \tag{8.2}$$

where $0 < \beta \leq 1$.

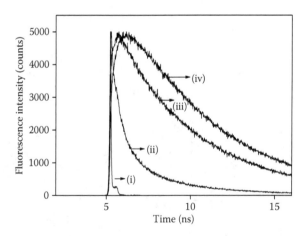

FIGURE 8.5 Fluorescence decays of C-153 in [Bmim][PF$_6$]/TX-100/water microemulsions ($w_0 = 5$) at (i) instrument response function, (ii) 470, (iii) 525, and (iv) 630 nm. (From Seth, D., Chakraborty, A., Setua, P., and Sarkar, N., *J. Chem. Phys.*, 126, 224512/1, 2007. With permission.)

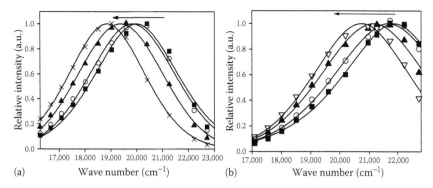

(a) Wave number (cm⁻¹) (b) Wave number (cm⁻¹)

FIGURE 8.6 TRES of (a) C-153 in [Bmim][PF$_6$]/TX-100/water ([Bmim][PF$_6$]-in-water) microemulsions ($R = 0.24$) at (i) 0 ps (■), (ii) 200 ps (○), (iii) 1000 ps (▲), and (iv) 4000 ps (×). (From Seth, D., Chakraborty, A., Setua, P., and Sarkar, N., *Langmuir*, 22, 7768, 2006. With permission.) (b) C-151 in [Bmim][PF$_6$]/TX-100/water (water-in-[Bmim][PF$_6$]) microemulsions ($w_0 = 5$) at (i) 0 (■), (ii) 200 (○), (iii) 1000 (▲), and (iv) 4000 (▽) ps. (From Seth, D., Chakraborty, A., Setua, P., and Sarkar, N., *J. Chem. Phys.*, 126, 224512/1, 2007. With permission.)

The stretched exponential model gives the worst fitting to our $C(t)$ data which is shown in Figure 8.7.

The triexponential model gives a result similar to that of the biexponential fitting. Therefore, we choose biexponential model to fit all of our $C(t)$ curves. The decay of $C(t)$ with time (Figure 8.8) was fitted by the biexponential function:

$$C(t) = a_1 e^{-t/\tau_1} + a_2 e^{-t/\tau_2} \tag{8.3}$$

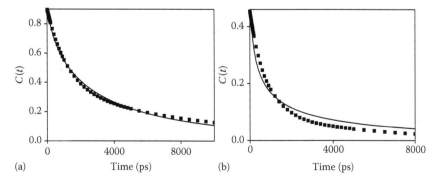

(a) Time (ps) (b) Time (ps)

FIGURE 8.7 Stretched exponential fitting of decay of the solvent correlation function ($C(t)$) of C-153 and C-151 in [Bmim][PF$_6$]/TX-100/water ([Bmim][PF$_6$]-in-water) microemulsion at $R = 0.17$. (From Seth, D., Chakraborty, A., Setua, P., and Sarkar, N., *Langmuir*, 22, 7768, 2006. With permission.)

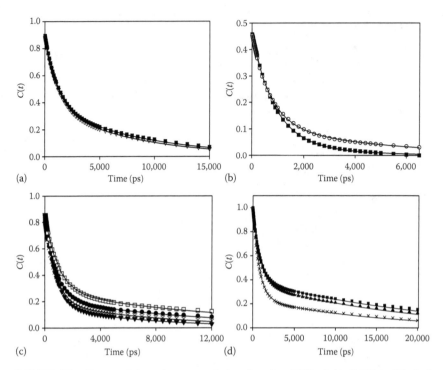

FIGURE 8.8 Decay of the solvent correlation function ($C(t)$) of (a) C-153 in [Bmim] [PF$_6$]/TX-100/water ([Bmim][PF$_6$]-in-water) microemulsions at (i) $R = 0.17$ (■) and (ii) $R = 0.41$ (▽). (From Seth, D., Chakraborty, A., Setua, P., and Sarkar, N., *Langmuir*, 22, 7768, 2006. With permission.) (b) C-151 in [Bmim][PF$_6$]/TX-100/water ([Bmim][PF$_6$]-in-water) microemulsions at (i) $R = 0.17$ (○) and (ii) $R = 0.41$ (■). (From Seth, D., Chakraborty, A., Setua, P., and Sarkar, N., *Langmuir*, 22, 7768, 2006. With permission.) (c) C-153 in [Bmim][PF$_6$]/TX-100/water (water-in-[Bmim][PF$_6$]) microemulsions at (i) $w_0 = 2$ (□), (ii) $w_0 = 5$ (●), (iii) $w_0 = 8$ (×), and (iv) $w_0 = 10$ (▼). (From Seth, D., Chakraborty, A., Setua, P., and Sarkar, N., *J. Chem. Phys.*, 126, 224512/1, 2007. With permission.) (d) C-153 in [Bmim][BF$_4$]/TX-100/cyclohexane microemulsions at $w = 0$ (■), $w = 0.2$ (▲), and $w = 1.5$ (×). (From Chakrabarty, D., Seth, D., Chakraborty., A., and Sarkar, N., *J. Phys. Chem. B*, 109, 5753, 2005.)

where τ_1 and τ_2 are the two solvation times with amplitudes of a_1 and a_2, respectively. The decay parameters of $C(t)$ in different microemulsions are summarized in Tables 8.8 through 8.11.

8.3.2.2 Rotational Dynamics

Time-resolved fluorescence anisotropy, $r(t)$, is calculated using the following equation

$$r(t) = \frac{I_{\parallel}(t) - GI_{\perp}(t)}{I_{\parallel}(t) + 2GI_{\perp}(t)} \qquad (8.4)$$

TABLE 8.8
Decay Parameters of ($C(t)$) for C-151 and C-153 in ([Bmim][PF$_6$]-in-Water) Microemulsions and TX-100 Water Mixture

System	Δv (cm^{-1})[a]	a_1	a_2	τ_1 (ns)	τ_2 (ns)[b]
C-151 System 1	860	0.33	0.12	0.70	4.56
C-151 System 2	885	0.33	0.11	0.74	3.13
C-151 System 3	950	0.39	0.07	0.84	2.38
C-153 system 1	1565	0.51	0.39	1.10	8.51
C-153 system 2	1595	0.50	0.41	1.06	7.92
C-153 system 3	1550	0.51	0.38	1.10	7.82
C-151 in TX-100/water mixture	690	0.21	0.09	0.72	0.72
C-153 in TX-100/water mixture	1220	0.24	0.33	0.93	2.87

[a] $\Delta v = v_0 - v_\infty$.
[b] Error in experimental data of ±5%.

where G is the correction factor for detector sensitivity to the polarization direction of the emission and $I_\parallel(t)$ and $I_\perp(t)$ are the fluorescence decays parallel and perpendicular to the polarization of the excitation light, respectively. The rotational relaxation time of C-153 in [Bmim][PF$_6$]-in-water microemulsions is fitted to a biexponential function. With an increase in R the average rotational relaxation time remains unchanged (Table 8.12). In the case of C-151, the rotational relaxation time in [Bmim][PF$_6$]-in-water microemulsions is fitted to a triexponential function. It is found that with an increase in R the average rotational relaxation

TABLE 8.9
Decay Parameters of $C(t)$ for C-153 in Tween 20/[Bmim][PF$_6$]/ Water-Containing Microemulsions at Different I Values (I = [Bmim] [PF$_6$]-to-Tween 20 Weight Ratio)

System	Δv (cm^{-1})[a]	a_1	a_2	τ_1 (ns)	τ_2 (ns)	$<\tau_s>$ (ns)[b,c]
C-153 at I = 0.05	1060	0.94	0.06	0.80	7.2	1.18
C-153 at I = 0.18	1170	0.86	0.14	0.77	14.7	2.72
C-153 at I = 0.25	1180	0.84	0.16	0.76	14.5	2.96
C-153 in tween 20/water binary mixture	1050	0.92	0.08	0.73	8.4	1.34

[a] $\Delta v = v_0 - v_\infty$.
[b] Error in experimental data of ±5%.
[c] $<\tau_s> = a_1\tau_1 + a_2\tau_2$.

TABLE 8.10

Decay Parameters of C-153 and C-151 in Water-in-[Bmim][PF$_6$] Microemulsions at Different w_0 Values

System	Δv (cm^{-1})[a]	a_1	a_2	τ_1 (ns)	τ_2 (ns)[b]
C-153 at $w_0 = 0$	1370	0.52	0.32	1.12	14.60
C-153 at $w_0 = 2$	1330	0.54	0.28	0.965	13.65
C-153 at $w_0 = 5$	1290	0.58	0.22	0.805	11.15
C-153 at $w_0 = 8$	1280	0.63	0.19	0.740	8.50
C-153 at $w_0 = 10$	1390	0.70	0.17	0.730	6.90
C-151 at $w_0 = 0$	1450	0.30	0.27	1.15	10.80
C-151 at $w_0 = 2$	1290	0.29	0.22	0.945	8.15
C-151 at $w_0 = 5$	1370	0.36	0.16	0.850	7.15
C-151 at $w_0 = 8$	1330	0.40	0.10	0.890	6.85
C-151 at $w_0 = 10$	1310	0.36	0.14	0.705	3.75
C-151 in [Bmim][PF$_6$]	530	0.03	0.15	0.400	1.10

[a] $\Delta v = v_0 - v_\infty$.
[b] Error in experimental data of $\pm 5\%$.

time of C-151 gradually increases (Figure 8.9a). The variation of the rotational relaxation time of C-151 and C-153 in different ways suggested different locations of the probes in the [Bmim][PF$_6$]-in-water microemulsion. This is shown schematically in Scheme 8.1a.

The anisotropy decay parameters of C-153 and C-151 in water-in-[Bmim] [PF$_6$] microemulsions are listed in Table 8.14 and shown in Figure 8.9b. The anisotropy decays of both probes are fitted to a biexponential function. The rotational relaxation times of both probes are gradually decreasing with an increase in w_0. In case of C-153 with a gradual increase in w_0 from 2 to 10 the rotational relaxation time decreases from 4.53 to 3.68 ns whereas, in the case of C-151 it decreases

TABLE 8.11

Decay Parameters of $C(t)$ and Missing Component of C-153 in [Bmim] [BF$_4$]/TX-100/Cyclohexane Microemulsions at Different w Values

W	a_1	τ_1 (ns)	a_2	τ_2 (ns)	$<\tau_s>$ (ns)[a,b]	Missing Component
0	0.74	0.731	0.26	12.96	3.91	24%
0.2	0.60	0.815	0.40	18.94	8.06	15%
1	0.62	0.888	0.38	16.99	7.01	22%
1.5	0.62	0.839	0.38	16.64	6.84	24%

[a] Error in experimental data of $\pm 5\%$.
[b] $<\tau_s> = a_1\tau_1 + a_2\tau_2$.

TABLE 8.12

Rotational Relaxation Parameters of C-151 and C-153 in [Bmim] [PF$_6$]-in-Water Microemulsions and TX-100/Water Mixture

System	r_0	a_{1r}	τ_{1r} (ns)	a_{2r}	τ_{2r} (ns)	a_{3r}	τ_{3r} (ns)	$<\tau_r>$ (ns)[a,b]
C-151 System 1	0.40	0.19	0.06	0.43	0.92	0.38	3.65	1.79
C-151 System 2	0.40	0.16	0.05	0.47	0.95	0.37	4.0	1.93
C-151 System 3	0.40	0.19	0.05	0.47	1.03	0.34	4.67	2.08
C-153 System 1	0.35	0.47	0.87	0.53	3.80	—	—	2.42
C-153 System 2	0.36	0.44	0.89	0.56	3.64	—	—	2.43
C-153 System 3	0.34	0.36	0.74	0.64	3.52	—	—	2.52
C-151 in TX-100/ water mixture	0.37	0.47	0.63	0.53	3.27	—	—	2.03
C-153 in TX-100/ water mixture	0.35	0.41	0.75	0.59	3.20	—	—	2.19

[a] $<\tau_r> = a_{1r}\tau_{1r} + a_{2r}\tau_{2r}$.
[b] Error in experimental data of ±5%.

from 4.83 to 2.63 ns. The rotational relaxation times of C-151 and C-153 in neat [Bmim][PF$_6$] are 3.89 and 3.57 ns, respectively.

In [Bmim][PF$_6$]/tween 20/water ternary microemulsions the rotational relaxation time of C-153 in all microemulsions is fitted to a biexponential function. In all microemulsions rotational relaxation time remains more or less the same. The anisotropy decay parameters are listed in Table 8.13.

The representative anisotropy decays of C-153 in [Bmim][BF$_4$]/TX-100/ cyclohexane microemulsions at $w = 0$ and at $w = 1.5$ are shown in Figure 8.9c. The decay parameters are listed in Table 8.15. The rotational relaxation of C-153 in cyclohexane is fitted with a single exponential function and the rotational relaxation time is 135 ps. But in microemulsions all the rotational relaxation decays are fitted with biexponential function. As for example at $w = 0.2$ the observed rotational relaxation times are 0.348 and 2.15 ns. The average rotational relaxation time of C-153 increases from cyclohexane to microemulsions. Moreover, the average rotational relaxation time increases with an increase in w values.

8.3.3 VISCOSITY MEASUREMENT

We have also measure the viscosity of different microemulsions. With the gradual addition of [Bmim][PF$_6$] in [Bmim][PF$_6$]-in-water microemulsions bulk viscosity of solution gradually increases (Table 8.4). We have also measured the viscosity of water-in-[Bmim][PF$_6$] microemulsions at different w_0. With the gradual addition of water in TX-100/[Bmim][PF$_6$] mixture bulk viscosity of the solution gradually decreases. The bulk viscosity of TX-100-[Bmim][PF$_6$] mixtures at different w_0 is tabulated in Table 8.6. All viscosity measurements were carried out at 298 K.

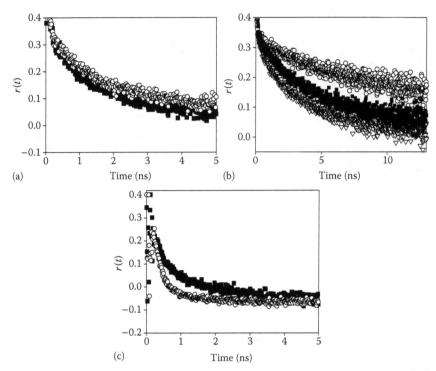

(a) Time (ns)   (b) Time (ns)   (c) Time (ns)

FIGURE 8.9 Decays of the time-resolved fluorescence anisotropy ($r(t)$) of (a) C-151 in [Bmim][PF$_6$]/TX-100/water ([Bmim][PF$_6$]-in-water) microemulsions at (i) $R = 0.17$ (■) and (ii) $R = 0.41$ (O). (From Seth, D., Chakraborty, A., Setua, P., and Sarkar, N., *Langmuir*, 22, 7768, 2006. With permission.) (b) C-151 in [Bmim][PF$_6$]/TX-100/water (water-in-[Bmim] [PF$_6$]) microemulsions at (i) $w_0 = 0$ (O), (ii) $w_0 = 5$ (■), and (iii) $w_0 = 10$ (∇). (From Seth, D., Chakraborty, A., Setua, P., and Sarkar, N., *J. Chem. Phys.*, 126, 224512/1, 2007. With permission.) (c) C-153 in [Bmim][BF$_4$]/TX-100/cyclohexane microemulsions at (i) $w = 0$ (O) and (ii) $w = 1.5$ (■). (From Chakrabarty, D., Seth, D., Chakraborty., A., and Sarkar, N., *J. Phys. Chem. B*, 109, 5753, 2005.)

TABLE 8.13
Rotational Relaxation Parameters of C-153 in [Bmim][PF$_6$]/Tween 20/Water Microemulsions and in Tween 20/Water Mixture

System	r_0	a_{1r}	τ_{1r} (ns)	a_{2r}	τ_{2r} (ns)	$<\tau_r>$ (ns)[a,b]
C-153 at $I = 0.05$	0.39	0.52	0.27	0.48	1.56	0.89
C-153 at $I = 0.18$	0.39	0.51	0.30	0.49	1.56	0.92
C-153 at $I = 0.25$	0.39	0.49	0.26	0.51	1.55	0.92
C-153 in tween 20/water binary mixture	0.39	0.55	0.29	0.45	1.43	0.80

[a] $<\tau_r> = a_{1r}\tau_{1r} + a_{2r}\tau_{2r}$.
[b] Error in experimental data of ±5%.

TABLE 8.14

Rotational Relaxation Parameters of C-153 and C-151 in Water-in-[Bmim] [PF$_6$] Microemulsions at Different w_0 Values

System	r_0	a_{1r}	τ_{1r} (ns)	a_{2r}	τ_{2r} (ns)	$<\tau_r>$ (ns)[a,b]
C-153 at $w_0 = 0$	0.39	0.39	0.190	0.61	7.86	4.87
C-153 at $w_0 = 2$	0.33	0.21	0.153	0.79	5.69	4.53
C-153 at $w_0 = 5$	0.32	0.20	0.503	0.80	5.56	4.55
C-153 at $w_0 = 8$	0.37	0.25	0.628	0.75	5.40	4.21
C-153 at $w_0 = 10$	0.39	0.28	0.608	0.72	4.87	3.68
C-151 at $w_0 = 0$	0.35	0.34	0.076	0.66	7.84	5.20
C-151 at $w_0 = 2$	0.40	0.32	0.118	0.68	7.06	4.83
C-151 at $w_0 = 5$	0.39	0.25	0.09	0.75	4.40	3.32
C-151 at $w_0 = 8$	0.38	0.23	0.20	0.77	3.64	2.85
C-151 at $w_0 = 10$	0.38	0.24	0.18	0.76	3.40	2.63
C-151 in [Bmim][PF$_6$]	0.40	0.21	0.08	0.79	4.90	3.89

[a] $<\tau_r> = a_{1r}\tau_{1r} + a_{2r}\tau_{2r}$.
[b] Error in experimental data of $\pm 5\%$.

8.4 DISCUSSIONS: SOLVATION AND ROTATIONAL DYNAMICS IN FOUR DIFFERENT RTILs-MICROEMULSIONS SYSTEMS

8.4.1 [BMIM][PF$_6$]/TX-100/WATER ([BMIM][PF$_6$]-IN-WATER) MICROEMULSIONS

From the steady-state results we can conclude that C-153 is located at the interface of the microemulsions. The rotational relaxation time of C-153 also does not change with an increase in R. We have also measured the fluorescence lifetime of C-153 in different microemulsions at the corresponding emission peak of the dye,

TABLE 8.15

Rotational Relaxation Time of C-153 in [Bmim][BF$_4$]/ TX-100/Cyclohexane Microemulsions at Different w Values

w	r_0	a_{1r}	τ_{1r} (ns)	a_2	τ_{2r} (ns)	$<\tau_r>$ (ns)[a,b]
0	0.40	0.95	0.310	0.05	5.01	0.545
0.2	0.40	0.87	0.348	0.13	2.15	0.582
1	0.37	0.70	0.421	0.30	2.36	1.00
1.5	0.39	0.64	0.396	0.36	2.64	1.20

[a] $<\tau_r> = a_{1r}\tau_{1r} + a_{2r}\tau_{2r}$.
[b] Error in experimental data of $\pm 5\%$.

and we have observed that the fluorescence lifetime of C-153 at the emission maximum is single exponential and remains unchanged for all microemulsions (Table 8.4). These data also suggest that C-153 is located at the interface of the microemulsions. The most interesting characteristic in these microemulsions is that the solvent reorganization time is not very sensitive to an increase in R or an increase in the size of droplet. This observation is different from what we have seen in AOT microemulsions [106–115]. In previous solvation dynamics studies on AOT microemulsions the dye molecules migrated to the water/polar solvent pool of the microemulsions as the hydrodynamic radii or the water content of the microemulsions are increased [106–115]. For this reason the solvent relaxation time gradually becomes faster with an increase in the water content of AOT microemulsions. In this work we do not observe a systematic variation of the solvent relaxation time, like in AOT microemulsions. This is probably due to the interfacial location of C-153 in the microemulsions and with the gradual addition of RTILs the position of C-153 remains more or less the same. C-153 is a hydrophobic probe, so it is more likely that C-153 resides at the interface of RTILs and TX-100 in the microemulsions. Recently, Corbeil and Levinger [116] and Hazra et al. [117] also showed a similar trend in the solvation time in SDS, CTAB, and TX-100 quaternary microemulsions. In our experiment, the fast component of the solvation time remained almost same and we have observed a small change in the slow component with an increase in R from 0.17 to 0.41 (Table 8.8). We have also observed the solvation time of C-153 in a 15 wt% TX-100/water binary system, and it is described by two components of 930 ps (24%) and 2.87 ns (33%). The slowing down of both components of the solvation time in [Bmim][PF$_6$]-in-water microemulsions compared to TX-100/water binary systems is due to the addition of [Bmim][PF$_6$]. Due to the addition of [Bmim][PF$_6$] the structure of TX-100/water binary system has been changed. Consequently, the probes molecules feel different microenvironment. In all RTILs-containing microemulsions, we have observed a bimodal solvation time and time constants and their relative weights are more or less the same in all microemulsions. The time constant of fast component ~1100 ps (51%) and relative weight of the slow component, ~39% are the same in all microemulsions. We have observed only a small change in the time constant of the long component. Chakrabarty et al. [46] observed the solvation time of C-153 in neat [Bmim][PF$_6$], the average solvation time is 3.35 ns with time constant 536 ps (83%) and 17.10 ns (17%). Hence, the observed dynamics in RTILs-containing microemulsions is not the same as the dynamics in TX-100/water mixtures or in neat [Bmim][PF$_6$]; rather it implies the dynamics inside the microemulsions. Recently, Chakraborty et al. [118] reported the solvation time of C-153 in neat [Bmim][PF$_6$] as ~3 ns. Maroncelli et al. [39,40] reported the solvation time in neat [Bmim][PF$_6$] ~1.8 ns using 4-aminophthalimide and 1.0 ns using C-153 as the probe. Karmakar and Samanta [34] reported the solvation time of Nile red in neat [Bmim][PF$_6$] as ~1.0 ns. Thus, the solvation times reported by various groups are different. Even we have also observed different solvation times of C-153 in neat [Bmim][PF$_6$] [46,118] in different experiments and various drying conditions. This is due to the fact that the qualities of the RTILs used by different groups are different and the presence of small amount of impurities like water and chloride ion vastly changes the viscosity of RTILs [119]. Recently, Ito et al. [42] reported that different results reported by different groups are due to different methods/setup used to represent the data. In this experiment the solvation dynamics is hindered in the pool of the microemulsions compared to the neat [Bmim][PF$_6$], but retardation is very small compared to the severalfold retardation of the solvation dynamics of conventional solvents inside the core of the microemulsions [106–117,120]. In pure water solvent relaxation occurs on the femtosecond time scale. Jimenez et al. [99] reported that the solvent relaxation of C-343 in water consists of an initial decay of 55 fs (50%), attributed to the librational motion. Solvent relaxation of C-102 in water is bimodal with time constants of <50 ps (26%) and 310 fs (74%) [99–102]. Due to the very fast intermolecular vibrational and librational motions of water, solvent relaxation in water is very fast, but in microemulsions the solvent relaxation time is retarded severalfold compared to that in pure water.

The nature of solvation in neat RTILs is totally different from other polar solvents. In polar solvents such as water, alcohols, etc. the solvent molecules reorient themselves around the photoexcited solute molecule. Whereas, in neat RTILs motions of cations and anions around photoexcited dye molecules are responsible for solvation. Chapman and Maroncelli [121] showed that ionic solvation is slower compared to other polar solvent [99–102] and is dependent on the viscosity of the medium. Bart et al. [122,123] also showed that ionic solvation is slow and biphasic in nature, which is in contrast to the monophasic nature of solvent response function in some polar solvents. They also showed that the solvent relaxation in molten salt depends on the solute. Due to different size of cations and anions of RTILs the mobility of cation is different from anion and consequently we have observed biphasic dynamics compared to monophasic dynamics observed in some polar solvent. Samanta et al. [32–38] have studied solvation dynamics measurement in different RTILs using different probe solutes. They have reported biphasic dynamics occurring on picosecond and nanosecond time scale arising from the anions and the collective diffusion of the cations and anions, respectively. Maroncelli et al. [39–45] observed biphasic solvation in different RTILs. Recently, Lang et al. studied solvation dynamics using fluorescence upconversion technique [49]. Apart from experimental studies a few theoretical studies [50–55] regarding dynamical properties of RTILs were reported. Shim et al. [50] used molecular dynamics simulations and divide the solvent response of the systems into cation and anion components. Their results indicate that the subpicosecond response is entirely due to anionic motions. Later on they proposed that the rapid subpicosecond dynamics arising mainly from inertial translation and slow relaxation ascribed to ion transport [51,52]. Kobrak et al. [54] proposed that collective cation–anion motions are responsible for subpicosecond response. In neat RTILs motions of ions are responsible for the fast dynamics and the collective diffusional motions are responsible for the slow dynamics. So in neat RTILs we get high value of solvent relaxation time compared to other solvent, and in RTILs-containing microemulsions we have observed a small increase in the solvation time compared to that in neat RTILs.

We have also chosen one more hydrophilic and flexible probe, C-151, to study solvation dynamics in this microemulsions. Steady-state absorption and fluorescence spectra of C-151 do not change with an increase in R, but the rotational relaxation time of C-151 gradually increases with an increase in R. Thus, the results show that with an increase in [Bmim][PF$_6$] content of the microemulsions more C-151 molecules move toward the core of the microemulsions. The solvation time of C-151 in these microemulsions is also biexponential in nature. Both components of the solvent relaxation time of C-151 gradually change with an increase in R from 0.17 to 0.41. With an increase in the [Bmim][PF$_6$] content in the microemulsions the time constant of the fast component and its relative amplitude gradually increase and vice versa for the slow component. The change in time constants of solvent relaxation time with an increase in [Bmim][PF$_6$] content of the microemulsions is due to an increase in the sizes of the microemulsions. An increase in the size of the microemulsions leads to an increase in the free motion of the ions, and as a result the solvent relaxation time has been changed. Recently, by quantum chemical studies Cave et al. [124] and Sulpizi et al. [125] showed that the flexibility of C-151 is only in the −NH$_2$ torsion angle, and traditional TICT formation is unlikely upon photoexcitation of the C-151 molecule. They showed that the real difference between C-151 and C-153 is the change of two ring-closing propyl groups for amino protons that will be strong hydrogen bond donors to water, RTIL anions, and ether oxygen on the TX-100 molecules. Thus, one can expect that C-151 would be substantially more sticky than C-153 if H-bonding were to occur. With increasing concentration of the RTIL the rotational relaxation time of C-151 increases substantially, which indicate strong specific interaction of C-151 with the RTIL. Thus, with an increase in R the microenvironment around C-151 changes substantially, which leads to different interactions that would affect the solvation dynamics, whereas in the case of C-153 with an increase in the concentration of the RTIL the rotational relaxation time does not change as that of C-151 does, which indicates that C-153 resides at the interfacial position of the microemulsions. Therefore, the addition of RTIL would not affect the solvent relaxation time of C-153.

In the present study using the TCSPC setup, we are missing the fast component of the solvation dynamics (<90 ps). We can apply the method of Fee and Maroncelli [126] to calculate the missing component. We have calculated a "time zero spectrum" using the above procedure. The time zero frequency can be estimated using the following relation from absorption and fluorescence spectra:

$$v_p(t=0) \approx v_p(abs) - [v_{np}(abs) - v_{np}(em)] \qquad (8.5)$$

where the subscripts "p" and "np" refer to the polar and nonpolar spectra, respectively. The percentage of the missing component is $(v_{cal}(0) - v(0))/(v_{cal}(0) - v(\infty))$. If we use C-153 as our experimental probes, using this procedure, we calculate a total Stokes' shift of ~1740 cm^{-1} at $R = 0.17$, but we have observed a total Stokes' shift of ~1565 cm^{-1}. Thus, we have missed ~10% of total spectral shift. In all microemulsions at different R we are missing ~10% of the solvent relaxation dynamics. We have also calculated the missing component of C-153 in a 15 wt% TX-100/water binary mixture. Here we have calculated using the above procedure a total Stokes' shift of ~2120 cm^{-1}, but we have observed a total Stokes' shift of ~1220 cm^{-1}. Therefore, we are missing about 43% of the total solvent relaxation dynamics. From the above missing component data we can easily conclude that the location of C-153 in all microemulsions at different R is more or less the same, and the microenvironment of dye molecule C-153 in the microemulsions is totally different from TX-100/water binary mixture. When we used C-151 as our experimental probe, we calculated a total Stokes' shift of ~2000 cm^{-1} at $R = 0.24$ using Fee and Maroncelli [126] procedure, but we have observed a total Stokes' shift of ~885 cm^{-1}. Thus, we are missing ~56% of the total spectral shift. With an increase in [Bmim][PF$_6$] content of the microemulsions the missing component increases. We have also calculated the missing parts of the total solvent relaxation in the TX-100/water binary mixture. Here we have calculated using the above procedure a total Stokes' shift of ~2280 cm^{-1}, but we have observed a total Stokes' shift of ~690 cm^{-1}. Thus, we are missing ~70% of the total solvent relaxation dynamics. From these data we can say that the microenvironment of C-151 in microemulsions is quite different from that in the TX-100/water binary mixture. Chakrabarty et al. [46] reported that 44% of the total solvent relaxation dynamics in neat [Bmim][PF$_6$] occurs on subpicosecond time scales using C-153 as the experimental probe. Chowdhury et al. [47] also reported that more than half of the solvent relaxation dynamics was completed within 100 ps. However, we have observed that ~10% of the total solvent relaxation dynamics was missed using C-153 as our experimental probes. This suggests that the solvent relaxation of RTILs is hindered in microemulsions compared to neat solvent.

We have used this time zero frequency calculated from the method of Fee and Maroncelli [126] to construct $C(t)$. For both probes decay of $C(t)$ starts from a value which is less than unity (Figure 8.8).This is due to the fact that it is not possible to capture the ultrafast dynamics in these systems by our setup. In the case of C-153 we have missed ~10% of the solvent relaxation dynamics, so decay of $C(t)$ starts from near about 0.9. The solvent relaxation time of C-153 in system 1 consists of two components, 1.10 ns (51%) and 8.51 ns (39%), and the rest, 10%, of the dynamics are missed. In the case of C-151 we have missed more than half of the dynamics. The decay of $C(t)$ of C-151 in system 1 starting form 0.45 indicates that we are missing 55% of the total dynamics. The solvent relaxation time of C-151 in system 1 consists of two components, 0.70 ns (33%) and 4.56 ns (12%).

With an increase in the amount of RTIL the bulk viscosity of the solution gradually increases. The rotational relaxation time of C-151 also gradually increases with an increase in R, and with an increase in R the time constant of solvent relaxation gradually changes. The rotational relaxation and solvent relaxation times of C-153 remain almost unaltered with an increase in R. In the case of C-151 the behavior of rotational relaxation time simply follows the hydrodynamic theory, but in the case of C-153 due to the location of the probe in the interfacial region the rotational relaxation time is almost unaffected, which indicates that the microscopic viscosity sensed by the probe is different from macroscopic viscosity.

Therefore, it is plausible that the bulk viscosity of the solvent mixture in this case is actually relatively meaningless due to the biphasic nature of the solution, and it is very difficult to conclude anything from the bulk viscosity data.

In microemulsions RTILs feel a restricted environment in the core of the microemulsions. Thus, we have observed slower solvation time of C-153 in microemulsions compared to neat RTILs, but the retardation in the solvent relaxation time is much less compared to that of neat RTILs. With an increase in the [Bmim] [PF$_6$] content the size of microemulsions increases. The small change in time constants of solvent relaxation with an increase in R is due to an increase in the size of the microemulsions. With an increase in size, the free motion of ions of RTILs in the core of the microemulsions has increased, and this leads to change in time constants of solvent relaxation.

8.4.2   [Bmim][PF$_6$]/TX-100/Water (Water-in-[Bmim][PF$_6$]) Microemulsions

The emission peak of C-153 in [Bmim][PF$_6$]–TX-100 binary system ($w_0 = 0$) (with weight ratio of [Bmim][PF$_6$]-to-TX-100 = 1.5) is at 518 nm. With the gradual addition of water, i.e., with an increase in w_0 the emission peak of C-153 has been gradually red shifted in this mixture. The emission peak of C-153 at $w_0 = 10$ is at 529 nm. The red shift in the emission spectra indicates that with the gradual addition of water most of the C-153 molecules have been shifted toward the pool of the microemulsions. In the case of C-151 we have also observed a similar trend in the emission spectra. In [Bmim][PF$_6$]–TX-100 binary system (at $w_0 = 0$) C-151 shows an emission peak at 472 nm. With an increase in water loading in this binary system, the emission peak of C-151 has been gradually red shifted. This indicates that the most of C-151 molecules have been shifted toward the pool of the microemulsions. The emission peak of C-151 at $w_0 = 10$ of microemulsions is at 481 nm.

We have also studied the solvation dynamics of C-151 in neat [Bmim][PF$_6$]. The solvent relaxation of C-151 in neat [Bmim][PF$_6$] consists of two components with time constants of 400 ps (3%) and 1.10 ns (15%). Due to small size of the anion it is reasonable to assume that the fast component is due to the translational motion of the anion. Since the amplitude of the slow component is much larger compared to the fast component, the slow component not only arises due to the motion of the bigger cations, but also probably contributes collective motion of cations and anions. Cammarata et al. [127] showed the existence of symmetric 1:2 type H-bonded complexes: [PF$_6$]$^-$···HOH···[PF$_6$]$^-$ in [Bmim][PF$_6$]/water mixture. Baker et al. [128] showed that slowing down of the subnanosecond component with the addition of solubilized water is due to the formation of 1:2 type H-bonded complexes: [PF$_6$]$^-$···HOH···[PF$_6$]$^-$. It suggests that –NH$_2$ group of C-151 may form H-bond with the PF$_6^-$ ion and contributing to the slow component of the solvent relaxation.

For both probes we have observed a bimodal solvation dynamics in this water-in-[Bmim][PF$_6$] microemulsions at different w_0 values. In the case of C-153 the fast and slow components of solvent relaxation are decreased from 0.965 to 0.730 ns and 13.65 to 6.90 ns, respectively, with change in w_0 from 2 to 10. The decrease in both components of solvation time with an increase in w_0 (Table 8.10) indicates that with an increase in water loading more C-153 molecules have been shifted toward the pool of the microemulsions. With an increase in w_0 the time constant of the fast component gradually decreases and the amplitude of fast component increases. The time constant and amplitude of slow components are gradually decreasing with an increase in w_0. The slow components are due to water molecules present at the surface of the microemulsions, i.e., due to bound water molecules. The water molecules residing at the central region of the water pool, i.e., the free water molecules, are contributing to the fast component of solvation dynamics. These data also support the fact that with an increase in w_0 more C-153 molecules have been shifted to the pool of the microemulsions. The average solvation time of C-153 in neat [Bmim][PF$_6$] is ~3.35 ns with time constants of 536 ps (83%) and 17.10 ns (17%) [46]. These time constants are totally different from the time constants observed in microemulsions. The solvent relaxation of C-153 at $w_0 = 0$, i.e., in [Bmim][PF$_6$]–TX-100 binary system is also bimodal with time constants of 1.120 ns (52%) and 14.60 ns (32%), respectively. At $w_0 = 2$ the solvation time of C-153 can be described by two time constants of 965 ps (54%) and 13.65 ns (28%). Thus dynamics at different w_0 represents the dynamics inside the core of the microemulsions.

In the case of C-151 the fast and slow components of the solvation time are decreasing from 0.945 to 0.705 ns and 8.15 to 3.75 ns, respectively, with an increase in w_0 from 2 to 10. The decrease in time constants of solvation time indicates that more number of C-151 molecules are shifted to the core of the microemulsion with an increase in w_0. With an increase in w_0 the contribution of fast component is gradually increasing upto $w_0 = 8$. The solvation time of C-151 at $w_0 = 2$ consists of time constants of 945 ps (29%) and 8.15 ns (22%). At $w_0 = 8$ the solvation time consists of time constants of 890 ps (40%) and 6.85 ns (10%) and at $w_0 = 10$ the time constants are 705 ps (36%) and 3.75 ns (14%), respectively, i.e., at $w_0 = 10$, the amplitude of fast component decreases with respect to the amplitude of fast component of C-151 at $w_0 = 8$. The solvation relaxation of C-151 in neat [Bmim][PF$_6$] consists of time constants of 400 ps (3%) and 1.10 ns (15%), respectively. The relative difference in the fast and slow components of solvation time of C-153 and C-151 and the amplitudes are probably arising due to the different hydrophobicities of the dyes. Thus the slow dynamics in the microemulsion is essentially the dynamics inside the core of the microemulsions. For both probes with an increase in w_0 the amplitude of the first components increases and the average solvation time decreases. In pure water solvent relaxation time occurs on the femtosecond time scale. Due to very fast intermolecular vibrational and librational motions of water, the solvent relaxation in water is very fast [99–102], but in microemulsions the solvent relaxation time is retarded several times compared to pure water. Chakrabarty et al. [129,130] observed that with the gradual addition of cosolvent in RTILs, the average solvation and rotational relaxation times are gradually decreasing due to the

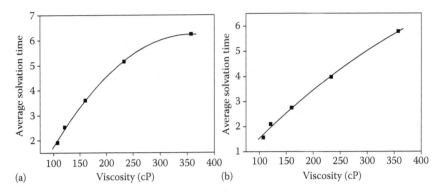

FIGURE 8.10 Plot of average solvation time vs. bulk viscosity at (a) C-153 and (b) C-151. (From Seth, D., Chakraborty, A., Setua, P., and Sarkar, N., *J. Chem. Phys.*, 126, 224512/1, 2007. With permission.)

decrease in viscosity of the medium. Here we have also observed that with the gradual addition of water in [Bmim][PF$_6$]–TX-100 binary system the average solvation and rotational relaxation times of both probes have decreased, whereas in our present work, with the addition of water in [Bmim][PF$_6$]–TX-100 binary system microemulsions are formed. In this work the decrease in the solvation time with the addition of water is mainly due to the water dynamics inside the microemulsions. We have plotted the average solvation time with viscosity at different w_0 values in Figure 8.10. From the plot it is inferred that the average solvation time shows a nonlinear relationship with bulk viscosity. In the previous studies of the solvation dynamics in neat RTILs, Maroncelli et al. [39–42] showed that the solvation time can be reasonably represented by relations of the form $\tau_{solv} \propto \eta^p$ with $p \cong 1$. It is seen that in our case the average solvation time vs viscosity plot shows a curvature and was fitted by the second order polynomial. Due to heterogeneity in our system microviscosity sensed by the probe molecules inside the microemulsion is different from the bulk viscosity. So in these systems it is not possible to maintain linear correlation between the average solvation time and bulk viscosity.

In the present study using the TCSPC setup, we are missing the fast component of the solvation dynamics (<90 ps). We can apply the method of Fee and Maroncelli [126] to calculate the missing component. We have calculated a time zero spectrum using the above procedure. Using this procedure in the case of C-153 we have calculated a total Stokes' shift of ~1615 cm^{-1} at $w_0 = 2$ whereas we have observed a total Stokes' shift of ~1330 cm^{-1}. Thus we have missed ~18% of the total dynamics. With an increase in w_0 the percentage of missing component is almost unchanged. In the case of C-151 we have calculated a total Stokes' shift of ~2500 cm^{-1} at $w_0 = 2$ but we have observed a total Stokes' shift of ~1290 cm^{-1}. Thus we are missing ~50% of the total dynamics. With an increase in w_0 the amount of missing component remains more or less the same. In the case of C-151 in neat [Bmim][PF$_6$] using Fee and Maroncelli [126] procedure we have calculated a total Stokes' shift of ~2960 cm^{-1} and the observed Stokes' shift is ~530 cm^{-1}. Thus we are missing 82% of total spectral shift of C-151 in neat [Bmim][PF$_6$].

We have used these time zero frequencies calculated from Fee and Maroncelli [126] procedure to construct $C(t)$. For both probes decay of $C(t)$ starts from a value which is less than unity (Figure 8.8), because it is not possible to capture the ultrafast dynamics in these systems by using our setup. The solvent relaxation time of C-153 at $w_0 = 2$ consists of two components of 0.965 ns (54%) and 13.65 ns (28%) and the rest 18% dynamics are missed. In the case of C-151 we have missed about 50% of the total dynamics. The solvent relaxation time of C-151 at $w_0 = 2$ consists of two components of 0.945 ns (29%) and 8.15 ns (22%), and the rest 49% dynamics are missed. On the other hand the solvent relaxation time of C-151 in neat [Bmim][PF$_6$] consists of two components of 400 ps (3%) and 1.10 ns (15%), i.e., 82% dynamics is missed.

We have also studied the rotational relaxation of C-153 and C-151 in microemulsions at different w_0 values to understand the location of probes. We found that with an increase in w_0 the rotational relaxation times of both probes are gradually decreasing. With an increase in w_0 the free water molecules in the core of the microemulsions are gradually increasing and probe molecules gradually shifted toward the core of the microemulsions. As a result we have observed a gradual decrease in the average solvation time for both probes. The rotational relaxation times also supported this fact that probe molecules are gradually shifting toward the core of the microemulsions with an increase in w_0. With an increase in w_0 the probe molecules experience a lesser-restricted environment in the microemulsions, consequently the rotational relaxation times of both probes are gradually decreasing. This is also true for other microemulsions. We have also measured the rotational relaxation time of C-151 in neat [Bmim][PF$_6$] and the value is ~3.89 ns with time constant of 0.08 (21%) and 4.90 (79%) ns. The rotational relaxation time of C-153 in neat [Bmim][PF$_6$] is ~3.57 ns [46]. The biphasic rotational relaxations in RTILs are also supported from the molecular dynamics simulation where two different time scales of diffusions were observed [131,132]. With the gradual addition of water, i.e., with an increase in w_0 the bulk viscosity of the solution gradually decreases. The rotational relaxation time of C-151 and C-153 also gradually decreases with an increase in w_0. We have plotted the average rotational relaxation time of the probes as a function of the bulk viscosity of the solution (Figure 8.11). Debye–Stokes–Einstein (DSE) hydrodynamic model [133] predicted a linear relationship between orientational relaxation time and the viscosity. In our case the rotational relaxation time vs viscosity plot shows a curvature and not the straight line as predicted by DSE model. Magee [134] modifies and brings nonlinearity in this model. According to Magee [134] macroscopic viscosity is useful to predict the relaxation times provided nonlinear relationship is acknowledged. The modified relation between macroscopic viscosity and orientational relaxation time is

$$\tau = \frac{\tau'\eta}{\eta + \eta'} \tag{8.6}$$

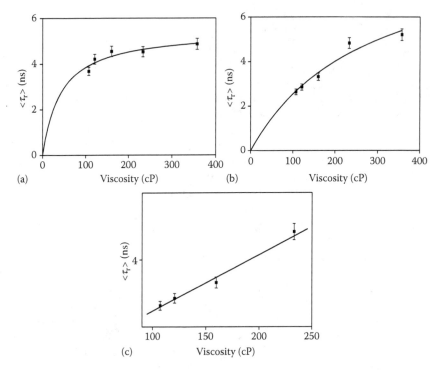

FIGURE 8.11 Plot of average rotational relaxation time vs bulk viscosity at (a) C-153, (b) C-151, and (c) C-151. (From Seth, D., Chakraborty, A., Setua, P., and Sarkar, N., *J. Chem. Phys.*, 126, 224512/1, 2007. With permission.)

where
η' is the limiting viscosity
τ' is the limiting relaxation time

When $\eta \ll \eta'$ the relaxation time varies linearly with bulk viscosity. When $\eta \gg \eta'$ the relaxation time is independent of bulk viscosity, i.e., independent of macroscopic viscosity, and this type of behavior is observed in highly viscous media. In the case of C-153 the values of τ' and η' are 5.4 ns and 42.3 cP, respectively, and limiting viscosity is always much lower than the macroscopic viscosity so we have observed nonlinearity in the plot of orientational relaxation time vs viscosity. In the case of C-151 the values of τ' and η' are 9.5 ns and 275 cP, respectively, i.e., limiting viscosity is higher than the macroscopic viscosity from $w_0 = 2$ to $w_0 = 10$. So we have observed linearity in the plot from $w_0 = 2$ to $w_0 = 10$ (Figure 8.11c), but when we include data of $w_0 = 0$ (viscosity = 357.5 cP) the overall plot shows a nonlinearity (Figure 8.11b). The similar type of nonlinear relation between the orientational relaxation time and viscosity was also observed by Chuang and Eisenthal [135].

8.4.3   [Bmim][PF$_6$]/Tween-20/Water ([Bmim][PF$_6$]-in-Water) Microemulsions

From steady-state result it is observed that C-153 is located in a position in the microemulsions, where the polarity sensed by the C-153 molecule is less than the bulk water. The average solvent relaxation time of C-153 in tween 20/[Bmim] [PF$_6$]/water-containing microemulsions at $R = 0.216$ is 1.18 ns. With a gradual increase in the amount of [Bmim][PF$_6$], i.e., with an increase in R the solvent relaxation time gradually increases. The solvent relaxation time of C-153 at $R = 1.08$ is 2.96 ns, i.e., a 2.5 times increase in the solvent relaxation time on going from $R = 0.216$ to $R = 1.08$. We also observed the solvent relaxation time of C-153 in tween 20/water mixture and it is 1.34 ns. In common reverse micelle or microemulsions with an increase in w ([polar solvent]/[surfactant]) the solvent relaxation time generally decreases [112–115]. It is due to the fact that with an increase in w value the dye molecules migrate toward the polar solvent pool of the microemulsions and also the size of the microemulsions increases, so that mobility of the solvent in the pool of the microemulsions increases, which leads to decrease in the solvent relaxation time. In some cases where probe molecule resides at the surface of the microemulsions, the solvent relaxation time is practically unaltered with change in w value [116,117]. This result, i.e., with an increase in R the solvent relaxation time increases, is somewhat the opposite trend that is observed in common microemulsions.

In water-containing micelles, the solvent relaxation time is retarded several times compared to pure water. The extended hydrogen bonding between water and surfactant molecules, counterions, and micellar head groups is responsible for slowing down the solvent relaxation compared to the neat water. Recent computer simulation studies [136–139] showed that the hydrogen bonding between water molecules and surfactant is much stronger than the hydrogen bond between two water molecules. These two types of water molecules are responsible for slow and fast component of the solvation relaxation in water-containing micelles.

Let us discuss about our present result of solvent relaxation time in the microemulsions. It is unusual that with an increase in R the solvent relaxation time increases, but we have observed it in our experiment. It may be possible that with the addition of [Bmim][PF$_6$] the microscopic viscosity sensed by the C-153 molecules increases. From steady-state spectral feature and rotational relaxation time of C-153 in microemulsions, it is expected that location of C-153 may be in the interface of the microemulsions. With the addition of [Bmim][PF$_6$] weightage of fast component gradually decreases and time constant of fast component remains almost the same (Table 8.9). Similarly, with the addition of [Bmim][PF$_6$] the time constant of slow components and their amplitude gradually increase. With the addition of R the number of [Bmim][PF$_6$] molecules in the core as well as in the surface of the microemulsions increases, i.e., more [Bmim][PF$_6$] molecules interact with the C-153 molecules. The solvent relaxation time of C-153 in neat [Bmim][PF$_6$] is ~3 ns [46,118]. So with an increase in [Bmim][PF$_6$] the solvent relaxation time of C-153 in microemulsions shows features of neat [Bmim] [PF$_6$] in the microenvironment of the microemulsions. Since the motion of ions are responsible for solvation in neat RTILs, with an increase in [Bmim][PF$_6$] concentration the motion of ions is hindered. Moreover, it is expected that with an increase in [Bmim][PF$_6$] concentration the microviscosity in the pool of the microemulsions increases. Since the slow component of solvent relaxation in RTILs is viscosity dependent, we have observed an increase in the time constant of the slow component of solvent relaxation in the microemulsions with an increase in R.

We have also measured the rotational relaxation time ($r(t)$) of C-153 in micro-emulsions at different R-value. It is observed that with an increase in R, i.e., with the addition of [Bmim][PF$_6$] the change in rotational relaxation time of C-153 is very small. The average rotational relaxation time of C-153 in tween 20/water mixture is 0.80 ns, whereas the average rotational relaxation time of C-153 in microemulsions at $R = 1.08$ is 0.92 ns (Table 8.13). This 120 ps increase in $r(t)$ is due to the addition of [Bmim][PF$_6$].

In the present study using the TCSPC setup, we are missing the fast compo-nent of the solvation dynamics (<90 ps). We can apply the method of Fee and Maroncelli [126] to calculate the missing component. We have calculated a time zero spectrum using the above procedure. The time zero frequency can be estimated using Equation 8.5 from the absorption and fluorescence spectra. The percentage of missing component is $(v_{cal}(0) - v(0))/(v_{cal}(0) - v(\infty))$. Using the above procedure we have observed that at $R = 0.216$ we are missing about 54% of the solvent relaxation dynamics. With an increase in R the missing component decreases. At $R = 1.08$ we are missing 41% of the solvent relaxation dynamics. So the missing component data also show a similar feature that with an increase in R the solvent relaxation of C-153 becomes more hindered.

8.4.4   [Bmim][BF$_4$]/TX-100/Cyclohexane Microemulsions

The emission spectra of C-153 in microemulsions are markedly different from the emission spectra in cyclohexane. The red shift in emission maximum indicates C-153 molecules have experienced more polarity compared to cyclohexane. It indicates that probe molecules are gradually encapsulated in [Bmim][BF$_4$] pool of the microemulsions. TX-100 can form microemulsions in cyclohexane [140,141]. Due to this reason we have observed a large red shift from cyclohexane to $w = 0$. Now as we add [Bmim][BF$_4$] in this system the polarity again increases and emission peak become further red shifted. But the observed emission peak at $w = 1.5$ (516 nm) is much less than emission peak in neat [Bmim][BF$_4$] (537 nm) [32]. It indicates that the polarity of the microemulsions is much lower compared to neat [Bmim][BF$_4$].

To comprehend the solvation dynamics results, a thorough understanding of the microemulsion structure and location of the probe within the microemulsions is necessary. Zhu et al. [140,141] characterized the water/TX-100/cyclohexane microemulsions. According to them TX-100 forms a nonspherical reverse micelles in cyclohexane, with cyclohexane penetrating the polar interior of the aggregates. Gao et al. [80] have showed that with the addition of [Bmim][BF$_4$] in TX-100/

cyclohexane the size of the droplets increases with w values similar to other microemulsions with water domains. DLS experiment shows the structure of the microemulsions is spherical and the hydrodynamic diameter increases from 10 to 94 nm as w varies from 0.2 to 1.5 (Figure 8.2).

The observed rotational relaxation time of C-153 in cyclohexane is 135 ps. In microemulsions the rotational relaxation time is bimodal in nature. The biexponential nature of rotational relaxation in TX-100/water reverse micelles has been reported [142,143]. Both rotational relaxation times in microemulsions are slower compared to cyclohexane. It strongly suggests that the probe molecules are residing at the core of the microemulsions. With an increase in w value the number of [Bmim][BF$_4$] molecules increases in the core of the microemulsions, thus microviscosity also increases. The average rotational relaxation time also increases due to the increase in the viscosity of the core due to the addition of highly viscous [Bmim][BF$_4$]. From the above discussion it is clear C-153 is located in the core of the [Bmim][BF$_4$]/TX-100 microemulsions.

The striking observation of this experiment is that we have observed slow solvation at $w = 0$ also. The observed solvation is bimodal with 3.91 ns as average solvation time. Zhu et al. [140,141] suggest that at $w = 0$ TX-100 forms reverse micelles in cyclohexane with cyclohexane penetrating the polar region. The observed slow solvation at $w = 0$ is possible due to the polymer motion. The cooperative motions arising from the local dynamics of the ethylene oxide parts of TX-100 may be responsible for the slow dynamics. The reorientation of the entire oligomer molecules may also be responsible for slow dynamics. Shirota et al. observed similar type of slow dynamics in different liquid poly(ethylene glycol)s [144,145].

At different w values of the TX-100 microemulsions we have observed slow bimodal solvation dynamics. The average solvation time varies from 8.06 to 6.84 ns as w is going from 0.2 to 1.5. Thus, the observed slow dynamics is not due to the polymer motion as observed at $w = 0$, rather it reflects the dynamics of the interior of the microemulsions. The diameters of the [Bmim][BF$_4$] pool of the microemulsions are 10, 47, and 94 nm at $w = 0.2$, 1, and 1.5, respectively. So the solvation time in the [Bmim][BF$_4$] pool of the microemulsions is almost independent of the [Bmim][BF$_4$] content or size of the microemulsions. The situation is completely different from AOT/water or TX-100/water reverse micelles, where solvation time decreases with an increase in w value [106–117]. The average solvation time of C-153 in neat [Bmim][BF$_4$] is 2.13 ns [32]. So the solvation dynamics is hindered in the pool of the microemulsions compared to neat [Bmim][BF$_4$]. But retardation is very small (only 4 times) compared to severalfold retardation of solvation dynamics of conventional solvents inside the core of the reverse micelles [106–117]. Recently, Chowdhury et al. [47] have observed that half or more of the solvation is completed within 100 ps in [Bmim][BF$_4$] using fluorescence up-conversion measurement. In comparison to this result solvent relaxation is much slower in the core of the RTIL-containing microemulsions.

Solvation in neat RTILs is possible due to the motion of the ions. It was verified earlier that ionic solvation is slow [121–123]. In neat RTILs, local motions of the ions are responsible for the fast component and collective diffusive motions are responsible for the slow component. In microemulsion the motions of the ions are also responsible for the slow solvation. Moreover, RTILs feel a restricted environment in the core of the microemulsions and both the local and collective diffusive motions are slower compared to neat RTILs. Hence solvation dynamics become retarded in the pool of the microemulsions compared to the neat RTILs.

The increase in [Bmim][BF$_4$] contents (due to increase in w values) increases the size of the microemulsions. Thus the local ion concentration remains almost constant near C-153. Thus the solvation time remains almost constant with an increase in w value. The slight decrease in solvation time with an increase in w value is due to the increase in size of the microemulsions. The increase in size leads to the increase in free motions of the ions and decreases the solvation time.

In the present study using TCSPC setup we are missing the fast component of the solvation dynamics (<90 ps). We can apply the method of Fee and Maroncelli [126] to calculate the missing component. We have calculated "time zero spectrum" using the above procedure. This indicates a total Stokes' shift of ~2372 cm^{-1} at $w = 0.2$. But we have observed 2007 cm^{-1} as total Stokes' shift. Thus we have missed ~15% of the total spectral shift. The missing components for all the w values are tabulated in Table 8.11. From Table 8.11 it is clear that missing components are increasing with an increase in w value. Karmakar and Samanta [32] reported that more than 50% of the total solvent response in neat [Bmim][BF$_4$] occurs in subpicosecond time scales. Chowdhury et al. [47] also observed that half or more than half of the solvent response completed within 100 ps. So, the percentage of missing components decreases in microemulsions compared to neat [Bmim][BF$_4$]. It suggests that the solvent relaxation of RTIL is much slower in microemulsions compared to neat RTIL.

8.5 CONCLUSION

The dynamics of solvent and rotational relaxation of C-153 and C-151 was studied in [Bmim][PF$_6$]/TX–100/H$_2$O ternary microemulsions, i.e., [Bmim][PF$_6$]-in-water microemulsions. In the case of C-153 with an increase in [Bmim][PF$_6$] content in the microemulsions change in the solvent relaxation time is small. The rotational relaxation time of C-153 also remains the same with increase in R. This is due to the fact that with an increase in [Bmim][PF$_6$] content in the microemulsions the position of C-153 remains the same, and this suggests that C-153 resides at the interface of these microemulsions. In the case of C-151, with an increase in R the slow component of the solvation time gradually decreases, the fast component gradually increases, and the rotational relaxation time gradually increases. Therefore, with an increase in R the C-151 molecules gradually shifted to the core of the microemulsions, and in the case of RTILs-containing microemulsions the solvation time is very much dependent on the probe location.

The interaction of water with [Bmim][PF$_6$] in [Bmim][PF$_6$]/TX-100/water, i.e., water-in-[Bmim][PF$_6$] ternary microemulsion, has been probed by the solvent and rotational relaxation of C-153 and C-151. There is a monotonic decrease of both solvation and rotational relaxation times with increase in water content of the microemulsions. With increase in water content of the microemulsions the probe molecules are gradually shifted to the core of the microemulsions and experience the free water-like environment, causing the decrease in the solvent and rotational relaxation time. Both the average solvation and rotational relaxation times have nonlinear relation with bulk viscosity. The decrease in solvent and rotational relaxation time in [Bmim][PF$_6$]/TX-100/water ternary system is similar to that of [Bmim][PF$_6$]–water binary system studied by us earlier [129,130]. But in this present work the decrease in solvation and rotational relaxation times with the gradual addition of water is an effect of water dynamics inside the ternary microemulsions. The retardation of solvation time of water in the core of the microemulsion is several thousand times compared to pure water whereas in the case of RTILs the retardation of solvation time is only 2–3 times compared to neat RTILs.

The solvent relaxation time of C-153 in tween 20/[Bmim][PF$_6$]/water containing, i.e., [Bmim][PF$_6$]-in-water microemulsions shows an unusual feature. With increase in R (R = [Bmim][PF$_6$]-to-tween 20 molar ratio) the solvent relaxation time gradually increases. Steady-state result and rotational relaxation time of C-153 in different microemulsions show that C-153 is located at the surface of the microemulsions. With increase in [Bmim][PF$_6$] content in the microemulsions the microviscosity sensed by the C-153 molecules in the core of the microemulsions increases and also the motions of ions of [Bmim][PF$_6$] become hindered. So we have observed an increase in the solvent relaxation time of C-153 with increase in R.

The gradual red shift in emission spectra of C-153 with increase in [Bmim][BF$_4$] content in the [Bmim][BF$_4$]/TX-100/cyclohexane microemulsions indicates C-153 molecules may reside in the core of the microemulsions. The bimodal rotational relaxation time also supports this conjecture. The average rotational relaxation time increases with increase in w values due to the addition of highly viscous [Bmim][BF$_4$]. We have also observed slow solvation dynamics at $w = 0$ (3.91 ns). The observed slow dynamics may be due to the polymer motions and the segmental dynamics of ethylene oxide chain. The observed solvation time varies from 8.06 to 6.84 ns as w is increasing from 0.2 to 1.5. The solvation time inside the core of microemulsions is retarded (~4 times) compared to solvation time in neat [Bmim][BF$_4$]. Although the retardation is very small compared to several-fold retardation of solvation time in conventional solvent inside reverse micelles. Ionic motions are also responsible for solvation inside microemulsions as in neat RTILs. The microemulsions impose restriction in the ionic motions and retarded the solvation time.

ACKNOWLEDGMENTS

Nilmoni Sarkar is indebted to the Department of Science and Technology (DST), and Council of Scientific and Industrial Research (CSIR), Government of India, for generous research grants. Debabrata Seth is thankful to CSIR for research fellowships. Scheme 8.1 is reprinted from Refs. [95,96]. All figures are reprinted from Refs. [94–96] with kind permission from the American Chemical Society (ACS), the United States and reused with permission from Debabrata Seth, Anjan Chakraborty, Palash Setua, and Nilmoni Sarkar, *Journal of Chemical Physics*, 126, 224512 (2007). Copyright 2007, American Institute of Physics.

SYMBOLS AND TERMINOLOGIES

RTILs	room temperature ionic liquids
[Bmim][PF$_6$]	1-butyl-3-methylimidazolium hexafluorophosphate
[Bmim][BF$_4$]	1-butyl-3-methylimidazolium tetrafluoroborate
TX-100	triton X-100
VOCs	volatile organic compounds
AOT	dioctylsulfosuccinate, sodium salt
CTAB	cetyltrimethyl-ammonium bromide
SDS	sodium dodecyl sulfate
DMF	*N,N*-dimethyl formamide
TCSPC	time correlated single photon counting
FFEM	freeze-fracture electron microscopy
DLS	dynamic light scattering
SANS	small-angle neutron scattering
TEM	transmission electron microscope
w	(water or other polar solvents)/(surfactant)
R	[Bmim][PF$_6$]/TX-100 molar ratio
TRES	time-resolved emission spectra
DSE	Debye–Stokes–Einstein
TICT	twisted intramolecular charge transfer

REFERENCES

1. Seddon, K. R. 2003. Ionic liquids: A taste of the future. *Nat. Mater.* 2, 363–365.
2. Rogers, R. D. and Seddon, K. R. 2003. Ionic liquids—solvents of the future? *Science* 302, 792–793.
3. Sheldon, R. 2001. Catalytic reactions in ionic liquids. *Chem. Commun.* 2399–2407.
4. Parvulescu, V. I. and Hardacre, C. 2007. Catalysis in ionic liquids. *Chem. Rev.* 107, 2615–2665.
5. Seoud, O. A. E., Koschella, A., Fidale, L. C., Dorn, S., and Heinze, T. 2007. Applications of ionic liquids in carbohydrate chemistry: A window of opportunities. *Biomacromolecules* 8, 2629–2647.
6. Chowdhury, S., Mohan, R. S., and Scott, J. L. 2007. Reactivity of ionic liquids. *Tetrahedron* 63, 2363–2389.
7. Plechkova, N. V. and Seddon, K. R. 2008. Applications of ionic liquids in the chemical industry. *Chem. Soc. Rev.* 37, 123–150.
8. Welton, T. 1999. Room-temperature ionic liquids. Solvents for synthesis and catalysis. *Chem. Rev.* 99, 2071–2084.
9. Dupont, J., de Souza, R. F., and Suarez, P. A. Z. 2002. Ionic liquid (molten salt) phase organometallic catalysis. *Chem. Rev.* 102, 3667–3692.
10. Welton, T. 2004. Ionic liquids in catalysis. *Coord. Chem. Rev.* 248, 2459–2477.

11. Baker, G. A., Baker, S. N., Pandey, S., and Bright, F. V. 2005. An analytical view of ionic liquids. *Analyst* 130, 800–808.
12. Ray, P. C. and Rakshit, J. N. 1911. Nitrites of the alkylammonium bases: Ethylammonium nitrite, dimethylammonium nitrite, and trimethylammonium nitrite. *J. Chem. Soc.* 1470–1475.
13. Walden, P. 1914. *Bull. Acad. Imper. Sci.* (St. Petersburg), 1800.
14. Swatloski, R. P., Holbrey, J. D., and Rogers, R. D. 2003. Ionic liquids are not always green: Hydrolysis of 1-butyl-3-methylimidazolium hexafluorophosphate. *Green Chem.* 5, 361–363.
15. Baker, G. A. and Baker, S. N. 2005. A simple colorimetric assay of ionic liquid hydrolytic stability. *Aust. J. Chem.* 58, 174–177.
16. Earle, M. J., Esperanca, J. M. S. S., Gilea, M. A., Lopes, J. N. C., Rebelo, L. P. N., Magee, J. W., Seddon, K. R., and Widegren, J. A. 2006. The distillation and volatility of ionic liquids. *Nature* 439, 831–834.
17. Wasserscheid, P. 2006. Volatile times for ionic liquids. *Nature* 439, 797–797.
18. Ludwig, R. and Kragl, U. 2007. Do we understand the volatility of ionic liquids? *Angew. Chem. Int. Ed.* 46, 6582–6584.
19. Smiglak, M., Reichert, W. M., Holbrey, J. D., Wilkes, J. S., Sun, L., Thrasher, J. S., Kirichenko, K., Singh, S., Katritzky, A. R., and Rogers, R. D. 2006. Combustible ionic liquids by design: Is laboratory safety another ionic liquid myth? *Chem. Commun.* 2554–2556.
20. Aki, S. N. V. K., Brennecke, J. F., and Samanta, A. 2001. How polar are room-temperature ionic liquids? *Chem. Commun.* 413–414.
21. Muldoon, M. J., Gordon, C. M., and Dunkin, I. R. 2001. Investigations of solvent–solute interactions in room temperature ionic liquids using solvatochromic dyes. *J. Chem. Soc. Perkin Trans.* 2, 433–435.
22. Carmichael, A. J. and Seddon, K. R. 2000. Polarity study of some 1-alkyl-3-methylimidazolium ambient-temperature ionic liquids with the solvatochromic dye, Nile Red. *J. Phys. Org. Chem.* 13, 591–595.
23. Fletcher, K. A. and Pandey, S. 2003. Solvatochromic probe behavior within ternary room-temperature ionic liquid 1-butyl-3-methylimidazolium hexafluorophosphate + ethanol + water solutions. *J. Phys. Chem. B* 107, 13532–13539.
24. Fletcher, K. A., Baker, S. N., Baker, G. A., and Pandey, S. 2003. Probing solute and solvent interactions within binary ionic liquid mixtures. *New J. Chem.* 27, 1706–1712.
25. Baker, S. N., Baker, G. A., and Bright, F. V. 2002. Temperature-dependent microscopic solvent properties of 'dry' and 'wet' 1-butyl-3-methylimidazolium hexafluorophosphate: Correlation with $E_T(30)$ and Kamlet–Taft polarity scales. *Green Chem.* 4, 165–169.
26. Pandey, S., Fletcher, K. A., Baker, S. N., and Baker, G. A. 2004. Correlation between the fluorescent response of microfluidity probes and the water content and viscosity of ionic liquid and water mixtures. *Analyst* 129, 569–573.
27. Reichardt, C. 2005. Polarity of ionic liquids determined empirically by means of solvatochromic pyridinium *N*-phenolate betaine dyes. *Green Chem.* 7, 339–351.
28. Hyun, B. -R., Dzyuba, S. V., Bartsch, R. A., and Quitevis, E. L. 2002. Intermolecular dynamics of room-temperature ionic liquids: Femtosecond optical Kerr effect measurements on 1-alkyl-3-methylimidazolium bis((trifluoromethyl)sulfonyl) imides. *J. Phys. Chem. A* 106, 7579–7585.
29. Xiao, D., Rajian, J. R., Cady, A., Li, S., Bartsch, R. A., and Quitevis, E. L. 2007. Nanostructural organization and anion effects on the temperature dependence of the optical Kerr effect spectra of ionic liquids. *J. Phys. Chem. B* 111, 4669–4677.

30. Chakrabarty, D., Chakraborty, A., Hazra, P., Seth, D., and Sarkar, N. 2004. Dynamics of photoisomerisation and rotational relaxation of 3,3′-diethyloxadicarbocyanine iodide in room temperature ionic liquid and binary mixture of ionic liquid and water. *Chem. Phys. Lett.* 397, 216–221.

31. Karmakar, R. and Samanta, A. 2003. Intramolecular excimer formation kinetics in room temperature ionic liquids. *Chem. Phys. Lett.* 376, 638–645.

32. Karmakar, R. and Samanta, A. 2002. Solvation dynamics of Coumarin-153 in a room-temperature ionic liquid. *J. Phys. Chem. A* 106, 4447–4452.

33. Karmakar, R. and Samanta, A. 2002. Steady-state and time-resolved fluorescence behavior of C153 and Prodan in room-temperature ionic liquids. *J. Phys. Chem. A* 106, 6670–6675.

34. Karmakar, R. and Samanta, A. 2003. Dynamics of solvation of the fluorescent state of some electron donor-acceptor molecules in room temperature ionic liquids, [BMIM] [(CF$_3$SO$_2$)$_2$N] and [EMIM][(CF$_3$SO$_2$)$_2$N]. *J. Phys. Chem. A* 107, 7340–7346.

35. Saha, S., Mandal, P. K., and Samanta, A. 2004. Solvation dynamics of Nile red in a room temperature ionic liquid using streak camera. *Phys. Chem. Chem. Phys.* 6, 3106–3110.

36. Mandal, P. K., and Samanta, A. 2005. Fluorescence studies in a pyrrolidinium ionic liquid: Polarity of the medium and solvation dynamics. *J. Phys. Chem. B* 109, 15172–15177.

37. Samanta, A. 2006. Dynamic Stokes' shift and excitation wavelength dependent fluorescence of dipolar molecules in room temperature ionic liquids. *J. Phys. Chem. B* 110, 13704–13716.

38. Paul, A. and Samanta, A. 2007. Solute rotation and solvation dynamics in an alcohol-functionalized room temperature ionic liquid. *J. Phys. Chem. B* 111, 4724–4731.

39. Ingram, J. A., Moog, R. S., Ito, N., Biswas, R., and Maroncelli, M. 2003. Solute rotation and solvation dynamics in a room-temperature ionic liquid. *J. Phys. Chem. B* 107, 5926–5932.

40. Arzhantsev, S., Ito, N., Heitz, M., and Maroncelli, M. 2003. Solvation dynamics of Coumarin 153 in several classes of ionic liquids: Cation dependence of the ultrafast component. *Chem. Phys. Lett.* 381, 278–286.

41. Ito, N., Arzhantsev, S., Heitz, M., and Maroncelli, M. 2004. Solvation dynamics and rotation of Coumarin 153 in alkylphosphonium ionic liquids. *J. Phys. Chem. B* 108, 5771–5777.

42. Ito, N., Arzhantsev, S., and Maroncelli, M. 2004. The probe dependence of solvation dynamics and rotation in the ionic liquid 1-butyl-3-methyl-imidazolium hexafluorophosphate. *Chem. Phys. Lett.* 396, 83–91.

43. Arzhantsev, S., Jin, H., Ito, N., and Maroncelli, M. 2006. Observing the complete solvation response of DCS in imidazolium ionic liquids, from the femtosecond to nanosecond regimes. *Chem. Phys. Lett.* 417, 524–529.

44. Arzhantsev, S., Jin, H., Baker, G. A., and Maroncelli, M. 2007. Measurements of the complete solvation response in ionic liquids. *J. Phys. Chem. B* 111, 4978–4989.

45. Jin, H., Baker, G. A., Arzhantsev, S., Dong, J., and Maroncelli, M. 2007. Solvation and rotational dynamics of Coumarin 153 in ionic liquids: Comparisons to conventional solvents. *J. Phys. Chem. B* 111, 7291–7302.

46. Chakrabarty, D., Hazra, P., Chakraborty, A., Seth, D., and Sarkar, N. 2003. Dynamics of solvent relaxation in room temperature ionic liquids. *Chem. Phys. Lett.* 381, 697–704.

47. Chowdhury, P. K., Halder, M., Sanders, L., Calhoun, T., Anderson, J. L., Armstrong, D. W., Song, X., and Petrich, J. W. 2004. Dynamic solvation in room-temperature ionic liquids. *J. Phys. Chem. B* 108, 10245–10255.

48. Headley, L. S., Mukherjee, P., Anderson, J. L., Ding, R., Halder, M., Armstrong, D. W., Song, X., and Petrich, P. W. 2006. Dynamic solvation in imidazolium-based ionic liquids on short time scales. *J. Phys. Chem. A* 110, 9549–9554.
49. Lang, B., Angulo, G., and Vauthey, E. 2006. Ultrafast solvation dynamics of Coumarin 153 in imidazolium-based ionic liquids. *J. Phys. Chem. A* 110, 7028–7034.
50. Shim, Y., Duan, J., Choi, M. Y., and Kim, H. J. 2003. Solvation in molecular ionic liquids. *J. Chem. Phys.* 119, 6411–6414.
51. Shim, Y., Choi, M. Y., and Kim, H. J. 2005. A molecular dynamics computer simulation study of room-temperature ionic liquids. I. Equilibrium solvation structure and free energetics. *J. Chem. Phys.* 122, 44510/1–44510/12.
52. Shim, Y., Choi, M. Y., and Kim, H. J. 2005. A molecular dynamics computer simulation study of room-temperature ionic liquids. II. Equilibrium and nonequilibrium solvation dynamics. *J. Chem. Phys.* 122, 44511/1–44511/12.
53. Shim, Y., Jeong, D., Manjari, S., Choi, M. Y., and Kim, H. J. 2007. Solvation, solute rotation and vibration relaxation, and electron-transfer reactions in room-temperature ionic liquids. *Acc. Chem. Res.* 40, 1130–1137.
54. Kobrak, M. N. and Znamenskiy, V. 2004. Solvation dynamics of room-temperature ionic liquids: Evidence for collective solvent motion on sub-picosecond timescales. *Chem. Phys. Lett.* 395, 127–132.
55. Kobrak, M. N. 2006. Characterization of the solvation dynamics of an ionic liquid via molecular dynamics simulation. *J. Chem. Phys.* 125, 64502/1–64502/11.
56. Mandal, P. K., Sarkar, M., and Samanta, A. 2004. Excitation-wavelength-dependent fluorescence behavior of some dipolar molecules in room-temperature ionic liquids. *J. Phys. Chem. A* 108, 9048–9053.
57. Mandal, P. K., Paul, A., and Samanta, A. 2006. Excitation wavelength dependent fluorescence behavior of the room temperature ionic liquids and dissolved dipolar solutes. *J. Photochem. Photobio. A* 182, 113–120.
58. Hu, Z. and Margulis, C. J. 2006. Heterogeneity in a room-temperature ionic liquid: Persistent local environments and the red-edge effect. *Proc. Natl. Acad. Sci. U S A* 103, 831–836.
59. Hu, Z. and Margulis, C. J. 2006. A study of the time-resolved fluorescence spectrum and red edge effect of ANF in a room-temperature ionic liquid. *J. Phys. Chem. B* 110, 11025–11028.
60. Hu, Z. and Margulis, C. J. 2007. Room-temperature ionic liquids: Slow dynamics, viscosity, and the red edge effect. *Acc. Chem. Res.* 40, 1097–1105.
61. Paul, A., Mandal, P. K., and Samanta, A. 2005. On the optical properties of the imidazolium ionic liquids. *J. Phys. Chem. B* 109, 9148–9153.
62. Paul, A., Mandal, P. K., and Samanta, A. 2005. How transparent are the imidazolium ionic liquids? A case study with 1-methyl-3-butylimidazolium hexafluorophosphate, [bmim][PF$_6$]. *Chem. Phys. Lett.* 402, 375–379.
63. Shirota, H., Funston, A. M., Wishart, J. F., and Castner, E. W. Jr. 2005. Ultrafast dynamics of pyrrolidinium cation ionic liquids. *J. Chem. Phys.* 122, 184512/1–184512/12.
64. Sando, G. M., Dahl, K., and Owrutsky, J. C. 2006. Vibrational relaxation in ionic liquids and ionic liquid reverse micelles. *Chem. Phys. Lett.* 418, 402–407.
65. Shirota, H. and Castner, E. W. Jr. 2005. Why are viscosities lower for ionic liquids with –CH$_2$Si(CH$_3$)$_3$ vs –CH$_2$C(CH$_3$)$_3$ substitutions on the imidazolium cations? *J. Phys. Chem. B* 109, 21576–21585.
66. Shirota, H., Wishart, J. F., and Castner, E. W. Jr. 2007. Intermolecular interactions and dynamics of room temperature ionic liquids that have silyl- and siloxy-substituted imidazolium cations. *J. Phys. Chem. B* 111, 4819–4829.

67. Brands, H., Chandrasekhar, N., and Unterreiner, A. -N. 2007. Ultrafast dynamics of room temperature ionic liquids after ultraviolet femtosecond excitation. *J. Phys. Chem. B* 111, 4830–4836.

68. Iwata, K., Kakita, M., and Hamaguchi, H. 2007. Picosecond time-resolved fluorescence study on solute–solvent interaction of 2-aminoquinoline in room-temperature ionic liquids: Aromaticity of imidazolium-based ionic liquids. *J. Phys. Chem. B* 111, 4914–4919.

69. Vieira, R. C. and Falvey, D. E. 2007. Photoinduced electron-transfer reactions in two room-temperature ionic liquids: 1-Butyl-3-methylimidazolium hexafluorophosphate and 1-octyl-3-methylimidazolium hexafluorophosphate. *J. Phys. Chem. B* 111, 5023–5029.

70. Luisi, P. L. and Straube, B. E. (Eds.). 1984. *Reverse Micelles.* New York, Plenum Press.

71. Riter, R. E., Undiks, E. P., Kimmel, J. R., and Levinger, N. E. 1998. Formamide in reverse micelles: Restricted environment effects on molecular motion. *J. Phys. Chem. B* 102, 7931–7938.

72. Riter, R. E., Kimmel, J. R., Undiks, E. P., and Levinger, N. E. 1997. Novel reverse micelles partitioning nonaqueous polar solvents in a hydrocarbon continuous phase. *J. Phys. Chem. B* 101, 8292–8297.

73. Anderson, J. L., Pino, V., Hagberg, E. C., Shears, V. V., and Armstrong, D. W. 2003. Surfactant solvation effects and micelle formation in ionic liquids. *Chem. Commun.* 2444–2445.

74. Fletcher, K. A. and Pandey, S. 2004. Surfactant aggregation within room-temperature ionic liquid 1-ethyl-3-methylimidazolium bis(trifluoromethylsulfonyl)imide. *Langmuir* 20, 33–36.

75. Tang, J., Li, D., Sun, C., Zheng, L., and Li, J. 2006. Temperature dependant self-assembly of surfactant brij 76 in room temperature ionic liquid. *Colloids Surf. A* 273, 24–28.

76. Patrascu, C., Gauffre, F., Nallet, F., Bordes, R., Oberdisse, J., de Lauth-Viguerie, N., and Mingotaud, C. 2006. Micelles in ionic liquids: Aggregation behavior of alkyl poly(ethyleneglycol)-ethers in 1-butyl-3-methyl-imidazolium type ionic liquids. *Chem. Phys. Chem.* 7, 99–101.

77. He, Y., Li, Z., Simone, P., and Lodge, T. P. 2006. Self-assembly of block copolymer micelles in an ionic liquid. *J. Am. Chem. Soc.* 128, 2745–2750.

78. Triolo, A., Russina, O., Keiderling, U., and Kohlbrecher, J. 2006. Morphology of poly(ethylene oxide) dissolved in a room temperature ionic liquid: A small angle neutron scattering study. *J. Phys. Chem. B* 110, 1513–1515.

79. Zheng, L., Guo, C., Wang, J., Liang, X., Chen, S., Ma, J., Yang, B., Jiang, Y., and Liu, H. 2007. Effect of ionic liquids on the aggregation behavior of PEO–PPO–PEO block copolymers in aqueous solution. *J. Phys. Chem. B* 111, 1327–1333.

80. Gao, H., Li, J., Han, B., Chen, W., Zhang, J., Zhang, R., and Yan, D. 2004. Microemulsions with ionic liquid polar domains. *Phys. Chem. Chem. Phys.* 6, 2914–2916.

81. Eastoe, J., Gold, S., Rogers, S. E., Paul, A., Welton, T., Heenan, R. K., and Grillo, I. 2005 Ionic liquid-in-oil microemulsions. *J. Am. Chem. Soc.* 127, 7302–7303.

82. Li, J., Zhang, J., Gao, H., Han, B., & Gao, L. 2005. Nonaqueous microemulsion-containing ionic liquid [bmim][PF$_6$] as polar microenvironment. *Colloid Polym. Sci.* 283, 1371–1375.

83. Gao, Y., Zhang, J., Xu, H., Zhao, X., Zheng, L., Li, X., and Yu, L. 2006. Structural studies of 1-butyl-3-methylimidazolium tetrafluoroborate/TX-100/ p-xylene ionic liquid microemulsions. *Chem. Phys. Chem.* 7, 1554–1561.

84. Li, N., Gao, Y., Zheng, L., Zhang, J., Yu, L., and Li, X. 2007. Studies on the micropolarities of bmimBF$_4$/TX-100/toluene ionic liquid microemulsions and their behaviors characterized by UV-visible spectroscopy. *Langmuir* 23, 1091–1097.

85. Gao, Y., Han, S., Han, B., Li, G., Shen, D., Li, Z., Du, J., Hou, W., and Zhang, G. 2005. TX-100/water/1-butyl-3-methylimidazolium hexafluorophosphate microemulsions. *Langmuir* 21, 5681–5684.

86. Gao, Y., Li, N., Zheng, L., Zhao, X., Zhang, S., Han, B., Hou, W., and Li, G. 2006. A cyclic voltammetric technique for the detection of micro-regions of bmimPF$_6$/tween 20/H$_2$O microemulsions and their performance characterization by UV-Vis spectroscopy. *Green Chem.* 8, 43–49.

87. Li, J., Zhang, J., Han, B., Gao, Y., Shen, D., and Wu, Z. 2006. Effect of ionic liquid on the polarity and size of the reverse micelles in supercritical CO$_2$. *Colloids Surf. A* 279, 208–212.

88. Cheng, S., Zhang, J., Zhang, Z., and Han, B. 2007. Novel microemulsions: Ionic liquid-in-ionic liquid. *Chem. Commun.* 2497–2499.

89. Liu, J., Cheng, S., Zhang, J., Feng, X., Fu, X., and Han, B. 2007. Reverse micelles in carbon dioxide with ionic-liquid domains. *Angew. Chem. Int. Ed.* 46, 3313–3315.

90. Feng, Y. and John T. 2006. Surfactant ionic liquid-based microemulsions for polymerization. *Chem. Commun.* 2696–2698.

91. Li, Z., Zhang, J., Du, J., Han, B., and Wang, J. 2006. Preparation of silica microrods with nano-sized pores in ionic liquid microemulsions. *Colloids Surf. A* 286, 117–120.

92. Gao, Y., Li, N., Zheng, L., Bai, X., Yu, L., Zhao, X., Zhang, J., Zhao, M., and Li, Z. 2007 Role of solubilized water in the reverse ionic liquid microemulsion of 1-butyl-3-methylimidazolium tetrafluoroborate/TX-100/benzene. *J. Phys. Chem. B* 111, 2506–2513.

93. Li, N., Cao, Q., Gao, Y., Zhang, J., Zheng, L., Bai, X., Dong, B., Li, Z., Zhao, M., and Yu, L. 2007. States of water located in the continuous organic phase of 1-butyl-3-methylimidazolium tetrafluoroborate/triton X-100/triethylamine reverse microemulsions. *Chem. Phys. Chem.* 8, 2211–2217.

94. Chakrabarty, D., Seth, D., Chakraborty, A., and Sarkar, N. 2005. Dynamics of solvation and rotational relaxation of Coumarin 153 in ionic liquid confined nanometer-sized microemulsions. *J. Phys. Chem. B* 109, 5753–5758.

95. Seth, D., Chakraborty, A., Setua, P., and Sarkar, N. 2006. Interaction of ionic liquid with water in ternary microemulsions (triton X-100/water/1-butyl-3-methylimidazolium hexafluorophosphate) probed by solvent and rotational relaxation of Coumarin 153 and Coumarin 151. *Langmuir* 22, 7768–7775.

96. Seth, D., Chakraborty, A., Setua, P., and Sarkar, N. 2007. Interaction of ionic liquid with water with variation of water content in 1-butyl-3-methyl-imidazolium hexafluorophosphate ([bmim][PF$_6$])/TX-100/water ternary microemulsions monitored by solvent and rotational relaxation of Coumarin 153 and Coumarin 490. *J. Chem. Phys.* 126, 224512/1–224512/12.

97. Seth, D., Setua, P., Chakraborty, A., and Sarkar, N. 2007. Solvent relaxation of a room-temperature ionic liquid [bmim][PF$_6$] confined in a ternary microemulsion. *J. Chem. Sci.* 119, 105–111.

98. Adhikari, A., Sahu, K., Dey, S., Ghosh, S., Mandal, U., and Bhattacharyya, K. 2007. Femtosecond solvation dynamics in a neat ionic liquid and ionic liquid microemulsion: Excitation wavelength dependence. *J. Phys. Chem. B* 111, 12809–12816.

99. Jimenez, R., Fleming, G. R., Kumar, P. V., and Maroncelli, M. 1994. Femtosecond solvation dynamics of water. *Nature* 369, 471–473.

100. Vajda, S., Jimenez, R., Rosenthal, S. J., Fidler, V., Fleming, G. R., and Castner, E. W. Jr. 1995. Femtosecond to nanosecond solvation dynamics in pure water and inside the γ-cyclodextrin cavity. *J. Chem. Soc. Faraday Trans.* 91, 867–873.
101. Kahlow, M. A., Kang, T. J., and Barbara, P. F. 1988. Transient solvation of polar dye molecules in polar aprotic solvents. *J. Chem. Phys.* 88, 2372–2378.
102. Gustavsson, T., Cassara, L., Gulbinas, V., Gurzadyan, G., Mialocq, J. -C., Pommeret, S., Sorgius, M., and van der Meulen, P. 1998. Femtosecond spectroscopic study of relaxation processes of three amino-substituted Coumarin dyes in methanol and dimethyl sulfoxide. *J. Phys. Chem. A* 102, 4229–4245.
103. Alfassi, Z. B., Huie, R. E., Milman, B. L., and Neta, P. 2003. Electrospray ionization mass spectrometry of ionic liquids and determination of their solubility in water *Anal. Bioanal. Chem.* 377, 159–164.
104. Shvedene, N. V., Borovskaya, S. V., Sviridov, V. V., Ismailova, E. R., and Pletnev, I. V. 2005. Measuring the solubilities of ionic liquids in water using ion-selective electrodes. *Anal. Bioanal. Chem.* 381, 427–430.
105. Maroncelli, M. and Fleming, G. R. 1987. Picosecond solvation dynamics of Coumarin 153: The importance of molecular aspects of solvation. *J. Chem. Phys.* 86, 6221–6239.
106. Nandi, N., Bhattacharyya, K., and Bagchi, B. 2000. Dielectric relaxation and solvation dynamics of water in complex chemical and biological systems. *Chem. Rev.* 100, 2013–2046.
107. Bhattacharyya, K. and Bagchi, B. 2000. Slow dynamics of constrained water in complex geometries. *J. Phys. Chem. A* 104, 10603–10613.
108. Sarkar, N., Das, K., Datta, A., Das, S., and Bhattacharyya, K. 1996. Solvation dynamics of Coumarin 480 in reverse micelles. Slow relaxation of water molecules. *J. Phys. Chem.* 100, 10523–10527.
109. Das, S., Datta, A., and Bhattacharyya, K. 1997. Deuterium isotope effect on 4-aminophthalimide in neat water and reverse micelles. *J. Phys. Chem. A* 101, 3299–3304.
110. Pal, S. K., Mandal, D., Sukul, D., and Bhattacharyya, K. 1999. Solvation dynamics of 4-(dicyanomethylene)-2-methyl-6-(p-dimethylaminostyryl)-4H-pyran (DCM) in a microemulsion. *Chem. Phys. Lett.* 312, 178–184.
111. Lundgren, J. S., Heitz, M. P., and Bright, F. V. 1995. Dynamics of acrylodan-labeled bovine and human serum albumin sequestered within Aerosol-OT reverse micelles. *Anal. Chem.* 67, 3775–3781.
112. Hazra, P. and Sarkar, N. 2002. Solvation dynamics of Coumarin 490 in methanol and acetonitrile reverse micelles. *Phys. Chem. Chem. Phys.* 4, 1040–1045.
113. Hazra, P., Chakrabarty, D., and Sarkar, N. 2002. Intramolecular charge transfer and solvation dynamics of Coumarin 152 in Aerosol-OT, water-solubilizing reverse micelles, and polar organic solvent solubilizing reverse micelles. *Langmuir* 18, 7872–7879.
114. Hazra, P., Chakrabarty, D., and Sarkar, N. 2003. Solvation dynamics of Coumarin 153 in aqueous and non-aqueous reverse micelles. *Chem. Phys. Lett.* 371, 553–562.
115. Hazra, P. and Sarkar, N. 2001. Intramolecular charge transfer processes and solvation dynamics of Coumarin 490 in reverse micelles. *Chem. Phys. Lett.* 342, 303–311.
116. Corbeil, E. M. and Levinger, N. E. 2003. Dynamics of polar solvation in quaternary microemulsions. *Langmuir* 19, 7264–7270.
117. Hazra, P., Chakrabarty, D., Chakraborty, A., and Sarkar, N. 2003. Solvation dynamics of Coumarin 480 in neutral (TX-100), anionic (SDS), and cationic (CTAB) water-in-oil microemulsions. *Chem. Phys. Lett.* 382, 71–80.

118. Chakraborty, A., Seth, D., Chakrabarty, D., Setua, P., and Sarkar, N. 2005. Dynamics of solvent and rotational relaxation of Coumarin 153 in room-temperature ionic liquid 1-butyl-3-methylimidazolium hexafluorophosphate confined in brij-35 micelles: A picosecond time-resolved fluorescence spectroscopic study. *J. Phys. Chem. A* 109, 11110–11116.

119. Seddon, K. R., Stark, A., and Torres, M. -J. 2000. Influence of chloride, water, and organic solvents on the physical properties of ionic liquids. *Pure. Appl. Chem.* 72, 2275–2287.

120. Shirota, H. and Segawa, H. 2004. Solvation dynamics of formamide and N,N-dimethylformamide in Aerosol OT reverse micelles. *Langmuir* 20, 329–335.

121. Chapman, C. F. and Maroncelli, M. 1991. Fluorescence studies of solvation and solvation dynamics in ionic solutions. *J. Phys. Chem.* 95, 9095–9114.

122. Bart, E., Meltsin, A., and Huppert, D. 1994. Solvation dynamics in molten salts. *J. Phys. Chem.* 98, 10819–10823.

123. Bart, E., Meltsin, A., and Huppert, D. 1994. Solvation dynamics of Coumarin 153 in molten salts. *J. Phys. Chem.* 98, 3295–3299.

124. Cave, R. J., Burke, K., and Castner, E. W. Jr. 2002. Theoretical investigation of the ground and excited states of Coumarin 151 and Coumarin 120. *J. Phys. Chem. A* 106, 9294–9305.

125. Sulpizi, M., Carloni, P., Hutter, J., and Rothlisberger, U. 2003. A hybrid TDDFT/MM investigation of the optical properties of aminocoumarins in water and acetonitrile solution. *Phys. Chem. Chem. Phys.* 5, 4798–4805.

126. Fee, R. S. and Maroncelli, M. 1994. Estimating the time-zero spectrum in time-resolved emission measurements of solvation dynamics. *Chem. Phys.* 183, 235–247.

127. Cammarata, L., Kazarian, S. G., Salter, P. A., and Welton, T. 2001. Molecular states of water in room temperature ionic liquids. *Phys. Chem. Chem. Phys.* 3, 5192–5200.

128. Baker, S. N., Baker, G. A., Munson, C. A., Chen, F., Bukowski, E. J., Cartwright, A. N., and Bright, F. V. 2003. Effects of solubilized water on the relaxation dynamics surrounding 6-propionyl-2-(N,N-dimethylamino)naphthalene dissolved in 1-butyl-3-methylimidazolium hexafluorophosphate at 298 K. *Ind. Eng. Chem. Res.* 42, 6457–6463.

129. Chakrabarty, D., Chakraborty, A., Seth, D., Hazra, P., and Sarkar, N. 2004. Dynamics of solvation and rotational relaxation of Coumarin 153 in 1-butyl-3-methylimidazolium hexafluorophosphate [bmim][PF_6]–water mixtures. *Chem. Phys. Lett.* 397, 469–474.

130. Chakrabarty, D., Chakraborty, A., Seth, D., and Sarkar, N. 2005. Effect of water, methanol, and acetonitrile on solvent relaxation and rotational relaxation of Coumarin 153 in neat 1-hexyl-3-methylimidazolium hexafluorophosphate. *J. Phys. Chem. A* 109, 1764–1769.

131. Margulis, C. J., Stern, H. A., and Berne, B. J. 2002. Computer simulation of a "Green Chemistry" room-temperature ionic solvent. *J. Phys. Chem. B* 106, 12017–12021.

132. Margulis, C. J. 2004 Computational study of imidazolium-based ionic solvents with alkyl substituents of different lengths. *Mol. Phys.* 102, 829–838.

133. Debye, P. 1929. *Polar Molecules.* London, Dover Publications.

134. Magee, M. D. 1974. Dielectric relaxation time, a non-linear function of solvent viscosity. *J. Chem. Soc. Faraday Trans. 2* 70, 929–938.

135. Chuang, T. J. and Eisenthal, K. B. 1971. Studies of effects of hydrogen bonding on orientational relaxation using picosecond light pulses. *Chem. Phys. Lett.* 11, 368–370.

136. Balasubramanian, S. and Bagchi, B. 2001. Slow solvation dynamics near an aqueous micellar surface. *J. Phys. Chem. B* 105, 12529–12533.

137. Balasubramanian, S., Pal, S., and Bagchi, B. 2002. Hydrogen-bond dynamics near a micellar surface: Origin of the universal slow relaxation at complex aqueous interfaces. *Phys. Rev. Lett.* 89, 115505–115508.

138. Pal, S., Balasubramanian, S., and Bagchi, B. 2002. Temperature dependence of water dynamics at an aqueous micellar surface: Atomistic molecular dynamics simulation studies of a complex system. *J. Chem. Phys.* 117, 2852–2859.

139. Pal, S., Balasubramanian, S., and Bagchi, B. 2003. Identity, energy, and environment of interfacial water molecules in a micellar solution. *J. Phys. Chem. B* 107, 5194–5202.

140. Zhu, D. -M., Feng, K. -I., and Schelly, Z. A. 1992. Reverse micelles of triton X-100 in cyclohexane: Effects of temperature, water content, and salinity on the aggregation behavior. *J. Phys. Chem.* 96, 2382–2385.

141. Zhu, D. -M. and Schelly, Z. A. 1992. Investigation of the microenvironment in triton X-100 reverse micelles in cyclohexane, using methyl orange as a probe. *Langmuir* 8, 48–50.

142. Dutt, G. B. 2004. Rotational relaxation of hydrophobic probes in nonionic reverse micelles: Influence of water content on the location and mobility of the probe molecules. *J. Phys. Chem. B* 108, 805–810.

143. Dutt, G. B. 2004. Does the onset of water droplet formation alter the microenvironment of the hydrophobic probes solubilized in nonionic reverse micelles? *J. Phys. Chem. B* 108, 7944–7949.

144. Shirota, H. and Segawa, H. 2003. Time-resolved fluorescence study on liquid oligo(ethylene oxide)s: Coumarin 153 in poly(ethylene glycol)s and crown ethers. *J. Phys. Chem. A* 107, 3719–3727.

145. Frauchiger, L., Shirota, H., Uhrich, K. E., and Castner, E. W. Jr. 2002. Dynamic fluorescence probing of the local environments within amphiphilic starlike macromolecules. *J. Phys. Chem. B* 106, 7463–7468.

9 Microemulsion Systems and Their Potential as Drug Carriers

Raid G. Alany, Gamal M. M. El Maghraby,
Karen Krauel-Goellner, and Anja Graf

CONTENTS

9.1   HISTORY, TERMINOLOGY, AND DEFINITION

Microemulsion (ME) systems have attracted commercial interest for many years. Long before Hoar et al. described MEs for the first time in 1943, women in Australia used a transparent dispersion of eucalyptus oil, water, soap flakes, and spirit to wash wool. Indeed Hoar and Schulman were the first to introduce the term "microemulsion" to describe transparent, fluid systems obtained by titration of a conventional emulsion with medium chain alcohol such as pentanol or hexanol. Since then, the term ME has been used to describe multicomponent systems comprising a nonpolar, an aqueous, surfactant, and cosurfactant components. ME systems can be formulated without a cosurfactant, i.e., using a single surfactant, although cosurfactants or an auxiliary surfactant is widely used. It is critical to point out that the term ME was (and occasionally is) used in the literature to describe various surfactant association systems (micelles and reverse micelles), mesophases and liquid crystalline systems (lamellar, hexagonal, and cubic), and even coarse emulsions that are micronized using external energy (submicron emulsions). To avoid such confusion, Danielsson and Lindmann [1] introduced the following definition: "Microemulsion is a system of water, oil and amphiphile which is optically isotropic and a thermodynamically stable liquid solution." By this definition, the following systems were excluded:

- Systems that are surfactant-free
- Liquid crystalline systems (mesophases)
- Dilute surfactant systems with water or oil alone (micellar and nonmicellar)
- Kinetically stabilized conventional emulsions including micronized coarse emulsions

TABLE 9.1
Comparison of Coarse Emulsions and MEs

Property	ME	Coarse Emulsion
Disperse phase droplet size	Less than $0.2\,\mu m$	$0.2{-}10\,\mu m$
Visual appearance	Transparent to translucent	Turbid to milky
Stability	Thermodynamically stable	Thermodynamically unstable
Formation	Spontaneous	Requires energy input
Microstructure	Droplet or bicontinuous	Droplet

Source: Tenjarla, S., *Crit. Rev. Ther. Drug Carrier Sys.*, 16, 461, 1999.

The term microemulsion is often incorrectly used in the literature to describe oil and water dispersions of small droplet size produced by prolonged ultrasound mixing, high-shear homogenization, and microfluidization, i.e., submicron emulsions. The major differences between a micro- and coarse emulsion are shown in Table 9.1 [2].

9.2   STRUCTURE OF ME SYSTEMS

The term micro"emulsion" has led to a common belief that most are droplet-based systems, whereas many are bicontinuous. An ME system can be one of three types depending on the relative ratios of the constituting components: oil-in-water (o/w ME) systems comprise water as the continuous medium; water-in-oil (w/o ME), where oil is the continuous medium and water and oil bicontinuous ME in which almost equal amounts of water and oil coexist [3]. The simplest representation of the ME microstructure is with reference to the droplet model in which an interfacial film comprising an amphiphile (surfactant/cosurfactant) molecules surrounds the dispersed droplets (Figure 9.1). The interfacial orientation of the amphiphile varies depending on the ME type. The structure of the ME being o/w or w/o depends to a great extent on the volume fraction of water, oil and amphiphile as well as the nature of the interfacial film. The nature of the interfacial film is usually governed by the geometry of the amphiphile molecules comprising the film. Droplet w/o ME form when the water volume fraction is

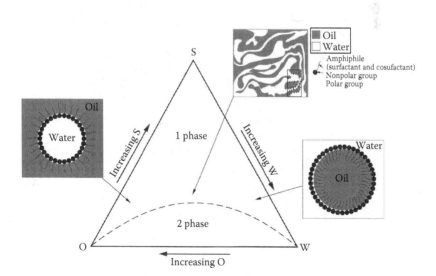

FIGURE 9.1   Theoretical ternary phase diagram outlining the region of existence of one-phase and two-phase systems. Note the illustrative representation of the droplet w/o, droplet o/w, and bicontinuous MEs. O, oil component; W, water component; S, amphiphile component (surfactant/cosurfactant).

low, whereas o/w ME droplets from when the oil volume fraction is low and water is present in abundance. On the other hand, in systems with almost equal amounts of water and oil, a bicontinuous ME is likely to exist (Figure 9.1). In these systems, both oil and water entwine as nano-domains that are separated by an amphiphile-stabilized interface with a net curvature close to zero. (Upward and downward curvatures of the interfacial film are the same and as such the amphiphile favors none of the two immiscible phases.)

9.3 ROLE OF SURFACTANTS IN THE FORMATION OF ME SYSTEMS (THERMODYNAMIC AND FORMULATION CONSIDERATIONS)

A simplified thermodynamic model has been proposed to explain the formation of an ME system as follows:

$$\Delta G_f = \gamma \Delta A - T \Delta S \qquad (9.1)$$

where
   ΔG_f is the free energy of ME formation
   γ is the interfacial tension at the oil–water interface
   ΔA is the change in the interfacial area (associated with reducing droplet size)
   S is the system entropy
   T is the absolute temperature

The formation of an ME is a process that is promoted by the entropy term ΔS due to the increased randomness associated with the dispersion of one of the two immiscible phases as small droplets in the second phase. This leads to higher entropy and delivers free energy, which in turn is necessary to form a stable ME. The interfacial tension is lowered by the migration of surfactant molecules to the interface of the two immiscible phases. Once the critical micelle concentration is reached and micelles are formed, no further decrease in the interfacial tension is possible by adding more surfactant. However, by adding a second surfactant, a further reduction of the interfacial tension can be achieved which results in the thermodynamic stability of an ME. The ME formation process is also associated with a reduction in droplet size. This results in an increased ΔA due to an increase in surface area due to droplet size reduction. This is compensated by a very low-interfacial tension (decreased γ), normally achieved by using relatively high-amphiphile concentrations. The net outcome is a negative value for ΔG_f that translates into a spontaneous ME formation [4]. There are, however, a number of other effects such as solvation enthalpy, entropy changes due to alteration of the water structure and enthalpy changes caused by dissolution of the micelles that are not considered in this simplified model.

   The formation of ME systems can be explained with reference to the "self-assembly theory of micelle and bilayer forming surfactant molecules" where the volume of the surfactant is denoted v, its head group surface area a, and its length l [5,6]. If the critical packing parameter (CPP = v/al) has values between *zero* and *one* o/w ME forms, whereas w/o ME forms when CPP is greater than *one*. When using surfactants with critical packing parameters close to *one* (i.e., CPP ≈ *one*) and at nearly equal volumes of water and oil, the mean curvature of the interfacial film approaches *zero* and droplets coalesce into a bicontinuous structure (Figure 9.1). However, it should be emphasized that this approach is based on sole geometrical considerations that are related to surfactant molecules and as such does not account for many processes including those outlined in Equation 9.1.

The ratio of the hydrophilic and the hydrophobic groups of the surfactant molecules, that is their hydrophile–lipophile balance (HLB), has been used for the selection of surfactants to formulate ME and accordingly the HLB of the candidate surfactant blend should match the required HLB of the oily component for a particular system, furthermore a match in the lipophilic part of the surfactant used with the oily component is favorable [7].

Shinoda and Kuineda introduced the concept of the phase inversion temperature (PIT) or so-called HLB temperature [8]. They described the recommended formulation conditions to produce ME with surfactant concentration of about 5%–10% w/w being (a) optimum HLB or PIT of a surfactant, (b) optimum-mixing ratio of surfactants, i.e., HLB or PIT of the mixture, and (c) optimum temperature for a given nonionic surfactant. They concluded that (a) the closer the HLBs of the two surfactants, the larger the cosolubilization of the two immiscible phases, (b) the larger the size of the solubilizer, the more efficient is the solubilization process, and (c) that mixtures of ionic and nonionic surfactants are more resistant to temperature changes than nonionic surfactants alone [8].

9.4 ROLE OF COSURFACTANTS/COSOLVENTS IN THE FORMATION AND STABILIZATION OF ME SYSTEMS

Cosurfactants are molecules with weak surface-active properties that are combined with the surfactants to enhance their ability to reduce the interfacial tension and promote the formation of an ME [3]. On the other hand cosolvents are weak amphiphiles that tend to distribute between the aqueous phase, the oil phase and the interfacial layer. They promote ME formation by rendering the oily phase less hydrophobic, the aqueous phase less hydrophilic, and the interfacial film more flexible and less condensed [9,10].

Most single chain surfactants do not sufficiently reduce interfacial tension at the oil/water interface to form MEs, furthermore they may lack the right molecular attributes (i.e., HLB) to act as cosolvents. To overcome such a hurdle, cosurfactant/cosolvent molecules are introduced to sufficiently lower the oil/water interfacial tension, fluidize the rigid hydrocarbon region of the interfacial film, and induce ideal curvature of the interfacial film. Typically, molecules with small to medium hydrocarbon chains (C3–C8) with a polar head group (hydroxyl, amine group, sulfoxide, or *n*-oxides) that can effectively diffuse between the immiscible phases and the interfacial film are used [11].

9.5 FORMULATION OF ME SYSTEMS

9.5.1 SELECTION OF COMPONENTS

Conceivably, the most significant problem associated with the formulation of pharmaceutical MEs is related to excipient's compatibility and acceptability. Pharmaceutically acceptable ME systems should be prepared using at least generally regarded as safe (GRAS) and ideally pharmaceutical grade ingredients, i.e., ones already approved by regulatory bodies for pharmaceutical use and are devoid of undesirable effects [12]. Most of published work focuses on using hydrocarbons such as, hexane, heptane, dodecane, or cyclohexane as the oily component. Whilst these oily components are of interest for fundamental research, they are not suitable pharmaceutical ingredients. On the other hand, common pharmaceutical oily components such as vegetable oils tend to be more difficult to microemulsify. Amongst the most widely used oily components in the formulation of pharmaceutical ME systems are triglycerides (vegetable oils mostly), mixtures of mono and diglycerides (considered by some researchers as cosurfactants) as well as di- and triglycerides, and fatty acid esters such as isopropyl myristate (IPM), isopropyl palmitate, and ethyl oleate [2]. The effect of the oily component on the phase behavior of o/w ME forming systems formulated with nonionic surfactants was reported [23]. Zwitterionic and nonionic surfactants are commonly used to formulate pharmaceutical ME systems [2]. Amongst the nonionic surfactants are sucrose esters [13], polyoxyethylene alkyl ethers [14], poly glycerol fatty acid esters [15], polyoxyethylene hydrogenated castor oil [16], and sorbitan esters [17]. Systems based on zwitterionic phospholipids, particularly lecithin, have been also investigated due to their biocompatible nature [18–22].

The choice of pharmaceutical cosurfactants is challenging, as most of the cosurfactants utilized in fundamental research are not suitable due to bioincompatibility issues. Amongst the pharmaceutically acceptable cosurfactants are ethanol [24] medium chain mono and diglycerides [25–28], 1,2-alkanediols [29,30] and sucrose ethanol mixtures [31] alkyl monoglucosides, and geraniol [32].

9.5.2 PHASE BEHAVIOR STUDIES

Before an ME can be used as a drug delivery carrier, the phase behavior of the candidate ingredients should be established. This is critical due to the distinct range of multiphase systems—that results when the comprising components are mixed (coarse emulsions, vesicles, lyotropic liquid crystals, and micellar systems to name few) Moreover, it is common to form ME systems in equilibrium with excess water or oil phases as described by Winsor [1–4]. This is mostly seen with water- or oil-rich systems. The most acceptable experimental method to establish the phase behavior of such systems is to construct a ternary phase diagram using Gibbs triangle (Figure 9.1). This is ideally done under isocratic and isobaric conditions to avoid any potential phase changes upon temperature and pressure fluctuations. Figure 9.1 outlines a theoretical ternary phase diagram for an ME-forming systems where the corners represent 100% of the comprising components and the sidelines are binary systems. In general, a ternary phase diagram can be constructed by one of two methods [33]:

- Preparing a large number of samples with different composition
- Titrating a binary or a pseudobinary mixture with the third component

If the prepared mixtures reach equilibrium rapidly, both methods yield matching results. On the other hand, for mixtures that do not reach rapid equilibrium, the first method is favorable due to the continuous composition changes associated with the titration method. Simply, the change in the components ratio during titration is likely to occur rapidly; therefore, it is possible to overlook phase changes upon titration [33].

It is common for a formulation to contain more than three components. Accordingly, the phase behavior cannot be fully represented using Gibbs triangle. The phase behavior of a quaternary mixture at fixed temperature and pressure can be instead represented using a phase tetrahedron. Full characterization of such a systems is time consuming and requires a large number of experiments [34]. A common approach to represent these systems is by fixing the mass ratio of two components (surfactant/cosurfactant, oil/surfactant, oil/cosurfactant, oil/drug, etc.) and as such considered a single component. Such an approach is actually an oversimplification of the system, yet is acceptable for the purpose of phase behavior studies. Such systems are denoted as "pseudoternary" and may comprise four (quaternary), five (quinary), or even six components yet are represented using a Gibbs triangle.

Expert systems have been explored as a pioneering approach to minimize the experimental effort associated with constructing phase diagrams. Artificial neural networks, fuzzy logic, and genetic algorithms were specifically used to predict the phase behavior of multicomponent ME forming system [17,36] as well as for selection of system ingredients [37].

9.6   TECHNIQUES USED TO CHARACTERIZE ME AND RELATED SYSTEMS

Since MEs have become commercially valuable for tertiary oil recovery, in the food and pharmaceutical industry, their characterization has become a pertinent feature of many publications dealing with this type of colloidal systems [4,7,38]. Knowing the microstructure of an ME is of great importance as the phase behavior will influence the ability to solubilize compounds and can determine drug release [39] or influence the manufacture of polymeric nanoparticles from these systems [40].

In the pharmaceutical field, MEs belong to the group of colloidal drug delivery systems and can therefore be subjected to several physicochemical analytical techniques used to characterize conventional colloids. MEs can be differentiated from coarse emulsions or other two-phase systems by visual inspection (ME = transparent or translucent, two-phase systems = turbid). Visual inspection is therefore commonly used for the investigation of phase diagrams to establish the initial phase boundaries. For exact determination of the phase boundaries, other techniques have to be used which give detailed information about the phase behavior and the microstructure of the pseudoternary systems to be able to differentiate between liquid crystalline systems and the different ME types (w/o droplet, bicontinuous, o/w droplet, or solution-type). To determine the nature of the continuous phase of the ME, one can use conductivity measurements, with which a water continuum will lead to an increased conductivity. In addition, viscosity measurements can give good insight into the phase behavior of MEs, in that w/o-droplet systems generally show a higher viscosity due to the higher viscosity of the oil-continuous pseudophase. Viscosity measurements are further important in differentiating between liquid–crystalline structures and MEs as liquid–crystalline systems exhibit high viscosities whereas MEs are usually Newtonian fluids with low viscosity [35]. Furthermore, determination of the refractive index (RI) can give useful information about the dispersed and continuous pseudophases. The measured RI of the ME system should be close to the RI of the pure component forming the continuous phase. A more recent approach to the characterization of MEs involves the utilization of differential scanning calorimetry (DSC) [41]. Information related to the phase-behavior of an ME can be gained from the thermal behavior, particularly the freezing behavior of the water component, which will show a similar freezing behavior compared to pure water if it is in the continuous phase. On the microstructure level, a wide variety of techniques can be exploited to study MEs. The progression from a pure ME to an ME in equilibrium with lamellar liquid crystals for instance can be monitored by polarizing light microscopy (ME = isotropic/nonbirefringent, lamellar liquid crystal = anisotropic/ birefringent) and by viscosity measurements as mentioned above (ME = Newtonian flow with low, viscosity, liquid crystal = non-Newtonian with high viscosity). Further microstructure investigations can be carried out using pulsed-field gradient self-diffusion NMR [42], where the rates at which the molecules diffuse can give insight into the type of ME (droplet, bicontinuous, or solution), but data interpretation can be difficult and NMR equipment is not readily available. A more direct visualization of internal ME structures can be gained by electron microscopy (EM) techniques, such as freeze fracture transmission EM (FF-TEM) and cryo field emission scanning and transmission EM (cryo-FESEM and cryo-TEM). The recent improvements of cryo preparation techniques for EM have proven to be very useful in the study of ME allowing these systems to be visualized closer to their natural state after a rapid-freezing step as compared to the time-consuming freeze fracture-replica sample preparation in FF-TEM [43,44]. The different techniques used in the characterization of ME and related systems along with relevant examples will be described below.

9.6.1 Visual Inspection

Visual inspection of pseudoternary systems is the first step in establishing the region boundaries in a phase diagram (Figure 9.1). The individual systems once formulated (usually in transparent vials) are visually inspected for clarity, signs of phase separation, and birefringence, subsequently left to equilibrate (for 24 h up to 1 week) and then re-inspected. When using nonionic surfactants, care should be taken that temperature fluctuations are avoided over the storage period, as nonionic surfactants are temperature sensitive. Visually, MEs are clear and thus easy to differentiate from a coarse emulsion or other two-phase systems. The clarity is due to their very small "droplet" size (usually in the range of 10–100 nm), which does not scatter visible light as compared to bigger aggregates like droplets in a coarse emulsion. A Tyndall-effect (opacity) might be seen, though. Figure 9.2 outlines the visual appearance of various pseudoternary systems ME (clear), a coarse emulsion (turbid), and an ME system in equilibrium with lammellar liquid crystals.

MEs can also coexist with excess water or oil phase or both (Winsor systems I–III) [45]. To aid the determination of any phase separation, centrifugation of the ME samples followed by visual inspection is recommended. Djordjevic et al. [46] and Mehta et al. [47] described the use of centrifugation to study the physical stability of ME systems. Djordjevic et al. investigated ME consisting of IPM, PEG-8 caprylic/capric glycerides (surfactant), polyglyceryl-6 dioleate (cosurfactant), and water as delivery systems for the drug diclofenac diethylamine. Centrifugation for 30 min at 13,000 rpm could show that drug incorporation had

FIGURE 9.2 Visual appearance of the psuedoternary systems comprising caprylic/capric triglycerides (Crodamol GTCC), caprylic/capric mono-/diglycerides (Capmul MCM), polysorbate 80 (Crillet 4), sorbitan mono-oleate (Crill 4), and water. Note the ME system with its characteristic clear appearance (left vial), ME system with liquid crystals (middle vial), and coarse emulsion (right vial).

no influence on the stability of the ME. Mehta et al. centrifuged ME samples made up of oleic acid, phosphate buffer, Tween 80, and ethanol at 2000 rpm for 30 min to demonstrate that incorporation of the drug rifampicin had no influence on the phase behavior. When approaching the phase boundary between the ME and liquid crystals regions, visual examination may fail to detect a small amounts of the formed liquid crystals as a distinct phase. For such systems, polarized light microscope examination is necessary.

9.6.2 POLARIZING LIGHT MICROSCOPY

For the characterization of ME, one can take advantage of the fact that these colloidal systems are isotropic and can be differentiated from liquid crystalline systems, which are anisotropic and show birefringence when viewed using polarizing light microscopy. The optical properties of isotropic material are not direction dependent, and with only one RI the propagation direction of light passing through this material is not influenced. Anisotropic material like liquid crystals (excluding cubic mesophases), on the other hand, has optical properties that vary with the orientation of the incident light. Liquid crystals demonstrate a range of refractive indices depending on the propagation direction of light through the substance. Since an anisotropic material is also a beam splitter, it divides light rays into two parts: the ordinary and extraordinary light beam. The two light beams are reunited along the same optical path, hence information related to the type of liquid crystal (smectic or nematic) can be extracted. To observe the effects of birefringence and isotropy, one needs linearly polarized light. Linearly polarized light consists of light waves that only propagate in one direction, and it is generated by the use of a polarizing filter or polarizer. Above the objective, a second polarizer is positioned, called the analyser, which is usually positioned at 90° to the polarizer (crossed). White light from an ordinary light source of a light microscope is shone onto the polarizer through which only light propagating into a certain direction can pass. This leads to plane-polarized light. The plane-polarized light subsequently travels through the sample specimen and is split into an ordinary and extraordinary light beam if the specimen sample is anisotropic and shows birefringence (such as lamellar, hexagonal, or reverse hexagonal liquid crystals). On the other hand, if the sample is isotropic (like ME systems or coarse emulsions), light will travel through undisturbed.

Polarizing light microscopy is invaluable to detect small amounts of a liquid crystalline phase coexisting with an ME before phase separation takes place. Hence, it allows for the fine-tuning of phase boundaries in a phase diagram. The issue of differentiating the various types of liquid crystals will not be further discussed here, as it is beyond the aims of this chapter. Nevertheless, the differentiation between ME systems containing liquid crystals and pure ME systems is critical for phase behavior studies. These phase transitions are often critical when MEs are to be used as drug delivery systems or for other commercial purposes. Constantinides et al. investigated w/o ME systems with medium or long-chain glycerides as the oil phase and used polarizing light microscopy to confirm the existence of the "isotropic behavior of MEs" [48]. Similarly, Djordjevic et al. claimed that "observing whether the sample rotates the plane of polarization of polarized light is a very useful tool to distinguish between isotropic microemulsions and anisotropic lamellar or hexagonal phases," when studying caprylocaproyl macrogolglyceride-based MEs [46]. For determining the phase boundaries in a phase diagram of n-heptane, the nonionic surfactant Igepal CA520, and water, Grätz et al. used the technique to differentiate between lamellar liquid crystals and ME-forming systems [49]. For a more concise analysis of liquid crystalline phases, the reader may be referred to the following references [50,51].

9.6.3 ELECTRICAL CONDUCTIVITY

Electrical conductivity (σ) has been traditionally used as a standard technique to study the phase behavior of coarse emulsions and reference to it can be found in most undergraduate pharmaceutical sciences textbook [52]. Similarly, MEs can exist as w/o or o/w systems, therefore conductivity measurements can be applied for their characterization. The underlying principle for phase determination by conductivity is the ability of water to conduct an electric current, which is measured in S cm^{-1} or μS cm^{-1}. If water forms the continuous phase of an ME, the system will show a high conductivity, on the other hand, the system will exhibit low conductivity if oil becomes the continuous phase. The change from a w/o system to an o/w system is generally associated with a significant increase in conductivity. If the conductivity of pure/deionized water is too low in the system of interest, addition of a small amount of an electrolyte (e.g., NaCl) may facilitate the measurements; however, the results have to be in interpreted in comparison with the conductivity readings of the pure electrolyte solution. It has to be remembered, however, that incorporation of an electrolyte may affect the phase behavior of MEs, particularly if they contain ionic surfactants, in which case MEs with and without electrolytes have to be compared. Several studies have investigated the relationship between electrical conductivity and phase transitions in ME systems allowing the determination of percolation thresholds (formation of long-range connectivity in random systems). Although percolation phenomena often are not of great interest in the characterization of MEs when used as drug delivery systems, some basic facts will be presented as this may make the interpretation of conductivity data easier.

   Lagües and Sauterey [53] and Mehta and Bala [54] described the phenomenon of phase transition from a w/o ME to an o/w ME with an increasing water concentration as follows:

1. At low-water concentration, the conductivity is low and reflects the mean charge of the droplets. The water droplets are isolated from each other in a nonconducting continuous oil phase.

2. A steep increase in conductivity at intermediate water concentrations can be explained by percolation transition, and every ME mixture will exhibit a specific critical water volume ratio/concentration ϕ_c at which percolation occurs. The increased conductivity leading up to the ϕ_c is caused by an increased number of (still) individual water droplets. The conductivity measured around and above the ϕ_c is due to dynamic droplet clusters or transient water channels, and microscopically droplets do not exist anymore at this stage. The critical water concentration ϕ_c needed to induce percolation usually ranges from 0.1 to 0.26 depending on the ME components, specifically the type of cosurfactant, and the temperature.

3. Around a water concentration of 0.4–0.45 bicontinuous structures are formed, but this can only be seen as a small plateau of the viscosity graph rather than with conductivity measurements [55].

4. Second transition from a bicontinuous into an o/w system can then be seen at high-water concentrations (~0.6). The concentration at which this inversion occurs (inversion concentration) is reflected by a steep increase in conductivity.

One can argue whether structural differences between the stage of "dynamic droplet structures" around the percolation threshold and bicontinuous MEs really exist. In the following example, a study by Alany et al. [35], one can apply some of the above-mentioned ideas to the analysis of conductivity data. Phase diagrams with the components ethyloleate, Crillet 4, and Crill 1 (in a ratio of 6:4 as the surfactant-mix), with and without butanol as a cosurfactant (in a ratio of 7:3 surfactant:cosurfactant) and water, were constructed and samples along a cut at a constant surfactant-to-oil ratio (4:6) through the respective phase diagram were investigated by conductivity measurements.

In Figure 9.3, one can see how the conductivity increased with increasing water fraction, especially in the cosurfactant-containing system. The initial increase

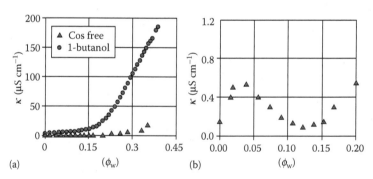

(a)   (b)

FIGURE 9.3 (a) Specific conductance (κ) as a function of the water volume fraction (ϕ_w) for cosurfactant-free and cosurfactant-containing (1-butanol) systems. (b) Specific conductance (κ) as a function of the water volume fraction (ϕ_w) for the cosurfactant-free system (expanded scale). (Reproduced from Alany, R.G., Davies, N.M., Tucker, I.G., and Rades, T. *Drug Dev. Ind. Pharm.*, 27(1), 33–41, 2001. With permission.)

is only very gradual and could be explained by the increasing amounts of droplets in the ME systems. Around a water fraction of ~0.15, the increase in conductivity seemed to become more pronounced, however, due to the formation of dynamic droplet clusters or transient water channels. This indicates that the percolation threshold has been reached at a water fraction of 0.15. Examining the conductivity curve on an expanded scale at water fractions between 0.0 and 0.2 in the cosurfactant-free system, one can notice an initial increase in conductivity up to water fraction of 0.04. This is due to initial hydration of ethylene oxide surfactant head groups. The following dip in conductivity is likely due to aggregation of surfactant molecules to from w/o ME droplets. The following increase in conductivity is associated with formation of lamellar liquid crystals [35].

The percolation threshold of an ME system can also be recorded along a temperature gradient by keeping the composition of the ingredients constant. The percolation threshold then becomes the percolation temperature or phase-inversion temperature. Examples can be found in the following references [56,57].

Electrical conductivity plays an important role in the characterization of colloidal systems due to its ease of use and data interpretation, and the low cost of the measurement equipment. Conductivity measurements can be used to simply determine the phase behavior of a system (o/w, w/o), to investigate microstructural changes along a concentration or temperature gradient and to determine the percolation threshold.

9.6.4 Flow Properties

The flow properties of a colloidal system are very much dependent on its microstructure, as determined by the molecular arrangement and interaction of its components. ME systems show flow typical of a Newtonian liquids, for which the shear stress is directly proportional to the shear rate. Since viscosity measurements are dynamic experiments, they will give information on dynamic properties of the ME. These will depend on the microstructure, type of aggregates, or interactions within the ME, which in turn are determined by the concentration of the various components and the temperature. The dispersion of one component in another, e.g., water in oil, will generally increase the bulk viscosity in comparison to the individual components (oil and water) [58]. For at true colloidal dispersion, viscosity will increase with increasing volume fraction of dispersed phase according to the formula generated by Einstein:

$$\eta_r = 1 + 2.5\Phi \qquad (9.2)$$

The formula describes the increase in relative viscosity η_r in relation to an increase in the volume fraction Φ of the dispersed phase existing as hard spheres [59,60]. This formula allows for the calculation of various shapes the dispersed phase can have (sphere, rodlike, wormlike, etc.). A useful modification of the previous equation was proposed by Attwood et al. [61]. The investigated ME system was formulated using liquid paraffin, water, Span 60, and Tween 80 and made reference to the following equation:

$$\eta_{rel} = a^{\phi/(1 - \kappa\phi)} \qquad (9.3)$$

where

η_{rel} is the relative viscosity
a is a viscosity constant with a theoretical value of 2.5 for solid spheres
ϕ is the volume fraction of the disperse phase
κ is the hydrodynamic interaction coefficient

In the same study, the effect of increasing the surfactant concentration on the overall viscosity was investigated for an o/w ME system and yielded values for the viscosity constant (a) of 3.19 to 4.17. The authors concluded that allowance for the hydration of the polyoxyethylene chain of the used surfactant reduced the value of the viscosity constant (a) toward the theoretical value of 2.5 for a solid sphere. They further concluded that changing the ratio of the nonionic surfactants did not significantly affect the viscosity of the system.

When characterizing MEs by viscosity measurements, the Newtonian flow behavior of an ME is a decisive feature in the differentiation from other colloidal systems, especially when other characteristics such as spontaneous formation, clarity, etc. already point at the existence of an ME. The Newtonian flow behavior of ME is based on their equilibrium structure. As long as the increasing stress does not alter the internal structure, flow increases linearly with increasing shear stress. Formation of liquid crystalline structures on the other hand will result in the emergence of non-Newtonian flow behavior. When displaying the results of viscosity measurements, however, one commonly finds graphs showing the viscosity as function of a component fraction, often water volume fraction, or temperature. This appears to be a straightforward method for determining percolation phenomena or phase transitions depending on water volume fraction or temperature.

To differentiate a cubic phase from an ME, D'Antona et al. compared the viscosity of the samples over an increasing shear rate [63]. The samples consisted of 70% glycerol monooleate, 2% ethanol, 28% water for the cubic phase, and 70% glycerol monooleate, 20% ethanol and 10% water for the ME sample. As expected, the ME did show Newtonian flow and the viscosity was not influenced by the increasing shear rate. On the other hand, the sample containing a cubic phase showed a decrease in viscosity upon increasing shear rate, i.e., displayed a typical non-Newtonian shear thinning flow.

Viscosity measurements of MEs can also be useful in terms of developing an ME formulation with characteristics desirable for application. Constantinides and Scalart for instance compared the viscosity of w/o MEs with long- versus medium-chain glycerides as oil phase [48]. The medium-chain glyceride MEs showed a lower viscosity than the long-chain glycerides MEs promising better spreading ability for topical applications and handling during manufacture. The medium-chain glyceride MEs also showed a better bioavailability for the model peptide calcein, probably due to the less restriction of the release from the lower viscosity medium-chain glyceride ME.

Incorporation of a drug can also have an influence on the viscosity of an ME system. Djordjevic et al. detected an increase in viscosity when incorporating the drug diclofenac diethylamine at a concentration of 1.16% into an ME of the composition water, IPM, PEG-8 caprylic/capric glycerides, and polyglyceryl-6 dioleate [46], although incorporation of the drug did not influence the overall flow behavior of the ME. Diclofenac diethylamine can cause an increase in viscosity as it has a very low-critical association concentration or in other words tends to self-associate or so-called aggregate easily and thereby may interact with the interfacial film of the ME [64].

9.6.5 Differential Scanning Calorimetry

DSC has been only recently added to the range of characterization techniques for MEs, but has long been a standard technique for solid-state characterization of [65]. DSC belongs to the thermal analysis methods and probes the pharmaceutical thermal behavior of a sample as a function of an externally applied temperature. During heating or cooling of the sample, endothermic (melting, sublimation, chemical degradation) or exothermic (crystallization) events can be monitored in a thermogram. For DSC, a sample and reference (usually an empty sample pan) are heated or cooled at a certain rate and the heat flow required to keep sample and reference at the same temperature is measured. Plots therefore show the differential rate of heating (expressed as watts/second, calories or joules/second) over the temperature. The area under a peak resulting from a thermal event is proportional to heat obtained from or absorbed by the sample material and displayed in the unit cal/(s·g) or J/(s·g). Since every compound exhibits a characteristic pattern of endo- or exothermic events (such as a characteristic melting point upon heating or cooling) impurities, mixtures of two or more components, polymorphic changes, or phase transitions can easily be detected with DSC.

When using DSC for investigating ME systems, the freezing and melting behavior of water can give valuable insight into the phase behavior of these systems. For investigation of liquid samples like MEs, hermetically sealed sample pans should be used to avoid evaporation of water. Water can be detected as being free (bulk) or bound (interfacial) water. Garti et al. studied the freezing behavior of water in an ME consisting of n-alkanes (C_{12}–C_{16})/sucrose esters/1-butanol and water [66]. The authors managed to assign a freezing temperature of $-4^{\circ}C$ for free water and $-10^{\circ}C$ for bound water. The freezing peak for free water increased upon increasing water concentration in the sample and became the dominant peak in the thermogram above water concentration of 20%. When plotting the area under the freezing peak over the water content, an increase in the peak area was observed which then leveled off into a plateau above a water content of 20%. The authors determined this as a phase transition from a w/o to a bicontinuous or o/w system in agreement with other characterization techniques. Podlogar et al. [41] investigated an ME system consisting of isopropyl mysistate/Tween 40/Imwitor 308 and water by DSC measurements amongst other characterization techniques. A freezing peak for water emerged at a water content of ~15% at a temperature of $-45^{\circ}C$ pointing at the existence of interfacial water. With increasing water content, this peak shifted toward the freezing temperature of free water ($-17^{\circ}C$ under the given measurement conditions) and showed the closest approximation at a water content of 65%. A pronounced shift of the water freezing peak toward the freezing temperature of free water was seen at 35% water content, which the authors related to a phase transition from w/o to bicontinuous or o/w ME [41].

9.6.6 PULSED-FIELD GRADIENT SPIN-ECHO NMR

Unlike most other characterization techniques introduced in this chapter that only allow conclusions to be drawn on the bulk level, self-diffusion NMR gives detailed information on colloidal systems by focussing on the molecular level. This complex technique will therefore be introduced in detail. NMR spectroscopy, based on physical properties of the molecular spin, is a very powerful method for the measurement of self-diffusion of small molecules in complex solution [67] with direct insight into general aspects of the solution structure [68].

This technique, more correctly called Fourier-transform pulsed-gradient spin-echo (FT-PGSE) NMR, is very convenient [69]. It rapidly provides accurate data simultaneously for all components in the sample in a single set of measurements. Direct information can be gained without the need for isotopic labelling of the sample [69].

Self-diffusion measurements give uniquely detailed and sensitive information on molecular organization and phase structure by reflecting binding and association phenomena as well as diffusion and flow, thereby making it a particularly suitable method for the characterization of colloidal systems [69]. In other words, self-diffusion of a molecular species as a "net result of the thermal motion-induced random-walk process" [69] (Brownian motion) is influenced by the microenvironment and thus the mobility of the molecule may be used to asses the type of ME formed. Although individual molecules in an ME have very rapid short-range dynamics, self-diffusion NMR probes for the long-range dynamics, e.g., in the micrometer scale. FT-PGSE-NMR therefore measures the diffusion of the whole droplets in a dispersion medium rather than the diffusion of molecules inside the droplet.

Self-diffusion experiments are based on a linear magnetic field gradient along one sample axis. Under the influence of a nonuniform magnetic field, otherwise equivalent nuclei at different locations give rise to different NMR frequencies. Thus, the gradient encodes the spin in position and quantitatively maps the NMR frequency with a location along the gradient-applied axis. The mapping of NMR frequencies has become available with Fourier transformation and is explained below. PGSE-NMR self-diffusion experiments are based on the Hahn spin-echo experiment, which has been developed further, by Stejskal and Tanner in 1965 with the introduction of gradient pulses in an otherwise homogeneous field. Field homogeneity, however, is more an experimental presumption and in reality residual field inhomogeneity is compensated by high-resolution NMR standards.

The basic principle of PGSE-NMR in a generally homogeneous magnetic field works by imposing position-dependent phase shifts to the nuclear spins through linear gradient pulsed of the amplitude g, duration δ, and the pulse-interval Δ, in addition to and following the 90° and 180° radiofrequency (RF) pulses. These are required for the formation of a spin-echo. Initially, the 90° RF pulse is applied to the magnetic field resulting in transverse magnetization of the spins. This is followed by the first gradient pulse leading to the phase dispersion in the rotating frame (dephasing). Subsequently, a second RF pulse (180° pulse) is applied at time t after the first pulse and leads to phase inversion. The second gradient pulse applied thereafter refocuses the spins resulting in the spin-echo. Refocusing of the spins is only complete and a true echo obtained in the absence of any phase shifts of the individual spin vectors. Here, an initial 90° RF pulse is applied resulting in phase dispersion in the rotating frame. However, a second RF pulse (180° pulse at time t after the first pulse) refocuses resonant spins yielding the formation of a spin-echo at time $2t$, only if the NMR precession frequency is not changing during the whole experiment. Diffusion affects this experiment in that the random molecular movement of the nuclei, due to Brownian motion/self-diffusion, leads to a random change of the location (random phase shift of individual spin vectors) [70] and variation of the NMR precession frequency of all nuclei in the sample, resulting in a more or less incomplete refocusing of the echo. This principle provides the means to detect diffusion phenomena in colloidal systems.

Advantages of the pulsed-gradient modification are the possibility of variation of the observation time, providing more experimental flexibility and very sharp echo detection in a relatively homogeneous magnetic field.

The frequency separation of multicomponent measurements to obtain self-diffusion coefficients for each component in only one measurement is possible due to the development of Fourier transformation and its introduction to NMR techniques.

The second half of the echo is applicable to Fourier transformation which separates the individual contribution of the components to the echo [70] and in turn enables the simultaneous determination of several self-diffusion coefficients by the mapping of their individual frequencies.

During one diffusion experiment, several echoes are collected upon increasing gradient strengths. When averaged over the entire sample, the NMR signal decreases with increasing gradient strength, with the rate of decrease being proportional to the diffusion coefficient, D.

Stejskal and Tanner devised a quantitative relationship of the echo attenuation as a function of gradient strength, the Stejskal–Tanner equation [71]:

$$I/I_0 = \exp[-D(\gamma\delta g)2(\Delta - \delta/3)] \qquad (9.4)$$

where

I is the signal intensity at an applied gradient strength g
I_0 is the signal intensity in the absence of the gradient
D is the diffusion coefficient
γ is the gyromagnetic ratio of the monitored spin (usually ^{1}H)
δ is the pulse duration
Δ is the pulse-interval (diffusion time between the gradient pulses)

From this equation, the diffusion coefficient D can be derived from a semilogarithmic plot or more accurately by modelling the data to an exponential plot. Exponential modelling allows deconvolution of overlapping signals and thus derivation of the diffusion coefficients for all underlying components.

A typical NMR spectrum of an ME mixture consisting of ethyloleate, a surfactant-mix of polyoxyethylene 20 sorbitan mono-oleate and sorbitan mono-laurate (6:4) and water, obtained with an increasing gradient strength over 17 increments is shown in Figure 9.4.

In general, for unrestricted Gaussian diffusion, the diffusion coefficient (D) is a simple translation from the mean square displacement in space ($\langle \Delta r^2 \rangle$) during the observation time (Δt) [71]:

$$\langle \Delta r^2 \rangle = 6D\Delta t \tag{9.5}$$

In the case of spherical diffusing species (particles or droplets), D is related to the Stokes–Einstein relation, where k_B represents the Boltzmann constant, T the temperature in Kelvin, η the viscosity of the medium, and r_H the hydrodynamic radius of the sphere:

$$D = k_B T / 6\pi \eta r_H \tag{9.6}$$

The Stokes–Einstein relation therefore allows for calculation of the hydrodynamic radius (r_H) of droplets in an ME.

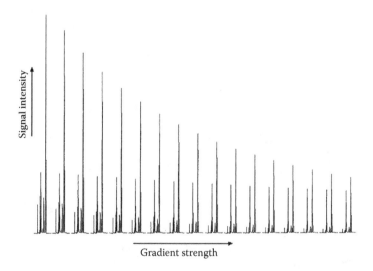

FIGURE 9.4 Typical FT-PGSE spectrum of an ME showing the decreasing proton signal intensity with increasing gradient strength. (From Krauel, K., *Formulation and Characterisation of Nanoparticles Based on Biocompatible Microemulsions*, School of Pharmacy, University of Otago, Dunedin, 2005.)

Furthermore, as can be seen from this relation, the diffusion coefficient is temperature dependent and thus good temperature control is required when performing PGSE-NMR measurements. Short-term fluctuations in temperature or temperature gradient will affect the echo decay and may cause the induction of convection [72], which will overall compromise the precision of the experiment.

For self-diffusion NMR measurements, ME samples are filled into standard NMR tubes. Tetramethylsilane dissolved in D_2O and sealed into a capillary is commonly used as reference. ^1H-NMR spectra of the pure components and the MEs are acquired and Fourier transformed. For analysis, peaks of the ME formulation can be assigned to the individual components by comparison to spectra obtained from the pure ME components. To obtain the data for exponential modelling using the Stejskal–Tanner equation (Equation 9.4), one or more peaks of the component are either integrated or analysis is performed on the peak height. Self-diffusion coefficients ranging from about 10^{-9} to $10^{-12}\,m^2\,s^{-1}$ are typically observed in solutions at ambient temperature [73].

Through comparison of the diffusion coefficients of the components in the mixtures with those of the pure components, determination of the type of MEs is possible according to the guidelines established by Lindman and Stilbs [74]. In w/o droplet type MEs, the water diffusion is determined by the diffusion of the droplet and therefore will be slower than the diffusion of the oil in the continuous oil phase whereas in an o/w droplet type ME the reverse can be observed. Diffusion of the surfactant molecules is slow in both cases as they are associated with the droplets. Furthermore, one can say that the diffusion coefficients of molecules in the pure component generally are about two orders of magnitude higher than the diffusion coefficients of the respective droplet. In a bicontinuous ME, where both oil and water form large domains, the diffusion coefficients of these two components are almost of the same order of magnitude as the ones observed for the pure components, and the surfactant is the slowest-diffusing component. One can differentiate a solution from a bicontinuous arrangement by considering the diffusion of the surfactant molecules. Surfactant molecules will diffuse freely and therefore faster in a true solution as the diffusion will not be obstructed by any long-range order arrangements of the surfactant molecules unlike in a bicontinuous ME.

$$D_{oil} > D_W = D_{surfactant} \rightarrow \text{water-in-oil droplet microemulsion}$$

$$D_W > D_{oil} = D_{surfactant} \rightarrow \text{oil-in-water droplet microemulsion}$$

$$D_W = D_{oil} > D_{surfactant} \rightarrow \text{bicontinuous microemulsion}$$

It has to be noted, however, that these guidelines apply to ideal cases only. In most situations, particularly with pharmaceutically acceptable MEs, the order of magnitude of the diffusion coefficients will vary due to the complexity of the ME components. In addition, molecular and droplet diffusion will be affected by obstruction within the ME [75] caused by a high viscosity of the sample which may obstruct the diffusion strongly making it difficult to detect, or an increasing droplet volume fraction, with which the rate of droplet diffusion will decrease proportionally. Peak separation and assigning peaks to a certain component can also be difficult. Furthermore, solvation effects will complicate diffusion measurements, as molecules of the dispersion medium may for instance associate with the droplets and therefore diffuse more slowly than expected, or surfactant molecules may show a high solubility in the dispersion medium (water or oil) and diffuse faster than expected. A known example on this phenomenon is the ethylene oxide surfactants and their tendency to dissolve in the oil phase [58].

Besides comparing the self-diffusion coefficients D, one can also compare relative diffusion coefficients $D_{rel} = D/D_0$, where D_0 denotes the self-diffusion coefficient of the pure component and D the corresponding value in the ME. D_{rel} sometimes is also described as the obstruction factor A_D. Values for D_{rel} between 0.01 and 0.001 for the water or oil component point at the existence of a w/o or o/w droplet ME system. Values above 0.01 can be found for bicontinuous systems and D_{rel} for the water and oil component should be in the same order of magnitude. D_{rel} values of 1 describe unobstructed flow behavior in a solution.

Self-diffusion NMR is being applied more and more in the characterization of MEs and a few examples will be given in the following paragraph. Biruss et al. investigated an ME consisting of 22% (w/w) polyoxyethylene lauryl ether (Brij-30), 22% (w/w) ethanol, 45% (w/w) eucalyptus oil and 10% (w/w) distilled water for the topical delivery of various hormonal drugs [76].

Self-diffusion NMR measurements showed that the ME was of the bicontinuous type as D_{rel} values for oil and water were in the same order of magnitude (0.32 and 0.21, respectively). The authors could also show that the surfactant Brij-30 existed in a free state, exhibiting fast diffusion behavior ($D = 2.2 \times 10^{-10}$ m^2 s^{-1}) and a micellized state, exhibiting slow obstructed diffusions behavior (8.3×10^{-11} m^2 s^{-1}). When measuring the diffusion behavior of the hormonal compounds 17-β-estradiol, progesterone, cyproterone acetate, and finasteride, it was found that 17-β-estradiol, the smallest molecule, exhibited slower diffusion behavior than the other substances. This was besides being the smallest molecule, which the authors attributed to the formation of hydrogen bonds between the hydroxyl group of the molecule and water and ethanol of the ME. The slowed diffusion behavior of finasteride ($D = 9.5 \times 10^{-11}$ m^2 s^{-1}), when compared to the diffusion of the substance in pure methanol ($D = 7.4 \times 10^{-10}$ m^2 s^{-1}), was attributed to the increased solubility of the compound in the lipophilic phase of the ME. MEs consisting of various mixtures of Labrasol, Plurol isostearique (surfactants), isostearyic isostearate (oil), and water as topical delivery systems for the drugs lidocaine, lidocaine hydrochloride, prilocaine, and prilocaine hydrochloride were studied by Kreilgaard et al. [77]. The characterization of MEs by self-diffusion NMR revealed the existence of o/w (isostearylic isostearate 3%, Labrasol 24%, Plurol isostearique 8%, water 65% (w/w)), bicontinuous (isostearylic isostearate 10%, Labrasol 35%, Plurol isostearique 35%, water 20% (w/w)), and w/o (isostearylic isostearate 70%, Labrasol 11.5%, Plurol isostearique 11.5%, water 7% (w/w)) ME systems.

Lidocaine showed slow diffusion in all ME systems suggesting association with the lipophilic phase. Prilocaine hydrochloride also showed slow diffusion behavior and the authors explained this by association of the drug with the surfactant layers. Drugs showing fast diffusion behavior in the various ME systems also showed superior in vitro transdermal permeation rates.

Apart from determining the microstructure of a colloidal system, once diffusion coefficients have been obtained, this information can also be used to calculate the hydrodynamic radius of micelles or droplets in the system by using the Stokes–Einstein relation mentioned above (Equation 9.4). D'Antona et al. calculated a hydrodynamic radius of 2.48 nm for spherical structures in an ME composed of 1.8% Myverol (monoglyceride) (w/w), 42.9% ethanol (w/w), and 55.3% water. This agreed well with the 2.2 nm size of a monoglyceride molecule [63]. Size calculations can become difficult, however, when solvation effects exist in the colloidal system and light scattering techniques are more reliable to measure particle sizes in colloidal systems. The few examples given here show the variety of possible applications for self-diffusion NMR and an increased accessibility to NMR equipment will certainly increase the number of studies using self-diffusion NMR in the characterization of MEs.

9.6.7   Electron Microscopy Techniques

The techniques described in the previous sections are important in the characterization of colloidal systems and give valuable insights into the microstructure. However, they all probe indirectly for the microstructure and use mathematical models to interpret the results. FF-TEM has been one of the first electron microscopic techniques used to visualize colloidal systems, and advances in cryo preparation techniques and EM in general have established these techniques as invaluable tools in the characterization of MEs. EM allows the direct observation of the system of interest and the obtained micrographs can satisfy our sense of vision. To view a colloidal sample with EM, a transformation from the liquid into the solid state of the sample has to be performed. Otherwise, volatile components of the sample may evaporate into the vacuum of the microscope stage, or the sample may flow due to the forces of the electron beam or be damaged by forces of the electron beam. Cryo preparatory techniques and coatings mainly achieve this transformation and "protection." Coatings can also improve the low contrast between the aqueous and organic components in the ME. Sample preparation can be invasive, and alter the microstructure of an ME, therefore caution has to be exercised in the interpretation of micrographs.

The following section will not explain the general principles of EM and the reader may be referred to relevant textbooks for this. Instead, it will introduce the application of EM to the characterization of MEs and the various sample preparation techniques that can be utilized.

9.6.7.1   Freeze Fracture Transmission Electron Microscopy

The basic steps involved in the sample preparation of an ME by freeze fracturing and subsequent visualization under a TEM are freezing, fracturing, etching, replicating, cleaning, and mounting. For the freezing step, the sample is usually mounted onto the sample holder, that is the sample placed between to copper plates in a sandwich fashion, and which will later be used for the fracturing. Freezing occurs by plunging the sample holders into a liquid cryogen, e.g., propane or ethane by hand or an automatic plunging device. Rapid-freezing rates are desired to achieve a vitrification of the sample, meaning that water in the sample freezes in an amorphous rather than a crystalline state, as the latter state would obstruct sample observation and possibly destroy the microstructures. Freezing rates of ~10^5 K/s can be achieved by using ethane as cryogen and plunging the sample with an automatic plunging device. Following this, the sample specimen is transferred under liquid nitrogen cooling into the freeze-etch device and fractured by separating the two copper plates. The freeze-etch chamber is held at $-150°C$ and a pressure below 10^{-6} torr. For the etching step, the temperature in the chamber is slightly raised to allow for evaporation of ice from the sample surface and the cold microtome, used to fracture the sample, functions as a cold trap on which the evaporating water molecules can condense. Subsequently, a replica of the sample is prepared to preserve the topographical details, as the sample would otherwise melt, when it is transferred into the TEM chamber at room temperature. To create a replica, the sample is coated with thin layers of platinum and carbon. A recent improvement in replica preparation has been the coating with a mix of tantalum and tungsten (Ta–W) instead of platinum, which decorates the oil fracture face more specifically [78]. Once a sample replica is created, it needs to be washed with solvents to dissolve the original sample, and mounted onto an EM-grid before it can be viewed in the TEM. One can imagine that this extensive sample and replica preparation procedure is prone to the creation of artifacts. The development of a controlled environment vitrification system (CEVS), in which all sample preparation steps can be carried out in one chamber and under controlled conditions, will certainly diminish the risk of creating artifacts. More information regarding the CEVS can be found in the following reference [79].

The interest to use FF-TEM in the characterization of MEs lies in the differentiation of the microstructures including droplet, bicontinuous, and liquid crystals [80]. A study by Alany et al. used FF-TEM, amongst other techniques, to investigate the different ME systems formed by various compositions of ethyloleate, Crillet 4 and Crill 1 (in a ratio of 6:4 as the surfactant-mix), with and without 1-butanol and water [35]. The FF-TEM micrograph (Figure 9.5a) of a cosurfactant-free ME containing 5% water showed a droplet structure supporting the results from conductivity and viscosity measurements. The same ME containing 25% water showed streaky structures typical for liquid crystals, and a system containing 35% water with 1-butanol as a cosurfactant showed a bicontinuous structure (Figure 9.5b and c, respectively). The liquid crystalline structures are particularly dominant in the micrographs as fracturing during sample preparation preferentially occurs along the surfactant film or surfactant–oil interfaces due to the molecular structure of these components (hydrocarbon chain + polar head) [82].

Some studies also use the obtained FF-TEM micrographs in combination with an image analyzer to determine the droplet size for the ME [84]. The droplet size of an ME usually is in the range of 10–100 nm. Considering the thickness of the

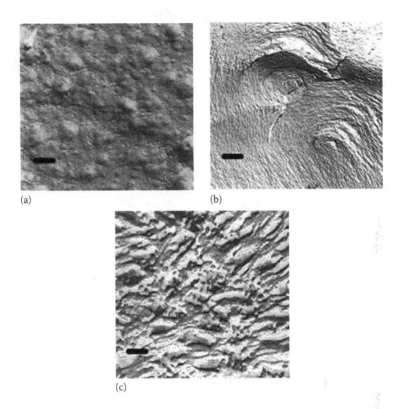

(a)   (b)

(c)

FIGURE 9.5 FFTEM micrographs for (a) droplet ME, (b) lamellar liquid crystals, and (c) bicontinuous ME. (Reproduced from Alany, R.G., Davies, N.M., Tucker, I.G., and Rades, T., *Drug Dev. Ind. Pharm.*, 27, 33, 2001. With permission.)

coating applied to an EM sample with average values of ~20 nm, one has to be aware that the droplet sizes may be overestimated by this procedure and light scattering techniques are suggested as more reliable means to determine the droplet size. Further FF-TEM micrographs of ME undergoing transition from globular to bicontinuous structures can be found in the following reference [58].

9.6.7.2 Cryo Transmission Electron Microscopy

Transmission EM is normally used to study the bulk of a sample. For the characterization of ME, TEM is used in a somewhat paradoxical way in that the sample is prepared for surface investigation only. For cryo-TEM, the sample is simply plunge-frozen (vitrified) and then viewed in the cryo-stage of a TEM. Cryo-TEM images of MEs still very much look like FF-TEM images; however, only the surface morphology can be investigated [58]. For TEM imaging, the sample has to be spread very thinly to be able to gain information from the transmission mode of the EM. This can be problematic for viscous and strongly interconnected

ME samples. A major advantage of cryo-TEM over FF-TEM, however, is the less invasive and time-consuming sample preparation as fracture and replica producing steps are not necessary. Cryo-TEM is a valuable technique not only for the investigation of MEs but colloidal structures in general, such as wormlike micelles [81], cubosomes [82], ISCOMS [83], etc. These samples are usually dilute and thin, hence one-layered samples can be prepared. Furthermore, investigations into improved sample preparation for FF-TEM and combinations of FF- and cryo-TEM can be found in the literature [84], and it can be anticipated that cryo-TEM may be added to the various characterization techniques for MEs in the near future.

9.6.7.3   Cryo Scanning Electron Microscopy

The advances made in the FF-TEM-freezing techniques have made cryo-SEM a novel tool in the visual characterization of MEs. The surface characteristics of MEs can be observed in the frozen state on the cryo-stage of the microscope without the need for replica production, and cryo-SEM may replace the time-consuming FF-TEM technique for ME characterization in the future, provided the instrumentation is available to the researcher. For cryo-SEM, a sample is filled into a suitable sample holder (copper rivets, etc.) and plunge frozen in a cryogen, which may be propane or ethane. As for FF-TEM, cooling rates should be fast to achieve vitrification of the often water-rich ME samples. The rivets are then mounted onto the sample slide under liquid nitrogen cooling, and together again plunge frozen in slushy nitrogen before being transferred into the cryo chamber of the microscope. In the cryo chamber, the sample can now be "fractured," i.e., some sample material can be scraped off from the surface with a knife followed by etching to allow for evaporation of condensed water and coating under vacuum at $-150°C$. The sample slide is then transferred onto the cryo-stage of the microscope, which again is cooled to $-150°C$, from where the sample can be viewed. Coating may not always be necessary if only a low voltage is used. This, however, may be a compromise between contrast quality and damage to the sample. One can also inspect the sample for heavy ice contamination before coating by transferring it directly onto the cryo-stage after the slushy nitrogen freezes. By slowly raising the temperature to $-85°C$, one can watch the ice evaporate, and then reset the temperature to $-150°C$ before moving the sample back into the cryo chamber for coating. Cryo-SEM is a rapid technique to transfer the liquid ME samples into solid material thereby enabling the visualization close to the natural state. Only few examples can be found in the literature so far, but with the establishment of this relatively new technique (at least in the area of colloidal systems) more results will surely be published in the future.

ME systems containing ethyloleate, sorbitan monolaurate, polyoxyethylene 20 sorbitan mono-oleate, butanol, and water in various compositions were investigated using this technique [85]. The cryo-SEM micrographs in Figure 9.6 are for (a) water droplets in a continuous oil matrix, (b) a bicontinuous ME system, and (c) a solution type ME [85].

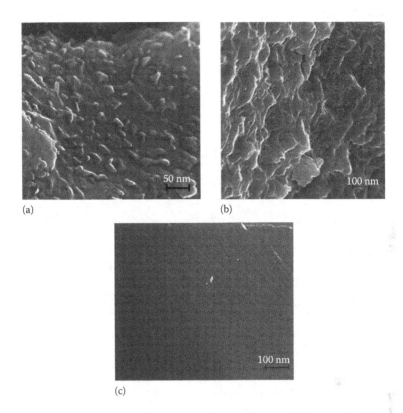

(a)   (b)

(c)

FIGURE 9.6 Cryo-FESEM micrographs of (a) w/o droplet ME (s:o 6:4, 10% water), (b) bicontinuous ME (s:o 6:4, 30% water), and (c) solution ME (s:o 9:1, 10% water). (Reproduced from Krauel, K., Girvan, L., Hook, S., and Rades, T., *Micron*, 38(8), 796–803, 2007. With permission.)

In addition, cryo-SEM investigations by Graf et al., of a system containing IPM, lecithin, the sugar-based surfactant decyl glucoside, ethanol and water, revealed uncharacteristic structures which were dominated by the most abundant phase in the ME system rather than the distinct ME structures mentioned in the system above [86]. All MEs in the latter study, however, were determined to be of the solution type by several other complementary characterization techniques and therefore lack a significant microstructure. This may explain the absence of droplet or bicontinuous structures in the cryo-SEM micrographs.

The various above-mentioned techniques should give the reader an introduction to the possibilities that are available to characterize ME systems. This section will hopefully make the reader aware that usually a multitude of techniques and comparison of the various results are necessary to fully investigate the microstructure of ME systems. We would like to point out that other techniques are available for the characterization of colloidal systems, such as x-ray and neutron scattering (SAXS and SANS), light scattering/diffraction, microcalorimetry, density, and dielectric birefringence which have not been discussed in this section, but are equally valuable for the characterization of MEs. It is up to the investigator to decide how thoroughly a particular system needs to be investigated. This will depend on the intended application of the ME system and the equipment available. Factors like storage stability, toxicity, and release characteristics, however, might be more important and a deeper understanding of the intrinsic microstructure of the system may not always be necessary.

9.7 MICROEMULSIONS AS DRUG DELIVERY SYSTEMS

9.7.1 ORAL DELIVERY

Poor drug bioavailability after oral administration can be attributed to different reasons including the enzymatic degradation of drug, poor permeability, solubility and dissolution, intestinal efflux, and presystemic metabolism in the intestine and liver. The use of MEs to overcome these problems is being extensively investigated. For the purpose of oral administration, MEs can be used as o/w or w/o MEs, ME preconcentrates, or as self-microemulsifying drug delivery systems (SMEDDS). SMEDDS comprise the nonaqueous components of ME that can readily disperse upon dilution in an aqueous environment with mild agitation to form MEs. They are often preferred over ME formulations for hydrolyzable drugs. In addition, their low volume enables packing into soft gelatin capsules for oral administration [87].

The first distinct benefit of oral ME and SMEDDS systems is the improvement in the oral bioavailability of hydrophilic drugs, namely, peptides and proteins. For example, improved oral delivery of insulin from ME system has been demonstrated [88]. Kraeling and Ritschel found that the oral pharmacological availability of insulin MEs as compared to intravenous insulin in beagle dogs was 2.1%, which further increased to 6.4% with the encapsulation of gelled MEs in hard gelatin capsules along with the protease inhibitor aprotinin and coating of the capsules for colonic release [89]. Furthermore, Cilek et al. tested the oral absorption of recombinant human insulin dissolved in the aqueous phase of w/o ME composed of Labrafil, lecithin, ethanol, and water in streptozotocin-induced diabetic male Wistar rats. The authors' demonstrated significant improvement in the oral bioavailability compared with an insulin solution [90]. In the same study, SMEDDS provided similar oral insulin bioavailability compared to formulations containing enzyme inhibitor. This may suggest that the ME formulation can protect the peptide drug from the metabolizing enzymes. In addition, other w/o ME systems were shown to improve the oral bioavailability of the linear water-soluble nonapeptide leuprolide acetate [91] and dipeptide N-acetylglucosaminyl-N-acetylmuramic acid [92]. Further, the intragastric administration of w/o ME of the epidermal growth factor was more effective in healing acute gastric ulcers in rats as compared to both intraperitoneal and intragastric administration as an aqueous solution [93]. The beneficial effects of MEs were attributed to the prevention of degradation in the gastrointestinal environment and the permeability enhancing effect of the lipid/surfactant components. ME systems have also been claimed to improve storage stability of proteins. For example, w/o ME-based media have been utilized for immobilization of water-soluble enzymes, such as lipase, in the internal, dispersed aqueous phase of w/o ME systems for biocatalytic conversion of water-insoluble substrates in the outer nonaqueous layer [94,95]. In all these applications, hydrophilic peptides or proteins were dissolved in the aqueous phase at or below their solubility levels.

The second benefit is the solubility and oral bioavailability improvement for lipophilic drugs. Early investigations revealed higher bioavailability of hydrophobic drugs after incorporation into SEDDS. For example, a SEDDS-containing lipophilic naphthalene derivative produced threefold higher values in C_{max} and AUC as compared to other dosage forms [96]. The most classical example for the application of ME formulations for improving the solubility and oral bioavailability of lipophilic drugs is the delivery of cyclosporine A (CsA). CsA is a potent immunosuppressive drug used in organ transplantation. It is a cyclic undecapeptide with very poor aqueous solubility [97]. Thus, extensive studies were carried out to improve the oral bioavailability of CsA, which eventually led to the formulation of Sandimmune which was improved further by producing an ME preconcentrate formulation of CsA, Sandimmune Neoral. The later improved the oral bioavailability and reduced intersubject variability [98,99]. However, for successful preparation of ME or SMEDDS for lipophilic drug candidates, drug precipitation after oral administration must be prevented. This can be achieved by increasing the solubilization capacity of the formulation so as to exceed the required drug concentration. Precipitation can be prevented also by incorporating antinucleant polymers that will help in formation of supersaturated system after dilution. These polymers include methyl cellulose, hydroxypropyl methylcellulose (HPMC), and polyvinylpyrrolidone [87]. A new, supersaturable self-emulsifying drug delivery system (S-SEDDS) of paclitaxel was developed employing HPMC as a precipitation inhibitor with a conventional SEDDS formulation. A pharmacokinetic study was conducted in male Sprague–Dawley rats to assess the exposure after an oral paclitaxel dose of 10 mg/kg in the SEDDS formulations with (S-SEDDS) and without HPMC. The paclitaxel S-SEDDS formulation shows 10-fold higher-maximum concentration (C_{max}) and fivefold higher oral bioavailability ($F = 9.5\%$) compared with that of the orally dosed Taxol formulation ($F = 2.0\%$) and the conventional SEDDS formulation without HPMC ($F = 1\%$) [100].

The mechanisms of improved oral bioavailability of drugs from ME systems can be summarized as follow:

- Protection of drugs (such as peptides and proteins) from degradation by the metabolizing enzymes [90].
- Increasing the gastrointestinal membrane permeability. The transcellular permeability can be increased by the surfactant components, which can disrupt the structural organization of the lipid bilayer of the absorptive membrane [101,102]. In addition, depending on the composition of the

SMEDDS, they may open the tight junctions and thus increasing the paracellular permeability [103].

* Enhancing the solubility and dissolution of lipophilic drugs [97,100].
* Increasing the translymphatic drug transport for lipophilic drugs. Depending on the lipid composition, ME can increase lymphatic transport which can reduce hepatic metabolism of drugs that have significant first pass effect [104]. Accordingly, to maximize the oral bioavailability of a hydrophobic drug from the ME formulations, the components must be selected taking into consideration the biopharmaceutical properties of the drug.
* Inhibition of the P-glycoprotein (P-gp) efflux pump, which is responsible for secretion of the absorbed molecules entering the enterocytes back to the gastrointestinal lumen. The P-gp activity can be inhibited by some pharmaceutical excipients like polyethylene glycol, Tween 80, and Cremophor EL [105]. Accordingly, incorporation of any of these excipients into the ME components can increase the bioavailability for drugs, which are known substrates of P-gp efflux pumps.

Examples of commercialized SMEDDS formulations include cyclosporine (Neoral), ritonavir (Norvir), and saquinavir (Fortovase) [106,107]. Only few SEDDS and SMEDDS formulations have been commercialized because of limitations in the usage level of excipients, e.g., surfactants and cosolvents, and the unpredictable improvement of oral bioavailability due to possibility of drug precipitation upon aqueous dilution in vivo. Predictive ability and quick methods for assessment of such problems could be very useful to the formulator in selecting lead formulations.

9.7.2   TRANSDERMAL DELIVERY

Transdermal drug delivery offers many advantages over other traditional routes of drug delivery. These include avoidance of the hepatic first pass metabolism, ease of administration with a good control over the rate of drug delivery. In addition, it allows for immediate termination of therapy when needed. Unfortunately, the barrier nature of the skin makes it difficult for most drugs to be delivered into and through it [108]. Alternative strategies have been employed to enhance dermal and transdermal delivery. These include the use of chemical penetration enhancers [109], preparation of supersaturated drug delivery systems [110], electrically driving molecules into or through the tissue employing iontophoresis [111], physically disrupting the skin structure, for example, by electroporation or sonophoresis [112,113], or by encapsulating the drug in vesicular delivery systems [114–116].

ME provides another promising alternative for dermal and transdermal delivery of both hydrophilic and lipophilic drugs [117,118]. The use of MEs in skin drug delivery has been reviewed and it was noted that although many reports of cutaneous drug delivery potential of topical MEs have been published recently, most of the studies have not been very systematic or consecutive which hampered drawing general conclusions on the interrelations between the ME properties or composition and the drug delivery rate [119,120]. This section will attempt to present these investigations in the most possible systematic way with an overview on the mechanisms of action of MEs as skin drug delivery systems.

The components of the ME not only affect the phase behavior, but may also influence its efficacy in transdermal delivery. Trotta and coworkers conducted a series of investigations using MEs with variable composition of oily phase for skin drug delivery. In the first study, they probed the transdermal delivery of hematoporphyrin from MEs containing constant surfactant and water concentrations but with variable cosurfactant–oil components. The cosurfactant–oil components were selected from benzyl alcohol, IPM, decanol, hexadecanol, oleic acid, and monoolein. All ME formulations showed significant increase in drug flux compared to the aqueous drug solution. The skin delivery depended on the cosurfactant–oil composition with formulation containing IPM–decanol (1:1) producing the highest flux [121]. In another study, the effect of varying ratios of IPM and benzyl alcohol on the transdermal delivery of felodipine from ME was investigated. The results revealed increased drug flux with increased proportions of benzyl alcohol. This may be attributed to the enhancing effect of benzyl alcohol or to the increased drug-loading capacity of the formulation [122]. In a more recent study, the effect of oily phase on the transdermal delivery of ketoprofen from ME was monitored. Oleic acid containing formulation produced the highest flux followed by IPM, Myvacet with the triacetin-containing formulation showing the least flux [123]. This effect did not depend on the solubilizing power of the oil, but can be attributed to the penetration enhancing effect.

The effect of the aqueous phase composition on the transdermal delivery from ME was also monitored. Replacing propylene glycol with transcutol in the aqueous phase increased the transdermal flux of nifedipine [124]. The influence of the ME water content was also evaluated, and it was found that the transdermal flux of glucose was higher in MEs containing 35% and 68% water compared to that containing 15% water [125]. However, it should be noted that the study employed only a water-soluble molecule and thus general conclusion about the effect of water content is difficult to report before monitoring its effect on transdermal delivery of drugs with varying lipophilicity.

The cosurfactant is another important component in ME formulations. Ethanol, isopropanol, propylene glycol, and glycerol are the most commonly used cosurfactants for transdermal MEs. The effects of cosurfactant incorporation in ME on the transdermal drug delivery have not been extensively studied. Ethanol and isopropanol containing MEs were investigated for skin delivery of estradiol in one study [126]. The study revealed greater tnasdermal fluxes of estradiol from ethanol-containing MEs compared to isopropanol-containing MEs, but it should be noted that the authors used different oil for each cosurfactant. In a recent study, El Maghraby investigated the effect of cosurfactant on transdermal delivery of hydrocortisone from eucalyptus oil ME. The author compared the effects of ethanol, isopropanol, and propylene glycol. He employed the same concentration of oil and water (20% each) with 60% of the system being either pure surfactant Tween 80) or a 1:1 mixture of surfactant/cosurfactant. All ME formulations increased the transdermal drug flux significantly, compared to the saturated aqueous solution of the drug. This effect was evident even with the basic formulation, which contained no cosurfactant. Investigating the effects of incorporation of cosurfactants on the hydrocortisone skin delivery from ME, the permeation parameters obtained in presence and absence of cosurfactants were compared. Incorporation of any of the tested cosurfactants in the ME significantly increased the transdermal flux of the drug compared to the cosurfactant-free formulation. Comparing between different cosurfactants, ethanol produced the greatest enhancement in transdermal delivery followed by propylene glycol and isopropanol [127]. Further investigations on the effects of cosurfactants are required in which drugs with different physicochemical properties are to be tested.

The mechanisms of improved skin delivery from ME systems can be summarized as follows:

- High-drug loading capacity of ME can provide higher-concentration gradient and thus increasing the driving force across the skin [119].
- Penetration enhancing effect of the ME components may provide an explanation [128]. This was not considered as the main causative effect by other investigators [127].
- Possibility that the ME components can enter the skin as monomers with the result that the solubility of the drug in the skin is increased [119]. This process will increase the partitioning of the drug into the skin creating high-drug concentration within the upper layers of the skin and thus high-driving force for transdermal drug delivery.
- Possibility of direct drug transfer from the ME droplet to the stratum corneum was considered. This consideration was based on the fact that the presence of the drug in the microstructure of the ME, which has a very small droplet size, will give large surface area for drug transfer to the skin.
- Fact that the ME had very low-interfacial tension will allow for excellent contact with the skin surface, with the vehicle filling even the wrinkles and microscopic gaps. This should enhance the vehicle skin drug transfer [119].
- Supersaturation process increases the thermodynamic activity and the driving force for the transdermal drug transfer [129]. This process can result form the fact that ME systems usually undergo phase transition upon dilution with aqueous phase or evaporation of any volatile constituents, which can influence the drug loading with the possibility of formation of supersaturated systems.

Finally, it should be noted that most of the suggested mechanisms are based on speculations and systematic investigations are sill required.

9.7.3   OCULAR DELIVERY

The simplicity and convenience of topical ocular solutions rendered them the most common dosage form for treatment of ocular diseases. These conventional dosage forms account for 90% of the available ophthalmic formulations [130]. However, rapid precorneal loss caused by drainage and high-tear fluid turnover is amongst the major problems associated with topical ophthalmic drug delivery. Only 5% of the applied drug in conventional eye drops penetrates the cornea and reaches the intraocular tissues with the rest of the dose undergoing transconjunctival absorption or drainage via the nasolacrimal duct before transnasal absorption. This results in loss of drug into the systemic circulation providing undesirable systemic side effects [131,132]. Accordingly, the challenging target is to develop topical ocular delivery systems with high-ocular retention, increased corneal drug absorption, and reduced systemic side effects whilst maintaining the simplicity and convenience of the dosage form as eye drops. Alternative strategies have been pursued to partially or fully achieve such target. These included the use of bioadhesive hydrogels [133] formulation of in situ gel forming systems [134], preparation of collagen shields [135], application of particulate, and vesicular drug delivery systems such as nanoparticles, liposomes and niosomes [136,137], or employing micellar solutions [138]. MEs on the other hand offer a promising alternative as they comprise aqueous and oily components and can therefore accommodate both hydrophilic and lipophilic drugs. Moreover, they are transparent, thermodynamic stable and possess ultralow-interfacial tension and as such offer excellent wetting and spreading properties. Further advantages result from possible improvement of solubility and stability of incorporated drugs with potential increase in bioavailability; hence, these systems off a promising alternative to conventional eye formulations. Few investigators [139–145] have considered the use of ME for ocular drug delivery. Their work was mostly focused on o/w MEs or on ME loaded into contact lenses. A w/o ME system comprising Crill 1, Crillet 4 with ethyl oleate, and water has been developed [35]. This system was found to undergo phase change upon dilution with water with viscosity of the ME increasing before transformation to the more viscous liquid crystalline system. This system has been recently investigated for ocular delivery. The concept was based on formulating liquid ME with low-water concentration. The ME can undergo phase transition after ocular application due to dilution with the resident tears with subsequent increase in viscosity, prolonged ocular retention, and improved ocular bioavailability [146,147].

Five different formulations with constant oil, surfactant ratio, and varying aqueous component at concentrations of 5% w/w (ME 5%), 10% w/w (ME 10%), 26% w/w (LC), 85% w/w (o/w EM), and 100% w/w (solution) were evaluated. The preocular retention of the selected formulations was investigated in rabbit eye using gamma scintigraphy. The results revealed that the retention of ME formulations was significantly greater than an aqueous solution with no significant difference between ME systems (containing 5% and 10% water) [147]. To evaluate the ability of these formulations to improve the ocular bioavailability of a model hydrophilic drug, a pharmacodynamic study was designed and conducted [147]. The formulations under investigated were loaded with pilocarpine HCl and applied to the rabbit eye before monitoring the miotic response [147]. The miotic response and duration of action were greatest in case of ME and LC formulations indicating high-ocular bioavailability. It was thus concluded that these phase transition ME systems can be potentially delivered into eye drops as they provide the fluidity necessary for successful instillation, with the viscosity being increased after application. Furthermore, they offered prolonged ocular retention with improved therapeutic efficacy. In addition, the safety of these phase transition formulations has been evaluated with respect to their surfactant content and microstructure. Employing the modified hen's egg chorioallantoic membrane test (HET-CAM), which specifically monitors conjunctival irritation; these formulations were found to be practically none-irritant [147]. In a more recent study, the safety of these systems was further probed by evaluating their effect on the precorneal tear film (PCTF) of rabbit eyes [148]. The effects were evaluated by assessment of the PCTF lipid layer stability and integrity using interferometry, tear evaporation rate measurements, and indirect estimation of tear volume. Ocular application of the ME formulations changed the appearance of the PCTF lipid layer, indicating lipid layer disruption. The recovery time was longer in case of ME compared with an aqueous solution (SOL). The tear evaporation rate was increased after application of both ME and LC systems compared with the SOL, with the LC system showing the greatest effect. Tear volume measurement results revealed minimal changes associated with the instillation of both ME systems. Whilst phase transition w/o ME systems did interact with the PCTF lipid layer in albino New Zealand rabbits, their effect on the volume of resident tears was found to be minimal. It is also encouraging to note that none of these systems induced any visible signs of irritation or discomfort when instilled into the eye of albino New Zealand rabbits [146–148].

In conclusion, ME systems can be considered as promising for ocular delivery; yet, further research is required to establish the safety of these surfactant-containing systems and ensure proper selection of their components.

9.7.4 PARENTERAL DELIVERY

The parenteral route is one of the most important paths for drug administration. Its importance stems from the process of generation and evaluation of new chemical entities, where the development of suitable parenteral formulations is required for preclinical studies. In some cases, it offers the only viable choice for the administration of poorly orally absorbed drugs such as amphotericin B and paclitaxel [149]. In addition, it is the most effective route in the case of emergency as it ensures very quick onset of action. However, design of parenteral formulations is challenging as only limited number of excipients can be safely employed. The challenge is even greater when formulating hydrophobic drugs due to solubility restrictions. Alternative strategies have been adopted to overcome this challenge. Micellar solubilization which has limited drug loading

capacity was employed for compounds such as vitamin D [150]. Phospholipids/
bile salts-mixed micelles were also used with some success. This was utilized for
amphotricin B [151]. However, the significant hemolytic effect of the bile salt is
a major drawback for these formulations [152]. Liposomes, which are vesicles,
made of phospholipids with or without cholesterol attracted some interest with
few liposomes-based formulations reaching the market due to poor stability and
high cost. The representative examples of liposome-based parenterals include
doxorubicin (Doxil) and Amphotericin B (AmBisome). In addition to these,
cyclodextrins were also used for the solubilization of hydrophobic drugs, but
they are very expensive with some problems in complexation [150].

Lipid emulsions showed superior solubilization capacity, but they suffer from
poor physical stability on long-term storage, risk of emboli formation, and rapid
growth of microorganisms [153]. Accordingly, MEs were investigated to overcome some of these problems but selection of their components is critical. The
selected excipients should be biocompatible, sterilizable, available as nonpyro-
genic grade, nonirritant to nerves, and nonhemolytic. The rest of this section will
present some potential applications for MEs as parenteral drug delivery systems.

Paclitaxel is a potent anticancer but with high lipophilicity and thus poor oral
bioavailability. It is administered as an intravenous infusion to patients as Taxol
[154]. Taxol is based on cosolvent system of Cremophor EL (polyoxyethylated
castor oil) and ethanol. Despite its chemical stability for 27 days after dilution
with the infusion fluid, this preparation showed evidence of paclitaxel precipitation 3 days after dilution [149]. In addition, the adverse effects associated with the
Cremophore EL severely limited the clinical utility of the product in several cases.
Parenteral paclitaxel ME was first developed by He et al. [155]. The system comprised lecithin, ethanol, poloxamer 188, and a very small amount of Cremophore
EL as compared to that used in Taxol. This ME formulation demonstrated significantly less hypersensitivity reactions compared to Taxol and was attributed to the
lower content of Cremophore EL. Pharmacokinetic evaluations revealed higher
AUC values in case of the ME formulation compared to Taxol. In a recent investigation, Cremophor-free w/o ME was developed for paclitaxel. The potential of
medium chain mono- and di-glycerides such as Capmul MCM and Myvacet in the
development of parenteral MEs was evaluated. The developed MEs were found to
be safer as compared to Taxol with respect to erythrocyte toxicity. The pharmacokinetic studies indicated that paclitaxel MEs exhibited—two- to fivefold higher-
circulation time, and—three- to eightfold wider tissue distribution as compared to
that of Taxol [156,157]. Zhang et al. [158] have successfully developed Cremophor-free paclitaxel SMEDDS. The formulation included tributyrin, tricaproin, ethanol, lecithin, and poloxamer 188. The SMEDDS formulation on dilution with
saline yielded MEs with a globule size of 16 nm. Comparative evaluation of the
pharmacokinetic parameters of paclitaxel SMEDDS and Taxol indicated that
paclitaxel SMEDDS yielded significantly higher AUC values and circulation time
compared to Taxol.

Amphotericin B is an extremely potent drug, used for treatment of systemic
fungal infections. Unfortunately, its oral bioavailability is very low due to negligible

gastrointestinal absorption. Accordingly, parenteral delivery is the only viable alternative, however, is associated with serious adverse effects [151]. Formulation of this drug in the form of mixed micellar formulation, liposomes or lipid complexes, reduced the adverse effects significantly but remained very expensive [151]. Amphotericin B was formulated as biocompatible ME based on polysorbate 80, lecithin, and IPM. The acute toxicity, efficacy, and in vivo tolerability of the amphotericin B ME were tested relative to the traditional mixed micellar formulation. The study revealed higher LD50 for the ME formulation reflecting a better safety profile [159].

In addition to being safer, lecithin-based amphotericin MEs were superior to marketed formulation with respect to ability to reduce fungal load and to increase the survival period of the infected mice [160].

In conclusion, MEs are promising candidates for the formulation of parenteral hydrophobic drugs. Once again, it is critical to ensure proper selection of the components and monitor their long-term safety.

9.8   CLOSING STATEMENT

MEs are emerging as promising drug carriers. They are attractive to formulation and drug delivery scientists due to their thermodynamic stability and simple preparation method. They can accommodate poorly soluble drugs (both hydrophilic and lipophilic) and protect those that are vulnerable to chemical and enzymatic degradation. They have the potential to increase the solubility of poorly soluble drugs, enhance the bioavailability of drugs with poor permeability, reduce patient variability, and offer an alternative for controlled drug release. A critical look at the literature shows that exciting research is taking place. It is only a matter of time before we witness the marketing of new ME-based dosage forms, following the successful introduction of Sandimmune Neoral, Norvir, and Fortovase.

ABBREVIATIONS

Cal	calories
CEVS	controlled environment vitrification system
cm	centimeter
CPP	critical packing parameters
DSC	differential scanning calorimetry
EM	electron microscopy
FESEM	field emission scanning electron microscopy
FF-TEM	freeze fracture transmission electron microscopy
FT-PGSE	Fourier transformed pulsed-gradient spin-echo
GRAS	generally regarded as safe
HET-CAM	hen's egg chorioallantoic membrane test
HLB	hydrophile–lipophile balance
J	joules
K	kelvin

ME	microemulsion
NaCl	sodium chloride
nm	nanometer
NMR	nuclear magnetic resonance
o/w	oil-in-water
PEG	polyethylene glycol
PIT	phase inversion temperature
PLM	polarizing light microscopy
RF	radio frequency
RI	refractive index
rpm	rotations per minute
s	seconds
S	Siemens
S	system entrophy
s:o	surfactant:oil
SANS	small angle neutron scattering
SAXS	small angle x-ray scattering
SEM	scanning electron microscopy
T	absolute temperature
Ta–W	tantalum–tungsten
TEM	transmission electron microscopy
w/o	water-in-oil
w/w	weight-per-weight
μS	microSiemens

SYMBOLS AND TERMINOLOGIES

γ	interfacial tension
ΔA	change in the interfacial area
ΔG_f	free energy of ME formation
σ	electrical conductivity
ϕ_c	critical water concentration
η	dynamic viscosity
υ	kinematic viscosity
ρ	density
τ	shear stress
D	shear rate
ϕ	volume fraction
g	pulse amplitude
Δ	pulse interval
δ	pulse duration
t	time
I	signal intensity
I_0	signal intensity at time zero
D	diffusion coefficient

γ gyromagnetic ratio
$\langle \Delta r^2 \rangle$ mean square displacement in space
Δt observation time
k_B Boltzmann constant
T temperature in kelvin
r_H hydrodynamic radius

REFERENCES

1. Danielsson, I. and Lindman, B. 1981. The definition of microemulsion. *Colloid and Surfaces*, 3, 391.
2. Tenjarla, S. 1999. Microemulsions: An overview and pharmaceutical applications. *Critical Reviews in Therapeutic Drug Carrier Systems*, 16(5), 461–521.
3. Attwood, D. 1994. Microemulsions. In Kreuter, J. (Ed.), *Colloidal Drug Delivery Systems* (pp. 31–71). New York: Marcel Dekker.
4. Lawrence, M. J. and Rees, G. D. 2000. Microemulsion-based media as novel drug delivery systems. *Advanced Drug Delivery Reviews*, 45(1), 89–121.
5. Mitchell, D. and Ninham, W. 1981. Micelles vesicles and microemulsions. *Journal of Chemical Society Faraday Transactions*, 77(2), 601–629.
6. Israelachvili, J. N., Mitchell, D., and Ninham, W. 1976. Theory of self-assembly of hydrocarbon amphiphiles into micelles and bilayers. *Journal of Chemical Society Faraday Transactions II*, 72, 1525–1568.
7. Prince, L. M. 1977. Formulation. In Prince, L. M. (Ed.), *Microemulsions: Theory and Practice* (pp. 33–49). New York: Academic Press.
8. Shinoda, K. and Kuineda, H. 1973. Condition to produce so-called microemulsions: Factors to increase mutual solubility of oil and water solubilisers. *Journal of Colloid and Interface Science*, 42, 381–387.
9. Aboofazeli, R., Lawrence, C. B., Wicks, S. R., and Lawrence, M. J. 1994. Investigations into the formation and characterization of phospholipid microemulsions. Part 3. Pseudo-ternary phase diagrams of systems containing water–lecithin–isopropyl myristate and either an alkanoic acid, amine, alkanediol, polyethylene glycol alkyl ether or alcohol as cosurfactant. *International Journal of Pharmaceutics*, 111, 63–72.
10. Kahlweit, M., Busse, G., and Faulhaber, B. 1995. Preparing microemulsions with alkyl monoglucosides and the role of *n*-alkanols. *Langmuir*, 11(9), 3382–3387.
11. Alany, R. G., Rades, T., Agatonovic-Kustrin, S., Davies, N. M., and Tucker, I. G. 2000. Effects of alcohols and diols on the phase behaviour of quaternary systems. *International Journal of Pharmaceutics*, 196(2), 141–145.
12. FDA. (Revised as of April 1, 2001). Food and Drugs, Chapter 1. Food and Drug Administration. In Department of Health and Human Services (Ed.), *Code of Federal Regulations*, Title 21 (Vol. 3): U.S. Government Printing Office via GPO Access.
13. Thevenin, M. A., Grossiord, J. L., and Poelman, M. C. 1996. Sucrose esters cosurfactant microemulsion systems for transdermal delivery—Assessment of bicontinuous structures. *International Journal of Pharmaceutics*, 137(2), 177–186.
14. Malcolmson, C., Satra, C., Kantaria, S., Sidhu, A., and Lawrence, M. J. 1998. Effect of oil on the level of solubilization of testosterone propionate into nonionic oil-in-water microemulsions. *Journal of Pharmaceutical Sciences*, 87(1), 109–116.

15. Ho, H. O., Hsiao, C. C., and Sheu, M. T. 1996. Preparation of microemulsions using polyglycerol fatty acid esters as surfactant for the delivery of protein drugs. *Journal of Pharmaceutical Sciences*, 85(2), 138–143.
16. Kuineda, H., Hasegawa, Y., John, A. C., Naito, M., and Muto, M. 1996. Phase behaviour of polyoxyethylene hydrogenated caster oil in oil/water systems. *Colloid and Surfaces*, 109, 209–216.
17. Agatonovic-Kustrin, S., Glass, B. D., Wisch, M. H., and Alany, R. G. 2003. Prediction of a stable microemulsion formulation for the oral delivery of a combination of antitubercular drugs using ANN methodology. *Pharmaceutical Research*, 20(11), 1760–1765.
18. Aboofazeli, R. and Lawrence, M. J. 1993. Investigations into the formation and characterization of phospholipid microemulsions. Part 1. Pseudo-ternary phase diagrams of systems containing water–lecithin–alcohol–isopropyl myristate. *International Journal of Pharmaceutics*, 93(May 31), 161–175.
19. Aboofazeli, R. and Lawrence, M. J. 1994. Investigations into the formation and characterization of phospholipid microemulsions. Part 2. Pseudo-ternary phase diagrams of systems containing water–lecithin–isopropyl myristate and alcohol: Influence of purity of lecithin. *International Journal of Pharmaceutics*, 106, 51–61.
20. Aboofazeli, R., Patel, N., Thomas, M., and Lawrence, M. J. 1995. Investigations into the formation and characterization of phospholipid microemulsions. 4. Pseudo-ternary phase diagrams of systems containing water–lecithin–alcohol and oil—the influence of oil. *International Journal of Pharmaceutics*, 125(1), 107–116.
21. Shinoda, K., Araki, M., Sadaghiani, A., Khan, A., and Lindman, B. 1991. Lecithin-based microemulsions; phase behaviour and microstructure. *Journal of Physical Chemistry*, 95, 989–993.
22. Saint-Ruth, H., Attwood, D., Ktistis, G., and Taylor, C. J. 1995. Phase studies and particle size analysis of oil-in-water phospholipid microemulsions. *International Journal of Pharmaceutics*, 116, 253–261.
23. Warisnoicharoen, W., Lansley, A. B., and Lawrence, M. J. 2000. Nonionic oil-in-water microemulsions: The effect of oil type on phase behaviour. *International Journal of Pharmaceutics*, 198(1), 7–27.
24. Park, K. M., Lee, M. K., Hwang, K. J., and Kim, C. K. 1999. Phospholipid-based microemulsions of flurbiprofen by the spontaneous emulsification process. *International Journal of Pharmaceutics*, 183(2), 145–154.
25. Constantinides, P. P. 1995. Lipid microemulsions for improving drug dissolution and oral absorption: Physical and biopharmaceutical aspects. *Pharmaceutical Research*, 12(Nov.), 1561–1572.
26. Constantinides, P. P., Lancaster, C. M., Marcello, J., Chiossone, D. C., Orner, D., Hidalgo, I., et al. 1995. Enhanced intestinal absorption of an RGD peptide from water-in-oil microemulsions of different composition and particle size. *Journal of Controlled Release*, 34, 109–116.
27. Constantinides, P. P. and Yiv, S. H. 1995. Particle size determination of phase-inverted water-in-oil microemulsions under different dilution and storage conditions. *International Journal of Pharmaceutics*, 115(Mar. 7), 225–234.
28. Constantinides, P. P., Welzel, G., Ellens, H., Smith, P. L., Sturgis, S., Yiv, S. H., et al. 1996. Water-in-oil microemulsions containing medium-chain fatty acids/salts: Formulation and intestinal absorption enhancement evaluation. *Pharmaceutical Research*, 13(2), 210–215.
29. Kahlweit, M., Busse, G., Faulhaber, B., and Eibl, H. 1995. Preparing nontoxic microemulsions. *Langmuir*, 11(11), 4185–4187.

30. Kahlweit, M., Busse, G., and Faulhaber, B. 1996. Preparing microemulsions with alkyl monoglucosides and the role of alkanediols as cosolvents. *Langmuir*, 12, 861–862.
31. Joubran, R., Parris, N., Lu, D., and Trevino, S. 1994. Synergetic effect of sucrose and ethanol on formation of triglyceride microemulsions. *Journal of Dispersion Science and Technology*, 15(6), 687–704.
32. Stubenrauch, C., Paeplow, B., and Findenegg, G. H. 1997. Microemulsions supported by octyl monoglucoside and geraniol. 1. The role of alcohol in the interfacial layer. *Langmuir*, 13, 3652–3658.
33. Bhargava, H. N., Narurkar, A., and Lieb, L. M. 1987. Using microemulsions for drug delivery. *Pharmaceutical Technology*, 11(Mar.), 46.
34. Bourrel, M., Schechter, R. S. 1988. The R-ratio. In Bourrel, M., Schechter, R. S. (Eds.), *Microemulsions and Related Systems* (pp. 1–30). New York, Basel: Marcel Dekker.
35. Alany, R. G., Davies, N. M., Tucker, I. G., and Rades, T. 2001. Characterising colloidal structures of pseudoternary phase diagrams formed by oil/water/amphiphile systems. *Drug Development and Industrial Pharmacy*, 27(1), 33–41.
36. Alany, R. G., Agatonovic-Kustrin, S., Rades, T., and Tucker, I. G. 1999. Use of artificial neural networks to predict quaternery phase systems from limited experimental data. *Journal of Pharmaceutical and Biomedical Analysis*, 19(3–4), 443–452.
37. Richardson, C. J., Mbanefo, A., Aboofazeli, R., Lawrence, M. J., and Barlow, D. J. 1997. Prediction of phase behavior in microemulsion systems using artificial neural networks. *Journal of Colloid and Interface Science*, 187(2), 296–303.
38. Flanagan, J. and Singh, H. 2006. Recent Advances in the delivery of food-derived bioactives and drugs using microemulsions. In Mozafari, R. M. (Ed.), *Nanocarrier Technologies: Frontiers of Nanotherapy* (pp. 95–112). Dordrecht, The Netherlands: Springer.
39. Koshnevis, P., Mortazavi, S. A., Lawrence, M. J., and Aboofazeli, R. 1997. In-vitro release of sodium salicylate from water-in-oil phospholipid microemulsions. *Journal of Pharmacy and Pharmacology*, 49(S4), 47.
40. Krauel, K., Davies, N. M., and Rades, T. 2005. Using different structure types of microemulsions for the preparation of poly(alkylcyanoacrylate) nanoparticles by interfacial polymerisation. *Journal of Controlled Release*, 106(1–2), 76–87.
41. Podlogar, F., Gasperlin, M., Tomsic, M., Jamnik, A., and Bester Rogac, M. 2004. Structural characterisation of water–Tween 40/Imwitor 308–isopropyl myristate microemulsion using different experimental methods. *International Journal of Pharmaceutics*, 276, 115–128.
42. Stilbs, P. 1986. A comparative study of micellar solubilisation for combination of surfactants and solubilisates using Fourier transform pulsed-gradient spin-echo NMR multicomponent self-diffusion technique. *Journal of Colloid and Interface Science*, 94(2), 463–469.
43. Krauel, K., Girvan, L., Hook, S. M., and Rades, T. 2007. Characterisation of colloidal drug delivery systems: From the naked eye to cryoFESEM. *Micron*, 38, 796–803.
44. Regev, O., Ezrahi, S., Aserin, A., Garti, N., Wachtel, E., Kaler, E. W., Khan, A., and Talmon, Y. 1996. A study of the microstructure of a four-component nonionic microemulsion by Cryo-TEM, NMR, SAXS, and SANS. *Langmuir*, 12(3), 668–674.
45. Nagarajan, R. and Ruckenstein, E. 2000. Molecular theory of microemulsions. *Langmuir*, 16, 6400–6415.
46. Djordjevic, L., Primorac, M., Stupar, M., and Krajisnik, D. 2004. Characterization of caprylocaproyl macrogolglycerides based microemulsion drug delivery vehicles for an amphiphilic drug. *International Journal of Pharmaceutics*, 271(1–2), 11–19.

47. Mehta, S., Kaur, G., and Bhasin, K. 2007. Incorporation of antitubercular drug isoniazid in pharmaceutically accepted microemulsion: Effect on microstructure and physical parameters. *Pharmaceutical Research*, 25(1), 227–236.
48. Constantinides, P. P. and Scalart, J.-P. 1997. Formulation and physical characterization of water-in-oil microemulsions containing long- versus medium-chain glycerides. *International Journal of Pharmaceutics*, 158(1), 57–68.
49. Graetz, K., Helmstedt, M., Meyer, H. M., and Quitzsch, K. 1998. Structure and phase behaviour of the ternary system water, *n*-heptane and the nonionic surfactant Igepal CA520. *Colloid and Polymer Science*, 276(2), 131–137.
50. Collings, P. J. 2002. *Liquid crystals: Natures's Delicate Phase of matter.* Princeton, NJ: Princeton University Press.
51. Collings, P. J. and Hird, M. 1997. *Introduction to Liquid Crystals: Chemistry and Physics.* London, UK: Taylor & Francis Ltd.
52. Sinko, P. J. and Martin, A. N. 2006. *Martin's Physical Pharmacy and Pharmaceutical Sciences: Physical Chemical and Biopharmaceutical Principles in the Pharmaceutical Sciences.* Philadelphia, PA: Lippincott Williams & Wilkins.
53. Laguës, M. and Sauterey, C. 1980. Percolation transition in water in oil microemulsions: Electrical conductivity measurements. *Journal of Physical Chemistry*, 84, 3503–3508.
54. Mehta, S. K. and Bala, K. 2000. Tween-based microemulsions: A percolation view. *Fluid Phase Equilibria*, 172(2), 197–209.
55. Mehta, S. K. and Bala, K. 1995. Volumetric and transport properties in microemulsions and the point of view of percolation theory. *Physical Review E*, 51(6), 5732.
56. Strey, R. 1994. Microemulsion microstructure and interfacial curvature. *Colloid and Polymer Science*, 272(8), 1005–1019.
57. Baroli, B., Lopez-Quintela, M. A., Delgado-Charro, M. B., Fadda, A. M., and Blanco-Mendez, J. 2000. Microemulsions for topical delivery of 8-methoxsalen. *Journal of Controlled Release*, 69(1), 209–218.
58. Kumar, P. and Mittal, K. L. 1999. *Handbook of Microemulsion Science and Technology.* New York: Marcel Dekker, Inc.
59. Einstein, A. 1906. New determination of molecular dimensions. *Annalen der Physik*, 19, 289–305.
60. Einstein, A. 1911. Eine neue Bestimmung der Moleküldimensionen. *Annalen der Physik*, 34, 591–592.
61. Attwood, D., Currie, L. R. J., and Elworthy, P. H. 1974. Studies of solubilised micellar solutions. *Journal of Colloid and Interface Science*, 46, 261–265.
62. Krauel, K. 2005. *Formulation and Characterisation of Nanoparticles Based on Biocompatible Microemulsions.* Dunedin: School of Pharmacy, University of Otago, p. 231.
63. D'Antona, P., Parker, W. O., Zanirato, M. C., Esposito, E., and Nastruzzi, C. 2000. Rheologic and NMR characterization of monoglyceride-based formulations. *Journal of Biomedical Materials Research*, 52(1), 40–52.
64. Kriwet, K. and Mueller-Goymann, C. C. 1995. Diclofenac release from phospholipid drug systems and permeation through excised human stratum corneum. *International Journal of Pharmaceutics*, 125(2), 231–242.
65. Craig, D. Q. M. and Reading M. 2006. *Thermal Analysis of Pharmaceuticals.* Boca Raton, FL: CRC Press.
66. Garti, N., Aserin, A., Tiunova, I., and Fanun, M. A. 2000. DSC study of water behaviour in water-in-oil microemulsions stabilized by sucrose esters and butanol. *Colloids and Surfaces A: Physicochemical and Engineering Aspects*, 170(1), 1–18.

67. Moulik, S. P. and Paul, B. K. 1998. Structure, dynamics and transport properties of microemulsions. *Advances in Colloid and Interface Science*, 78(2), 99–195.

68. Price, W. S. 1997. Pulsed-field gradient nuclear magnetic resonance as a tool for studying translational diffusion: Part I. Basic theory. *Concepts in Magnetic Resonance*, 9(5), 299–336.

69. Stilbs, P. 1987. Fourier transform pulsed-gradient spin-echo studies of molecular diffusion. *Progress in NMR Spectroscopy*, 19(1), 1–45.

70. Södermann, O. and Stilbs, P. 1994. NMR studies of complex surfactant systems. *Progress in NMR Spectroscopy*, 26, 445–482.

71. Stilbs, P. 1999. Diffusion studied using NMR spectroscopy. In Tranter, G., Holmes, J., Lindon, J. (Eds.), *Encyclopedia of Spectroscopy and Spectrometry* (pp. 369–374). London, UK: Academic Press.

72. Kärger, J., Pfeiffer, H., and Heink, W. 1988. Principles and application for self-diffusion measurements by nuclear magnetic resonance. *Advances in Magnetic Resonance*, 12, 2–89.

73. Stilbs, P. 1987. Fourier transform pulsed-gradient spin-echo studies of molecular diffusion. *Progress in Nuclear Magnetic Resonance Spectroscopy*, 19(1), 1–45.

74. Lindmann, B. and Stilbs, P. 1987. Molecular diffusion in microemulsions. In: Friberg, S. E., Bothorel, P. (Eds.), *Microemulsions: Structure and Dynamics* (pp. 119–152). Boca Raton, FL: CRC Press.

75. Jönsson, B., Lindman, B., Holmberg, K., and Kronberg, K. 1998. *Surfactants and Polymers in Aqueous Solution*. Chichester, UK: John Wiley & Sons.

76. Biruss, B., Kahlig, H., Valenta, C. 2007. Evaluation of an eucalyptus oil containing topical drug delivery system for selected steroid hormones. *International Journal of Pharmaceutics*, 328(2), 142–151.

77. Kreilgaard, M., Pedersen, E. J., Jaroszewski, J. W. 2000. NMR characterisation and transdermal drug delivery potential of microemulsion systems. *Journal of Controlled Release*, 69(3), 421–433.

78. Burauer, S., Belkoura, L., Stubenrauch, C., and Strey, R. 2003. Bicontinuous microemulsions revisited: A new approach to freeze fracture electron microscopy (FFEM). *Colloids and Surfaces A: Physicochemical Engineering Aspects*, 228, 159–170.

79. Bellare, J. R., Davis, H. T., Scriven, L. E., and Talmon, Y. 1988. Controlled environment vitrification system: An improved sample preparation technique. *Journal of Electron Microscopy Technique*, 10(1), 87–111.

80. Marchand, K. E., Tarret, M., Lechaire, J. P., Normand, L., Kasztelan, S., and Cseri, T. 2003. Investigation of AOT-based microemulsions for the controlled synthesis of MoSx nanoparticles: An electron microscopy study. *Colloids and Surfaces A: Physicochemical and Engineering Aspects*, 214(1–3), 239–248.

81. Magdassi, S., Ben Moshe, M., Talmon, Y., and Danino, D. 2003. Microemulsions based on anionic gemini surfactant. *Colloids and Surfaces A: Physicochemical and Engineering Aspects*, 212(1), 1–7.

82. Spicer, P. T., Small, W. B., Lynch, M. L., and Burns, J. L. 2002. Dry powder precursors of cubic liquid crystalline nanoparticles (cubosomes). *Journal of Nanoparticle Research*, 4(4), 297–311.

83. Sanders, M., Brown, L., Deliyannis, G., and Pearse, M. 2005. ISCOM(TM)-based vaccines: The second decade. *Immunology and Cell Biology*, 83(2), 119–128.

84. Agarwal, V., Singh, M., McPherson, G., John, V., and Bose, A. 2004. Freeze fracture direct imaging of a viscous surfactant mesophase. *Langmuir*, 20(1), 11–15.

85. Krauel, K., Girvan, L., Hook, S., and Rades, T. 2007. Characterisation of colloidal drug delivery systems from the naked eye to Cryo-FESEM. *Micron*, 38(8), 796–803.

86. Graf, A., Ablinger, E., Peters, S., Zimmer, A., Hook, S. M., and Rades, T. 2008. Microemulsions containing lecithin and sugar-based surfactants: Nanoparticle templates for delivery of proteins and peptides. *International Journal of Pharmaceutics*, 350(1–2), 351–360, 2008.

87. Narang, A. S., Delmarre, D., and Gao, D. 2007. Stable drug encapsulation in micelles and microemulsions. *Advanced Drug Delivery Reviews*, 345, 9–25.

88. Cho, Y. W. and Flynn, M. 1989. Oral delivery of insulin. *Lancet*, 2, 1518–1519.

89. Kraeling, M. E. and Ritschel, W. A. 1992. Development of a colonic release capsule dosage form and the absorption of insulin. *Methods and Findings in Experimental and Clinical Pharmacology*, 14, 199–209.

90. Cilek, A., Celebi, N., Tirnaksiz, F., and Tay, A. 2005. A lecithin-based microemulsion of rh-insulin with aprotinin for oral administration: Investigation of hypoglycemic effects in non-diabetic and STZ-induced diabetic rats. *International Journal of Pharmaceutics*, 298, 176–185.

91. Zheng, J. Y. and Fulu, M. Y. 2006. Decrease of genital organ weights and plasma testosterone levels in rats following oral administration of leuprolide microemulsion. *International Journal of Pharmaceutics*, 307, 209–215.

92. Lyons, K. C., Charman, W. N., Miller, R., and Porter, C. J. 2000. Factors limiting the oral bioavailability of *N*-acetylglucosaminyl-*N*-acetylmuramyl dipeptide (GMDP) and enhancement of absorption in rats by delivery in a water-in-oil microemulsion. *International Journal of Pharmaceutics*, 199, 17–28.

93. Celebi, N., Turkyilmaz, A., Gonul, B., and Ozogul, C. 2002. Effects of epidermal growth factor microemulsion formulation on the healing of stress-induced gastric ulcers in rats. *Journal of Controlled Release*, 83, 197–210.

94. Schuleit, M. and Luisi, P. L. 2001. Enzyme immobilization in silica-hardened organogels. *Biotechnology and Bioengineering*, 72, 249–253.

95. Madamwar, D. and Thakar, A. 2004. Entrapment of enzyme in water-restricted microenvironment for enzyme-mediated catalysis under microemulsion based organogels. *Applied Biochemistry and Biotechnology*, 118, 361–369.

96. Shah, N. H., Carvajal, M. T., Patel, C. I., Infeld, M. H., and Malick, A. W. 1994. Self emulsifying drug delivery systems (SEDDS) with polyglycolized glycerides for improving in vitro dissolution and oral absorption of lipophilic drugs. *International Journal of Pharmaceutics*, 106, 15–23.

97. Ritschel, W.A. 1996. Microemulsion technology in the reformulation of cyclosporine: The reason behind the pharmacokinetic properties of Neoral. *Clinical Transplantation*, 10, 364–373.

98. Vondercher J, and Meizner A. 1994. Rationale for the development of Sandimmune Neoral. *Transplantation Proceedings*, 26, 2925–2927.

99. Meinzer A, Mueller E, and Vonderscher J. 1995. Microemulsion—a suitable galenical approach for the absorption enhancement of low soluble compounds? *B T Gattefosse*, 88, 21–26.

100. Gao, P., Rush, B. D., Pfund, W. P., Huang, T., Bauer, J. M., Morozowich, W., Kuo, M. S., and Hageman, M. J. 2003. Development of a supersaturable SEDDS (S-SEDDS) formulation of paclitaxel with improved oral bioavailability. *Journal of Pharmaceutical Sciences*, 92, 2386–2398.

101. Swenson ES. and Curatolo WJ. 1992. Means to enhance penetration. *Advanced Drug Delivery Reviews*, 8, 39–92.

102. Kim, H. J., Yoon, K. A., Hahn, M., Park, E. S., and Chi, S. C. 2000. Preparation and in vitro evaluation of self-microemulsifying drug delivery systems containing idebenone. *Drug Development and Industrial Pharmacy*, 26, 523–529.

103. Sha, X., Yan, G., Wu, Y., Li, J., and Fang, X. 2005. Effect of self-microemulsifying drug delivery systems containing Labrasol on tight junctions in Caco-2 cells. *European Journal of Pharmaceutical Science*, 24, 477–486

104. Dahan, A. and Hoffman, A. 2006. Use of a dynamic in vitro lipolysis model to rationalize oral formulation development for poor water soluble drugs: Correlation with in vivo data and the relationship to intra-enterocyte processes in rats. *Pharmaceutical Research*, 23, 2165–2174.

105. Hugger, E. D., Novak, B. L., Burton, P. S., Audus, K. L., and Borchardt, R. T. 2002. A comparison of commonly used polyethoxylated pharmaceutical excipients on their ability to inhibit P-glycoprotein activity in vitro. *Jouranl of Pharmaceutical Sciences*, 91, 1991–2002.

106. Cooney, G. F., Jeevanandam, V., Choudhury, S., Feutren, G., Mueller, E. A., and Eisen, H. J. 1998. Comparative bioavailability of Neoral and Sandimmune in cardiac transplant recipients over 1 year. *Transplantation ProcEedings*, 30, 1892–1894.

107. Porter, C. J. and Charman, W. N. 2001. In vitro assessment of oral lipid based formulations. *Advanced Drug Delivery Reviews*, 50 (Suppl. 1), S127–S147.

108. Barry, B. W. 1983. *Dermatological Formulations: Percutaneous Absorption*. Marcel Dekker: New York and Basel.

109. Goodman, M. and Barry, B. W. 1988. Action of penetration enhancers on human skin as assessed by permeation of model drugs 5-fluorouracil and estradiol. I. Infinite dose technique. *The Journal of Investigative Dermatology*, 91, 323–327.

110. Megrab, N. A Williams, A. C., and Barry, B. W. 1995. Oestradiol permeation across human skin and silastic membranes: Effects of propylene glycol and supersaturation. *Journal of Controlled Release*, 36, 277–294.

111. Miller, L. L., Kolaskie, C. J., Smith, G. A., and Rivier, J. 1990. Transdermal iontophoresis of gonadotropin releasing hormone and two analogues. *Journal of Pharmaceutical Sciences*, 79, 490–493.

112. Kost, J., Pliquett, U., Mitragotri, S., Yamamoto, A., Langer, R., and Weaver, J. 1996. Synergistic effect of electric field and ultrasound on transdermal transport. *Pharmaceutical Research*, 13, 633–638.

113. Banga, A. K., Bose, S., and Ghosh, T. K. 1999. Iontophoresis and electroporation: Comparisons and contrasts. *International Journal of Pharmaceutics*, 179, 1–19.

114. Mezei, M. and Gulasekharam, V. 1980. Liposomes—a selective drug delivery system for topical route of administration. 1. Lotion dosage form. *Life Sciences*, 26, 1473–1477.

115. El Maghraby, G. M., Williams, A. C., and Barry, B. W. 1999. Skin delivery of oestradiol from deformable and traditional liposomes: Mechanistic studies. *Journal of Pharmaceutical Sciences*, 51, 1123–1134.

116. El Maghraby, G. M., Williams, A. C., and Barry, B. W. 2006. Can drug bearing liposomes penetrate intact skin? *The Journal of Pharmacy and Pharmacology*, 58, 415–429.

117. Boltri, L., Morel, S., Trotta, M., and Gasco, M. R. 1994. In vitro transdermal permeation of nifedipine from thickened microemulsion. *Journal de Pharmacie de Belgique*, 49, 315–320.

118. Delgado-Charro, M. B., Iglesias-Vilas, G., Blanco-Mendez, J., Lopez-Quintela, M. A., and Guy, R. H. 1997. Delivery of a hydrophilic solute through the skin from novel microemulsion systems. *European Journal of Pharmaceutics and Biopharmaceutics*, 43, 37–42.

119. Kreilgaard, M. 2002. Influence of microemulsions on cutaneous drug delivery. *Advanced Drug Delivery Reviews*, 54, S77–S98.

120. Kogan, A. and Garti, N. 2006. Microemulsions as transdermal drug delivery vehicles. *Advances in Colloid and Interface Science*, 123–126, 369–385.
121. Trotta, M., Gasco, M. R., Caputo, O., and Sancin, P. 1994. Transcutaneous diffusion of hematoporphyrin in photodynamic therapy: In vitro release from microemulsions. *STP Pharma Sciences* 4, 150–154.
122. Trotta, M., Morel, S., and Gasco, M. R. 1997. Effect of oil phase composition on the skin permeation of felodipine from o/w microemulsions. *Pharmazie*, 52, 50–53.
123. Rhee, Y. S., Choi, J. G., Park, E. S., and Chi, S. C. 2001. Transdermal delivery of ketoprofen using microemulsions. *International Journal of Pharmaceutics*, 228, 161–170.
124. Boltri, L., Morel, S., Trotta, M., and Gasco, M. R. 1994. In vitro transdermal permeation of nifedipine from thickened microemulsions. *Journal de Pharmacie de Belgique*, 49, 315–320.
125. Osborne, D. W., Ward, A. J., and O'Neill, K. J. 1991. Microemulsions as topical drug delivery vehicles: In-vitro transdermal studies of a model hydrophilic drug. *Journal of Pharmacy and Pharmacology*, 43, 450–454.
126. Peltola, S., Saarinen-Savolainen, P., Kiesvaara, J., Suhonen, T. M., and Urtti, A. 2003. Microemulsions for topical delivery of estradiol. *International Journal of Pharmaceutics*, 254, 99–107.
127. El Maghraby, G. M. 2008. Transdermal delivery of hydrocortisone from eucalyptus oil microemulsion: Effects of cosurfactants, *International Journal of Pharmaceutics*, 355(1–2), 285–292, 2008.
128. Dreher, F., Walde, P., Walther, P., and Wehrli, E. 1997. Interaction of a lecithin microemulsion gel with human stratum corneum and its effect on transdermal transport. *Journal of Controlled Release,* 45, 131–140.
129. Kemken, J., Ziegler, A., and Muller, B. W. 1992. Influence of supersaturation on the pharmacodynamic effect of bupranolol after dermal administration using microemulsion as vehicle, *Pharmaceutical Research*, 9, 554–558.
130. Le Bourlais, C., Acar, L., Zia, H., and Sado, P. A. 1998. Ophthalmic drug delivery systems—recent advances. *Progress in Retinal and Eye Research*, 17(1), 33–55.
131. Meseguer, G., Gurny, R., Buri, P. 1994. In vivo evaluation of dosage forms: Application of gamma scintigraphy to non-enteral routes of administration. *Journal of Drug Targeting*, 2, 269–288.
132. Lang, J. C. 1995. Ocular drug delivery conventional ocular formulations. *Advanced Drug Delivery Reviews*, 16, 39–43.
133. Durrani, A. M., Farr, S. J., and Kellaway, I. W. 1995. Influence of molecular weight and formulation pH on the precorneal clearance rate of hyaluronic acid in rabbit eye. *International Journal of Pharmaceutics*, 118, 243–250.
134. Miller, S. C. and Donovan, M. D. 1982. Effect of poloxamer 407 gel on the miotic activity of pilocarpine nitrate in rabbits. *International Journal of Pharmaceutics*, 12, 147–152.
135. Unterman, S. R., Rootman, D. S., Hill, J. M., Parelman, J. J., Thompsom, H. W., and Kaufman, H. E. 1988. Collagen shield drug delivery: therapeutic concentrations of tobramycin in the rabbit cornea and aqueous humor. *Journal of Cataract and Refractive Surgery*, 14, 500–5004.
136. Fitzgerald, P., Hadgraft, J., Kreuter, J., and Wilson, C. G. 1987. A gamma scintigraphic evaluation of microparticulate ophthalmic delivery systems: Liposomes and nanoparticles. *International Journal of Pharmaceutics*, 40, 81–84.
137. Aggarwal, D. and Kaur, I. P. 2005. Improved pharmacodynamics of timolol maleate from a mucoadhesive niosomal ophthalmic drug delivery system. *International Journal of Pharmaceutics*, 290, 155–159.

138. Pepić, I., Jalšenjak, N., and Jalšenjak, I. 2004. Micellar solutions of triblock copolymer surfactants with pilocarpine. *International Journal of Pharmaceutics*, 272, 57–64.

139. Gasco, M. R., Gallarate, M., Trotta, M., Bauchiero, L., Gremmo, E., and Chiappero, O. 1989. Microemulsions as topical delivery vehicles: Ocular administration of timolol. *Journal of Pharmaceutical and Biomedical Analysis*, 7(4), 433–439.

140. Gallarate, M., Gasco, M. R., Trotta, M., Chetoni, P., and Saettone, M. F. 1993. Preparation and evaluation in vitro of solutions and o/w microemulsions containing levobunolol as ion-pair. *International Journal of Pharmaceutics*, 100(1–3), 219–225.

141. Siebenbrodt, I. and Keipert, S. 1993. Poloxamer-systems as potential ophthalmics II. *Microemulsions. European Journal of Biopharmaceutics*, 39, 25–30.

142. Lia, C., Abrahamsona, M., Kapoora, Y., and Chauhan, A. 2007. Timolol transport from microemulsions trapped in HEMA gels. *Journal of Colloid and Interface Sciences*, 315(1), 297–306.

143. Malina I. and Radomska-Soukharev A. 2006. Microemulsions with timolol as potential eye drops. *Pharmazie*, 61(7), 65.

144. Hasse, A. and Keipert, S. 1997. Development and characterization of microemulsions for ocular application. *European Journal of Pharmaceutics and Biopharmaceutics*, 43, 179–183.

145. Gulsen, D. and Chauhan, A. 2005. Dispersion of microemulsion drops in HEMA hydrogel: A potential ophthalmic drug delivery vehicle. *International Journal of Pharmaceutics*, 292, 95–117.

146. Alany, R. G., Rades, T., Nicoll, J., Tucker, I. G., and Davies, N. M. 2006. W/O microemulsions for ocular delivery: Evaluation of ocular irritation and precorneal retention. *Journal of Controlled Release*, 111(1–2), 145–152.

147. Chan J., El Maghraby, G. M. M., Craig, J. P., and Alany R.G. 2007. Ocular delivery of pilocarpine hydrochloride from phase transition microemulsions: In vitro in vivo evaluation. *International Journal of Pharmaceutics*, 328(1): 65–71.

148. Chan J., El Maghraby, G. M. M., Craig, J. P., and Alany R.G. 2008. Effect of phase transition water-in-oil microemulsions on the precorneal tear film of albino New Zealand rabbits. *Clinical Ophthalmology*, 2(1), 1–9.

149. Vyas, D. M. 1995. Paclitaxel (Taxol) formulation and prodrugs. In: Farina, V. (Ed.), *The Chemistry and Pharmacology of Taxol and Its Derivatives*. Vancouver, British Columbia: Elsevier Science, 165–245.

150. Strickley, R. G. 2004. Solubilizing excipients in oral and injectable formulations. *Pharmaceutical Research*, 21, 201–230.

151. Dupont, B. 2002. Overview of the lipid formulations of amphotericin B. *The Journal of Antimicrobial Chemotherapy*, 49 S1, 31–36.

152. Zhang, Y., Jiang, X. G., and Yao, J. 2001. Lowering of sodium deoxycholate-induced nasal ciliotoxicity with cyclodextrins. *Acta Pharmacologica Sinica*, 22, 1045–1050.

153. Bennett, S. N., McNeil, M. M., Bland, L. A., Arduino, M. J., Villarino, M. E., and Perrotta, D. M. 1995. Postoperative infections traced to contamination of an intravenous anesthetic, propofol. *New England Journal of Medicine*, 333, 147–154.

154. Trissel, L. A. 1997. Pharmaceutical properties of paclitaxel and their effects on preparation and administration. *Pharmacotherapy*, 17, 133S–139S.

155. He L., Wang G., and Zhang Q. 2003. An alternative paclitaxel microemulsion formulation: hypersensitivity evaluation and pharmacokinetic profile. *International Journal of Pharmaceutics*, 250, 45–50.

156. Nornoo, A. O. and Chow, D. S. L. 2008. Cremophor-free intravenous microemulsions for paclitaxel II. Stability, in vitro release and pharmacokinetics. *International Journal of Pharmaceutics*, 349, 117–123.

157. Nornoo, A. O., Osborne, D. W., and Chow, D. S. L. 2008. Cremophor-free intravenous microemulsions for paclitaxel I: Formulation, cytotoxicity and hemolysis. *International Journal of Pharmaceutics*, 349, 108–116.

158. Zhang, X. N., Tang, L. H., Gong, J. H., Yan, X. Y., and Zhang, Q. 2006. An alternative paclitaxel self-emulsifying microemulsion formulation: Preparation, pharmacokinetic profile, and hypersensitivity evaluation. *PDA Journal of Pharmaceutical Science and Technology*, 60, 89–94.

159. Moreno, M. A., Frutos, P., and Ballesteros, M. P. 2001. Lyophilized lecithin based oil-water microemulsions as a new and low toxic delivery system for Amphotericin B. *Pharmaceutical Research*, 18, 344–351.

160. Brime, B., Molero, G., Frutos, P., and Frutos, G. 2004. Comparative therapeutic efficacy of a novel lyophilized amphotericin B lecithin-based oil–water microemulsion and deoxycholateamphotericin B in immunocompetent and neutropenic mice infected with Candida albicans. *European Journal of Pharmaceutical Sciences*, 22, 451–458.

10 Physicochemical Characterization of Pharmaceutically Applicable Microemulsions: Tween 40/Imwitor 308/ Isopropyl Myristate/ Water

Mirjana Gašperlin and Marija Bešter-Rogač

CONTENTS

10.1   INTRODUCTION

Microemulsions are clear, stable, isotropic mixtures of oil, water, and surfactant, frequently in combination with a cosurfactant. They are currently of pharmaceutical interest because of their considerable potential as drug delivery vehicles able to incorporate a wide range of drug molecules [1,2]. Depending on their composition, the microemulsions exhibit a number of structures that crucially influence their properties and applications [3,4]. A substantial amount of research work has been carried out to understand the physicochemical properties of microemulsions for their technical or pharmaceutical applications. In our previous work [5] the pseudoternary system water-surfactant (Tween 40 [TW40])/cosurfactant (Imwitor 308 [IMW])/isopropyl myristate (IPM) was investigated, using the minimum content of surfactant/cosurfactant mixture for maximum water solubilization, since higher amounts of surfactants can cause irritation. It was found that the microstructure of such complex systems needs to be characterized by a group of experimental methods [3] in order to obtain an insight into the influence of the internal structure on drug release [4]. However, on entering the physiological environment, the microemulsion can change its structure and therefore, if loaded with drugs, its ability to release them [6]. Hence, the determination of phase stability diagrams (or phase maps) and location of the different structures formed with these water–oil–surfactant/cosurfactant systems are very important. The structures of microemulsions are in the nanometer range, the so-called colloidal domain. They can be formed as elongated, rodlike micelles, W/O (water-in-oil) or O/W (oil-in-water) spherical droplets, and bicontinuous or lamellar structures. In the water-rich region, O/W droplets are the most frequent form while, in the microemulsions with more similar contents of water and oil, bicontinuous structures are formed. A great variety of methods are available to study colloidal systems [3] but a serious limitation with some of them lies in the requirement to dilute the microemulsion systems in order to eliminate particle–particle interactions. Most of the work reported in the pharmaceutical literature has been conducted using concentrated microemulsion systems, where many approaches have met with limited success [7–9]. The structural characterization of concentrated systems is extremely problematic and could be carried out using a wide range of different techniques, but the complementarity of methods [10,11] is generally required in order to fully characterize these systems. A further problem is that it is quite difficult to find pharmaceutically applicable surfactants that form microemulsions and allow continuous variation of the oil-to-water ratio.

In this work the pseudoternary system water surfactant TW40/cosurfactant IMW/IPM was investigated. This system forms microemulsions and allows continuous variation of the oil-to-water ratio over a wide range. Methods were chosen that could operate over the whole investigated range of the system composition; conductivity, surface tension, density, small angle x-ray scattering (SAXS), and differential scanning calorimetry (DSC) were used. In addition we have studied the in vitro release of nonsteroidal anti-inflammatory drug ibuprofen from microemulsions of various compositions on a dilution line with constant ratio of surfactant mixture and IPM 4:1 to confirm the influence of microstructure on release rate and consequently to lay the basis for predicting drug release under in vivo conditions where the microemulsion composition is continuously varying.

10.2 EXPERIMENTAL

10.2.1 MATERIALS

IPM was obtained from Fluka Chemie GmbH, Switzerland and was used as the lipophilic phase. TW40—polyoxyethylene (20) sorbitan monopalmitate (Fluka Chemie GmbH, Switzerland) was used as a surfactant and IMW—glyceryl caprylate (Condea Chemie GmbH, Germany) as a cosurfactant. Twice distilled water was used as the hydrophilic phase. Ibuprofen was obtained by Lek d.d., Slovenia.

10.2.2 PREPARATION OF MICROEMULSIONS

Surfactant and cosurfactant were blended in a 1:1 mass ratio to give a surfactant mixture. Stock solutions were prepared by mixing appropriate amounts of surfactant mixture and IPM or water to obtain the desired starting point mixture. Compositions of the starting systems are given in Table 10.1. The appropriate amount of water or IPM was added to give the desired microemulsion composition along

TABLE 10.1

Compositions of the Starting Systems of Surfactant Mixture and IPM or Water

Denotations of Systems	TW40 + IMW (wt %)	IPM (wt %)	Water (wt %)
W0	100	0	0
W1	80	20.0	0
W2	62.8	37.2	0
W3	39.8	60.2	0
IPM0	100	0	0
IPM1	80.0	0	20.0
IPM2	63.1	0	36.9
IPM3	40.0	0	60.0

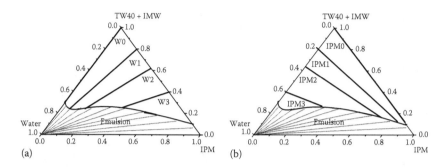

FIGURE 10.1 Pseudoternary phase diagram of the system of water/TW40/IMW/IPM at 25°C, in which the weight ratio of TW40 to IMW was fixed at 1:1. (a) W0, W1, W2, and W3 designate the water dilution lines for which the TW40/IMW mixture:IPM ratio is constant at the compositions given in Table 10.1. (b) IPM0, IPM1, IPM2, and IPM3 designate the IPM dilution lines for which TW40/IMW mixture:water ratio is constant with the compositions given in Table 10.1.

the corresponding dilution curve. Components were blended for 5 min at room temperature (25°C ± 1°C) using a magnetic stirrer. The corresponding dilution curves are drawn on a pseudoternary diagram (W lines in Figure 10.1a, IPM lines in Figure 10.1b).

In microemulsion samples, selected for release study, ibuprofen was added to the already prepared microemulsions in 1 wt % concentration. After the addition samples were mixed with magnetic stirrer at room temperature for additional 15 min.

10.2.3 Density Measurements

The density of microemulsions and solvents was measured with a Paar digital precision density meter DMA 60 with an external measuring cell DMA 602 (Anton Paar, Austria). An ultrathermostat controlled the temperature at 25.00°C ± 0.01°C. The accuracy of density measurements was within $\pm 5 \times 10^{-6}$ kg dm^{-3}.

10.2.4 Surface Tension Measurements

Surface tension was measured at 25.0°C ± 0.5°C with a Kruss processor tensiometer K21 (Kruss GmbH, Germany) using Wilhelmy's plate method. A square platinum plate was cleaned, washed with twice distilled water and heated in a reductive flame to purge all impurities. This cleaning procedure was repeated before every measurement. During the measurement, the plate is dipped into the liquid. The tensiometer measures the pulling force of the liquid on the plate and, with known plate size, calculates the surface tension.

10.2.5   Electrical Conductivity Measurements

Conductance was measured using an Iskra conductivity meter MA 5964 (Iskra, Slovenia) with a homemade conductivity cell with cell constant of $0.7265\,cm^{-1}$. Conductivity and surface tension measurements require large amount of samples and for this reason these measurements were carried out during titration of the starting mixtures with water or IPM.

10.2.6   Small-Angle X-Ray Scattering Measurements

SAXS experiments were performed with an evacuated Kratky compact camera system (Anton Paar, Austria) with a block collimating unit attached to a conventional x-ray generator (Bruker AXS, Karlsruhe, Germany) equipped with a sealed x-ray tube (Cu-anode target type) producing Ni-filtered Cu $K\alpha$ x-rays with a wavelength of $0.154\,nm$. The operating power was $35\,kV \times 10\,mA$. The samples were transferred to a standard quartz capillary placed in a thermally controlled sample holder centered in the x-ray beam. The scattering intensities were measured with a linear, position sensitive detector (PSD 50 m, M. Braun, Garsching, Germany) detecting the scattering pattern within the whole scattering range simultaneously. All measurements were performed at $25°C$. For each sample, five SAXS curves with a sampling time between 3,600 and 15,000 s were recorded and subsequently averaged to ensure reliable statistics.

10.2.7   Differential Scanning Calorimetry

DSC measurements were performed with a DSC Pyris 1 with Intracooler 2P, both from Perkin Elmer, USA. Nitrogen, with a flow rate of $20\,mL\,min^{-1}$, was used as purge gas. Approximately 5 to 15 mg of sample was weighed precisely into a small aluminium pan and quickly sealed hermetically to prevent water evaporation. The empty sealed pan was used as a reference. Samples were cooled from $30°C$ to $-56°C$. Cooling rate was $10\,K\,min^{-1}$.

10.2.8   In Vitro Release

In vitro release of ibuprofen from microemulsions was tested using a dialysis method [12] in USP dissolution apparatus 2 (DT-6, Erweka, Germany). The temperature was set to $37°C \pm 0.5°C$ and paddle revolution speed to 50 rotations min^{-1}. Two milliliters of drug-loaded microemulsions with 1 wt % of ibuprofen were directly instilled into dialysis bag (Spectrapor/Por 4 membrane, MWCO12-14.000, Spectrum, USA) and placed into 500 mL of phosphate buffer pH 7.4 ± 0.05. To maintain sink conditions 1 wt % of SDS was added in acceptor medium. At predetermined time intervals (1, 1.5, 2, 3, 4, 5, 6, and 8h) 5 mL of sample was withdrawn and filtered through $0.45\,\mu m$ membrane filter (Sartorius, Germany). The experiment was carried out in four parallels. Results are reported as arithmetic means ± standard error. Drug release was determined by two parameters: The amount of released drug after 8 h (mg) and the release rate expressed as the slope of the linear regression line (mg h^{-1}).

Ibuprofen content was assayed by Agilent 1100 series HPLC system (Agilent, USA). The stationary phase was a 250 × 4mm ID column packed with 5μm Nucleosil C8, the mobile phase acetonitrile:phosphate buffer (pH 7.4) = 28.5:71.5, the flow rate 1.2mL min⁻¹, injection volume 20μL, and UV detection at 225 nm.

10.3 RESULTS AND DISCUSSION

The pseudoternary diagram for the investigated system at 25°C is shown in Figure 10.1. The system is homogenous along the marked dilution W and IPM lines. In all investigated microemulsions the weight ratio of TW40 to IMW was fixed at 1:1. W0, W1, W2, and W3 (Figure 10.1a) designate the water dilution lines for which TW40/IMW mixtures:IPM ratios were constant at the values given in Table 10.1. IPM0, IPM1, IPM2, and IPM3 (Figure 10.1b) denote the IPM dilution lines with constant TW40/IMW mixtures:water ratios (Table 10.1). The nonhomogenous region was not investigated in this work.

10.3.1 DENSITY MEASUREMENTS

The measured densities for samples along W lines are shown in Figure 10.2a as a function of the weight percent of water in the microemulsion, and for IPM lines as a function of the weight percent of IPM (Figure 10.2b).

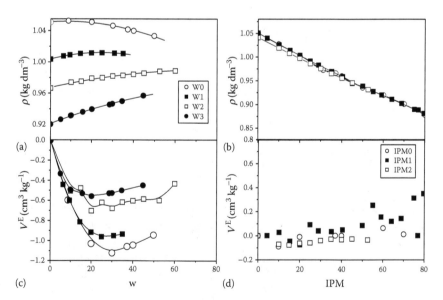

FIGURE 10.2 Densities along the dilution lines shown on the phase diagram (Figure 10.1): (a) densities of the W samples as a function of the water weight ratio; (b) densities of the IPM samples as a function of IPM weight ratio, and the excess volume, V^E, as a function of weight ratio of (c) water and (d) IPM.

The density increases monotonically in water-poor samples along the W1, W2, and W3 lines, at higher proportions of water remaining constant (W2) or decreasing slightly (W1) (Figure 10.2a). Density is almost constant along the W0 line for samples with low content of water and decreases with increasing amounts of water in the microemulsion.

The volume of the microemulsion, V_{exp}, was obtained from the measured density. The excess volume, $V^E = V_{exp} - V_{id}$, was calculated, where the ideal volume, V_{id}, is given by the following densities: 0.997047 (water), 1.05088 (W0), 1.00345 (W1), 0.96620 (W2), 0.91999 (W3) kg dm^{-3}, assuming additivity of volumes. The values of V^E along the W lines are presented in Figure 10.2c. The volumes are in fact seen not to be additive. Considerable contraction of volume, V^E, occurs from ~10 wt % water (W0) up to ~30 wt% water (W3) in the system. At higher water contents, V^E changes little, suggesting that the attractive interactions are constant in that region. Above ~50 wt% water in W2 samples, V^E increases again, indicating that the attractive interactions are declining. It was not possible to determine the density in W0 and W1 samples with higher water content because of their high viscosity.

Densities along IPM lines decreased (Figure 10.2b) with increasing IPM in the microemulsions. No dramatic change was observed. Using the densities of pure IPM (0.84962) and of the starting solutions 1.05580 (IPM0), 1.05100 (IPM1), and 1.04153 (IPM2) kg dm^{-3}, the excess volume, V^E, was again calculated (Figure 10.2d). There is no marked deviation from additivity. Densities along the IPM3 line were not measured due to the high viscosity of samples.

10.3.2 SURFACE TENSION

The dependence of surface tension as a function of water weight ratio is shown for W lines (Figure 10.3a) and for samples along IPM lines (Figure 10.3b). From Figure 10.3a it is seen that the surface tension decreases linearly with increasing

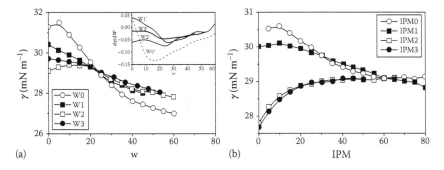

FIGURE 10.3 Surface tension along the dilution lines which are shown on the phase diagram (Figure 10.1). (a) The dependence of surface tension on the water weight ratio. Inset: The first derivative of the surface tension. (b) The dependence of surface tension on the IPM ratio.

content of water—except for W0 and W2 samples up to ~10 wt% of water. The first derivatives exhibit slight minima at the positions of the more marked breaks (Figure 10.3a, inset). This suggests that structural changes occur at these compositions (~25 wt % of water in W1 and W2, ~20 wt % of water in W3).

At higher proportions of water the surface tension in general decreases linearly, although, in W1 and W2, slight breaks are observable at ~35 wt% and 40 wt% water, respectively, suggesting the structural changes again at these compositions.

The surface tension along IPM0 and IPM1 lines decreases with increasing proportions of IPM (Figure 10.3b). Pure IPM has a lower surface tension (28.9 mN m^{-1}) than water (72.0 mN m^{-1}) and the decrease could be explained by this difference. In contrast, surface tension along IPM2 and IPM3 increases with the addition of IPM up to ~20 wt%–30 wt%. It can be assumed that the added IPM is incorporated in the internal structure in such a way that more water is on the outside of the "oil drops," causing the increase in surface tension.

10.3.3 Electrical Conductivity

The conductance increases with the addition of water along the W0, W1, W2, and W3 curves (Figure 10.4a), reflecting the higher conductivity of water compared with that of the IPM and surfactant mixture. A maximum in the first derivative at ~35 and 40 wt % water in the microemulsion is observed for W1 and W2 samples (Figure 10.4a, inset) confirming the presence of percolation behavior (bicontinuous microstructure) in this region [5,13–18]. The conductivity of microemulsions containing more than 45 wt % water in W1 and 50 wt % water in W2 decreases significantly, probably due to the observed higher viscosity [18]. It can be assumed that, in W3, the bicontinuous microstructure is presented also at higher water contents, although the maximum in the derivative is not achieved.

The dependence of conductivity on the IPM content along IPM lines shows quite a different picture (Figure 10.4b). Conductivity decreases along IPM1 and

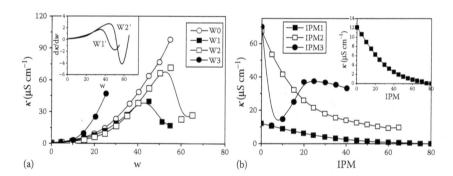

FIGURE 10.4 (a) Electrical conductivity along the water dilution lines as a function of the water weight ratio. Inset: The first derivative of the electrical conductivity along the W1 and W2 lines. (b) Electrical conductivity along the IPM dilution lines as a function of the IPM weight ratio. Inset: Conductivity along the IPM1 dilution line.

IPM2, reflecting the lower conductivity of IPM compared to that of the water and surfactant mixture, but no maxima or minima were observed in the first derivatives. Because the conductivity of the IPM and surfactant mixture is very low, measurements were not made in the solutions without water (IPM0).

The conductivities of samples lying on the IPM3 line exhibit an entirely different behavior: Here the initial, small addition of IPM causes a strong decrease but, with the further addition of IPM, the electrical conductivity increases. The small amount of added IPM converts the already highly viscous mixture to an even more viscous microemulsion, resulting in even lower conductance. This suggests that the added IPM incorporates into the microstructure, extruding water thus causing the rise in conductivity. Above ~20 wt% IPM in the system, the conductivity decreases slightly.

10.3.4 SAXS Measurements

Examples of SAXS experiments on the TW40/IMW/IPM/water system performed on samples along two investigated dilution lines (W1 and IPM1) are shown in Figure 10.5. It is evident that the scattering intensity, I, increases markedly with the addition of water (Figure 10.5a) or IPM (Figure 10.5b). Specifically, the scattering functions increase and slowly shift toward the lower values of the scattering vector q. Along W1 dilution line (Figure 10.5a) in the case of samples with more than 55 wt % of water, a sharp peak develops in the low q region. Sharp peaks are characteristic for the scattering pattern of the lamellar phases [19–24] and apart from W1 line they were observed only along the W2 line for samples containing more than ~60 wt % of water.

In Figure 10.6 the values of scattering vector q at the maximum scattering intensity, q_{max}, in dependence of the amount of water (Figure 10.6a) and IPM

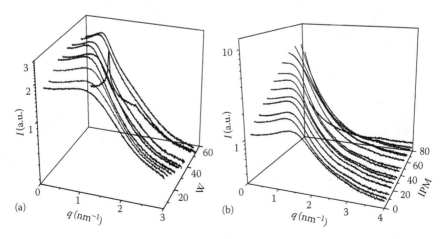

FIGURE 10.5 Scattering curves of the tested samples (a) along W1 and (b) along IPM1 dilution line.

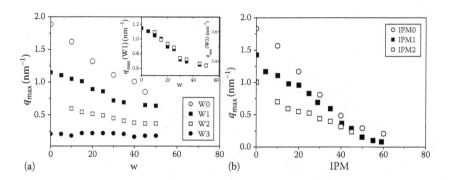

FIGURE 10.6 Values of the scattering vector at the maximum scattering intensity, q_{max}, along the dilution lines which are on the phase diagram (Figure 10.1). (a) The dependence of q_{max} on the water weight ratio. Inset: The dependence obtained for the samples along W1 and W2 dilution lines. (b) The dependence of q_{max} on the IPM ratio.

(Figure 10.6b) for all investigated samples are shown. At first glance in all cases q_{max} is shifted smoothly toward lower values with increasing amount of water or IPM in the sample. However, a careful inspection of the dependence of W1 and W2 lines reveals (inset of Figure 10.6a), that the position of the maximum scattering intensity shifted smoothly toward lower values of the scattering vector q up to ~30 wt % of water, than it changes only slightly. The positions of sharp peaks, indicating the lamellar phases, are not drawn.

10.3.5 DSC Measurement

The state of water in the microemulsion system, and consequently the structure, can be determined by DSC. In most cases a distinction can be made between bulk and bound water [25–27]. Three representative cooling curves for systems water/TW40/IMW/IPM are shown in Figure 10.7.

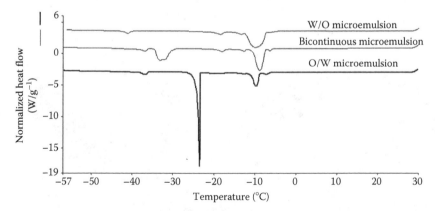

FIGURE 10.7 DSC cooling curves for three representative structures of the system water/TW40/IMW/IPM, where the weight ratio of TW40 to IMW was fixed at 1:1.

For the W/O microemulsions, the largest peak appeared at ca. −8°C, indicating solidification of IPM. The second, much smaller peak at ca. −42°C indicates freezing of the surfactant/cosurfactant mixture. In these systems water is interacting strongly with the other components, so that no separate freezing peaks of water are detected, since the water molecules freeze with the surfactant mixture. With the bicontinuous type of microemulsion we observed that water interacts less with the surfactant molecules and freezing of this "bound" water is seen as a new, larger peak in the temperature range from ca. −50°C to −30°C, the temperature depending on the strength of the interaction between surfactant and water molecules. The peak is comparatively broad. In the systems where water is present in the continuous phase, a distinctive, large sharp peak appears at ca. −20°C, which indicates the freezing of supercooled water [5,6,28]. However, some interactions between water and surfactant are still present and this is the reason why we observed a slight deviation in peak position from that of pure water. This information is the key to distinguishing between different microemulsion types by DSC.

The temperatures of the freezing peaks of water observed along the W0, W1, W2, and W3 dilution lines are plotted in one diagram (Figure 10.8a). The results are in accordance with previously published data where a detailed analysis of water freezing behavior along dilution lines has been shown [5,6]. Taking into account the experimental error, the water freezing peak for microemulsions containing up to 20 wt%–25 wt% water is located at −39°C ± 3°C, then moving up to ca. −18°C ± 3°C at higher water content. Thus, up to ~20 wt%, water interacts strongly with its environment but, in the range 20–40 wt % of water, the interactions in the microemulsion weaken. At higher water content, the water is in the continuous phase. Along the W0 line (system without IPM), similar interactions of water molecules with the surfactant/cosurfactant mixtures can be presumed: water is strongly bonded until 30 wt % of water and above 50 wt% it is in the continuous phase.

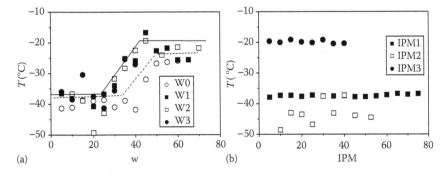

FIGURE 10.8   (a) Temperatures of the freezing peak of water observed on DSC cooling lines along W0, W1, W2, and W3 dilution lines. Lines are only to guide the eye. (b) The temperatures of the freezing peak of water observed on DSC cooling lines along IPM1, IPM2, and IPM3 dilution lines.

Along the IPM1 and IPM3 lines the temperatures of the freezing peaks of water were observed at ca. $-20°C$ and $-38°C$, respectively (Figure 10.8b). For IPM2 a separate peak for freezing of water molecules is detected, indicating that these samples are bicontinuous.

10.3.6   PSEUDOTERNARY PHASE DIAGRAM

The results obtained from density, surface tension, and conductivity (Figures 10.2 through 10.4) agree well with those from SAXS and DSC measurements. The observed excess volumes V^E (Figure 10.2c) lead to the conclusion that the attractive forces increase up to ~10 wt% water in W3 and ~20 wt% in W2 and W1 lines. The first derivatives of the surface tension exhibit minima at the same compositions (Figure 10.3a inset). With the addition of water, the conductance increases monotonically up to ~10 wt% for W1 and ~20 wt% water for W2 and W3 samples, reflecting the higher conductivity of water compared to that of the IPM and surfactant mixtures. No separate freezing peak of water was observed in these samples. It can be assumed that, due to the large amount of surfactant present, water interacts strongly, lowering the freezing point to very low temperatures and giving a very low freezing enthalpy, presumably below the limit of detection. This suggests a W/O type of microemulsion. The excess volumes above ~20 wt% water (Figure 10.2c) change only slightly, indicating similar attractive forces. Surface tension (Figure 10.3a) also decreases in this range of composition. The electrical conductivity measurements (Figure 10.4a) indicate percolation behavior [15] in this region. These results support the conclusion that, at more than ~15 wt% of water in W1 samples and ~20 wt% in W2 and W3, bicontinuous phases are present. From the DSC measurements (Figure 10.8a) it can be concluded that the water in the system containing more that ~45–50 wt% water in W1 and W2 samples is in the outer phase.

Many of these samples, especially those with even higher water content, were very viscous and the density could not be measured. The surface tension decreases linearly over this range of water content, but, in W1 and W2, breaks at ~35 wt% and 40 wt% water are observed, suggesting that the structure changes at these compositions. No distinctive break could be observed in W3 in this region. The conductivity of microemulsions containing more than ~45 wt% water in W1 and 55 wt% water in W1 decreased significantly, probably due to the higher viscosity. W3 lines evidently do not reach this range.

These findings are supported by the SAXS measurements of samples along W1 and W2 lines (inset in Figure 10.6a). When the water content in the samples is increasing the position of the maximum scattering intensity (q_{max}) is moving toward lower values of the scattering vector up to ~30 wt% of water, than it changes only slightly. In our previous work it has been shown that as the amount of water increases, the droplets tend to swell and W/O type can be predicted [5]. In the region where from conductivity measurements the transition to bicontinuous structures is detected, the position of the maximum scattering intensity changes only slightly ($q_{max} \approx$ constant). Recently it has been shown that the

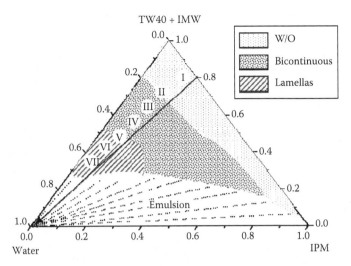

FIGURE 10.9 Pseudoternary phase diagram of the system water/TW40/IMW/IPM at 25°C, where the weight ratio of TW40 to IMW was fixed at 1:1. The compositions of the microemulsions used in the drug release study are denoted.

evaluation of the scattering data does not give a clear indication for the bicontinuous phase [29]. The coincidence of the results from SAXS measurements and other techniques demonstrated in this work shows that SAXS can give a rough qualitative estimation of restructuring in the system. Moreover, in the case of samples with more than 55 wt% of water the lamellar phases can be predicted from SAXS measurements. The observed high viscosity of the samples resulting in the decreasing electric conductivity supports the existence of the lamellar phases.

Whereas along W lines useful information about microstructure was obtained, this cannot be claimed for corresponding IPM lines. Along these lines no reliable changes can be observed. We can only assume that the samples along each line experience similar environment or at least there is no dramatical changes.

Finally, from all obtained results the ranges of the discussed structures could be drawn in the ternary phase diagram (Figure 10.9). It is necessary to stress that the crossing between two microstructures could not be regarded as a sharp border but more or less as a "transitional" region, which is obviously detected differently by various techniques.

10.3.7 IBUPROFEN RELEASE

According to the constructed phase diagram (Figure 10.9), the W1 dilution line was selected for in vitro drug release studies while only on this line all discussed microstructures were clearly identified. The composition of selected samples is shown in Table 10.2 and in the phase diagram. Sample I is of W/O type, II represents the start

TABLE 10.2

Compositions of the Microemulsions along Dilution Line W1 Used for Ibuprofen Release Study

ME Sample	TW40 + IMW (wt %)	IPM (wt %)	Water (wt %)
I	72.0	18.0	10.0
II	62.4	15.6	22.0
III	54.4	13.6	32.0
IV	47.2	11.8	41.0
V	39.5	9.9	50.5
VI	32.0	8.0	60.0
VII	24.0	6.0	70.0

of the percolation phenomenon, III and IV bicontinous phase, V is located in the border area between the bicontinuous and water continuous phase, while VI and VII are water continuous with lamellar structure.

The release experiment was performed in vitro using a dialysis bag method, which was developed by Kang et al. [12]. The method is simple, fast, and reproducible. Although in vitro studies hardly match the in vivo conditions, they remain an important part of formulation development. So designed experiments reveal the important parameters derived from carrier systems (vehicle viscosity, partition of drug molecules between microemulsion phases, droplet size, and microstructure) which influence bioavailability of incorporated drugs.

As a model nonsteroidal anti-inflammatory drug ibuprofen was chosen. It is very effective for the systemic treatment of rheumatoid arthritis, osteoarthritis, and ankylosing spondylitis. It is almost insoluble in water (1.74×10^{-4} mol L^{-1}) [30], while its solubility in IPM is 0.160 ± 0.015 g mL^{-1} [31]. As a weak acid ($pK_a = 4.4$) its solubility increases with increasing pH. Ibuprofen shows a considerable surface activity [32] that could contribute to its enhanced permeation through biological membranes [33]. The release profiles of tested ME samples are shown in Figure 10.10a.

The kinetic of ibuprofen release was evaluated by fitting the experimental data to different order kinetics. The data were transformed for linear regression analysis for zero-order, first-order, and Higuchi ($\sqrt{t}$) kinetics. The calculated Pearson coefficients (in the range of 0.9945–0.9990) indicate the best fits for zero-order kinetics, which is valid, when the diffusion across the membrane controls the release. This kinetics was also confirmed in the investigations of ibuprofen-loaded microemulsions and microemulsion-based hydrogels [31]. Comparing the amount of released ibuprofen after 8 h as well as the release rate the slowest release (Figure 10.10b) was observed for ME I; with the increasing amount of water in the delivery system both parameters are increasing.

Evidently, samples can be divided into three groups. First group are samples I and II which represent W/O type of microemulsion and the starting of percolation

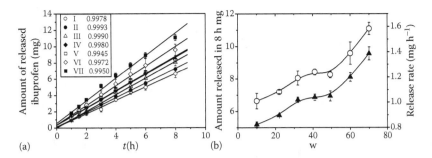

FIGURE 10.10   (a) Release profiles of ibuprofen from microemulsions of different composition, given in Table 10.2. Lines: zero-order release kinetics. Pearson coefficients are given next to the symbols in the legend. (b) Correlation between release parameters; (O) amount of released ibuprofen after 8 h and (▲) ibuprofen release rate as function of the water weight ratio in the microemulsion (see Table 10.2).

phenomena. For the second group (III, IV, V) containing 32, 41, and 49 wt % of water, respectively, almost constant release rate is observed. According to the phase diagram on Figure 10.9, the samples are in bicontinuous region. The third group are water continuous samples (VI, VII), where on the basis of conductivity and SAXS measurements lamellar structures are predicted.

The surface activity of ibuprofen is raising the possibility that on solubilization in microemulsions it will not only remain in oil phase but also be incorporated in surfactant film where it may act as a cosurfactant and consequently interact with other microemulsion components that could affect the release.

The results are in agreement with the recent release study of nonsteroidal drug ketoprofen [6]. Despite the fact that the ketoprofen release was studied by different techniques and therefore the release rate values are not comparable directly, the same tendency was observed. Additionally, it has been shown [34] that incorporation of ketoprofen does not alter the microemulsion system significantly; however, its presence prevents the formation of stronger interaction and formation of gel-like structure in water rich region. It was also found out that stronger interactions between microemulsion components in W/O as well in the bicontinuous phase lead to slower ketoprofen release. Because of similar molecule structure of ibuprofen the same could be assumed also for it. We can conclude that release behavior of ibuprofen is influenced with the microstructure and can be predicted to a certain extent, using a combination of several tested methods for physical characterization of microemulsions.

10.4   SUMMARY

Density, surface tension, electrical conductivity, DSC, and SAXS measurements have been applied to investigate pharmaceutically applicable microemulsions comprising water-surfactant (TW40)/cosurfactant (IMW)/IPM over the entire phase diagram. The results provide a qualitative description of the microstructures present in the microemulsion regime of the phase diagram, together with evidence for transitions between them.

ACKNOWLEDGMENTS

Financial support from the Slovenian Research Agency is gratefully acknowledged (P1-0201 and P1-0189). MB-R would like to express her gratitude to the Alexander von Humboldt Foundation, Germany for the donation of the small x-ray scattering system. This work was partially supported by COST Action D43.

SYMBOLS AND TERMINOLOGIES

ρ	density (kg dm^{-3})
γ	surface tension (mN m^{-1})
κ	electrical conductivity (μS cm^{-1})
I	scattering intensity at SAXS experiments (a.u.)
q	scattering vector (nm^{-1})
q_{max}	value of the scattering vector at the position of the maximum scattering intensity (nm^{-1})
V_{exp}	real volume of 1 kg of microemulsion, obtained from its density (cm^3 kg^{-1})
V_{id}	ideal volume of 1 kg of microemulsion, calculated from the densities of all components and assuming additivity of volumes (cm^3 kg^{-1})
V^{E}	excess volume, $V_{exp} - V_{id}$ (cm^3 kg^{-1})
m	mass (mg, g)
wt %	weight percent
T	temperature (°C)
t	time (h)
W/O	water-in-oil
O/W	oil-in-water
SAXS	small angle x-ray scattering
DSC	differential scanning calorimetry
IPM	isopropyl myristate and weight percent of IPM in the sample (used in figures)
w	weight percent of IPM and water in the sample (used in figures)
TW40	Tween 40
IMW	Imwitor 308
W0, W1, W2, W3	denotation of the microemulsions obtained by the stepwise addition of water to the starting mixtures (with compositions given in Table 10.1)

IPM0, IPM1, IPM2, IPM3   denotation of the microemulsions obtained by the stepwise addition of IPM to the starting mixtures (with compositions given in Table 10.1)

REFERENCES

1. Bagwe, R. P., Kanicky, J. R., Palla, B. J., Patanjali, P. K., and Shah, D. O. 2001. Improved drug delivery using microemulsions: Rationale, recent progress, and new horizonts. *Crit. Rev. Ther. Drug*, 18(1), 77–140.
2. Tenjarla, S. 1999. Microemulsions: An overview and pharmaceutical applications. *Crit. Rev. Ther. Drug*, 16(5), 461–521.
3. Kahlweit, M. 1999. Microemulsions. *Annu. Rep. Prog. Chem., Sect. C. Phys. Chem.*, 95, 89–115.
4. Lawrence, M. J. and Rees, G. D. 2000. Microemulsion-based media as novel drug delivery systems. *Adv. Drug Deliv. Rev.*, 45, 89–121.
5. Podlogar, F., Gašperlin, M., Tomšič, M., Jamnik, A., and Bešter-Rogač, M. 2004. Structural characterisation of water–Tween 40/Imwitor 308–isopropyl myristate microemulsions using different experimental methods. *Int. J. Pharm.*, 276, 115–128.
6. Podlogar, F., Bešter-Rogač, M., and Gašperlin, M. 2005. The effect of internal structure of selected water—Tween 40–Imwitor 308–IPM microemulsions on ketoprofene release. *Int. J. Pharm.*, 302, 68–77.
7. Aboofazeli, R. and Lawrence, M. J. 1993. Investigations into the formation and characterization of phospholipid microemulsions. I. Pseudo-ternary phase diagrams of systems containing water–lecithin–alcohol–isopropyl myristate. *Int. J. Pharm.*, 93, 161–175.
8. Aboofazeli, R., Barlow, D. J., and Lawrence, M. J. 2000. Particle size analysis of phospholipid microemulsions. I. Total intensity light scattering. *AAPS Pharm. Sci*, 2(2), 13, DOI: 10.1208/ps020213.
9. Aboofazeli, R., Barlow, D. J., and Lawrence, M. J. 2000. Particle size analysis of concentrated phospholipid microemulsions II. Photon correlation spectroscopy. *AAPS PharmSci*, 2(3), 19, DOI: 10.1209/ps020319.
10. Regev, O., Ezrahi, S., Aserin, A., Garti, N., Wachte, E., Kaler, E. W., Khan, A., and Talmon, Y. 1996. A study of the microstructure of a four-component nonionic microemulsion by Cryo-TEM, NMR, SAXS, and SANS. *Langmuir*, 12, 668–674.
11. Fanun, M. 2007. Structure probing of water/mixed nonionic surfactants/caprylic-capric triglyceride system using conductivity and NMR. *J. Mol. Liq.*, 133, 22–27.
12. Kang, B. K., Lee, J. S., Chon, S. K., Jeong, S. Y., Yuk, S. H., Khang, G., Lee, S. B., and Cho, S. H. 2004. Development of self-microemulsifying drug delivery system (SMEDDS) for oral bioavailability enhencement of simvastatin in beagle dogs. *Int. J. Pharm.*, 274, 63–73.
13. Borkovec, M., Eicke, H. -F., Hammerich, H., and Das Gupty, B. 1988. Two percolation processes in microemulsions. *J. Phys. Chem.*, 92, 206–211.
14. Giustini, M., Palazzo, G., Colafemmina, G., DellaMonica, M., Giomini, M., and Geglie, A. 1996. Microstructure and dynamics of the water-in-oil CTAB/*n*-pentanol/*n*-hexane/water microemulsion: A spectroscopic and conductivity study. *J. Phys. Chem.*, 70, 959–971.
15. Gradzielsky, M. and Hoffman, H. 1999. Rheological properties of microemulsions. In *Handbook of Microemulsion Science and Technology*, Kumar, P. and Mittal, K.L., Eds. Marcel Dekker, New York, Chapter 13, pp. 411–436.

16. Allouche, J, Tyrode, E., Sadtler, V., Choplin, L., and Salager, J. -L. 2004. Simultaneous conductivity and viscosity measurements as a technique to track emulsion inversion by the phase-inversion-temperature method. *Langmuir*, 20, 2134–2140.

17. Garcia-Rio, L., Mejuto, J. C., Perez-Lorenzo, M., Rodriguez-Alvarez, A., and Rodriguez-Dafonte, P. 2005. Influence of anionic surfactants on the electric percolation of AOT/isooctane/water microemulsions. *Langmuir*, 21, 6259–6264.

18. Plaza, M., Tadros, Th. F., Solans, C., and Pons, R. 2002. Characterization of microemulsions formed in a water/ABA block copolymer poly(hydroxystearic acid)-poly(ethylene oxide)-poly(hydroxystearic acid)/1,2-hexanediol/isopropyl myristate system. *Langmuir*, 18, 5673–5680.

19. Castelletto, V., Fisher, J., Hamley, I.,and Yang, Z. 2002. SAXS study of the swelling and shear orientation of the lamellar phase formed by a diblock copolymer. *Colloid Surf. A*, 211, 9–18.

20. Fairhurst, C. E., Holmes, M. C., and Leaver, M. S. 1997. Structure and morphology of the intermediate phase region in the nonionic surfactant C16EO6/water system. *Langmuir*, 13, 4964–4975.

21. Ficheux, M.-F., Bellocq, A.-M., and Nallet, F. 2001. Elastic properties of polymer-doped dilute lamellar phases: A small-angle neutron scattering study. *Eur. Phys. J.*, E4, 315–326.

22. Grillo, I., Levitz, P., and Zemb, T. 2000. Insertion of small anionic particles in negatively charged lamellar phases. *Langmuir*, 16, 4830–4839.

23. Hyde, S. T. 1995. On swelling and structure of composite materials. Some theory and applications of lyotropic mesophases. *Colloid Surf. A*, 103, 227–247.

24. Yamashita, I., Kawabata, Y., Kato, T., Hato, M., and Minimikawa, H. 2004. Small angle X-ray scattering from lamellar phase for beta-3,7-dimethyloctylglucoside/water system: Comparison with beta-*n*alkylglucosides. *Colloid Surf. A*, 250, 485–490.

25. Erzahi, S., Anserin, A., Fanun, M., and Garti, N. 2001. Subzero behaviour of water in microemulsions. In: *Thermal Behaviour of Dispersed Systems*. Marcel Dekker, New York, Chapter 3, pp. 59–120.

26. Schulz, P. C., Soltero, J. F. A., and Puig, J. E. 2001. DSC analysis of surfactant-based microstructures. In: *Thermal Behaviour of Dispersed Systems*. Marcel Dekker, New York, Chapter 4, pp. 121–182.

27. Garti, N.,Anserin, A., Ezrahi, S., Tiunova, I., and Berkovic, B. 1996. Water behaviour in nonionic surfactant systems I: Subzero temperature behavior of water in nonionic microemulsions studied by DSC. *J. Colloid Interface Sci.*, 178, 60–68.

28. Milon, J. J. G. and Braga, S. L. 2003. Supercooling water in cylindrical capsules. 15th symposium on thermophysical properties, June 22–27, 2003, Boulder, CO.

29. Freiberger, N., Moitzi, C., de Campo, L., and Glatter, O. 2007. An attempt to detect bicontinuity from SANS data. *J. Colloid Interface Sci.*, 312, 59–67.

30. Perlovich, L., Kukrov, V. S., Kinch, N. A., and Bauer-Brandl, A. 2004. Solvation and hydration characteristic of ibuprofen and acetylsalicylic acid. *AAPS Pharm. Sci.*, 6(1), 3, DOI: 10.1208/ps060103.

31. Chen, H., Chang, X., Du, D., Li, J., Xu, H., and Yang, X. 2006. Microemulsion-based hydrogel formulation of ibuprofen for topical delivery. *Int. J. Pharm.*, 315, 52–58.

32. Rao, S. C., Schoenwald, R. D., Barfknecht, C. F., and Laban, S. L. 1992. Biopharmaceutical evaluation of ibufenac, ibuprofen and their hydroxyethoxy analogs in the rabitt eye. *J. Pharmacokinet. Biopharm.*, 20, 357–388.

33. Al-Saidan, S. M. 2004. Transdermal self-permeation enhancement of ibuprofen. *J. Control. Release*, 100, 199–209.
34. Tomšič, M., Podlogar, F., Gašperlin, M., Bešter-Rogač, M., and Jamnik, A. 2006. Water–Tween 40/Imwitor 308–isopropyl myristate microemulsions as delivery systems for ketoprofen: Small-angle x-ray scattering study. *Int. J. Pharm.*, 327, 170–177.

11 Places of Microemulsion and Emulsion in Cancer Therapy: In Vitro and In Vivo Evaluation

Ercüment Karasulu, Burçak Karaca,
Levent Alparslan, and H. Yesim Karasulu

CONTENTS

11.1 INTRODUCTION TO CANCER CHEMOTHERAPY

Cancer is one of the most challenging health problems of the developing world. Cancer is actually a heterogenic group of disease which is mainly characterized by altered cell growth and apoptotic systems within the cell. All carcinogens or mutagens cause some damage to the DNA of the cell. This alteration may be in a wide pattern ranging from a single base pair mutation to chromosomal aberrations. The DNA mutations that are not repaired before the next cell division cycle pass on to the daughter cells and thus result in the accumulation of DNA mutations in the nucleus. All scientific studies on carcinogenesis show that cancer is a genetic disease caused by means of molecular change [1,2].

That is why most conventional cytotoxic drugs work on DNA by either causing extensive DNA damage that will lead to apoptosis or block DNA synthesis.

The development of cytotoxic agents for cancer is one of the cornerstones of treatment of cancer in recent years [3,4]. There are three main treatments established for cancer: surgery, irradiation, and cytotoxic agents. For several types of cancer, especially, at the early stage of the disease, cytotoxic therapy provides either cure of the disease or prolong survival combined with other treatment modalities or alone. Also in the advanced or metastatic stage of the disease, it may be helpful in symptom control of the patient, which is another important aspect of cancer treatment. However, there are still some important limitations of cytotoxic drugs mainly because of the lack of understanding of the detailed mechanism of action and the rationale use of them in cancer treatment. Most of them have very narrow therapeutic indexes, resulting in severe toxic side effects. It is a well-known fact that most conventional cytotoxic drugs not only cause the death of the tumor cell but also may damage normal cells, especially cells that divide rapidly like bone marrow cells. Another limitation of the conventional cytotoxic treatment is the lack of its selectivity to different types of cancer. Each cancer has its own unique genetic profile such as finger prints that differ among cancers. So new target molecules are being widely studied for cancer treatment, which will clearly result in improving therapeutic outcome and lessen the unwanted side effects. These side effects may sometimes be life threatening [2,4].

Therefore, while developing new drugs for cancer treatment, the selectivity to destroy cancer cells while sparing normal cells is an important issue. Similarly, although many of the cytotoxic drugs are capable of causing remissions in many types of cancer, the response duration is generally short because of drug resistance systems and the patient ultimately dies of his/her cancer. So there is a need for developing new cytotoxic drugs combining the features of safety and efficiency more than the conventional ones [4–6].

11.1.1 RATIONALE OF CANCER CHEMOTHERAPY

Invention of cytotoxic drugs is a recent event in the treatment of cancer. During World War II, soldiers who were accidentally exposed to great amounts of nitrogen mustard developed bone marrow hypoplasia. This coincidental observation led to the first clinical studies with nitrogen mustard in patients with lymphomas in Yale University in 1942. However, especially in the last decade, as our understanding of molecular basis of cancer that controls cell growth, differentiation, and apoptosis increased, a great progress was also seen in the field of cancer chemotherapy [1–4].

The rationale of cancer chemotherapy is based on cell growth, differentiation, and apoptotic mechanisms of the cell. Carcinogenesis is actually a multistep process that occurs within years. This is because the human cell has many defense mechanisms for protecting DNA. The number of cells in an individual is controlled by cell division and apoptotic systems. DNA mutations that affect the genes that control either cell cycle (cell division) or apoptosis result in an excessive number of mutated cells and that is cancer. Nowadays, there are a few cancer cell population kinetics hypotheses (i.e., Norton-Simon, Gompertzian model) and

TABLE 11.1
Drugs according to Cell Cycle

Cell Cycle	Agents
Cell cycle independent	Nitrogen mustard, aziridines, nitrosoureas, alkyl alkane sulfonates, nonclassic alkylating agents, anthracyclins, actinomycins, anthracenedions
Cell cycle specific	
S phase	Bleomycin, antimetabolites, camptothecins, epipodophyllotoxins
G2 phase	Bleomycin, epipodophyllotoxins
M phase	Vinca alkaloids, taxanes

Source: This table was prepared based on multiple books.

all of them are in agreement that there is an exponential growth of tumor cells and proliferating cells have a heterogeneous profile. However, none of the cell population kinetic hypotheses can accurately explain the clinical behavior of different types of cancer. The genetic heterogeneity of each tumor has led to new studies on the genetic identity of tumors, for example, microarray gene technologies and more specific and tailored therapies are now being discussed. So the selectivity of cancer chemotherapy accompanied by higher doses with lesser side effects is one of the major goals to be achieved in cancer treatment [2,5–7].

Actually cytotoxic drugs may be divided into two large groups according to their activity on the cell cycle (Table 11.1). The first group is cell cycle-independent drugs (alkylating agents, antitumor antibiotics) that kill tumor cells in both in dividing and resting phases. However, cell cycle-dependent drugs are mostly effective on dividing cells. This group includes antimetabolites, antitubulin agents, and topoisomerase inhibitors. In the recent years, some targeted drugs are also becoming important in cancer chemotherapy targeting some molecules that are either overexpressed or have a role in cell proliferation or differentiation, like monoclonal antibodies or kinase inhibitors [2,6,7].

11.1.2 DRUG RESISTANCE SYSTEMS

As we have already mentioned above, nonselectivity of conventional drugs may result in severe side effects. However, the main obstacle for conventional cytotoxic treatment is drug resistance. Since the tumor mass enlarges day by day, it becomes a huge mass of heterogenic cell groups. While applying conventional cytotoxic drugs, cancer cells within the mass will be exposed to different amount of drugs and thus they will develop different kinds of genetic mutations and some of them will become resistant to the treatment. This event is the answer why the tumor first responds to treatment but relapses in a short time [2,6,8].

There are many mechanisms why cancer cells become resistant to treatment. Finding the answer to this will be an excellent solution for many cancers. Cancer cells may become resistant either by an efflux system for the drug or by decreasing the intake of the drug, or increasing the number of molecules that play a role in cell proliferation and/or antiapoptotic systems within the cell, or by altering drug metabolism or DNA repair process. One of the most studied reasons of drug resistance systems is the increased efflux of the drug [8]. This is due to the over expression of P-glycoprotein (P-gp), which is encoded by the multidrug resistance (MDR) 1 gene. This glycoprotein is a member of the ATP-dependent transport system working as a transmembrane transport system for the whole body. The overexpression of the gene results in resistance to many cytotoxic drugs including doxorubisin, vinblastine, and taxol. The Pgp works by binding to chemotherapeutic drugs and ATP is hydrolyzed and a conformational change is seen with the protein. At last, the drug is released from the extracellular part of the cell. There are other mechanisms for different kinds of cytotoxic drugs. For instance, some drugs, like methotrexate, use specific transport systems to enter the cell. Any of the mutations seen in the genes coding these receptors may result in decrease of the intake of the drug into the cell. Methotrexate resistance is a result of the mutations of folate carrier systems [7,9,10].

After the process of carcinogenesis begins within a cell, it is normal that we see the overexpression of some molecules due to gene amplification that play a role in cell proliferation and/or antiapoptotic systems. In the recent years, these molecules are widely studied as target molecules for new anticancer drugs focusing to block them. Also, an increase in DNA repair, like increased alkyltransferase activity in the cell, may lead to resistance to alkylating agents [10,11]. Drug resistance is generally classified in two forms:

1. Biochemical or absolute resistance: This form of resistance represents a situation that the tumor cell cannot be killed at any dose of the drug.
2. Relative resistance: The tumor cell may be killed with higher level of drugs.

Absolute resistance is a very important issue regarding why most cancer types are not cured by cytotoxic treatment. However, relative resistance may be solved by some drug strategies by manipulating doses of the drugs. But, here we face another challenge, that is, severe toxic side effects of the cytotoxic drugs. Therefore, modern oncology looks for newly developed drugs that may be used at higher doses to overcome the resistance problem which is specific for cancer cells so that they are not fatal to the patients.

11.2 DRUG DELIVERY SYSTEMS: WHY ARE THEY NECESSARY FOR CANCER TREATMENT?

There is still a need for increased specificity of anticancer agents to target tumors in spite of marked progress in chemotherapy, which is often associated with systemic toxicity and/or severe side effects and damage to healthy tissues in the vicinity of the tumors. Moreover, drug resistance is one of the other major problems of cancer treatment. Hence biomedical scientists are still trying to find solutions for this problem by inventing new drugs or new systems for "old drugs." That is why the drug delivery systems are getting more importance in the field of oncology [8,11,12].

The use of colloidal and/or nanocarrier delivery systems such as liposomes, nanoparticles, microspheres, polymers, emulsions, and microemulsions has been increasing for cancer chemotherapy and they are aimed to package and redirect cytotoxic agents due to their pharmacological advantage. These systems have proven themselves in the delivery of many drugs and they are being used clinically in many cancer types. These nanocarriers are able to increase the therapeutic efficacy of cytotoxic drugs by changing the pharmacokinetics and biodistribution of the drugs. This leads to decreased drug toxicity and increased tumor response. The aim of all cytotoxic drugs is to have an effective tumor response with minimum side effect profile. Drug delivery systems may be accepted as one of the ways to achieve this goal in the treatment of cancer [13–20].

Enhanced drug accumulation within the tumor cell and increased sensitivity of the drug gives us the opportunity to use these drugs at higher doses and therefore drug delivery systems may also be the solutions for drug resistance. Because of the very narrow therapeutic indexes of many cytotoxic agents, most of them are to be used at subtherapeutical doses [7,8,11].

Due to the reasons mentioned above, this chapter focuses on the current status of microemulsions and emulsions in cancer therapy and discusses both the challenges and advantages of using them in cancer chemotherapy.

11.3   MICROEMULSION AND EMULSION-BASED SYSTEMS IN CANCER THERAPY

Efficient cancer chemotherapy requires a high degree of selective localization of cytotoxic drugs in a tumor. Selective targeting therapy is a promising approach for cancer treatment that requires the drug–carrier complex to be stable in plasma circulation and to have the ability to reach the target cells where the drug is specifically delivered. Among the various drug delivery systems for cancer therapy, microemulsion and emulsions are promising carriers due to their biocompatibility, stability, ability to solubilize large quantities of lipophilic compounds. These systems have been attracting much interest in industrial laboratories as well as among academic researchers due to their unique characteristics [21–23]. Despite the vast amount of literature available on microemulsions and emulsions as carriers for anticancer drugs, few studies have examined the relationships between composition and performance.

Microemulsion is defined as a dispersion consisting of oil, surfactant, cosurfactant, and aqueous phase, which is a single optically isotropic and thermodynamically stable liquid solution with a droplet diameter usually within the range of 10–140 nm. There are three different basic structural types of microemulsions

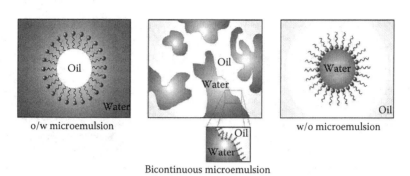

o/w microemulsion   Bicontinuous microemulsion   w/o microemulsion

FIGURE 11.1 Diagrammatic representation of microemulsion structures: (a) w/o micro-emulsion droplet; (b) o/w microemulsion droplet; (c) irregular bicontinuous structure. (Adapted from Lawrence, M.J. and Rees, G.D., *Adv. Drug Deliv. Rev.*, 45, 89, 2000. With permission.)

(Figure 11.1): water-in-oil (w/o), oil-in-water (o/w), and finally bicontinuous structures [21–25]. In practice, the key difference between emulsions and micro-emulsions is that the former, whilst they may exhibit excellent kinetic stability, are fundamentally thermodynamically unstable and will be eventually phase separate. The other important difference concerns their appearance; emulsions are cloudy while microemulsions are clear or translucent. Additionally, there are distinct differences in their method of preparation, since emulsions require a large input of energy while microemulsions do not. The latter point has obvious implications in considering the relative cost of commercial production of the two types of system [21–23,26,27].

Microemulsions offer several potential advantages as drug delivery systems that arise from their solubilization capacity, transparencies, high stability, and simplicity of manufacture. However, the most critical problem regarding micro-emulsion-based drug carriers is the toxicity of the components. Therefore, recent efforts have been focused on how to decrease or eliminate the toxicity or irritation of the microemulsion formulations [28,29]. A microemulsion system free of alcohols was investigated as a potential drug delivery system [30–32]. An interesting study was performed by Sha et al. [28]. The aim of this study was to investigate the effect of two novel self-microemulsifying drug delivery systems (SMEDDS) containing Labrosol with different dilutions on tight junctions. The cytotoxicity of SMEDDS and effect of surfactants on the mitochondrial activity of Caco-2 cells were evaluated by using the MTT (3-[4,5-dimethylthiazol-2-yl]-2,3-diphenyl tetrazolium bromide) assay. Changes in subcellular localization of the tight junction proteins, ZO-1 and F-actin, were examined by confocal laser scanning microscopy. Results demonstrated that negatively charged SMEDDS with different dilutions had no effect on the transepithelial electrical resistance, but significantly increased the permeability of the paracellular marker. The other significant problem of microemulsion systems is the lack of biological tolerance of the excipient such as a surfactant and a cosurfactant. Karasulu et al. [33] have examined microemulsions of methotrexate (M-MTX) and a solution of methotrexate (Sol-MTX) on a model biological environmental. For this purpose, a gastrointestinal cell culture model, the Caco-2 cell line, was used to investigate the cytotoxic effects of the polymeric carrier and its effect on the cell monolayer integrity. Results for the colorimetric assay revealed that for all empty microemulsion concentrations, the cell monolayers remained more than 95% viable, when compared to control indicating that this system appears to possess very low cytotoxicity (Figure 11.2). Caco-2 viability experiments were performed with M-MTX and Sol-MTX at the same concentrations. After incubation of the cells with Sol-MTX for 3 days, the Caco-2 cell proliferation was significantly inhibited (determined by Tukey's test $p < 0.05$) in a dose-dependent manner to an extent of $38.11\% \pm 3.90\%$ at the highest concentration of $40\,ng/75\,\mu L$. As it can be clearly seen from Figure 11.2, the differences between the viability of cells for M-MTX and Sol-MTX were found to be significantly different when ANOVA was applied according to a 2×8 factorial randomized design (p: 0.016; for α: 0.05, power: 0.695). When MTX was loaded into the microemulsion system, the toxicity to the cells was significantly lower when compared with the Sol-MTX ($p < 0.05$). However, this effect was not dose dependent. At the lowest M-MTX concentration of $0.5\,ng/75\,\mu L$, MTX had clearly no antiproliferative effect. If the dilution rates of M-MTX applied on Caco-2 cells

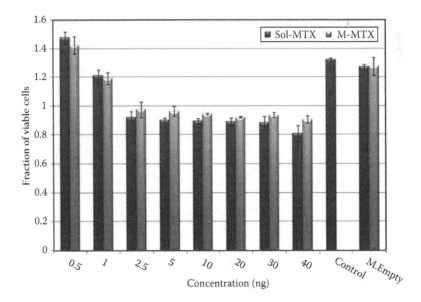

FIGURE 11.2 Cytotoxicity assays of Caco-2 cells treated with Sol-MTX, M-MTX, and empty microemulsion (no drug loaded) at different concentrations. The viability was measured by the MTT test. The values represent the mean of three independent experiments (mean ± SD, $n = 3$). (Adapted from Karasulu, H.Y., Karabulut, B., Goker, E., Guneri, T., and Gabor, F., *Drug Deliv.*, 14, 225, 2007. With permission.)

were between 2.5 and 40 ng/75 µL, the Caco-2 cell proliferations were inhibited, but no significant difference could be determined by the Dunnett test at these concentrations ($p > 0.05$). Results for the colorimetric assay revealed that for M-MTX concentrations the cell monolayers remained more than 72.11% viable when compared to control, indicating that this system appears to possess a very low cytotoxicity compared with Sol-MTX (61.89%). So, the microemulsion formulation of MTX had little cytotoxic effect on Caco-2 cells when compared with Sol-MTX. Therefore, by using M-MTX in the therapy, low cytotoxic effect on normal cells and low side effects may be expected.

Microemulsions are frequently used for administration of lipophilic drugs usually dissolved in the oil phase of the o/w microemulsion. Due to its biocompatibility and the long-term stability, microemulsions with internal phase diameters of submicron order can be used for intravenous (i.v.) administration. It has been described that i.v. administration of microemulsions with droplet size in the range of 100–200 nm is largely captured by blood cells in the liver [34,35]. In the Fomariz et al. study [36], a parenteral microemulsion stabilized by a phosphatidylcholine/hydrophilic surfactant mixture was prepared. The variation of the droplet diameter and rheological properties was studied in the presence and absence of doxorubicin incorporated into the microemulsion. The results suggested that doxorubicin interacts with the microstructure of the microemulsion increasing significantly the drug solubility. It was possible to conclude that the investigated microemulsion can be a very promising vehicle as a drug carrier for the administration of doxorubicin. In the literature, encapsulation of chemotherapeutic agents within colloidal systems usually improves drug efficiency and leads to decrease in toxicity because the carrier exits the blood circulation in tissues where capillary junctions have been disrupted and are not tightly bound, for example, tumor growth areas. A number of drugs have now been successfully encapsulated in liposomes and microemulsions and in most cases appear to improve therapeutic efficacy and largely decrease toxicity [37–39]. However, microemulsion is known to be a very stable system compared with liposomes. Furthermore, its oily core, instead of the aqueous core of liposomes, allows highly efficient incorporation of lipophilic drugs and it may also have great potential as a parenteral vehicle for sparingly used substances because of its high-solubilization capacity. Recently, a phospholipid-based microemulsion has attracted a great deal of interest as a pharmaceutically acceptable microemulsion. An interesting study was performed by Hwang et al. [40]. The primary aim of this study was to develop a parenteral formulation of all-trans-retinoic acid (ATRA) by overcoming its solubility limitation by utilizing phospholipid-based microemulsion system as a carrier. The pharmacokinetic profile of ATRA on human cancer HL-60 and MCF-7 cell lines was also similar (Table 11.2) between free ATRA and a microemulsion formulation of ATRA, suggesting that anticancer activity was not impaired by loading in microemulsion. This study herein demonstrated that phospholipid-based microemulsions may provide an alternative parenteral formulation of ATRA.

Other important research with injectable microemulsions of vincristine (M-VCR) were carried out and its pharmacokinetics, acute toxicity, and antitumor effects were evaluated [41]. The pharmacokinetics, acute toxicity, and antitumor effects of M-VCR were studied in C57BL/6 mice-bearing mouse murine histocytoma M5076

TABLE 11.2
**Non-Compartmental Pharmacokinetic Parameters of ATRA after
i.v. Administration of Sodium ATRA or ATRA Microemulsion,
Equivalent to 4 mg/kg as ATRA, to Rats ($n = 5$)**

	Formulation	
Parameters	Sodium ATRA	Microemulsion Formulation
$T_{1/2}$ (h)	1.31 ± 0.50	1.20 ± 0.39
AUC ($\mu g \cdot h/mL$)	8.27 ± 2.27	9.86 ± 1.39
MRT (h)	1.89 ± 0.13	1.73 ± 0.13
V_{ss} (mL/kg)	850.26 ± 145.91	651.18 ± 106.71[a]
CL (mL/h)	450.73 ± 178.47	376.84 ± 63.56

Source: Hwang, S.R., Lim, S-J., Park, J-S., and Kim, C-K., *Int. J. Pharm.*, 276, 175, 2004. With permission.

[a] $p < 0.05$, when compared with sodium ATRA.

tumors. The plasma AUC of M-VCR was significantly greater than that of free vincristine (F-VCR) (Table 11.3). M-VCR had lower acute toxicity and greater potential antitumor effects than F-VCR in M5076 tumor-bearing C57BL/6 mice. M-VCR is a useful tumor-targeting microemulsion drug delivery system.

TABLE 11.3
**Comparison of Pharmacokinetic Parameters
of M-VCR and F-VCR (VCR Dose: 2 mg/kg,
i.v. Administrated in C57BL/6 Mice)**

	F-VCR*	M-VCR*
C_0 ($\mu g/mL$)	1.11 ± 0.16	0.85 ± 0.12
K_{12} (L/h)	31.74 ± 4.24	0.30 ± 0.01
K_{21} (L/h)	6.78 ± 1.52	0.21 ± 0.08
K_e (L/h)	1.37 ± 0.28	0.07 ± 0.01
$T_{1/2\alpha}$ (h)	0.02 ± 0.01	1.25 ± 0.31[a,b]
$T_{1/2\beta}$ (h)	2.96 ± 0.43	25.76 ± 3.88[a,b]
V_1 (L/kg)	1.82 ± 0.37	2.35 ± 0.41
V_2 (L/kg)	8.51 ± 2.32	3.52 ± 0.59
V_{ss} (L/kg)	10.33 ± 3.45	5.88 ± 1.21
CL (L/h/kg)	3.28 ± 0.25	0.17 ± 0.09[a,b]

Source: Junping, W., Takayama, K., Nagai, T., and Maitan, Y., *Int. J. Pharm.*, 251, 13, 2003. With permission.

[a] Results were given as mean $\pm$ SD, $n = 3$.

[b] $p < 0.01$, compared with those of F-VCR.

An interesting microemulsion formulation based on oleic acid/Span 80/Tween 80/ isopropanol has been reported [42]. The microemulsion was prepared by loading an anticancer drug, Mitomycin C (MC), into this oil/water system and the stability studies of microemulsion in order to obtain physical and physicochemical properties were also examined. An electrochemical detection for the interaction of double-stranded DNA (dsDNA) with MC loaded into the microemulsion was performed by using differential pulse voltammetry (DPV) with a disposable sensor, pencil graphite electrode (PGE), for the first time, in this study. The magnitude of guanine oxidation signal was monitored before and after interaction between MC and dsDNA. The effect of different experimental parameters such as MC concentration, MC interaction time with dsDNA, and dsDNA concentration was also studied to find the optimum analytical performance.

Targeted delivery of antitumor agents to solid tumors can be achieved simply by intra-arterial or direct injection into the tissue. Compared with their systemic administration, these methods of administration certainly increase the selectivity in their exposure to tumor cells [43–45].

Direct injection of various anticancer agents into the tumor has several advantages over systemic administration. For example, lower doses can be injected into the tumor site to reduce the side effects [45–47]. However, most anticancer agents are composed of small molecules and their intratumoral clearance is relatively rapid [48]. To overcome this problem, lipid carrier systems have been used because of their favorable characteristics as a biodegradable drug reservoir. Thus, local disposition characteristics of lipid carrier formulations after intratumoral injection have become an important issue in drug delivery [45]. Particle size and electrical charge are important factors determining the in vivo fate of lipid carriers. The pharmacokinetic parameters of two emulsion formulations with different sizes (large and small emulsions) and two liposome formulations with diverse charges (neutral and cationic liposomes) are listed in Table 11.4. k_1, the rate constant of

TABLE 11.4
Pharmacokinetic Parameters for Lipid Carriers, and MC after Intratumoral Injection in Tissue-Isolated Tumors

	k_1 ($\times 10^{-3}$ min^{-1})	k_2 ($\times 10^{-3}$ min^{-1})	k_3 ($\times 10^{-3}$ min^{-1})	R
Large emulsion	2.32 ± 1.82	6.64 ± 1.21	3.02 ± 0.55	0.862 ± 0.067
Small emulsion	24.42 ± 6.58	24.96 ± 13.07	19.96 ± 2.44	0.226 ± 0.044
Neutral liposome	14.81 ± 8.28	7.74 ± 2.23	3.66 ± 0.25	0.702 ± 0.125
Cationic liposome	0.081 ± 0.017	4.44 ± 1.62	0.96 ± 0.69	0.999 ± 0.00006
MC	22.81 ± 11.35	6.34 ± 0.74	9.06 ± 2.14	0.278 ± 0.065

Source: Nishikawa, M. and Hashida, M., *Adv. Drug Deliv. Rev.*, 40, 19, 1999. With permission.

Notes: k_1, rate constant of transfer from poorly perfused region to well-perfused regipn; k_2, venous appearance rate constant; k_3, rate constant of leakage from the surface; R, dosing ratio into poorly perfused region.

transfer from the poorly perfused region to the well-perfused region, is small and extensively varies among the formulations, indicating that the intratumoral behavior of lipid carriers is largely determined by this rate constant, and the transfer of large or cationic carriers is highly restricted. On the other hand, k_2, the venous appearance rate, is not so different, suggesting that the transfer of these carriers with diameters of 80–250 nm from tissue to vasculature is not a rate-limiting step [48–50]. An injectable microemulsion of arsenic trioxide (As_2O_3-M) was prepared for intratumoral injection and the suppressive effect of As_2O_3-loaded microemulsion on human breast cancer cells MCF-7 was compared with those of a solution of the drug [51]. The microemulsion was made up of soybean oil as the oil phase, a mixture of Brij 58 and Span 80 as surfactants, the absolute ethanol as co-surfactant, and bi-distilled water containing As_2O_3 solution as the aqueous phase. Microemulsion formulation contained 5×10^{-6} molar (M) As_2O_3. The pH of As_2O_3-M was adjusted to 7.35 ± 0.1 and the physicochemical stability of the formulation was observed. The formulation was physically stable for 12 months at room temperature when kept in ampoules, as well as after autoclaving at 110°C for 30 min. The antitumor effects of As_2O_3-M were examined on human breast cancer cells MCF-7. It was clearly demonstrated that As_2O_3-M had significant cytotoxic effect on breast cancer cell lines and the cytotoxic effect of As_2O_3-M was significantly more than that of regular As_2O_3 solutions. Even an ~3000 times diluted microemulsion formulation loaded with 5×10^{-6} M As_2O_3 showed cytotoxic effects. As a result, this diluted concentration (~1.6×10^{-9} M) was found 1000 times more effective than regular As_2O_3 solutions (5×10^{-6} M). Figure 11.3 shows that

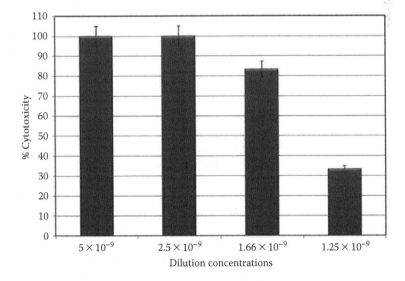

FIGURE 11.3 Cytotoxicity of tumor cells treated with As_2O_3–M. Cytotoxicity was assessed by trypan blue dye exclusion test following 72 h culture. Each point represented the mean ± SD. (Adapted from Karasulu, H.Y., Karabulut, B., Kantarcı, G., Ozgüney, I., Sezgin, C., Sanli, U.A., and Göker, E., *Drug Deliv.*, 11, 345, 2004. With permission.)

~1.6 × 10^{-9} M of the microemulsion form of arsenic trioxide exerted a highly cytotoxic effect and killed 80% of tumor cells. A significant difference was determined by using Scheffe's F-test ($p < 0.05$). According to in vitro cytotoxicity studies, it can be concluded that when As_2O_3 was incorporated into the microemulsion (As_2O_3-M), which is a new drug carrier system, it suppresses tumor cell growth on multiple tumor lines.

Lipid emulsions are considered to be superior to liposomes as they can be produced on an industrial scale, are stable during storage, are highly biocompatible, and have a high-solubilizing capacity as far as lipophilic drugs are concerned because they have an oil-in-particle form, so they can dissolve large amounts of drugs [52–54]. Furthermore, small lipid emulsions have been widely used as long-circulating carriers and have become useful parenteral drug delivery systems. It has been reported that particular lipid drug carriers mimic the metabolism of plasma lipoproteins or chylomicrons and are preferentially taken up by the liver [38–40,54]. In the Chansri et al. study [55], ATRA was incorporated into lipid emulsions in an attempt to alter its distribution characteristics and improve its inhibition of liver cancer metastasis. Lipid emulsions composed of egg phosphatidylcholine, cholesterol, and soybean oil were optimized carriers for ATRA delivery. The delivery of ATRA by emulsions can reduce the elimination of ATRA from the blood circulation and preferentially accumulate retention of ATRA in the liver can successfully suppress the progression of liver metastasis in mice injected with colon carcinoma cells. These findings indicate that the effective delivery and retention of ATRA in hepatocytes by emulsion is an efficient approach for the treatment of liver metastasis.

Paclitaxel is one of the most effective and most widely used anticancer agents. However, paclitaxel is difficult to formulate for parenteral administration because of its low aqueous solubility and Cremophor EL, the excipient used for its formulation, and has been shown to cause serious side effects. An interesting study reports an alternative administration vehicle involving a lipophilic paclitaxel derivate, paclitaxel oleate, incorporated in the core of a nanosized, sterically stabilized o/w lipid emulsion. This formulation has good physicochemical properties and demonstrates cytotoxic activity against HeLa cells. The formulated emulsion may be clinically useful not only for eliminating toxic effects of Cremophor EL, but also for improving the pharmacokinetic parameters of paclitaxel [56].

Among the emulsifying methods, the SMEDDS are worthy of notice. SMEDDS are isotropic mixtures of oil, a surfactant, and possibly one or more hydrophilic solvent or cosurfactants, which form fine o/w emulsions or microemulsions when exposed to aqueous media under condition of gentle agitation [57,58]. An interesting formulation of SMEDDS, which is a mixture of paclitaxel, tetraglycol, Cremophor ELP, and Labrafil 1944, was developed by Kang et al. [59]. This paclitaxel formulation contains poly (D,L-lactide-co-glycolide) (PLGA) in order to offer controlled release of paclitaxel. The obtained results indicated that the potential use of microemulsion using PLGA for the sustained release of lipophilic drugs such as paclitaxel was established. The release behavior of paclitaxel from microemulsion containing PLGA exhibited a biphasic pattern characterized by a fast initial release during the first 48 h, followed by a slower and continuous release for 144 h; in contrast the release of paclitaxel from microemulsion without PLGA was complete within 24 h. This result was identical with the result of antitumor activity in vitro of paclitaxel from microemulsion containing PLGA against human breast cancer cell line MCF7 and this formulation enhanced antitumor activity in vivo compared with microemulsion without PLGA against SKOV-3 human ovarian cancer cells in a nude mice model. Two cremophor-free microemulsions, lecithin:butanol:myvacet oil:water (LBMW) and capmul:myvacet oil:water (CMW) for paclitaxel (pac) were developed for i.v. administration [60,61]. In vivo pharmacokinetic studies in male Sprague–Dawley rats after i.v. administration revealed that paclitaxel in LBMW and CMW in the systemic circulation were five and two times longer and were eight and three times more widely distributed than paclitaxel from Taxol. LBMW and CMW offer a significant clinical advantage in terms of prolonged half-life and wide tissue distribution, indicating that paclitaxel delivered by these systems i.v. may result in prolonged exposure of paclitaxel to the tumor and subsequently an improved clinical efficacy [61]. However, after i.v. administration, colloidal drug carriers are usually recognized as foreign bodies and are rapidly taken up by circulating monocytes and macrophages in the lier, spleen, and bone marrow. The ability of colloidal particles to evade the mononuclear phagocyte system (MP) and exhibit long residence times in blood depends largely on carrier size and the physicochemical properties of the surface. Modifying the colloid surface with a hydrophilic and flexible polymer such as poly(ethylene glycol) (PEG) is widely used to prolong circulation time. The longevity of PEGylated colloids is attributed to the highly hydrated and flexible PEG chains, which reduce interactions with plasma proteins and cell surfaces [62,63]. For example, Rossi et al. [64] demonstrated that nanosized PEGylated emulsions prepared with commonly used pharmaceutical excipients can passively target neoplastic tissues. The degree of emulsion accumulation into the tumor was dependent on the PEG coating and tumor type, whereby C26 colon adenocarcinoma was more permeable to the emulsions than B16 melanoma. Moreover, a relationship between tighter molecular packing at the air/water interface and enhanced physical stability in the presence of albumin was established. However, prolonged circulation time in vivo was not observed.

11.4 CONCLUSIONS

Lipid dispersion formulations, such as oil in water emulsions and microemulsions, are attractive carriers for lipophilic and anticancer drugs. Antitumor agents show their therapeutic activities only in the restricted sites where intact agents reach after administration. In particular, several studies have reported the ability of emulsions to enhance the accumulation of anticancer agents into solid tumors compared to the free drug. Emulsions are considered superior due to their suitability for industrial scale production, stability on storage, biocompatibility, and the incorporation efficacy for lipophilic drugs. However, in practical terms, there remains a problem such as the rapid release of drug on its own or by attack from plasma proteins, high-density lipoproteins, and other biological components. On the other hand, the favorable drug delivery and solvent properties, together with the ease of preparations and the infinite physical stability of these unique oil–water–surfactant mixtures, make microemulsions very promising vehicles for future formulations of anticancer drugs.

A number of comparative researches exist in literatures, some of which have evaluated the utility of microemulsion and emulsion formulations against alternative delivery systems for antitumor agents. Furthermore, it has proven possible to formulate preparations suitable for most routes of administration. There is still, however, a considerable amount of fundamental work in order to figure out the physicochemical behavior of microemulsions that needs to be performed before they can live up to their potential as multipurpose drug delivery vehicles.

In conclusion, it is assumed that as a result of these recent developments, new formulations for new applications will emerge for cancer therapy. Therefore, microemulsion and emulsion formulations are considered to be promising formulation technique for anticancer drugs and drug targeting.

SYMBOLS AND TERMINOLOGIES

ATRA	all-trans-retinoic acid
V_{ss}	apparent volume of distribution at steady state
AUC	area under the drug concentration–time curve
CMW	capmul:myvacet oil:water
DPV	differential pulse voltammetry
dsDNA	double-stranded DNA
F-VCR	free vincristine
i.v.	intravenous
LBMW	lecithin:butanol:myvacet oil:water
MRT	mean resident time
As_2O_3-M	microemulsion of arsenic trioxide
M-MTX	microemulsion of methotrexate
M-VCR	microemulsions of vincristine
MC	mitomycin C
MDR	multidrug resistance
pac	paclitaxel
PGE	pencil graphite electrode
Pgp	P-glycoprotein
PLGA	poly (D,L-lactide-co-glycolide)
PEG	poly(ethylene glycol)
SMEDDS	self-microemulsifying drug delivery systems
Sol-MTX	solution of the methotrexate
SD	standard deviation
CL	total clearance

REFERENCES

1. Dy, G.K. and Adjei, A.A. 2005. Obstacles and opportunities in the clinical development of targeted therapeutics. *Prog. Drug. Res.*, 63, 9–41.
2. Alison, M.R. 2002. *The Cancer Handbook*. London, Nature Publishing Group.
3. Szekeres, T. and Novonthy, L. 2002. New targets and drugs in cancer chemotherapy. *Med. Princ. Pract.*, 11, 117–125.
4. Hanahan, D. and Weinberg, R.A. 2000. The hallmarks of cancer. *Cell*, 100, 57–70.
5. Reddy, A. and Kaelin, Jr. W.G., 2002. Using cancer genetics to guide the selection of anticancer drug targets. *Curr. Opin. Pharmacol.*, 2, 366–373.
6. King, R.J.B. 2000. *Cancer Biology* (2nd edition). London, Pearson Education Ltd.
7. Chabner, B.A. and Longo, D.L. 2001. *Cancer Chemotherapy and Biotherapy— Principles and Practice* (3rd edition). Philadelphia, PA: Lippincott, Williams & Wilkins.
8. Pinedo, H.M. and Giacone, G. 1998. *Drug Resistance in the Treatment of Cancer*. London, UK, Cambridge University Press.
9. Gottesman, M.M. 2002. Mechanisms of cancer drug resistance. *Annu. Rev. Med.*, 53, 615–627.
10. Ueda, K., Clark, D.P., Chen, C.J., Robinson, I.B., Gottesman, M.M., and Pastan, I. 1987. The human multidrug resistance (mdr1) gene. cDNA cloning and transcription initiation. *J. Biol. Chem.*, 262, 505–508.
11. Goldstein, L.J., Galski, H., Fojo, A., Willingham, M., Lai, S.L., Gazdar, A., Pirker, R., Green, A., Crist, W., Brodeur, G.M., et al. 1989. Expression of a multidrug resistance gene in human cancers. *J. Natl. Can. Inst.*, 81, 116–124.
12. Allen, T.M. and Cullis, P.R. 2004. Drug delivery systems: Entering the mainstream. *Science*, 303, 1818–1822.
13. Panayiotou, M., Pöhner, C., Vandevyver, C., Wandrey, C., Hilbrig, F., and Freitag, R. 2007. Synthesis and characterization of thermo-responsive poly(N,N'-diethylacrylamide) microgels. *React. Funct. Poly.*, 67, 807–819.
14. Sun, H., Yu, J., Gong, P., Xu, D., Zhang, C., and Yao, S. 2005. Novel core-shell magnetic nanogels synthesized in an emulsion-free aqueous system under UV irradiation for targeted radiopharmaceutical applications. *J. Magn. Magn. Mater.*, 294, 273–280.
15. Chawla, J.S. and Amjii, M.M. 2002. Biodegradable poly(ε-caprolactone) nanoparticles for tumor-targeted delivery of tamoxifen. *Int. J. Pharm.*, 249, 127–138.
16. Hamoudeh, M., Hamoudeh, M., Salim, H., Barbos, D., Paunoiu, C., and Fessi, H. 2007. Preparation and characterization of radioactive dirhenium decacarbonyl-loaded PLLA nanoparticles for radionuclide intra-tumoral therapy. *Eur. J. Pharm. Biopharm.*, 67, 597–611.
17. Lee, S.H., Zhang, Z., and Feng, S.-S. 2007. Nanoparticles of poly(lactide)-tocopheryl polyethylene glycol succinate (PLA-TPGS) copolymers for protein drug delivery. *Biomaterials*, 28, 2041–2050.
18. Pradhan, P., Giri, J., Banerjee, R., Bellare, J., and Bahadur, D. 2007. Preparation and characterization of manganese ferrite-based magnetic liposomes for hyperthermia treatment of cancer. *Magn. Magn. Mater.*, 311, 208–215.
19. Fournier, E., Passirani, C., Colin, N., Breton, P., Sagodira, S., and Benoit, J.-P. 2004. Development of novel 5-FU-loaded poly(methyllidene malonate 2.1.2)-based microspheres for the treatment of brain cancers. *Eur. J. Pharm. Biopharm.*, 57, 189–197.
20. Brigger, I., Chaminade, P., Marsaud, V., Appel, M., Besnard, M., Gurny, R., Renoir, M., and Couvreur, P. 2001. Tamoxifen encapsulation within polyethylene glycol-coated nanospheres. A new antiestrogen formulation. *Int. J. Pharm.*, 214, 37–42.

21. Lawrence, M.J. and Rees, G.D. 2000. Microemulsion-based media as novel drug delivery systems. *Adv. Drug Del. Rev.*, 45, 89–121.
22. Prankerd, R.J. and Stella, V.J. 1990. The use of oil-in-water emulsions as a vehicle for parenteral drug administration. *J. Parenter. Sci. Technol.*, 44, 139–143.
23. De Gennes, P.G. and Taupin, C. 1982. Microemulsions and flexibility of oil/water interfaces. *J. Phys. Chem.*, 86, 2294–2304.
24. Watari, H. 1997. Microemulsions in separation sciences. *J. Chromatogr. A.*, 780, 93–102.
25. Karasulu, H.Y. 2008. Microemulsions as novel drug carriers: The formation, stability, applications and toxicity. *Expert Opin. Drug Deliv.*, 5, 119–35.
26. Masahiro, N. 2000. Places of emulsions in drug delivery. *Adv. Drug Deliv. Rev.*, 45, 1–4.
27. Spernath, A. and Aserin, A. 2006. Microemulsions as carriers for drugs and nutraceuticals. *Adv. Colloid Interf. Sci.*, 128–130, 47–64.
28. Sha, X., Yan, G., Wu, Y., Li, J., and Fang, X. 2005. Effect of self-microemulsifying drug delivery systems containing labrosol on tight junctions in Caco-2 cells. *Eur. J. Pharm. Sci.*, 24, 477–486.
29. Lv, F.-F., Li, N., Zheng, L.-Q., and Tung, C.-H. 2006. Studies on the stability of the chloramphenicol in the microemulsion free of alcohols. *Eur. J. Pharm. Biopharm.*, 62, 288–294.
30. Pouton, C.W. 2000. Lipid formulations for oral administration of drugs: Non-emulsifying, self-emulsifying and self-microemulsfying drug delivery systems. *Eur. J. Pharm. Sci.*, 11, 93–98.
31. Porter, C.J.H., Kaukonen, A.M., Boyd, B.J., Edwards, G.A., and Charman, W.N. 2004. Susceptibility to lipase-mediated digestion reduces the oral bioavailability of danazol after administration as a medium chain lipid-based microemulsion formulation. *Pharm. Res.*, 21, 1405–1412.
32. Osborne, D.W., Middleton, C.A., and Rogers, R.L. 1988. Alcohol-free microemulsions. *J. Disper. Sci. Tech.*, 9, 415–423.
33. Karasulu, H.Y., Karabulut, B., Goker, E., Guneri, T., and Gabor, F. 2007. Controlled release of methotrexate from w/o microemulsion and its in vitro anti-tumor activity. *Drug Deliv.*, 14, 225–233.
34. Tarr, B.D., Sambandan, T.G., and Yalkowsky, S.H. 1987. A new parenteral emulsion for the administration of taxol. *Pharm. Res.*, 4, 162–165.
35. Prankerd, R.J. and Stella, V.J. 1990. The use of oil-in-water emulsions as a vehicle for parenteral drug administration. *J. Parenter. Sci. Technol.*, 44, 139–49.
36. Formariz, T.P., Sarmento, V.H.V., Silva-Junior, A.A., Scarpa, M.V., Santilli, C.V., and Oliveira, A.G. 2006. Doxorubicin biocompatible o/w microemulsion stabilized by mixed surfactant containing soya phosphatidylcholine. *Colloids Surfaces B: Biointerf.*, 51, 54–61.
37. Owens, M.D., Baillie, G., and Halbert, G.W. 2001. Physicochemical properties of microemulsion analogues of low density lipoprotein containing amphiphatic apoprotein B receptor sequences. *Int. J. Pharm.*, 228, 109–117.
38. Azevedo, C.H.M., Carvalho, J.P., Valduga, C.J., and Maranhao, R.C. 2005. Plasma kinetics and uptake by the tumor of a cholesterol-rich microemulsion (LDE) associated to etoposide oleate in patients with ovarian carcinoma. *Gyn. Oncol.*, 97, 178–182.
39. Kawakami, S., Opanasopit, P., Yokoyama, M., Chansri, N., Yamamoto, T., Okano, T., Yamashita, F., and Hashida, M. 2005. Biodistribution characteristics of alltrans retinoic acid incorporated in liposomes and polymeric micelles following intravenous administration. *J. Pharm. Sci.*, 94, 2606–2615.

40. Hwang, S.R., Lim, S.-J., Park, J.-S., and Kim, C.-K. 2004. Phospholipid-based microemulsion formulation of all-trans-retinoic acid for parentral administration. *Int. J. Pharm.*, 276, 175–183.

41. Junping, W., Takayama, K., Nagai, T., and Maitan, Y. 2003. Pharmacokinetics and antitumor effects of vincristine carried by microemulsions composed of PEG-lipid. oleic acid, vitamin E and cholesterol. *Int. J. Pharm.*, 251, 13–21.

42. Karadeniz, H., Alparslan, L., Erdem, A., and Karasulu, E. 2007. Electrochemical investigation of interaction between mitomycin C and DNA in a novel drug-delivery system. *J. Pharm. Biomed. Anal.*, 45, 322–326.

43. Hunt, C.A., MacGregor, R.D., and Siegel, R.A. 1986. Engineering target in vivo drug delivery. 1. The physiological and physicochemical principles governing opportunities and limitations. *Pharm. Res.*, 3, 333–344.

44. Goldstein, D., Nassar, T., Lambert, G., Kadouche, J., and Benita, S. 2005. The design and evaluation of a novel target drug delivery system using cationic emulsion-antibody conjugates. *J. Control. Release*, 108, 418–432.

45. Nishikawa, M. and Hashida, M. 1999. Pharmacokinetics of anticancer drugs, plasmid DNA, and their delivery systems in tissue-isolated perfused tumors. *Adv. Drug Deliv. Rev.*, 40, 19–37.

46. Kawakami, S., Yamashita, F., and Hashida, M. 2000. Disposition characteristics of emulsions and incorporated drugs after systemic or local injection. *Adv. Drug. Deliv. Rev.*, 45, 77–88.

47. Lee, In-H., Park, Y.T., Roh, K., Chung, H., Kwon, I.C., and Jeong, S.Y. 2005. Stable pclitaxel formulations in oily contrast medium. *J. Contol. Release*, 102, 415–425.

48. Nomura, T., Koreeda, N., Yamashita, F., Takakura, Y., and Hashida, M. 1998. Effect of particle size and charge on the disposition of lipid carriers after intratumoral injection into tissue-isolated tumors. *Pharm. Res.*, 15, 128–132.

49. Nomura, T., Nakajima, S., Kawabata, K., Yamashita, F., Takakura, Y., and Hashida, M. 1997. Intratumoral pharmacokinetics and in vivo gene expression of naked plasmid DNA and its cationic liposome complex after direct gene transfer. *Cancer Res.*, 57, 2681–2686.

50. Nomura, T., Saikawa, A., Morita, S., Sakaeda (ne Kakutani), Yamashita, F., Honda, K., Takakura, Y., and Hashida, M. 1998. Pharmacokinetic characteristics and therapeutic effect of mitomycin C-dextran conjugates after intratumoral injection. *J. Control. Release*, 52, 239–252.

51. Karasulu, H.Y., Karabulut, B., Kantarcı, G., Ozgüney, I., Sezgin, C., Sanli, U.A., and Göker, E. 2004. Preparation of arsenic trioxide loaded microemulsion and its enhanced cytotoxicity on MCF-7 breast carcinoma cell line. *Drug Deliv.*, 11, 345–350.

52. Terek, M.C., Karabulut, B., Selvi, N., Akman, L., Karasulu, Y., Ozguney, I., Sanli, A.U., Uslu, R., and Ozsaran, A. 2006. Arsenic trioxide-loaded, microemulsion-enhanced cytotoxicity on MDAH 2774 ovarian carcinoma cell line. *Int. J. Gynecol. Cancer.*, 16, 532–537.

53. Fukushima, S., Kishimoto, S., Takeuchi, Y., and Fukushima, M. 2000. Preparation and evaluation of o/w type emulsions containing antitumor prostaglandin. *Adv. Drug Deliv. Rew.*, 45, 65–75.

54. Tomi, Y. 2002. Lipid formulation as a drug carrier for drug delivery. *Curr. Pharm. Des.*, 8, 467–474.

55. Chansri, N., Kawakami, S., Yamashita, F., and Hashida, M. 2006. Inhibition of liver metastasis by all-trans retinoic acid incorporated into o/w emulsions in mice. *Int. J. Pharm.*, 321, 42–49.

56. Lundberg, B.B., Risovic, V., Ramaswamy, M., and Wasan, K.M. 2003. A lipophilic paclitaxel derivative incorporated in a lipid emulsion for parentral administration. *J. Control. Release*, 86, 93–100.

57. Wu, W., Wang, Y., and Que, L. 2006. Enhanced bioavailability of silymarin by self-microemulsifying drug delivery system. *Eur. J. Pharm. Biopharm.*, 63, 288–294.

58. Gershanika, T. and Benita, S. 2000. Self-dispersing lipid formulations for improving oral absorption of lipophilic drugs. *Eur. J. Pharm. Biopharm.*, 50, 179–188.

59. Kang, B.K., Chon, S.K., Kim, S.H., et al. 2004. Controlled release of paclitaxel from microemulsion containing PLGA and evaluation and anti-tumor activity in vitro and in vivo. *Int. J. Pharm.*, 286, 147–156.

60. Nornoo, A.O., Osborne, D.W., and Chow, D.S. 2008. Cremophor-free intravenous microemulsions for paclitaxel I: Formulation, cytotoxicity and hemolysis. *Int. J. Pharm.*, 349, 108–116.

61. Nornoo, A.O. and Chow, D.S. 2008. Cremophor-free intravenous microemulsions for paclitaxel II. Stability, in vitro release and pharmacokinetics. *Int. J. Pharm.*, 349, 117–123.

62. Suh, J., Choy, K.L., Lai, S.K., Suk, J.S., Tang, B.C., Prabhu, S., and Hanes, J. 2007. PEGylation of nanoparticles improves their cytoplasmic transport. *Int. J. Nanomedicine*, 2, 735–741.

63. Vroman, B., Ferreira, I., Jérôme, C., Jérôme, R., and Préat, V. 2007. PEGylated quaternized copolymer/DNA complexes for gene delivery. *Int. J. Pharm.*, 344, 88–95.

64. Rossi, J., Giasson, S., Khalid, M.N., Delmas, P., Allen, C., and Leroux, J.C. 2007. Long-circulating poly(ethylene glycol)-coated emulsions to target solid tumors. *Eur. J. Pharm. Biopharm.*, 67, 329–338.

12 Enzyme Kinetics as a Useful Probe for Micelle and Microemulsion Structure and Dynamics

Werner Kunz, Didier Touraud, and Pierre Bauduin

CONTENTS

12.1 INTRODUCTION

Since the pioneering work of Martinek et al. [1] more than 30 years ago, numerous studies have been published, where enzymatic reactions were investigated in surfactant containing media. Most of them deal with reverse microemulsions, in which the enzyme is entrapped in water nanodroplets that are dispersed in an external oil pseudophase. Surfactants and cosurfactants stabilize the whole system by forming an interfacial layer between the aqueous and the hydrophobic part.

A recent review summarizes the most relevant papers published over the last 10 years [2]. Depending on the composition and the enzyme, a higher or lower activity of the biocatalysts is measured, when compared to the corresponding aqueous media without surfactant and oil phase. The specificity of the system makes it very difficult to deduce general rules. A cationic surfactant can be favorable to enzyme reactions in microemulsions or highly defavorable. The enzyme can be situated in the core of the water droplets or very near to the interface with the oil phase. Accordingly, substances that are incorporated in the surfactant layer are more or less "seen" by the enzyme, which cause very different sensitivity of the enzyme to these molecules.

On the other hand, once an enzyme and its affinity to the interface are well characterized, this enzyme can be a very sensitive probe for tiny changes in the composition of microemulsion. There are not many other techniques that have such a high sensitivity. Thermodynamic experiments such as calorimetry are also very sensitive, but they do not give a detailed insight into structures. By contrast, scattering techniques and NMR yield more detailed pictures, but they are less sensitive. Therefore, enzyme reactions can be a useful complement for the investigation of microemulsion structures. This is the topic of this chapter.

To begin with we want to show how sensitive enzyme reactions can indeed be. To this purpose, we will consider surfactant solutions below the critical micellar concentration (cmc). Then, Section 12.3, we will show that enzymes can also be used to characterize the composition of direct micellar solutions or microemulsion, a field that is less often considered. And finally we will show some examples of reverse microemulsions and how the enzyme activity correlates there with other experimental observations such as electrical conductivity.

12.2   ENZYMES IN DILUTE SURFACTANT SOLUTIONS

Horse liver alcohol dehydrogenase (HLAD) is a cheap and often used enzyme to study microemulsion systems [3]. Its structure is known in detail. It converts alcohols to aldehydes according to the following reaction scheme:

$$R-CH_2-OH+NAD^+ \xrightarrow{\quad HLADH \quad} R-CHO+NADH+H^+$$

with R being the corresponding hydrocarbon chain of n-alcohols and aldehydes, respectively. The progress of such a reaction can easily be followed spectroscopically by recording the adsorption of the produced NADH at 340 nm. What makes this reaction interesting is the fact that it is highly sensitive to the alcohol concentrations (see Figure 12.1, [4]).

In order to restrict uncertainties of the measured results due to variations of the purity of the enzyme, A is normalized enzymatic activity, defined as $A = V/V_0$. The initial velocity V of the enzymatic reaction is inferred from the slope of the adsorption intensity of NADH versus time during the first minutes, and V_0 is the initial velocity of a reference solution containing the same amount of enzyme from the same batch and an arbitrarily chosen amount of a reference alcohol concentration in the otherwise same buffer solutions (here 10^{-5} mol ethanol per gram of water) (see Ref. [3] for further details). As can be seen from Figure 12.1, the enzyme starts to be inhibited even at very small alcohol concentrations of the order of 5×10^{-5} mol/g.* Therefore, any significant change of the alcohol concentration close to the enzyme can be detected. In particular, this is the case when the alcohols can play the role of cosurfactants in the formation of micelles. But even far below

---

* According to the shape of the reactivity curve, it can be concluded that HLADH is subjected to an uncompetitive inhibition [3].

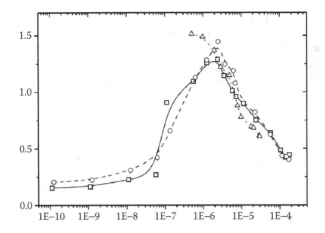

FIGURE 12.1 Typical normalized enzymatic activity A of HLADH in aqueous buffer solutions containing different alcohols. The curves are spline fits to guide the eyes. (From Schirmer, C., PhD Thesis, University of Regensburg, Germany, 2001. With permission; see also Ref. [3].)

the cmc specific interactions or aggregate formation may happen between a surfactant and an alcohol. Most of the experimental techniques are not sensitive enough to detect any such interactions. As can be seen from Figure 12.2, [4], even at surfactant concentrations as low as 10^{-6} mol/L, the HLADH activity exhibits significant

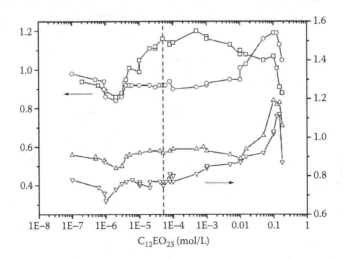

FIGURE 12.2 Normalized enzymatic activity of HLAD in aqueous 1-pentanol solutions as a function of added surfactant (polyoxyethylene(23)lauryl ether [$C_{12}EO_{23}$]). The vertical dashed line denotes the cmc of the pure surfactant. From top to bottom: alcohol concentrations of 10^{-3}, 5×10^{-3}, 7.5×10^{-3}, and 10^{-2} M. (From Schirmer, C., PhD Thesis, University of Regensburg, Germany, 2001.)

minima at all investigated 1-pentanol concentrations. It is beyond the scope of the present topic to discuss the origin of these variations at such low surfactant concentrations. However, this example shows how sensitive enzymatic reactions can be.* In Section 12.3, we will see how this fact can be used to characterize various direct microemulsions with different types of alcohols acting as cosurfactants.

12.3   ENZYMES IN DIRECT MICELLAR SYSTEMS

As a typical example of the different structuring of alcohols in direct micellar systems, we will discuss the ternary system water/$C_{12}EO_{23}$/alcohol with different alcohols and at different compositions. Two exemplary phase diagrams are given in Figure 12.3, [5]. It is well known that some alcohols are good cosurfactants, whereas others do not incorporate into the micellar structure. Therefore, it can be expected that water-soluble enzymes with a low affinity to interfaces, such as HLADH, will interact only with the amount of alcohols that is partitioned in the aqueous pseudophase. This is indeed the case, as can be concluded from Figure 12.4.

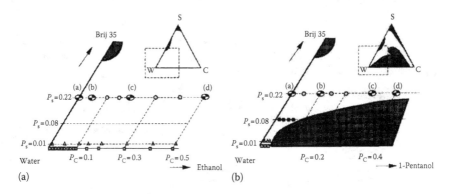

FIGURE 12.3 (a) Ternary system "aqueous buffer solution/ethanol/$C_{12}EO_{23}$" at 25°C. The water-rich part of the whole phase diagram, shown in the insert, as well as the different compositions used for the enzymatic studies and the neutron-scattering measurements are given. The bright zones denote macroscopically homogeneous phases of low viscosity. The dark zones give the compositions of either polyphasic or crystal-like systems. P_s is the quantity of the surfactant S ($C_{12}EO_{23}$), expressed in mass % of the whole composition, and P_c is the quantity of alcohol (C) in the same units. The points symbolized by □, △, and ○ denote the compositions for which enzymatic activities were measured. The corresponding results are given in (b). The points symbolized by the partially shaded circles and labeled a, b, c, and d refer to the compositions of the systems for which scattering experiments were carried out. (b) Corresponding diagram for the ternary system "aqueous buffer solution/1-pentanol/$C_{12}EO_{23}$" at 25°C. (From Meziani, A., Touraud, D., Zradba, A., Pulvin, S., Pezron, I., Clausse, M., and Kunz, W., *J. Phys. Chem. B*, 101, 3620, 1997.)

---

* In independent tests it was verified that the nonionic surfactant, which is also an alcohol, is not oxidized by the enzyme.

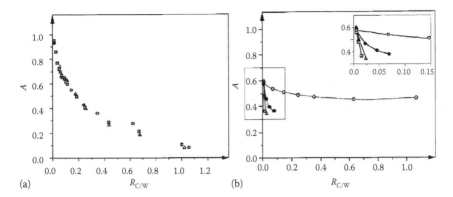

FIGURE 12.4   (a) Enzymatic activity of HLADH (relative to a pure aqueous buffer solution containing 10^{-2} mol/L of ethanol) as a function of the ratio of the mass % of ethanol to the mass % of water, $R_{c/w}$ $\square$, Δ, and $\bigcirc$ correspond to surfactant concentrations of $P_s = 0$, 0.01, and 0.22, respectively. The different $R_{c/w}$ ratios correspond to the series of the P_c values given in Figure 12.3a. (b) Enzymatic activity of HLADH (relative to a pure aqueous buffer solution containing 10^{-2} mol/L of ethanol) as a function of the ratio of the mass % of 1-pentanol to the mass % of water, $R_{c/w}$. $\square$, Δ, $\bullet$, and $\bigcirc$ correspond to surfactant concentrations of $P_s = 0$, 0.01, 0.08, and 0.22, respectively. The different $R_{c/w}$ ratios correspond to the series of the P_c values given in Figure 12.3b. (From Meziani, A., Touraud, D., Zradba, A., Pulvin, S., Pezron, I., Clausse, M., and Kunz, W., *J. Phys. Chem. B*, 101, 3620, 1997.)

Figure 12.4a, [5], shows that the enzymatic activity exponentially decreases with increasing amounts of ethanol present in the buffer solution. This is in agreement with Figure 12.1 and the fact that HLADH undergoes an uncompetitive inhibition by alcohol substrates. Remarkably, this inhibition is not influenced by the presence of the nonionic surfactant. Two reasons are responsible: first, this result confirms that $C_{12}EO_{23}$ does not significantly interact with the enzyme, a finding that was also verified in independent experiments, and, second, $C_{12}EO_{23}$ does not incorporate a significant amount of ethanol, even at very high surfactant concentrations (22 mass %).

This is in contrast to ternary systems containing 1-pentanol (see Figure 12.4b). It seems that this alcohol is less rapidly converted by HLADH, because the initial A values are only of the order of 60% compared to the reference ethanol system. However, this is not true. In further discussion, different enzyme batches may yield different absolute A values, without changing the shape of the curve. Further, the A values can be higher at still smaller alcohol concentration (Figure 12.1), where the concentration, at which the maximum A value is measured, is $c_{max} = 5 \times 10^{-5}$ mol/g. In any case, after this maximum, the addition of an increasing amount of surfactant suppresses more and more the uncompetitive inhibition by the alcohol. At 22 mass % of $C_{12}EO_{23}$ the normalized activity is stabilized at 50% and remains roughly constant over the whole concentration range. This can only be interpreted by a significant incorporation of 1-pentanol into the micellar structure. And indeed, it is well known that middle-chain alcohols are cosurfactants, whereas ethanol and propanol do not enter the micelles or even destroy them.

One of the drawbacks of enzymatic studies is that their activity depends on numerous factors that can vary from one experiment to the other. In addition, different batches of enzymes may not always give the same quantitative results. The consideration of normalized activities A reduces this uncertainty, but cannot fully avoid it. Therefore, it is interesting to investigate how far enzymatic reactions can deliver quantitative structural information on direct micellar systems.

In the present context, the partition coefficient of alcohols between the aqueous buffer and the micellar pseudophases is a convenient quantity to check this point. The partition coefficient p can be inferred from self-diffusion coefficients of the alcohol molecules according to following equation:

$$D_{app} = pD_{mic} + (1 - p)D_{free}$$

Here, the apparent self-diffusion coefficient D_{app} of the alcohol in the ternary system can be measured by the NMR spin-echo technique [6]. D_{mic} is the self-diffusion coefficient of the alcohol within or associated with the micelles and D_{free} the corresponding value in the aqueous pseudophase; it is approximated by measuring the self-diffusion coefficient in the binary D_2O–pentanol system. As D_{mic} is much smaller than D_{free}, the partition coefficient p can be approximated by

$$p = \frac{D_{free} - D_{app}}{D_{free} - D_{mic}} \approx \frac{D_{free} - D_{app}}{D_{free}}$$

It turns out that virtually no ethanol is partitioned into the micelles ($p \approx 0$), whereas NMR measurements suggest that p is about 0.4 in the case of 1-pentanol solutions with 8 mass % of alcohol (for more details, see Ref. [6]). The enzymatic activity measurements also lead to the conclusion that no ethanol is incorporated in the micelles. For the 1-pentanol system, a quantitative interpretation is more delicate.

Comparison of Figure 12.5 with Figure 12.4b, [5], shows rough, but not quantitative agreement between independent measurements, made with different badges of chemicals and enzymes and on different apparatuses. This is a typical example demonstrating that enzyme activity measurements cannot deliver quantitatively reproducible results. However, we can attempt to deduce p values in the following way, starting from an enzymatic activity study over several alcohol concentration decades, as done for Figure 12.5.

We suppose that the c_{max} value (i.e., the alcohol concentration, at which A has its maximum) always corresponds to the same optimum alcohol concentration in the aqueous (pseudo-) phase, at which the enzyme catalysis is at its maximum. If this is true, then the concentration of the alcohol incorporated in the micelles c_{mic} is given by $c_{mic} = c_{max} - c_{max}^0$, where c_{max}^0 is the optimum surfactant concentration at zero surfactant concentration. The alcohol partition coefficient can then be estimated as

$$p = c_{max} - c_{max}^0 / c_{max}$$

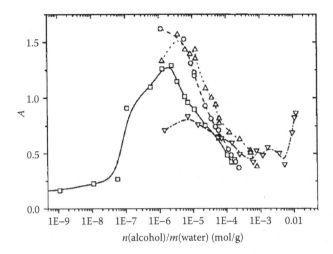

FIGURE 12.5   Enzyme activities in ternary systems water/$C_{12}EO_{23}$/1-pentanol for ($\square$) 0% surfactant, ($\bigcirc$) 1%, ($\triangle$) 8%, and ($\triangledown$) 22%. The lines are used for clarity. (From Schirmer, C., Liu, Y., Touraud, D., Meziani, A., Pulvin, S., and Kunz, W., *J. Phys. Chem. B*, 106, 7414–7421, 2002. With permission.)

For 1-pentanol at a surfactant concentration of 8% a p value of 0.6 is found, which is clearly higher than the value of 0.4 inferred from NMR. So all that can be inferred from the enzymatic measurements is that approximately one half of the alcohol molecules is incorporated in the micellar pseudophase.

Nevertheless and as further detailed in Ref. [3], the series of p values calculated from enzymatic reactivities for a series of alcohols is reasonable: for 8% of $C_{12}EO_{23}$ in the ternary system it is 0.5, 0.6, 0.9, and again 0.9, from 1-butanol to 1-heptanol, respectively. And for 22% of surfactant the series from ethanol to 1-octanol shows the following values: 0.0, 0.0, 0.8, 0.7, 0.9, 1, and 1. Except for 1-butanol, this result also makes sense.

Finally, we can compare both enzymatic activity and NMR results to a true structural investigation carried out with small-angle neutron scattering (SANS) experiments. Figure 12.6, [6], gives typical results, again for the ternary system water/$C_{12}EO_{23}$/1-pentanol and for water/$C_{12}EO_{23}$/1-pentanol [6]. The experimental spectra are well described by an ellipsoidal core shell model that is schematically represented in Figure 12.7, [6].

In order to fit this model to the experimental scattering data, the partition coefficient of the alcohol molecules must be adjusted. The following values were obtained for ternary systems with 8 mass % $C_{12}EO_{23}$: $p_{C4OH} = 0.15$, $p_{C5OH} = 0.38$ (the same value as obtained with NMR spin echo), $p_{C6OH} = 0.90$, $p_{C7OH} = 0.99$ as compared to $p_{C4OH} = 0.4$, $p_{C5OH} = 0.6$, $p_{C6OH} = 0.90$, $p_{C7OH} = 0.90$ inferred from the enzymatic reaction studies. Of course, the SANS descriptions are supposed to yield the more precise values. Nevertheless, in tendency, the enzyme studies are in qualitative agreement. It should be also added that for 22 mass % $C_{12}EO_{23}$ in the

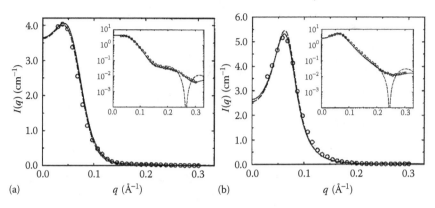

(a) q (Å^{-1})   (b) q (Å^{-1})

FIGURE 12.6 Absolute neutron scattering intensities I from SANS contrast experiments on ternary systems containing (a) 90.2 mass % D_2O, 7.9 mass % $C_{12}EO_{23}$, and 1.9 mass % 1-pentanol and (b) 89.7 mass % D_2O, 7.8 mass % $C_{12}EO_{23}$, and 2.5 mass % 1-heptanol at 25°C as a function of the wave number transfer q. The open circles are the experimental points, the dashed line is a theoretical result obtained by solving the Percus–Yevick equation with a spherical core/shell model, and the full line represents the results from a corresponding calculation with an ellipsoidal core/shell model. The insets show the same results in a logarithmic scale for the intensities. (From Preu, H., Zradba, A., Rast, S., Kunz, W., Hardy, E.H., and Zeidler, M.D., *Phys. Chem. Chem. Phys.*, 1, 3321, 1999. With permission.)

systems no modelling of the SANS spectra was possible, because of the shortcomings of the theory at such high surfactant concentrations.

12.4   ENZYMES IN REVERSE MICROEMULSIONS

As stated in the beginning, numerous studies concern enzymatic activities in reverse microemulsion systems [2]. Therefore, we will focus on a particular investigation,

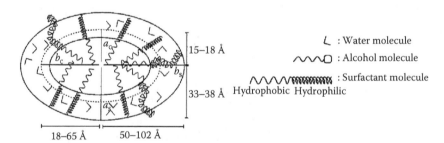

FIGURE 12.7 Schematic representation of the ellipsoidal core/shell model. The hydrophilic shell contains the more or less twisted and hydrated ethoxy groups of the surfactant molecules. In the diffuse layer nonhydrated ethoxy groups and some hydrophobic parts of the surfactant molecules form an intermediate area. In the inner core only hydrophobic tails of the surfactant molecules are present. Due to the steric hindrance, not all hydrophobic tails can completely stick in this hydrophobic core. (From Preu, H., Zradba, A., Rast, S., Kunz, W., Hardy, E.H., and Zeidler, M.D., *Phys. Chem. Chem. Phys.*, 1, 3321, 1999. With permission.)

in which it is attempted to use enzymes to characterize the rigidity of the interfacial surfactant film and some structural changes in the microemulsions [7,8].

The enzyme used this time is horseradish peroxidase (HRP), which catalyzes the oxidation of 2,2-azino-bis(3-ethylbenzothiazoline-6-sulfonic acid) diammonium salt (ABTS) by hydrogen peroxide. The reaction is followed spectroscopically by recording the absorption at 414 nm of the oxidized form of ABTS, which is detected during the first 4 min after the addition of HRP buffer solution to the other afore mixed components at 25°C. As in the studies of the direct micellar systems, the normalized activity $A = V/V_0$ is considered. The reference velocity V_0 is measured in a standard buffer solution (pH 5) under exactly the same conditions and buffer conditions, but without surfactant, alcohol, and oil.

The systems considered in these studies are composed of four major species: the buffer (regarded as one homogenous component for simplification), sodium dodecylsulfate (SDS), and dodecane as the oil phase. In contrast to AOT as surfactant, a cosolvent is needed to obtain reverse microemulsions. Here and as usual, various alcohols are used for this purpose; they are the forth species.

It should be noted that HRP is highly water soluble and insoluble in the oil phase. However, a certain, although small, affinity to interfaces is probable. SDS and ABTS are insoluble in water, whereas all alcohols considered here (from 1-butanol to 1-octanol) are completely miscible with dodecane, but will also show a partitioning between oil and water.

Further, it is of importance to stress that the activity of HRP is very pH dependent. A decrease of pH from 5 to 4.5 induces an increase of more than 100%. In independent measurements, it was found that alcohols are efficient inhibitors for HRP, whereas SDS leads to a superactivity, mainly because its addition to the buffer in significant amounts slightly reduces the pH. In Figure 12.8, [7], the phase diagrams of the regarded systems are given along with the composition paths, for which the enzyme activities were recorded.

This means that all SDS + alcohol contents are comparable in the different alcohol systems, except in the case of 1-butanol, where a slightly higher SDS + alcohol molarity had to be used in order to have a sufficiently extended monophasic range (see Figure 12.8a). The black areas (L_1, L_2) represent realms of existence of monophasic, isotropic, and thermodynamically stable liquids of low viscosity, whereas the bright areas correspond to other phases or mixtures that are not considered in this study [7].

Although the samples having compositions within the dark areas are macroscopically indistinguishable, they have different structures [7]. The butanol and pentanol systems show percolation and a clear transition from separated micelles to bicontinuous phases, whereas no percolation is found for hexanol, heptanol, and octanol systems. The main reason is that with increasing alcohol chain length the interfacial film consisting of SDS and alcohol becomes more and more rigid. We will see that the film rigidity is one of the crucial parameters for the enzymatic activity. Figure 12.9 gives the measured enzymatic activities in the 4-component systems along the composition paths in the monophasic regions.

To understand these results, several features of both the reverse microemulsions and of the enzymes must be taken into account. In the 1-butanol system, electric conductivity measurements revealed a percolation threshold at a relatively

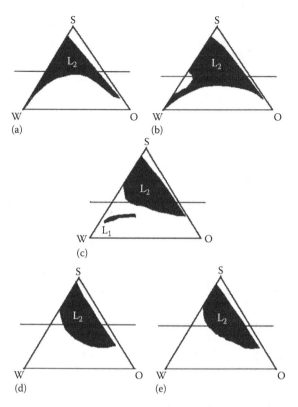

FIGURE 12.8 Phase diagrams of pseudoternary water/SDS/dodecane/*n*-alcohol systems (S stands for SDS/*n*-alcohols, W for water, and O for dodecane). (a) 1-butanol, (b) 1-pentanol, (c) 1-hexanol, (d) 1-heptanol, and (e) 1-octanol. In every case, the molar ratio K_m between surfactant and alcohol is 1:6.54. The full lines indicate the chosen compositions of the reaction media. The mass % of SDS + alcohol is: P_{SDS} + 1-butanol = 0.5, P_{SDS} + 1-pentanol = 0.4, P_{SDS} + 1-hexanol = 0.424, and P_{SDS} + 1-heptanol = 0.447. (From Bauduin, P., Touraud, D., Kunz, W., Savelli, M.P., Pulvin, S., and Ninham, B.W., *J. Coll. Inter. Sci.*, 292, 244–245, 2005. With permission.)

low water content (about 11%). At this composition, the surfactant is not yet fully hydrated. Dielectric relaxation measurements proved that about 18–20 water molecules are needed to completely hydrate SDS. At 11%, only about 10 molecules of water per SDS are present in the medium. Consequently, it can be assumed that there is a competition for water between SDS, the enzyme, and partially the butanol head groups. Therefore, the enzyme is certainly not fully hydrated. Despite of the low water content, percolation occurs at 11% buffer content, because the interfacial film is very flexible due to the presence of the short-chain alcohol. At about 25% the enzymatic activity has a maximum. This point corresponds nearly exactly to the value of 20 water molecules per SDS. A reasonable explanation is that here all SDS molecules are fully hydrated and consequently the remaining water can entirely hydrate the enzyme, as in pure bulk water. The reason for the

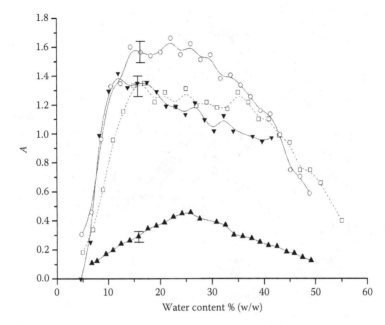

FIGURE 12.9 Enzymatic activities A in microemulsions (citrate buffer solution/SDS/ dodecane/n-alcohol [K_m, i.e., molar ratio between surfactant and alcohol is 1/6.54]) as a function of the buffer content. 1-butanol (▲), 1-pentanol (□), 1-hexanol (○), and 1-heptanol (▼). For the 1-octanol system no activity could be measured. Typical error bars are also included. (From Bauduin, P., Touraud, D., Kunz, W., Savelli, M.P., Pulvin, S., and Ninham, B.W., *J. Coll. Inter. Sci.*, 292, 244–245, 2005. With permission.)

low absolute enzymatic activity is mainly butanol that shows a relatively (compared to the other alcohols) high solubility in water. Further, butanol slightly enhances the pH of the buffer (confirmed by adding butanol to pure buffer solutions). A third reason is that the enzyme is also in slight contact with the inhibiting oil phase, because of the high flexibility of the interfacial film.

Concerning the pentanol system, three concentration ranges can be distinguished: after a sharp increase in A, a nearly constant activity is observed, followed by a pronounced decrease at buffer contents above 40%. This enzymatic behavior is strongly related to structural changes in the microemulsion, as illustrated in Figure 12.10, [7]. The initial step reflects the increasing hydration of all molecules, just as in the case of butanol. However, it reaches its maximum at about 15% of buffer content. At this composition, percolation occurs and, coincidently, the number of water molecules per SDS is about 20 so that also the hydration of enzyme is completed at this point.* Superactivity can be observed, because the concentration of pentanol in the aqueous pseudophase is much lower than with

---

* Note that the concentration of enzyme is in the nanomolar region only.

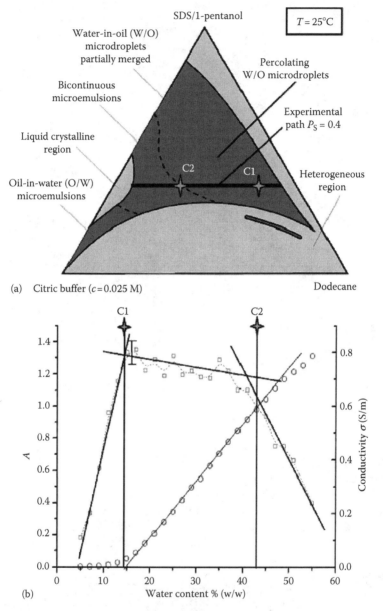

(a) Citric buffer ($c = 0.025$ M)

(b)

FIGURE 12.10 HRP enzymatic activity, defined as $A = V/V_0$, in the citrate buffer/SDS/1-pentanol/n-dodecane system. A tentative explanation of the correlation between enzymatic activities and electrical conductivity data is given. (a) Pseudoternary diagram of the citrate buffer solution ($c = 0.025$ M)/SDS/1-pentanol/n-dodecane ($K_m = 1/6.54$, $T = 25°C$). (b) HRP enzymatic activities (□) and conductivity (○) as a function of the buffer content (water content) along the experimental path. C1 and C2 appearing along the experimental path represent, respectively, the percolation threshold (14% buffer or water content) and the borderline between the effective medium and the bicontinuous system (43%). (From Bauduin, P., Touraud, D., Kunz, W., Savelli, M.P., Pulvin, S., and Ninham, B.W., *J. Coll. Inter. Sci.*, 292, 244–245, 2005. With permission.)

butanol so that the inhibition effect of the alcohol is less pronounced. In parallel, SDS causes a superactivitation, mainly because of its influence on the pH that is lowered in sulfate solutions. We will come back to the question of pH later on.

Upon further addition of water, the enzyme solution gets more diluted and the activity slightly decreases. At about 43% of buffer content, the electrical conductivity deviates from its linear increase, which can be interpreted as the transition to bicontinuous structures. At approximately that composition, the activity sharply abates, very probably because of the entire restructuration of the medium and a higher solubility of pentanol in the environment of the enzyme. Further details and a more rigorous discussion are given in Ref. [7]. A schematic summary of these effects and their relation with enzymatic activity are shown in Figure 12.11, [7].

For the longer-chain alcohols, the situation is slightly different. The too rigid structure induced by the cosurfactant prevents any percolation to occur. On the other hand, the very low solubility of alcohol causes a superactivity of the enzyme that is even higher than in the case of pentanol.

However, how to explain the lower superactivity in the case of the heptanol and the absence of any activity in the case of octanol? This is the consequence of a completely different feature: the reverse micelles have very rigid structures. During the preparation process of the reacting system, the enzyme containing solution is added, once the microemulsion system is already formed. By doing so, the enzyme is exposed to the highly denaturating oil phase (that also contains much of the inhibiting alcohol molecules) during a significant period of time, before it is incorporated in the aqueous pseudophase to reach the thermodynamic equilibrium. The more rigid the interfacial film, the longer the enzyme is in contact with the oil phase and the more it is inactivated. With other words, the decrease and, in the case of octanol, even absence of enzymatic activity is a consequence of the system's kinetics, which in turn is a measure for the film's rigidity.

As a conclusion of this part we can state that enzymatic reactivity is a very sensitive measure of (1) structural changes, (2) completion of hydration (of surfactant and cosurfactant), (3) partitioning of the various components, (4) interfacial film rigidity, and, not to forget, (5) pH.

Concerning the last point, pH, it is crucial to know if a simple glass electrode can be used to measure correctly the pH in a reverse microemulsion. To this purpose we compare the initial velocities of HRP in pure buffers at various pH to the velocities in reverse microemulsions. The results are shown in Figure 12.12. It turns out that when the same pH values are measured in both types of systems, the enzymes also show the same velocities. This means that on both the enzyme and the electrode surfaces exactly the same specific adsorption effects may occur—a fact that is not very probable. A better explanation is that the electrode still indicates the "right" pH value and that the enzyme is still as sensitive to pH changes as in the neat buffer solution without oil, surfactant, and alcohol.

The third curve denoted with (•) gives further information. It shows that the pH values in the buffers that are used to prepare the microemulsions are higher than that in the microemulsions themselves. It is easy to explain this effect simply by adding an equivalent amount of SDS to the buffer, without oil or alcohol. Then

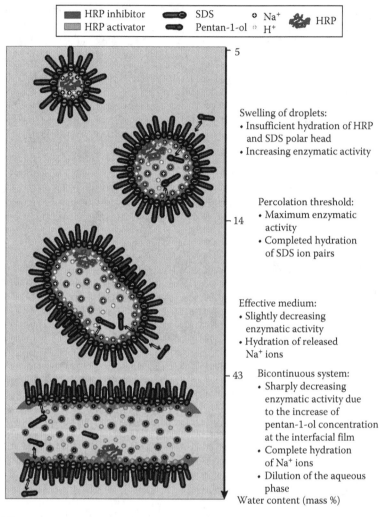

| HRP inhibitor | SDS | Na⁺ | |
| HRP activator | Pentan-1-ol | H⁺ | HRP |

5

Swelling of droplets:
• Insufficient hydration of HRP and SDS polar head
• Increasing enzymatic activity

14

Percolation threshold:
• Maximum enzymatic activity
• Completed hydration of SDS ion pairs

Effective medium:
• Slightly decreasing enzymatic activity
• Hydration of released Na⁺ ions

43

Bicontinuous system:
• Sharply decreasing enzymatic activity due to the increase of pentan-1-ol concentration at the interfacial film
• Complete hydration of Na⁺ ions
• Dilution of the aqueous phase

Water content (mass %)

FIGURE 12.11 Tentative interpretation of the enzymatic activity and electrical conductivity curves shown in Figure 12.10. (From Bauduin, P., Touraud, D., Kunz, W., Savelli, M.P., Pulvin, S., and Ninham, B.W., *J. Coll. Inter. Sci.*, 292, 244–245, 2005. With permission.)

the pH is indeed shifted to lower values, just by the presence of SDS, as already mentioned before.

From these results we can conclude that the pH (1) can be measured in reverse SDS microemulsions with some confidence and (2) is as a crucial factor for enzymatic activities in microemulsions as in pure buffer solutions. It must only be kept in mind that SDS molecules can change the pH values even in buffers.

Finally, we mention an even more complex system: further to SDS, a varying amount of dodecytrimethyammonium bromide (DTAB) is added to the system so

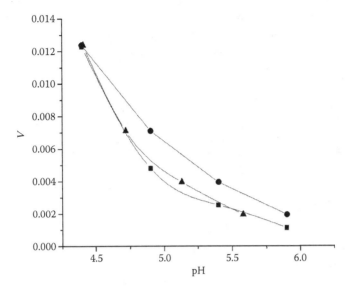

FIGURE 12.12   (▲) Initial (non-normalized) enzymatic velocities in citrate buffer solutions ($c = 0.025$M) as a function of pH and (■) in microemulsions (25% citrate buffer, 35% dodecane, 13.4% SDS, 26.6% 1-pentanol [$K_m = 1/6.54$]) as a function of the pH measured in that microemulsion with the help of a glass electrode. (●) Enzymatic velocities in the buffer solutions that were used to produce the microemulsions that show the behavior denoted with (■).

that the resulting microemulsions contain five main components now [8]: buffer, alcohol, two surfactants (SDS and DTAB), and *n*-dodecane.

   In the first experiment it was found that the relative activity *A* of HRP is nearly 0, when DTAB is added to a citrate buffer solution ($c = 0.025$ M, pH 5) at concentrations close to or higher than its cmc. This result supports the well-known strong inhibition effect of DTAB on HRP; however, catanionic mixtures with DTAB concentrations of up to 2.8 wt % (relative to the total system) in stable microemulsions still exhibit superactivities, provided that the amount of SDS is high enough (more than 86 wt % of the total amount of surfactant). This is despite the fact that DTAB is highly water soluble. Obviously, DTAB is strongly incorporated in the catanionic interfacial film and the decrease of the pH due to the presence of SDS still ensures superactivity.

   Of course, other conditions must be fulfilled, for example, a sufficient buffer content to fully hydrate the enzyme molecules and not a too high content of alcohol, preferably 1-hexanol so that the enzyme is not significantly inhibited by the cosurfactant.

   The fact that a mixture of cationic and anionic surfactants can maintain a high enzymatic activity is also interesting for the understanding of biological systems. This attenuation of the strong inhibition effect of one of the components is an example of how ionic interactions between oppositely charged surfactants can lead to partial preservation of biological activity, and this may occur in nature. Such an effect is of potential importance for disinfecting and bactericidal products based on cationic surfactants but this is not the topic of this chapter.

12.5   CONCLUSION

Micellar solutions and, in particular, reverse microemulsions are complex systems. Whereas the structures of the direct systems are quite well identified today, this is not the case for reverse systems. Enzymatic solutions are very sensitive systems. This is their force, but makes it also difficult to reproduce and to interpret the resulting enzymatic activities.

Despite this double complexity, it is possible to use enzymes in surfactant solutions to get a deeper insight in their structure and dynamics. In this chapter, we showed that enzymes can yield useful information on surfactant hydration, interfacial film rigidity, and partitioning of cosurfactants. Enzymes are also useful as an independent check for pH in reverse microemulsions. We could also show that for some of the properties of microemulsions such as cosurfactant partition coefficients, semi-quantitative results can be obtained.

SYMBOLS AND TERMINOLOGIES

A normalized enzymatic activity, defined as $A = V/V_0$
ABTS   2,2-azino-bis(3-ethylbenzothiazoline-6-sulfonic acid) diammonium salt
AOT   sodium bis(2-ethylhexyl) sulphosuccinate
Brij35®   polyoxyethylene(23)lauryl ether, also $C_{12}EO_{23}$ abbreviated
c_{max} concentration at which the maximum A value is measured
cmc   critical micellar concentration
D_{app} apparent self-diffusion coefficient of the alcohol in the ternary system
DTAB   dodecytrimethylammonium bromide
HLAD   horse liver alcohol dehydrogenase
HRP   horseradish peroxidase
K_m molar ratio between surfactant and cosurfactant in pseudoternary systems
NADH   nicotinamide adenine dinucleotide
P partition coefficient of alcohols between the aqueous buffer and the micellar pseudophases
P_s quantity of the surfactant S, expressed in mass % of the whole composition in ternary or pseudoternary systems
P_c quantity of the cosurfactant (C), expressed in mass % of the whole composition in ternary or pseudoternary systems, in the text C refers always to an alcohol
Q wave number transfer
$R_{c/w}$ cosurfactant (alcohol) to water mass % ratio

SANS   small-angle neutron scattering
SDS   sodium dodecylsulfate
V initial velocity of an enzymatic reaction
V_0 initial velocity of a reference solution

REFERENCES

1. Martinek, K., Levashov, A. V., Klyachko, N. I., and Berezin, I. V. 1977. Catalysis by water-soluble enzymes in organic solvents. Stabilization of enzymes against denaturation (inactivation) during their inclusion in inverted micelles of a surfactant. *Dokl. Akad. Nauk. SSSR*, 236, 920–923.
2. Biasutti, M. A., Abuin, E. B., Silber, J. J., Correa, N. M., and Lissi, E. A. 2008. Kinetics of reactions catalyzed by enzymes in solutions of surfactants. *Adv. Coll. Interface Sci.*, 136, 1–24.
3. Schirmer, C., Liu, Y., Touraud, D., Meziani, A., Pulvin, S., and Kunz, W. 2002. Horse liver alcohol dehydrogenase as a probe for nanostructuring effects of alcohols in water/nonionic surfactant systems. *J. Phys. Chem. B*, 106, 7414–7421.
4. Schirmer, C. 2001. PhD thesis, Verhalten von Enzymen und Farbstoffen in Lösungen Kationischer und nichtionischer Tenside, University of Regensburg, Germany.
5. Meziani, A., Touraud, D., Zradba, A., Pulvin, S., Pezron, I., Clausse, M., and Kunz, W. 1997. Comparison of enzymatic activity and nanostructures in water/ethanol/brij 35 and water/1-pentanol/Brij 35 systems. *J. Phys. Chem. B*, 101, 3620.
6. Preu, H., Zradba, A., Rast, S., Kunz, W., Hardy, E. H., and Zeidler, M. D. 1999. Small angle neutron scattering of D_2O–Brij 35 and D_2O–alcohol–Brij 35 solutions and their modelling using the Percus–Yevick integral equation. *Phys. Chem. Chem. Phys.*, 1, 3321–3329.
7. Bauduin, P., Touraud, D., Kunz, W., Savelli, M. -P., Pulvin, S., and Ninham, B. W. 2005. The influence of structure and composition of a reverse SDS microemulsion on enzymatic activities and electrical conductivities. *J. Coll. Inter. Sci.*, 292, 244–254.
8. Mahiuddin, S., Renoncourt, A., Bauduin, P., Touraud, D., and Kunz, W. 2005. Horseradish peroxidase activity in a reverse catanionic microemulsion. *Langmuir*, 21, 5259–5262.

13 Biocatalysis in Microemulsions

A. Xenakis, V. Papadimitriou, H. Stamatis, and F. N. Kolisis

CONTENTS

13.1 INTRODUCTION

Reverse micelles formed in water-in-oil (w/o) microemulsions, structurally inverse analogues to normal micelles, are capable of hosting proteins/enzymes in their so-called water pool. The biomolecule can be entrapped in the water pools, avoiding direct contact with the organic solvent, potentially limiting their

TABLE 13.1

Examples of Enzymes in Reverse Micellar Systems

Enzymes in Reverse Micelles	W/O Microemulsion Systems	References
Lipases	Various systems	[8,9]
Alcohol dehydrogenase	AOT-based system	[10]
Tyrosinase	AOT-based system	[11]
Cutinase	AOT-based system	[12,13]
Laccase	AOT-based system	[14]
DNA polymerase	Brij30, Triton X-100, SDS, CTAB, Brij58-based systems	[15]
α-Chymotrypsin	CTAB, AOT/bile salt-based system	[16,17]
Cholinesterase-like abzyme	AOT-based system	[18]
Lipoxygenase	AOT-based system	[19]
Hexokinase	Triton X-100-based system	[20,21]
Amyloglucosidase	Triton X-100-based system	[22]

denaturation. Biocatalysis, based on reverse micelles, is one of the earliest approaches introduced to solubilize enzymes in nonpolar organic solvents [1–7]. A whole range of enzymes, including hydrolases (lipases, esterases, glucosidases, and proteases), oxidoreductases (peroxidases, oxygenases, and dehydrogenases), transferases (kinases), etc., as well as catalytic antibodies, exhibited catalytic activity in reverse micelles as shown in Table 13.1.

13.1.1 ENTRAPMENT OF PROTEINS (ENZYMES) INTO W/O MICROEMULSIONS

There are three commonly used methods to incorporate enzymes in reverse micelles [6,23]. In the first method known as the "injection" method the enzyme solution is added to a solution of surfactant in a nonpolar organic solvent. The resulting mixture is vigorously shaken until an optically transparent solution is obtained. The second method consists of the addition of dry lyophilized protein to a surfactant solution in an organic solvent containing an aqueous phase. The third procedure is based on the phenomenon of spontaneous interfacial transfer of the protein in a two-phase system consisting of approximately equal volumes of the aqueous protein solution and surfactant containing organic solvent. Upon gentle shaking, the enzyme is transferred into the reverse micelles of the organic phase. The method is very useful for the separation, extraction, and purification of biomolecules (including enzymes and DNA).

The injection method is by far the most used method to microencapsulate enzymes, due to the simple procedure. One of the main drawbacks of the other two methods is the prolonged contact between the enzyme molecule and the organic solvent that contributes to the enzyme deactivation.

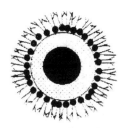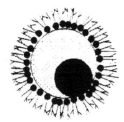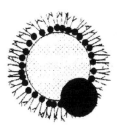

FIGURE 13.1   Representation of the localization sites of an enzyme in a reverse micelle. (From Papadimitriou, V., Enzymatic studies in microemulsions, PhD thesis, University of Athens, Athens, 1996.)

Several variables control the solubilization of proteins in microemulsions, including the pH and the ionic strength of the aqueous phase, the size of the protein, the size of the reverse micelles, and the nature of the surfactant [6,8,24]. The uptake of proteins into the aqueous microphase is a complex process. It was proposed that the hydrophobicity of the protein molecule plays an important role in its localization among the various microenvironments of the system. In fact, a hydrophilic protein can avoid direct contact with the continuous organic phase and remain localized in the water pool; a surface active enzyme (such as some lipases) can interact with the micellar interface, whereas a typical membrane protein can be in contact with the hydrophobic region of the micellar interface and also with the organic solvent [2,3,24] (Figure 13.1). The substrate molecules can also be partitioned between the aqueous microphase and the micellar interface, or the organic continuous phase, respectively.

13.1.2   CATALYTIC PROPERTIES OF ENZYMES IN W/O MICROEMULSIONS

The catalytic behavior and the stability of enzymes in reverse micelles are highly dependent on the composition and the structure of the microemulsion. The activity of entrapped enzymes strongly depends on the water content, the nature of the organic solvent, as well as the nature and the concentration of surfactant. Various surfactants, including the anionic AOT, the cationic CTAB, nonionics such as Triton, Brij, ethoxylated fatty alcohols, and zwitterionic phospholipids (phosphatidylcholine), were used for the preparation of reverse micellar systems-containing enzymes (Table 13.1). Most investigated systems used AOT as the surfactant because its phase behavior is well understood. The activity of some enzymes has been reported to depend on the surfactant concentration and in some cases it was attributed to the interaction of the enzymes with the micellar membrane [8,26,27]. Recent developments in this area include the use of modified surfactants or their mixtures with other additives and cosurfactants such as alcohols and sugars or the use of aprotic solvents for the reduction of the ionic interactions between the enzyme molecules and the micellar interface in order to improve the enzyme catalytic behavior and operational stability [8,17,28–34].

The kinetics of enzymatic reactions in microemulsions obey, as a rule, the classic Michaelis–Menten equation [6,26,35], but difficulties arise in interpreting the results because of the distribution of reactants, products, and enzyme molecules among the microphases of the microemulsion [8,36–38]. In addition, there are some enzymes in reverse micelles that exhibit enhanced activity as compared to that expressed in water; this has given rise to the concept of "super-activity" [6,26,39]. The superactivity has been explained in terms of the state of water in reverse micelles, the increased rigidity of the enzymes caused by the surfactant layer, and the enhanced substrate concentration at the enzyme micro-environment [36,40].

The determination of the enzyme activity as a function of the composition of the reaction medium is very important in order to find the optimal reaction conditions of an enzyme-catalyzed synthesis. However, the correlation between the reaction media properties and their effects on enzymatic reactions in reverse micelles is still unclear.

13.1.3 EFFECT OF WATER CONTENT

The effect of water content (expressed as the molar ratio of water to surfactant, w_o) of the microemulsion system on the catalytic behavior of enzymes has been studied most extensively in recent years. For many enzymes studied in reverse micelles, bell-shaped dependence of enzyme catalytic behavior (expressed as V_{max} or k_{cat}) as a function of w_o, usually with an optimum activity occurring when w_o is between 5 and 15, has been observed [8] (Figure 13.2).

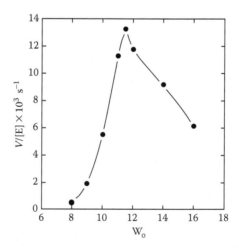

FIGURE 13.2 Hydrolysis rate of GPNA by α-chymotrypsin in a microemulsion of 50 mM AOT in isooctane as a function of w_o. (From Papadimitriou, V., Enzymatic studies in microemulsions, PhD thesis, University of Athens, Athens, 1996.)

It must be noted that the physicochemical properties of the water located inside the reverse micelles differ from those of bulk water, and the difference becomes progressively smaller as the w_o value increases. The optimum w_o value for catalysis depends on several aspects such as the enzyme concentration as well as its molecular dimensions. It has been proposed that an optimum of enzyme activity occurs around a value of w_o, at which the size of the droplet is similar to the size of the enzyme molecule [2]. However, there is an ongoing discussion in the literature as to whether the water content dependence of the activity is related to the size of the enzyme molecule or not [3,6,8,9,35,41].

13.1.4 BIOCATALYTIC APPLICATIONS OF ENZYME-CONTAINING MICROEMULSIONS

Reverse micelles have been associated with the idea of microreactors for enzymatic reactions, when substrates and/or products are lipophilic and low water content is desired. Microemulsions provide an enormous interfacial area through which the conversion of hydrophobic substrates can be catalyzed. Increasing the interfacial area is of great technological interest because it results in the increase in the number of substrate molecules available to react. Enzymes in w/o microemulsions offer considerable advantages as a reaction medium is used for biocatalytic transformations:

- Hydrophilic and hydrophobic substrates can be dissolved in high concentrations [8,42].
- Thermodynamic equilibrium in the enzyme-catalyzed condensation and hydrolysis reactions can be easily shifted by adjusting the water content of the system.
- Activity, stability, and stereoselectivity of the enzymes can be enhanced by controlling the composition of the reverse micellar reaction system [3,6,8,9,35,41].
- Multienzymatic reactions are feasible in reverse micelles [43].

Besides biocatalytic applications, it is interesting to note that nanostructures within a microemulsion are recognized as models of biological structures that facilitate the investigation of protein/enzyme structure—activity relationship under conditions that mimic biological environments in optically transparent solutions.

Since the beginning of enzyme catalysis in microemulsions in the late 1970s, several biocatalytic transformations of various hydrophilic and hydrophobic substrates have been demonstrated. Examples include reverse hydrolytic reactions such as peptide synthesis [44], synthesis of esters through esterification and transesterification reactions [42,45–48], resolution of racemic amino acids [49], oxidation and reduction of steroids and terpenes [50,51], electron-transfer reactions, [52], production of hydrogen [53], and synthesis of phenolic and aromatic amine polymers [54]. Isolated enzymes including various hydrolytic enzymes (proteases, lipases, esterases, glucosidases), oxidoreductases, as well as multienzyme systems [52], were employed.

13.2 ENZYMATICALLY CATALYZED REACTIONS IN MICROEMULSIONS

13.2.1 PROTEASES IN W/O MICROEMULSIONS

Since the appearance of "micellar enzymology" as a new field of research interest, proteases, being hydrophilic and relatively small, have been considered as good model systems for the effectuation of both hydrolytic and synthetic reactions in w/o microemulsions [44,55] (Table 13.2). These enzymes can be easily dissolved in the aqueous domains of w/o microemulsions and maintain their catalytically active conformation since they remain protected from the denaturing effect of both organic solvent and synthetic emulsifiers. In addition, the coexistence of aqueous, organic, and amphiphilic domains, which characterize w/o microemulsions, enables the contact of the enzymes with substrates of different polarities.

TABLE 13.2
Examples of Proteases in W/O Microemulsions

Enzyme	Reaction	W/O Microemulsion System	References
α-Chymotrypsin	Tripeptide synthesis	AOT/isooctane	[44]
α-Chymotrypsin	Dipeptide synthesis	CTAB/octane/hexanol and AOT/octane/ethyl acetate	[63]
α-Chymotrypsin	Dipeptide synthesis	TTAB/heptane/hexanol	[65]
Alcalase, trypsin	Precursor dipeptide synthesis	AOT/isooctane and CTAB/heptane/hexanol	[59]
α-Chymotrypsin	Precursor dipeptide synthesis	AOT/isooctane	[60]
α-Chymotrypsin	Oligopeptide synthesis	AOT-Brij30/n-heptane	[66]
Papain	Dipeptide synthesis	AOT and Tween 80/isooctane	[67]
α-Chymotrypsin	Ester hydrolysis	AOT/heptane	[55]
α-Chymotrypsin	Amide hydrolysis	AOT/isooctane	[68]
α-Chymotrypsin	Hydrolysis	AOT/n-heptane, CTAB/n-heptane/chloroform, SDS/toluene/pentanol	[1]
α-Chymotrypsin	Ester hydrolysis	AOT/isooctane	[70]
Trypsin	Ester, amide hydrolysis	AOT/isooctane and CTAB/chloroform/isooctane	[71]
α-Chymotrypsin and trypsin	Hydrolysis	AOT/isooctane and CTAB/pentanol/isooctane	[39]
Trypsin	Hydrolysis	Lecithin/isooctane/alcohol	[73]
Trypsin	Ester Hydrolysis	CTAB/chloroform/isooctane	[74]
Papain	Amide hydrolysis	AOT/isooctane	[77]
Papain	Amide hydrolysis	AOT and Tween 80/isooctane	[67]

The reaction conditions can be optimized by examining the effect of different factors such as water content, temperature, pH, surfactant concentration, reaction time, or product yield. Proteases are classified according to their catalytic mechanisms. Four mechanistic classes have been recognized by the International Union of Biochemistry and Molecular Biology: serine proteases (chymotrypsin, trypsin, elastase, subtilisin), cysteine proteases (papain, cathepsins, caspases), aspartic proteases (pepsins, cathepsins, rennins), and metallo proteases.

13.2.1.1 Peptide Synthesis

In aqueous solutions, proteases normally catalyze the hydrolysis of peptide bonds but in low water reaction mixtures like organic solvents containing minimal amounts of water, aqueous/organic biphasic systems, and reverse micelles, the thermodynamic equilibrium of the hydrolytic reaction can be shifted to the direction of peptide bond formation [44,56,57]. Reverse micellar environments have been proved more advantageous among other nonaqueous systems for biotransformations of substrates with varied polarities since they provide an increased interfacial area combined with an increased stabilization of the biocatalyst [23]. In parallel, the enzymatic method of peptide synthesis offers many advantages over the chemical method like stereospecificity, side chain protection, suppression of undesired side-products, and mild reaction conditions [58]. For all the reasons mentioned above, enzymatic peptide synthesis in reverse micelles has attained considerable interest by several research groups as a promising biotechnological alternative [23,44,59,60].

Proteases for peptide synthesis are selected on the basis of their specificity against amino acid residues and include the majority of the commercially available proteases of the four classes mentioned above [58]. Protease-catalyzed bond synthesis can be carried out either as an equilibrium-controlled process which is the direct reversal of the protease-catalyzed hydrolysis or a kinetically controlled process. In the latter case weakly activated carboxy components are employed [61].

The equilibrium-controlled peptide synthesis has the main advantage that all kinds of proteases can be used, independently of their catalytic mechanism. However, a serious disadvantage of the method is the low reaction velocity and low product yield of the peptide synthesis. As already mentioned, to overcome this problem, low water reaction mixtures like reverse micelles, which favor the shift of the reaction toward peptide synthesis, can be used.

In the kinetically controlled process, serine and cysteine proteases catalyze acyl transfer from activated carboxy components to various acceptors. The initially formed acyl–enzyme intermediate can be competitively deacylated by water (hydrolysis) or by an amine nucleophile (aminolysis). The yield of the peptide bond formation depends on two factors (1) the relative rate of hydrolysis and aminolysis, which is determined by the nucleophilicity of water versus that of the nucleophile and (2) the molar ratio of water and the nucleophile [62]. The ratio between aminolysis and hydrolysis of the acyl–enzyme intermediate is generally more favorable for cysteine proteases than for serine proteases [61]. Although the kinetic approach is faster and low enzyme concentrations are required, this method is limited to those proteases that form acyl–intermediates. If the kinetically controlled peptide bond formation takes place in low water reaction mixtures, the hydrolysis will be limited and the product yield can be improved. In the case of reverse micelles, two types of reactions can be considered (1) peptide synthesis takes place within the dispersed aqueous domains since both the enzyme and the substrates are hydrophilic. At the same time, hydrophobic peptide products (dipeptides, tripepdides, etc.) are transferred into the surrounding continuous phase (organic solvent) as soon as they are produced, not allowing the proteases to hydrolyze the peptide bonds any more and (2) as above, but one of the substrates is partitioned between water pool and organic solvent. In that case the water-insoluble peptide product is also expelled in the continuous phase [5,59].

Lüthi and Luisi were the first to report in 1984 the α-chymotrypsin-catalyzed synthesis of a hydrophobic tripeptide in AOT/isooctane w/o microemulsions [44]. In this case, N-benzyloxycarbonyl-L-alanyl-L-phenylalanyl-L-leucinamide was synthesized at pH 10, and at room temperature, using N-benzyloxycarbonyl-L-alanyl-L-phenylalanine methyl ester and leucinamide as reactants. The reaction yield was between 40% and 60% for w_0 values between 5 and 30 and the reaction product was analyzed by HPLC.

Jorba and coworkers reported the α-chymotrypsin catalyzed synthesis of the hydrophobic dipeptide N-acetyl-L-phenylalanine-L-leucinamide in two different reverse micellar systems. First, a system formulated with the cationic surfactant CTAB, n-octane, and either hexanol or octanol as cosurfactants [63]. In this case, the reaction yield was influenced by three parameters, namely w_0, pH, and the nature and concentration of emulsifiers. The second system studied was based on the well-known surfactant AOT and a mixture of n-octane/ethyl acetate as the organic phase. The addition of the polar solvent, ethyl acetate, in the reaction mixture resulted in the product yields up to 70% although the reaction rates were optimal when lower proportions of the polar solvent were added in the system [64].

In 1994, Serralheiro and coworkers described the α-chymotrypsin catalyzed synthesis of N-acetyl-L-phenylalanine-L-leucinamide in a reverse micellar membrane reactor operated in a batch mode [65]. The reverse micelles were formulated with TTAB, heptane, and hexanol. An ultrafiltration ceramic membrane was used to retain the enzyme and separate the peptide. The reactor was operated for four days without any loss of enzyme activity and the yield of the produced dipeptide was around 80%.

Protease-catalyzed synthesis of two precursor dipeptides of the hydrophilic tripeptide RGD (Arg-Gly-Asp) in three different types of w/o microemulsions, namely AOT/isooctane, Triton/ethyl acetate, and CTAB/heptane/hexanol, was reported by Chen and coworkers [59]. RGD is a tripeptide of pharmaceutical interest consisting of one neutral (Gly) and two hydrophilic amino acids (Arg and Asp). Among the proteases tested, alcalase and trypsin were selected as the most efficient catalysts and the reverse micellar system consisting of

AOT/isooctane was the most appropriate for the syntheses of the hydrophilic precursor dipeptides.

In 1999, the same group reported the synthesis of the precursor dipeptides, as mentioned above, catalyzed by free and immobilized α-chymotrypsin in AOT/ isooctane reverse micelles at product yields exceeding 80% [60]. The dipeptide products were successfully separated from the reaction medium by using a silica gel column.

α-Chymotrypsin catalyzed synthesis of oligopeptide derivatives in AOT-Brij30/ n-heptane mixed reverse micelles was reported by Xing et al. [66]. Mixed reverse micelles were proved more suitable for peptide synthesis since the product yields of 77% were obtained. Mixed reverse micelles containing two kinds of surfactants, namely AOT and Tween 80, were also considered from Fan, and coworkers for the improvement of papain's activity toward the synthesis of a model dipeptide such as N-benzyloxycarbonyl-glycine-L-phenylalanine methyl ester [67].

13.2.1.2   Peptide Hydrolysis

Proteases, when solubilized in the aqueous core of reverse micelles, can catalyze the hydrolysis of various small model peptides. In most of the cases, the substrates are partitioned between the micelles and the external solvent while the hydrolytic reaction is taking place within the dispersed aqueous domains. Since the beginning of micellar enzymology, α-chymotrypsin and trypsin have been extensively studied in different reverse micellar systems, employing various model peptides as substrates. In almost all cases the reactions followed Michaelis–Menten kinetics.

In 1979, Menger and Yamada were the first to report the α-chymotrypsin catalyzed hydrolysis of a synthetic ester, namely N-acetyl-L-tryptophan methyl ester, in AOT/heptane reverse micelles [55]. Circular dichroism studies of α-chymotrypsin containing reverse micelles at various water contents showed no major conformational changes of the protein. In this work, for the first time, increased k_{cat} values of the α-chymotrypsin-catalyzed hydrolysis in the microenvironment of reverse micelles were recorded. In 1981, Barbaric and Luisi studied the α-chymotrypsin-catalyzed hydrolysis of a model amide, N-glutaryl-L-phenylalanine p-nitornilide, in AOT/isooctane microemulsions [68]. The enzyme followed Michaelis–Menten kinetics with K_m and turnover values higher than those measured in aqueous solutions. The increased activity of the enzyme observed in reverse micelles was then characterized as superactivity. In addition, an increased stability of α-chymotrypsin at low water content was observed. In 1985, an extensive study on the activity of α-chymotrypsin toward a number of substrates in various w/o microemulsions was made by Fletcher and coworkers [1]. The main results obtained from the studies mentioned above are (1) α-chymotrypsin exhibits superactivity when solubilized in the aqueous core of reverse micelles; (2) the pH at which the enzyme shows maximum activity in reverse micelles remarkably shifts from the optimum pH in the aqueous solutions; and (3) the reaction rate is depended on the w_o value of the system. The so-called superactivity of α-chymotrypsin in reverse micelles formulated with a variety of surfactants and organic solvents and toward many synthetic peptide molecules was also observed by different research groups in the following years [16,17,69].

A few years later, the mechanism of a protease-catalyzed reaction in reverse micelles was for the first time elucidated by using stop-flow techniques and the α-chymotrypsin catalyzed hydrolysis of esters as model reaction [70]. It was shown that deacylation remains rate limiting although in reverse micelle the acylation process is slowed down considerably.

In 1988, Walde and coworkers studied the kinetic and structural properties of another serine protease, namely trypsin, in two reverse micellar systems, AOT/ isooctane and CTAB/chloroform/isooctane, employing three different model substrates, an amide and two esters [71]. The main aim of this work was to compare the behavior of trypsin in reverse micelles with that of α-chymotrypsin. In the case of trypsin, superactivity was not observed and in general no obvious similarities between the two enzymes were recorded. Some years later, reverse micelles formulated with biocompatible surfactants such as lecithin of variable chain lengths in isooctane/alcohol were studied in relation to their capacity to solubilize α-chymotrypsin and trypsin [72]. The hydrolytic behavior of the same serine proteases, namely α-chymotrypsin and trypsin, in both AOT and CTAB microemulsions was studied and related to the polarity of the reaction medium as expressed by the $\log P$ value and measured by the hydrophilic probe 1-methyl-8-oxyquinolinium betaine [39]. In this study a remarkable superactivity of trypsin in reverse micelles formed with the cationic surfactant CTAB was reported.

In 1997, Avramiotis and coworkers solubilized trypsin in lecithin-based microemulsions of different compositions and studied the hydrolysis of L-lysine-p-nitroanilide [73]. Trypsin exhibited the same general characteristics of serine proteases in reverse micelles as mentioned above (Michaelis–Menten kinetics, increased stability at low water content, increased activity, bell-shaped dependence of w_o). In this study, it was shown, by using fluorescence quenching techniques and conductivity measurements, that in the presence of trypsin the structural characteristics of the reverse micelles are affected and the system compartmentalization is induced.

In the following years some more studies appeared in the literature concerning trypsin activity in reverse micelles in relation to various characteristics of the reaction medium. Fadnavis et al. studied the pH dependence of hydrolytic activity of trypsin in CTAB reverse micelles toward a positively charged model ester substrate [74]. It was found that enzyme activity variations as a function of w_o are pH dependent. In 2005, Dasgupta and coworkers related the catalytic activity of trypsin in reverse micelles formulated with cationic surfactants with the concentration of the water-pool components and the aggregate size to delineate the independent role of both parameters [75]. Finally, in 2006, the influence of ethylene glycol on the thermostability of trypsin in AOT reverse micelles was examined and was found to exhibit a positive effect [76].

As already mentioned above, a vast majority of the studies concerning protease-catalyzed peptide hydrolysis in microemulsions were conducted with the serine proteases α-chymotrypsin and trypsin. Nevertheless, some studies of a cysteine protease, namely papain, were also performed in the microenvironment of reverse micelles. In 1994, Vicente et al., solubilized papain in AOT/isooctane reverse micelles and studied its catalytic activity and stability toward the hydrolysis of N-α-benzoyl-DL-arginine-p-nitroanilide. An increased stability was observed and related to the protective role of a surfactants shell layer around the biomolecule. In parallel, the enzyme activity was decreased in comparison to the aqueous solution and this behavior was attributed to papain unfolding within the reverse micelles [77]. Finally, the papain-catalyzed hydrolysis of N-α-benzoyl-L-arginine-p-nitroanilide in mixed reverse micelles containing AOT and Tween 80 was reported and an increased activity of the enzyme was observed [78]. This behavior was attributed to the presence of the nonionic surfactant Tween 80 and the consequent adjustment of the micropolarity in the vicinity of the enzyme molecule.

13.2.2 LIPASES IN REVERSE MICELLES

A particular case of enzymatic studies in microemulsions is that of lipases that act almost exclusively near interfaces in a classical heterogeneous procedure and are very stable and active in this medium.

Lipases (EC 3.1.1.3) represent a group of enzymes that are most widely used for various biotransformation reactions due to several factors such as low cost and availability, ease in handling and their regio- and enantioselectivity toward a wide range of substrates. Lipases are important industrial enzymes, which catalyze the hydrolysis of fats and oils and can also play a role in the synthesis of various useful compounds. As the activity of these hydrolytic enzymes is greatly increased at the lipid–water interface, a phenomenon known as interfacial activation [78,79], reverse micelles have been used for hosting these reactions due to the large interfacial area promoting contact between enzyme and substrates [8,9].

The biocatalytic behavior and stability of several lipases have been studied in reverse micellar systems formulated with ionic (such as AOT and CTAB) and nonionic (Tween, Triton) or natural (lecithin) surfactants. The catalytic efficiency of such encapsulated lipases seems to be dependent on the local molar concentration of water and other ions present in the vicinity of the enzyme [8,9,28], as well as the nature and the concentration of the surfactant used [80–82].

There are various reports in the literature concerning kinetic studies of the lipase-catalyzed hydrolysis or synthesis of esters in microemulsions [8,9,49,83,84]. A simple Michaelis–Menten kinetic model was proposed for the hydrolysis of triglycerides [85,86], while the esterifications of aliphatic alcohols with fatty acids follow a ping-pong bi–bi mechanism [87]. According to this mechanism the lipase reacts with the fatty acid to form a noncovalent enzyme–fatty acid complex, which is then transformed to an acyl–enzyme intermediate, while water, the first product, is released; this is followed by a nucleophile attack (by the alcoholic substrate) to form another binary complex that finally yields the ester and the free enzyme. The kinetic parameters K_m and V_{max} determined in these studies represent apparent values because the substrate concentration in the enzyme microenvironment is unknown. In order to determine the substrate concentration in the enzyme microenvironment, the substrate partitioning between the various microphases such as the water pool, the micellar interface, and the continuous organic phase is required [42,48,87].

13.2.2.1 Lipase Stability in Reverse Micelles

The operational and storage stability of microbial lipases reported in w/o microemulsions are markedly different depending on the lipase source, as well as the nature of surfactant and the composition of the revere micelle systems. The size of the microemulsion droplets seems to have an important role in the stability of the lipases. In many cases, the stabilities of lipases from *Candida rugosa, Rhizopus delemar, Pseudomonas* sp., *Chromobacterium viscosum*, and *Humicola lanuginosa* improved in small reverse micelles that had low w_0 values as compared to stabilities in reverse micelles with high w_0 values [86,88,89]. The control of several factors that affect lipases stability in reversed micelles are (1) the decrease of the water content of the system, (2) the addition of protective compounds (substrates, ligands, neutral salts, and polyols), (3) the decrease of the repulsive interactions between enzyme and the negatively or positively charged polar heads of the surfactant, and (4) the use of optimum pH and correct buffer concentration [8,9,29,49,89].

Spectroscopic studies (fluorescence and circular dichroism) indicate that the variation in lipases stability is related to their hydrophobicity and therefore to the different degrees of interactions between the enzymes and the micellar interface [88]. Two important factors contribute to inducing conformational changes on the enzyme molecules in reverse micelles: the unusual properties of the encapsulated water and the electrostatic interactions between ionic head groups of the surfactant and the polypeptide chains of the enzyme. Spectroscopic data on the structure of enzymes entrapped in reverse micelles are controversial, as regards the changes in the protein structure upon incorporation into reverse micelles (Table 13.3).

The information derived by several studies shows that proteins fall to the following different categories according to the perturbation of their structure induced by reverse micelles: (1) conformations which are not changed or only slightly changed after incorporation into reverse micelles, (2) conformations which are altered in reverse micelles but unaffected by the water content of the system, and (3) conformations which are altered and which also change as a function of the water content [91].

13.2.2.2 Lipase-Catalyzed Reactions in Reverse Micelles

Lipases are hydrolases that catalyze in vivo the hydrolysis of triacylglycerols to glycerol or glycerides and free fatty acids. However, in systems with low water content, such as in the case of reverse micelles, they can catalyze the synthesis of esters, either directly from acids and alcohols or through trans- and interesterifications.

TABLE 13.3
Structural and Spectral Characteristics of Various Lipases in AOT-Based Microemulsions

Lipases	CD and Fluorescence Data	References
Pancreatic	Large increase in β-sheet, decrease in α-helix content. Blue shift in the fluorescence emission maximum	[80]
C. rugosa isolipase A	Increase in α-helix content. Blue shift of the fluorescence emission maximum	[88]
C. rugosa isolipase B	No shift of the fluorescence emission maximum	[88]
Pseudomonas sp.	Decrease in the α-helix content	[92]
P. simplicissimum	No shift of the fluorescence emission maximum	[42]
R. delemar	Blue shift of the fluorescence emission maximum depending on the w_o value	[42]
C. rugosa	Decrease in the α-helix content at higher w_o	[92]
	Slight increase in the α-helix	[88]
Chromo. viscosum	Increase of the fluorescence intensity of acetone treated enzyme	[90]
Mucor miehei	The ellipticity of the solubilized lipase in the far-UV region markedly decreased with increasing water content	[27]
R. arrhizus	Small increase in α-helix, dramatic increase in β-sheet and in β-turns content. Blue shift of the fluorescence emission maximum depending on the w_o value	[82] [42]

Here, we describe some of the potentially important hydrolytic and synthetic reactions catalyzed by lipases in microemulsions.

Lipases encapsulated in reverse micelles have been used in the last 25 years as a reaction system for the enzymatic hydrolysis of triglycerides or other substrates. Anionic, cationic, nonionic, and zwitterionic surfactants have been used in various hydrolytic reactions. The first report of Malakhova et al. [93] on lipase hydrolytic activity in microemulsions was the investigation of hydrolysis of triglycerides by pancreatic lipase in an AOT/octane microemulsion system. Since then, various microemulsion systems, based on natural [94,95] or synthetic surfactants [28,96–104], have been used for the lipase-catalyzed hydrolysis of vegetable oils and triglycerides. The reverse micellar structural characteristics such as the size and shape together with the solubilizing ability, which depends on the nature of the surfactant used, strongly affect the enzyme activity. In most cases the lipase-catalyzed hydrolytic reactions have been performed in batch-type reactors. However, the development of an ultrafiltration ceramic membrane reactor for the simultaneous lipolysis/bioproducts separation in a reversed micellar medium was reported [105–107].

The hydration state of the microemulsions appears to be the most important factor that determines the reaction rate. Many reports on the activity of lipases from various sources as a function of w_o confirm a typical bell-shaped profile with maximum activity at w_o values of between 9 and 15 [85,98,108]. However, in lecithin-based reverse micelles, the maximum reaction rate for the lipase from *R. delemar* occurred at an extremely low water content ($w_o = 2.2$) [109]. Han et al. [86] showed that the initial water content significantly affects the equilibrium of the hydrolytic reactions in microemulsions. Moreover, the maximum conversion of esters to fatty acid and alcohol is generally increased as the initial water concentration increases.

Reverse micelles have been used as reaction media for lipase-catalyzed glycerolysis of triglycerides. This enzymatic process produces a mixture of mono- and diglycerides that are important emulsifiers in the food industries. It is interesting to note that such reactions were favored in glycerol-containing microemulsions of low water content ($w_o = 1$), while the relative quantities of monoglyceride, diglyceride, and fatty acid in the final product depended on the operational conditions such as substrate concentration and water content of the system [110].

Although studies on the application of lipases to catalyze hydrolytic reactions are important to understand the underlying principles, the synthetic reactions of lipases are industrially more important. Reverse micelles can alter the equilibrium position of reactions through reagent partitioning and in the case of hydrolysis/ condensation reactions by lowering water activity. The microemulsion medium facilitates solubilization of materials of widely different polarities. The lipophilic substrate (fatty acid) and products can be solubilized in the continuous organic phase, while the hydrophilic substrate such as glycerol or short-chain alcohol can be incorporated into reverse micelles (Figure 13.3). Lipase-catalyzed syntheses of glycerides and esters of alcohols as well as polyols have been reported. A few examples of lipase-catalyzed synthetic reactions in reversed micellar systems are given in Table 13.4

The formation of glycerol esters from glycerol and fatty acids in AOT/isooctane, CTAB-heptane/chloroform as well as phosphatidylcholine/hexane reverse micelles has been demonstrated by several groups [111,125–128].

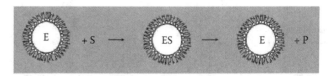

FIGURE 13.3 Model presentation of an enzymatically catalyzed reaction of lipophilic substrates in a w/o microemulsion.

TABLE 13.4
Lipase-Catalyzed Synthetic Reactions in Various W/O Microemulsions

Lipase Source	Surfactant	Reaction Type	References
R. delemar	AOT, Lecithin $C_{12}E_4$	Esterification of aliphatic alcohols, glycerol and ethylene glycol by fatty acids	[42,45,111,112,113]
R.arrhizus	AOT	Esterification of aliphatic alcohols by fatty acids	[42]
Rhizopus sp.	AOT, $C_{12}E_3$	Transesterification of palm oil by stearic acid	[114]
C. cylindracea	AOT Brij	Esterification of aliphatic alcohols by fatty acids	[113–117]
		Enantioselective synthesis of ibuprofen esters	
		Transesterification of triglycerides by fatty acids	
		Esterification of DL-menthol	
Rhizomucor miehei	AOT Lecithin	Synthesis of butyl butyrate Esterification of alcohols by fatty acids	[83,118–121]
P. simplicissimum	AOT	Esterification of alcohols and terpenols with fatty acids	[42,87]
P. cepacia	AOT Lecithin	Esterification of various aliphatic alcohols and glycerol by fatty acids	[46,48]
H. lanuginosa	AOT	Esterification of hydrophilic diols and aliphatic alcohols with fatty acids	[83,119,122]
Thermomyces lanuginosa	AOT	Esterification of alcohols	[103]
Penicillium coryophilum	AOT	Synthesis of butyl-oleate	[123]
Chromo. viscosum	AOT	Esterification of fatty acids	[30]
	CTAB	Esterification of glycerol	[96]
	AOT	Resolution of chiral alcohols	[89]
	SDS	Macrocyclic lactone synthesis	[115]
		Esterification of various alcohols	[124]
Lipozyme	AOT Na-oleate	Esterification of glycerol with fatty acids	[125,126]

The efficiency of lipases to catalyze the synthesis of glycerides in AOT/isooctane microemulsions depends strongly on the water content of the system. Lipases are active at very low w_0 values (w_0 = 1–3); this contrasts with the high optimum w_0 values found for the esterification of monofunctional aliphatic alcohols in similar microemulsions [111,127]. This difference is possibly due to the hydrophilic character of glycerol, which can also be located in the dispersed phase of the microemulsion droplets [111,125,128]. Reverse micelles systems have been also used for the lipase-catalyzed transesterification of triglycerides with free fatty acids [108,116]. Initial work designed to explore the applications of lipases in this area has been directed toward the production from inexpensive starting materials of high-value products. A typical case is the transesterification of palm oil to a cocoa butter analogue. This reaction requires a partial replacement of palmitoyl groups by stearoyl groups in 1(3)-position. Cocoa butter substitutes have been prepared in high yields using a 1,3-specific lipase in anionic and nonionic micro-emulsions [129].

The use of w/o microemulsions as media for the lipase-catalyzed synthesis of mono- and diesters of various hydrophilic diols which have nonionic surfactant properties suitable for cosmetic, pharmaceutical, and nutritional purposes was investigated [112,122,130,131]. In this system, relatively large quantities of hydrophilic diols can be solubilized and their selective esterification with various fatty acids can be achieved with very high reaction rates at room temperature. By controlling the reaction time and the concentration of the substrates it is possible to change the product distribution toward the preferential mono- or diesters of hydrophilic diols [122].

Reverse micellar systems have been successfully used as a media for the enzymatic esterification of alcohols (including terpenols) with various fatty acids (see Table 13.4). In these systems the synthesis of esters proceeds with high rates and high yields. The microstructure of the media, as dictated by hydrophobic/hydrophilic character and concentration of the substrates and the surfactant as well as the hydration state of the system, strongly influences the synthetic activity and the selectivity of lipases [8,9,45]. Stamatis et al. [42] reported studies on the activity of lipases from *R. delemar*, *Rhizopus arrhizus*, and *Penicillium simplicissimum* in esterification reactions of monofunctional aliphatic alcohols with fatty acids in AOT-based microemulsion systems. The results showed a remarkable selectivity of the lipases regarding the chain length and the structure of the alcohols. The nonspecific lipase from *P. simplicissimum* gave higher reaction rates for the esterification of long chain alcohols as well as secondary or branched alcohols (Figure 13.4). Primary alcohols had a low reac-tion rate and tertiary a very slow rate of esterification. The 1,3-specific lipases from *R. delemar* and *R. arrhizus* showed a preference for the esterification of short-chain primary alcohols. The differences in the selectivity of lipases toward the alcohol chain length can be attributed to various factors such as the shape and the size of the reverse micelles, the availability of the substrates, and the location of the enzyme within the microemulsion. Extensive studies using a large number of substrates with different polarities have shown that this selec-tivity appears to be related to the localization of the enzyme within the micellar microstructure [42].

W/O microemulsions have also been shown to be an effective system for the lipase-catalyzed enantioselective esterifications of racemic substrates [115,117]. It has been reported that the reaction rate for the esterification of the (−)-enantiomer of menthol with fatty acids catalyzed by *P. simplicissimum* lipase in AOT-based systems is seven times faster than the esterification rate of the (+)-enantiomer

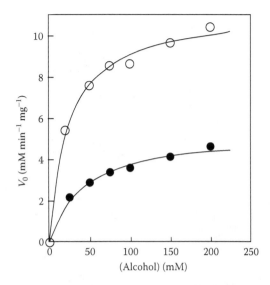

FIGURE 13.4 Dependence of the initial velocity V_0 of the esterification reaction of lauric acid with hexanol (●) and decanol (○) catalyzed by *P. simplicissimum* lipase in AOT/ isooctane microemulsion, as a function of the alcohol concentration. (From Stamatis, H. Enzymatic conversion of amphiphilic substrates in microemulsions, PhD thesis, University of Patras, Patras, 1996).

[47]. Moreover, the enantioselective synthesis of ibuprofen esters catalyzed by *Candida cylindracea* lipase was also studied in a similar system [115]. The enantioselectivity displayed by the lipase was much higher in AOT/isooctane microemulsions than in isooctane. The use of lipases to prepare large quantities of an optically enriched product has also been demonstrated in CTAB-based microemulsions [89]. Resolution of (±)-menthol by *C. cylindracea* lipase-catalyzed esterification with propionic anhydride in a nonionic reverse micellar system in an ultrafiltration membrane reactor has been recently described [117].

It is interesting to note that a combination of surfactants such as lecithin and AOT appears to be more advantageous for some biocatalytic reactions. An increased activity of *Chromo. viscosum* lipase in AOT or CTAB reverse micelles observed in the presence of nonionic surfactants (such as tetraethylene glycol dodecyl ether, $C_{12}E_4$) was attributed to the possible action of nonionic surfactant presumably as a protecting agent which prevents unfavorable ionic and hydrophobic interactions between anionic AOT molecules and the enzyme [99,104,133].

13.2.3 OXIDOREDUCTASES IN W/O MICROEMULSIONS

Oxidoreductases are enzymes that catalyze the reversible transfer of hydrogen atoms and electrons from one molecule (the reductant) to another (the oxidant). During the last four decades, a variety of oxidoreductases (peroxidases, oxidases,

TABLE 13.5

Examples of Oxidoreductases in W/O Microemulsions

Enzyme	Reaction	Microemulsion	References
HRP	Peroxidation	AOT/octane, AOT/benzene	[134]
Recombinant HRP	Peroxidation	AOT/octane	[138]
Tyrosinase	Oxidation	AOT/cyclohexane	[139]
Laccase	Oxidation of phenolics	AOT/isooctane	[140]
HRP and lactoper-oxidase	ABTS peroxidation	Nonionic surfactant/ cyclohexane	[141]
HRP, tyrosinase	ABTS peroxidation/ oleuropein oxidation	Lecithin/olive oil/ propanol	[142]
Cytochromes c and c_3	Pyrene oxidation	AOT/heptane	[144]
Hydrogenase	Hydrogen production	CTAB/chloroform/ octane	[145]
Hydrogenase, lipoamide dehydrogenase, and 20β-hydroxysteroid dehydrogenase	Reduction of ketosteroids	CTAB/hexanol/ octane	[146]
HLADH	Oxidation/reduction	AOT/cyclohexane	[147]
Xanthine oxidase	Oxidation of benzaldehydes	AOT/isooctane, Triton X-100/hexanol, CTAB/ hexanol/heptane	[149]
Yeast alcohol dehydrogenase	Reduction of 2-heptanone	Nonionic surfactant/ cyclohexane	[150]

hydrogenases, dehydrogenases, cytochromes, etc.) has been successfully incorporated within the aqueous core of reverse micelles and were shown to maintain both catalytic activity and conformational stability (Table 13.5). Since the early days of micellar enzymology, horseradish peroxidase (HRP) and cytochrome c were among the first enzymes studied in the restricted environment of reverse micelles. HRP is a small hemoprotein catalyzing by hydrogen peroxide, a variety of substrates. In a pioneer work which appeared in the literature in 1977, HRP was solubilized within the aqueous core of AOT reverse micelles and increased enzymatic activity characterized as superactivity was observed [134]. Since then, several studies based on the activity of plant and fungal peroxidases (HRP-like peroxidases, lignin peroxidases, etc.) and also mono- and/or polyphenol oxidases (tyrosinases, catechol oxidases, and laccases) in reverse micelles were published [38,135,136].

The increased biotechnological interest for enzymes that catalyze phenolic compounds in water restricted media mainly arises from the fact that these enzymes catalyze reactions leading to highly reactive radicals. These radicals, in aqueous solution, can start a nonenzymatic polymerization of the substrates (phenols, polyphenols, amines, polyamines) whereas in low water reaction media like w/o microemulsions, the polymerization process is considerably slowed down or almost inhibited. All related publications up to 2000 were summarized and presented in a comprehensive review by Rodakiewicz-Novak [137]. In 1997, an interesting study appeared in the literature which described the catalytic activity of a recombinant HRP in AOT/octane reverse micelles toward the peroxidation of different donor substrates [138]. The microheterogeneous system of reverse micelles provided, for the first time, evidence for the existence of dimeric structures of the recombinant HRP with altered substrate specificity compared with the native enzyme. In this study, it was also shown that the nature of peroxidase substrates affected the equilibrium between monomeric and dimeric HRP. During the last years, some more studies concerning encapsulation of peroxidases and polyphenol oxidases in various w/o microemulsions appeared in the literature thus proving the unceasing interest of that type of biocatalytic transformations in water restricted media. A thermostability study of tyrosinase in AOT/cyclohexane microemulsions toward the oxidation of 4-methylcatechol was presented by Rojo and collaborators in which the enzyme was protected by ligands against inactivation and its thermostability depended on the size of the reverse micelles [139]. This finding may allow the use of tyrosinase at higher temperatures with a considerable gain in its stability. The catalytic activity of a yellow laccase from *Pleurotus ostreatus* in AOT/isooctane microemulsions was investigated by Rodakiewicz-Novak et al. [140]. The enzyme was shown to retain catalytic activity although lower than in aqueous solution, toward substrates with medium or low solubility in water, both phenolic and nonphenolic. Recently, reverse micelles based on a nonionic synthetic surfactant and cyclohexane as the continuous organic phase were prepared and used for the encapsulation of HRP and lactoperoxidase [141]. The activity of the investigated peroxidases was measured by using hydrogen peroxide and 2,2′-azino-di-[3-ethyl-benzothiazoline-(6)-sulfonic acid] (ABTS). Activity and stability of the enzymes were not improved as compared to aqueous solution although a surfactant that does not interact electrostatically with proteins was used. In a more recent work, both ABTS oxidation catalyzed by HPR in the presence of hydrogen peroxide and oleuropein oxidation catalyzed by mushroom tyrosinase were investigated in microemulsions based on olive oil. The enzyme-catalyzed oxidations mentioned above were considered as model systems for the investigation of novel, olive oil, and lecithin-based microemulsions as potential biocompatible media for biotransformations [142]. Both enzymes were active in microemulsions based on virgin olive oil whereas their activity was considerably lowered in the environment of refined olive oil. This behavior indicates an increased sensitivity of both enzymes to the structural and chemical characteristics of the microenvironment.

As mentioned above, cytochromes were among the first enzymes that were successfully encapsulated in reverse micelles. In a study that appeared in 1979, cytochrome *c*, a small enzyme catalyzing several reactions and showing peroxidase activity by oxidation of various electron donors, was solubilized at room temperature as well as in a broad range of subzero temperatures in AOT/heptane microemulsions and absorbance spectra were obtained [143]. In 1982, two cytochromes from horse heart, namely cytochrome c and cytochrome c_3, were isolated from *Desulfovibrio vulgaris* and incorporated in AOT/heptane reverse micelles. Both enzymes were shown to retain their ability to mediate electron transfer. Hydrated electrons were produced by pulse radiolysis or alternatively formed in the laser photoionization of pyrene [144].

In the same year, Hilhorst and coworkers reported the solubilization of a microbial hydrogenase from *D. vulgaris* in CTAB/chloroform/octane microemulsions. It was shown that in the microenvironment of reverse micelles the enzyme was stabilized against inactivation as compared to aqueous solution. In addition, reducing equivalents for hydrogenase were produced from a photochemical system and the rate of hydrogen production was the same as in bulk water [145]. Next year, a combined enzyme system, consisting of hydrogenase, lipoamide dehydrogenase, and 20β-hydroxysteroid dehydrogenase, has been enclosed in reverse micelles formed with CTAB and a mixture of hexanol/octane. This system was shown to catalyze the stereo- and site-specific reduction of ketosteroids to their corresponding 20β-hydroxyform using an NADH-regenerating system [146]. After the reaction had been completed, the enzymes were regenerated in fairly good yield and the apolar product was isolated by precipitation of the surfactant with acetonitrile.

Horse liver alcohol dehydrogenase (HLADH) is an enzyme known to stereoselectively oxidize and reduce a wide range of alcohol and ketone substrates. In a study that appeared in 1987, HLADH was solubilized in reverse micelles formed with AOT/cyclohexane and the oxidation of ethanol and reduction of cyclohexanone in a coupled substrate/coenzyme recycling system was investigated [147]. Activity and stability studies showed that the enzyme remains active and stable for at least 2 weeks while the charged coenzyme is retained within the dispersed water droplets.

In 1994, a two-step conversion of an ester to an aldehyde was performed in reverse micelles formed with a variety of surfactants by using a combination of lipase and baker's yeast alcohol dehydrogenase [148]. The activity of the two enzymes was affected by the nature of the surfactants and the water content of the reverse micellar system. The aldehyde produced from the reaction partitions in the organic solvent thus preventing the enzyme's inhibition which is normally observed in aqueous solution.

Oxidation of different substituted benzaldehydes catalyzed by xanthine oxidase in different reverse micellar media was reported by Bommarius and collaborators [149]. The main aim of this study was to compare and quantify the enzymatic activity in water and microemulsions formulated with surfactants of different polarities. Kinetic constants were calculated in each of the reverse micellar media for substituted benzaldehydes with varying electronic and hydrophobic properties and were found similar to those calculated in aqueous solution.

The enzymatic reduction of a nonpolar substrate such as 2-heptanone to *S*-2-heptanol catalyzed by yeast alcohol dehydrogenase in w/o microemulsions formed with a technical nonionic surfactant and cyclohexane was studied by Orlich and Schomaecker [150]. The cofactor NADH required for the reduction was regenerated by a second enzyme, the formate dehydrogenase. The reaction rate of the reduction in microemulsions was increased up to 12 times compared to bulk water. A year later, the same research group used the reverse micellar system mentioned above for the solubilization of two alcohol dehydrogenases from yeast and horse liver, respectively, and also a microbial carbonyl reductase [151]. It was shown that under certain conditions activity and stability of the enzymes could be maintained for several weeks. Interestingly, experiments with semi-batch process including cofactor regeneration and product separation were successfully performed.

In another study concerning alcohol dehydrogenase activity in reverse micelles it was shown that under optimum conditions of pH, temperature, and water content, the enzyme was more active in AOT reverse micelles than in aqueous solution [152]. In addition, circular dichroism spectroscopy studies showed that the conformation of the enzyme was strongly affected by the reaction conditions.

The microenvironment of AOT/isooctane reverse micelles was considered as a biomimetic system for the solubilization of flavodoxin, aldehyde oxidoreductase, and a membrane-associated hydrogenase, all isolated from the sulfate reducing bacteria *Desulfovibrio gigas* [153,154]. The main perspective of this research group was the encapsulation of all the three microbial enzymes mentioned above in reverse micelles and the observation of the electron-transfer chain under the water restricted conditions.

To summarize, experimental work concerning the activity and the stability of a big variety of oxidoreductases in w/o microemulsions, formed with many different combinations of emulsifiers and organic solvents, has been published by several research groups during the last four decades. In general, oxidoreductases were shown to remain active if not superactive toward a variety of substrates in the restricted environment of reverse micelles giving rise to many interesting biotechnological applications in the near future.

13.3   TECHNICAL AND COMMERCIAL APPLICATIONS

13.3.1   Biocompatible Microemulsions for Applications in the Pharmaceutical, Cosmetics, and Food Technology Domains

The potential technical and commercial applications of microemulsions are mainly linked to their unique properties such as thermodynamic stability, optical clarity, and high solubilization capacity. However, the most critical problem regarding the use of microemulsions in the food, cosmetic, and pharmaceutical fields is the toxicity of their partial components. Formulation and characterization of nontoxic microemulsions based on naturally occurring amphiphiles and different oils have been studied for almost two decades. The first attempt to use natural biodegradable surfactants for the formulation of nontoxic microemulsions was reported by Shinoda et al. [155,156]. In this regard, soybean lecithin is a combination of naturally-occurring phospholipids extracted during the processing of soybean oil and therefore can be used for the construction of various nontoxic microemulsion formulations. Because of lecithin's high lipophilicity and strong tendency to form liquid crystalline structures, spontaneous formulation of reverse micelles is not feasible. In this case, addition of cosurfactants such as short-chain alcohols is necessary to facilitate the dispersion and stabilization of the aqueous phase within the continuous organic solvent. Since then, numerous studies on the microemulsification of various nonpolar solvents by lecithin of different chain lengths in the presence of alcohols were presented by different research groups and structural characteristics of these systems were reported [72,121,157–159]. Nevertheless, low molecular weight alcohols and alkane diols commonly used as cosurfactants in phospholipidic formulations are not generally accepted for the production of pharmaceutical and food products. In this regard, the use of alternative cosurfactants such as short straight-chain amines [160], alkanoic acids [161], glycol derivatives [162], and biodegradable chemical solvents such as N-methyl pyrrolidone [163] has been examined and proposed by many authors.

Phospholipids, as natural nontoxic amphiphiles, are in many respects regarded as ideal surfactant molecules for industrial applications in the food, cosmetic, and pharmaceutical domains. Besides, nonionic surfactants, such as polyethylene sorbitan n-acyl esters (Tweens), are generally considered as compounds with very low toxicity and therefore can be useful alternatives to natural surfactants for the construction of biocompatible microemulsions. In this respect, many studies of reverse micellar formulations based on nonionic surfactants have been published [164,165]. Nevertheless many attempts have been made by authors to synthesize and use new nontoxic surfactants other than lecithin and nonionic surfactants mentioned above. Sugar esters, mono- and diglycerides, fatty acid amides, and alkylglycosides are among the surfactants commonly used for the construction of stable and biocompatible microemulsion formulations [41]. Sucrose alkanoates are nonionic surfactants synthesized from sugars and natural fatty acids and thus considered nontoxic and biocompatible. In this regard, microemulsions based on sucrose alkanoates, hexanol, and decane were formulated and structurally characterized by Nakamura et al. in a study that appeared in 1999 [166]. Garti et al. also proposed the application of sugar esters for the construction of nonionic microemulsions in view of potential food applications [167]. Another biocompatible surfactant, namely n-dodecylammonium α-glutamate (GDA), was synthesized and the phase behavior, as well as the structural properties, of the GDA/n-pentanol/ water system was studied by various spectroscopic techniques [168]. In some cases, mixtures of biocompatible amphiphilic molecules were used for the microemulsification of apolar solvents. In a study which was published in 2000, a mixture of sorbitan monooleate and polysorbate 80 (nonionic emulsifier derived from polyoxylated sorbitol and oleic acid) was used together with medium chain glycerides and water to make biocompatible microemulsions for the encapsulation of a model peptide such as insulin [169]. In another study, a mixture of natural lipids, namely glycerol monooleate, diglycerol monooleate, and lecithin, was added in glycerol trioleate and water and the corresponding ternary and pseudoternary phase diagrams were investigated. In these systems the microemulsion regions occurred only at relatively high weight fraction of oil while a quite limited water uptake was observed [170]. In a recent work, microemulsions containing a mixture of pharmaceutical biocompatible components such as soya phosphatidylcholine, polyoxyethylene glycerol trihydroxystearate 40, and sodium oleate as surfactants and cholesterol as the continuous oil phase were prepared, structurally character-ized, and proposed as potential drug delivery systems [171].

In the majority of reverse micellar formulations used for the performance of enzyme-catalyzed reactions with biotechnological interest, the continuous organic phase usually consisted of saturated hydrocarbons such as hexane, octane, isooc-tane, cyclohexane, and benzene. The most significant problem associated with the preparation of microemulsions for application in the pharmaceutical and food domains is the difficulty associated with the acceptability of their components. In this respect, reverse micellar systems should be formed with highly biocompatible materials and consequently new "green" solvents must be introduced in the field. Microemulsions formulated with phospholipids as emulsifiers and mixtures of fatty acids and fatty acid ethyl esters as the organic solvent were prepared and characterized by scattering techniques by Ichikawa et al. [172,173]. The reversed micellar systems developed in these studies are expected to be used as solubiliza-tion media of various hydrophilic functional materials in the pharmaceutical and food industries. Although long and medium chain triglycerides and fatty acid esters (isopropyl myristate, isopropyl palmitate, etc.) have been used quite fre-quently for the preparation of microemulsions [174–176], their large size, slight amphiphilic, and semi-polar character, compared to hydrocarbon oils, makes the formulation of reverse micelles rather difficult. However, replacing triglycerides with cheap and commercially available oils such as soybean oil, sunflower oil, or jojoba oil [165,177] is an interesting alternative to prepare new stable and bio-compatible formulations. Recently olive oil, either virgin or refined, was intro-duced as a natural, nontoxic, and inexpensive product for the formulation of lecithin-based biocompatible microemulsions [142,178]. These systems were shown to provide an appropriate environment for the effectuation of various enzyme-catalyzed reactions although their ability to incorporate water is rather limited. Finally, essential oils, having low molecular weight and nonamphiphilic character, e.g., R(+)-limonene, could be an interesting alternative to triglyceride oils. Application of R(+)-limonene for the construction of food grade microemul-sions has been introduced a few years ago by Garti and coworkers [179,180]. More recently, an extensive structural study of lecithin-based w/o microemulsions formulated with R(+)-limonene and alcohols was presented [181]. These systems were successfully used as microreactors for the presentation of lipase-catalyzed esterifications.

13.3.2 PRODUCT RECOVERY AND ENZYME REUSE

Major problems that need to be solved before the employment of a microemulsion system in the industrial processes are the recovery of the products from the reaction mixture and the regeneration of the enzyme. The presence of surfactant impairs product recovery by usual methods such as extraction and distillation because surfactants stabilize emulsions and foam. Therefore, suitable immobilization methods and product separation strategies including phase separation techniques as well as the use of membrane reactors have to be developed.

Various membrane reactors have been proposed to allow recovery of products from microemulsions and permit to reuse the biocatalyst in various enzymatic processes such as hydrolysis of triglycerides [124,182], synthesis of esters [183,184], and peptide synthesis [65,185,186]. These reactors take advantage of the potential of synthetic semipermeable membranes, that usually are ultrafiltration membranes, to retain enzymes on the membrane and to recover products on the permeate side. The productivities as well as the enzyme stability in such reactors are usually high.

Lüthi and Luisi [44] have used a hollow fiber membrane reactor for peptide synthesis catalyzed by α-chymotrypsin in microemulsion. Chang et al. [110] described the immobilization of lipase on liposomes, which, in turn, were solubilized in AOT/isooctane reversed micelles and used for the continuous glycerolysis of olive oil in an ultrafiltration cell. The half-life of the *Chromo. viscosum* lipase was 7 weeks. The development of an ultrafiltration ceramic membrane bioreactor for the simultaneous lipolysis of olive oil and product separation in AOT/isooctane reversed micellar media has been also reported [106,107]. Cutinase performance was also evaluated in a ceramic membrane reactor [9]. An attempt to minimize the surfactant contamination problem was based on the use of an electro-ultrafiltration method which can decrease the gel formation in the membrane surface, improving the filtration flux, achieving the separation of the AOT reverse micelles [187].

Larsson et al. [188] proposed a simple technique for enzyme reuse and product recovery, taking advantage of the phase behavior of the surfactant system, to separate the microemulsion constituents into an oil-rich phase and a water-rich phase; the latter contained almost all the surfactant. The enzyme reuse and product recovery have been investigated using an HLADH catalyzing a coupled substrate–coenzyme regenerating cycle in AOT/isooctane or $C_{12}E_5$/heptane microemulsions. A small change in temperature induced a shift into a two-phase system: an oil-rich phase containing the products and a water-rich phase containing the surfactant and the enzyme. The oil-rich phase could be replaced by a new solution of substrate in oil, and then the temperature could be brought back to a value where the monophasic microemulsion was stable; hence, the reaction could be repeated. A similar temperature-induced phase separation procedure was applied for the enzyme recovery and separation of the product (ester) of an esterification reaction catalyzed by *Pseudomonas cepacia* lipase in AOT-based reverse micelles [127]. This procedure could extract a significant amount of active lipase whereas the product was almost completely separated from the surfactant system.

Another approach to simplify product recovery and enzyme recycle from microemulsion-based media is the use of gelled microemulsion systems. The formation of microemulsion-based organogels (MBGs), first reported in 1986 [189,190], involves the gelation of w/o microemulsions based on AOT or lecithin, into monophasic optically transparent rigid systems by mixing with an aqueous gelatin solution above the gelling temperature. Cooling to room temperature produces a transparent gel that has reproducible physical properties. The structural characteristics of the gelatin-containing MBG have been examined using various techniques such as small-angle neutron scattering, light scattering, and NMR [191,192]. These MBGs have been described as an extensive network of rigid rods of water and gelatin surrounded by a shell of surfactant. It was demonstrated that gelatin containing AOT or CTAB MBGs were particularly a useful form of immobilized enzymes [183,193–195]. These enzyme-containing MBGs are rigid and stable in various nonpolar organic solvents and may therefore be used for biotransformations in organic media. Using systems based on lipase containing MBGs formulated with gelatin, preparative-scale synthesis of different esters under mild conditions was possible, and both regio- and stereoselectivities have been demonstrated [124,133,183,184,193,195]. Under most conditions, the gel matrix fully retains the surfactant, gelatin, water, and enzyme components allowing the diffusion of nonpolar substrates or products between the surrounding organic phase and the gel (Figure 13.5).

Analogous to the MBG method, Fadnavis and Koteshwar [195] have used the phenomenon of reversible sol–gel transition to immobilize enzymes in the gelatin matrix by first entrapping the enzyme in the gelatin containing microemulsion at low temperature and then immobilizing it by cross-linking with glutaraldehyde. The immobilized enzymes (including lipase) termed as "gelozymes" are found to retain 70%–80% of their total activity.

The main drawbacks for biocatalytic applications of organogels formulated with gelatin are the poor mechanical and thermal stability and their low resistance to hydrophilic environments. An approach based on polymerization of tetraethoxysilane has been recently proposed in order to improve organogel stability [196]. Moreover, the use of biopolymers other than gelatin, such as agar k-carrageenan, and cellulose derivatives, to form MBGs as an enzyme immobilization matrix has also been described [130,197,198]. Enzyme-containing MBGs, to a great extent, overcome the major problems, which

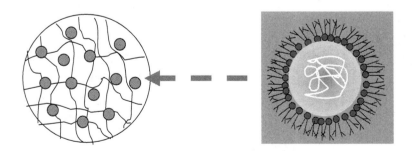

FIGURE 13.5 Enzyme-containing microemulsion droplet immobilized in an organogel.

must be solved for the employment of a microemulsion system in industrial processes, most notably the recovery of the products and the regeneration of the biocatalyst.

13.4 CONCLUSION

Microemulsions being isotropic and thermodynamically stable are very attractive reaction media for enzyme-catalyzed reactions. They offer an ideal microenvironment for hosting the biocatalyst whereas they facilitate solubilization of both hydrophobic and hydrophilic substrates in a macroscopically single phase. Various enzymes have been so far studied in different microemulsion systems. Of particular interest are the studies with proteases, lipases, and oxidoreductases. Proteases, such as α-chymotrypsin and trypsin, have been studied as regards their activity toward both proteolytic and peptide synthetic reactions. Lipases have been extensively studied as these enzymes undergo an interfacial activation step that is favored in reverse micelles. As a consequence, lipases in microemulsions offer a unique advantage to catalyze not only hydrolytic reactions of glycerides, but also synthetic reactions that cannot be carried out in conventional media. Oxidoreductases are also studied in these confined media as they catalyze reactions where the electron transfers and the generated highly reactive radicals can be inhibited. Numerous are the applications of the enzymes in reverse micelles. The regio- and stereoselectivity characteristics of lipase catalysis make possible the synthesis of a variety of new esters and glycerides. Reactor development and scale-up are important problems that need to be addressed for commercial utilization of these novel syntheses. Some key problems in the application of enzyme-containing reverse micelles to bioorganic reactions are the reuse of the biocatalyst and the separation of the product. Proposed solutions include the use of continuous reactors based on semipermeable membranes and temperature-induced phase separations. The use of biocompatible surfactants, such as lecithin, and oils, such as olive oil, to form reverse micelles would be useful for specific applications in pharmaceutical, cosmetic, and food industries.

SYMBOLS AND TERMINOLOGIES

ABTS	2,2′-azino-di-[3-ethyl-benzothiazoline-(6)-sulfonic acid]
AOT	bis(ethylhexyl) sulfosuccinate sodium salt
Brij30	polyethylene glycol dodecyl ether
$C_{12}E_4$	teteraethylene dodecyl ether
CTAB	cetyltrimethylammonium bromide
GDA	n-dodecylammonium α-glutamate
HLADH	horse liver alcohol dehydrogenase
HRP	horseradish peroxidase
k_{cat}	enzyme turnover number
K_m	Michaelis–Menten constant
$\log P$	partition coefficient of a solute between octane and water

MBGs	microemulsion-based organogels
SDS	sodium dodecyl sulfate
Triton X-100	polyethylene glycol p-(1,1,3,3-tetramethylbutyl)-phenyl ether
TTAB	tetradecyltrimethylammonium bromide
Tween 80	polyoxyethylene sorbitan monooleate
V_{max}	maximum enzyme velocity
w_o	molar ratio of water to surfactant

REFERENCES

1. Fletcher, P.D.I., Rees, G.D., Robinson, B.H., Freedman, R.B. 1985. Kinetic properties of a-chymotrypsin in water-in-oil microemulsions: Studies with a variety of substrates and microemulsion systems. *Biochem. Biophys. Acta* 832, 204–214.
2. Martinek, K., Levashov, A.V., Klyachko, N.L., Khmelnitsky, Y.L., Berezin, I.V. 1986. Micellar enzymology. *Eur. J. Biochem.* 155, 453–468.
3. Martinek, K., Klyachko, N.L., Kabanov, A.V., Khmelnitsky, Y.L., Levashov, A.V. 1989. Micellar enzymology: Its relation to membranology. *Biochim. Biophys. Acta* 981, 161–172.
4. Luisi, P.L., Bonner, F.J., Pellegrini, A., Wiget, P., Wolf, R. 1979. Micellar solubilization of proteins in aprotic solvents and their spectroscopic characterization. *Helv. Chim. Acta* 62, 740–753.
5. Luisi, P.L., Giomini, M., Pileni, M.P., Robinson, B.H. 1988. Reverse micelles as hosts for proteins and small molecules. *Biochim. Biophys. Acta* 947, 9–46.
6. Luisi, P.L., Magid, L. 1986. Solubilization of enzymes and nucleic acids in hydrocarbon micellar solutions. *CRC Crit. Rev. Biochem.* 20, 409–474.
7. Grandi, C., Smith, R.E., Luisi, P.L. 1981. Micellar solubilization of biopolymers in organic solvents. Activity and conformation of lysozymes in isooctane reverse micelles. *J. Biol. Chem.* 256, 837–843.
8. Stamatis, H., Xenakis, A., Kolisis, F.N. 1999. Bioorganic reactions in microemulsions: The case of lipases. *Biotechnol. Adv.* 17, 293–318.
9. Carvalho, C.M.L., Cabral, J.M.S. 2000. Reverse micelles as reaction media for lipases. *Biochimie* 82, 1063–1085.
10. Yang, H., Kiserowb, D.J., McGown, L.B. 2001. Effects of bile salts on the solubility and activity of yeast alcohol dehydrogenase in AOT reversed micelles. *J. Mol. Catal. B: Enzymatic* 14, 7–14.
11. Shipovskov, S., Ferapontova, E., Ruzgas, T., Levashov, A. 2003. Stabilisation of tyrosinase by reversed micelles for bioelectrocatalysis in dry organic media. *Biochim. Biophys. Acta* 1620, 119–124.
12. Papadimitriou, V., Xenakis, A., Cazianis, C.T., Stamatis, H., Egmond, M., Kolisis, F.N. 1996. EPR studies of cutinase in microemulsions. *Ann. NY Acad. Sci.* 799, 275–280.
13. Carvalho, C.M.L., Cabral, J.M.S., Aires-Barros, M.R. 1999. Cutinase stability in AOT reversed micelles: System optimization using the factorial design methodology. *Enzyme Microb. Technol.* 24, 569–576.
14. Liu, Z., Shao, M., Cai, R., Shen, P. 2006. Online kinetic studies on intermediates of laccase-catalyzed reaction in reversed micelle. *J. Colloid Interface Sci.* 294, 122–128.
15. Anarbaev, R.O., Khodyreva, S.N., Zakharenko, A.L., Rechkunova, N.I., Lavrik, O.I. 2005. DNA polymerase activity in water-structured and confined environment of reverse micelles. *J. Mol. Catal. B: Enzym.* 33, 29–34.

16. Spreti, N., Alfani, F., Cantarella, M., D'Amico, F., Germani, R., Savelli, G. 1999. Alpha-chymotrypsin superactivity in aqueous solutions of cationic surfactants. *J. Mol. Catal. B: Enzym.* 6, 99–110.

17. Freeman, K.S., Lee, S.S., Kiserow, D.J., McGown, L.B. 1998. Increased chymotrypsin activity in AOT/bile salt reversed micelles. *J. Colloid Interface Sci.* 15, 344–348.

18. Franqueville, E., Stamatis, H., Loutrari, H., Friboulet, A., Kolisis, F.N. 2002. Studies on the catalytic behavior of a cholinesterase-like abzyme in an AOT microemulsion system. *J. Biotechnol.* 97, 177–182.

19. Rodakiewicz-Nowak, J., Maslakiewicz, P., Haber, J. 1996. The effect of linoic acid on pH inside sodium bis(2-ethylhexyl)sulfosuccinate reverse micelles in isooctane and on the enzymatic activity of soybean lipogenase. *Eur. J. Biochem.* 238, 549–553.

20. Rodriguez, R.,Vargas, S., Fernandez-Velasko, D.A. 1998. Reverse micelle systems composed of water, Triton X100 and phospholipids in organic solvents. 1. Phase boundary titrations and dynamic light scattering analysis. *J. Colloid Interface Sci.* 197, 21.

21. Fernandez-Velasko, D.A., Rodriguez, R., Vargas, S., de Gomez-Puyou, M.T., Gomez-Puyou, A. 1998. Reverse micelle systems composed of water, Triton X100 and phospholipids in organic solvents. 2. Catalysis and thermostability of hexokinase. *J. Colloid Interface Sci.* 197, 29–35.

22. Shah, C., Sellappan, S., Madamwar, D. 2000. Entrapment of enzyme in water-restricted microenvironment—amyloglucosidase in reverse micelles. *Proc. Biochem.* 35, 971–975.

23. Klyachko, N.L., Levashov, A.V. 2003 Bioorganic synthesis in reverse micelles and related systems. *Curr. Opinion Colloid Interface Sci.* 8, 179–186.

24. Matzke, S.F., Creagh, A.L., Haynes, C.A., Blanch, H.W., Prausnitz, J.M. 1992. Mechanisms of protein solubilization in reverse micelles. *Biotechnol. Bioeng.* 40, 91–102.

25. Papadimitriou, V. 1996. Enzymatic studies in microemulsions. PhD thesis, University of Athens, Athens.

26. Kabanov, A.V., Levashov, A.V., Klyachko, N.L., Namyotkin, S.N., Pshezhetsky, A.V. 1988. Enzymes entrapped in reversed micelles of surfactants in organic solvents: A theoretical treatment of the catalytic activity regulation. *J. Theor. Biol.* 133, 327–343.

27. Naoe, K., Takeuchi, C., Kawagoe, M., Nagayama, K., Imai, M. 2007. Higher order structure of *Mucor miehei* lipase and micelle size in cetyltrimethylammonium bromide reverse micellar system. *J. Chromatography B* 850, 277–284.

28. Mitra, R.N., Dasgupta, A., Das, D., Roy, S., Debnath, S., Das, P.K. 2005. Geometric constraints at the surfactant headgroup: Effect on lipase activity in cationic reverse micelles. *Langmuir* 21, 12115–12123.

29. Moniruzzaman, M., Hayashi, Y., Talukder, M.M.R., Saito, E., Kawanishi, T. 2006. Effect of aprotic solvents on the enzymatic activity of lipase in AOT reverse micelles. *Biochem. Engin. J.* 30, 237–244.

30. Moniruzzaman, M., Hayashi, Y., Talukder, E., Kawanishi, T. 2007. Lipase-catalyzed esterification of fatty acid in DMSO (dimethyl sulfoxide) modified AOT reverse micellar systems. *Biocatal. Biotransform.* 25, 51–58.

31. Rairkar, M.E., Hayes, D.G., Harris, Z.J.M. 2007. Solubilization of enzymes in water-in-oil microemulsions and their rapid and efficient release through use of a pH-degradable surfactant. *Biotechnol. Lett.* 29, 767–771.

32. Spirovska, G., Chaudhuri, J.B. 1998. Sucrose enhances the recovery and activity of ribonuclease A during reversed micelle extraction. *Biotechnol. Bioeng.* 58, 374–379.

33. Eryomin, A.N., Metelitza, D.I. 1999. Effect of hydration degree of aerosolOT reversed micelles and surfactant concentration in heptane on spectral and catalytic properties of catalase. *Biochemistry (Moscow)* 64, 1049–1060.

34. Goncalves, A.M., Serrob, A.P., Aires-Barros, M.R., Cabral, J.M. 2000. Effects of ionic surfactants used in reversed micelles on cutinase activity and stability. *Biochim. Biophys. Acta* 1480, 92–106.

35. Orlich, B., Schomacker, R. 2002. Enzyme catalysis in reverse micelles. In Scheper Th (Ed.), *Advances in Biochemical Engineering/Biotechnology*, Vol. 75, pp. 185–209. Heidelberg: Springer-Verlag.

36. Martinek, K., Berezin, I.V., Khmelnitsky, Y.L., Klyachko, N.L., Levashov, A.V. 1987. Micellar enzymnology: Potentialities in applied areas (biotechnology). *Coll. Czechoslovak. Chem. Commun.* 52, 2589–2602.

37. Khmelnitsky, Y.L., Neverova, I.N., Polyakov, V.I., Grinberg, V.Y., Levashov, A.V., Martinek, K. 1990. Kinetic theory of enzymatic reactions in reversed micellar systems. Application of the pseudophase approach for partitioning substrates. *Eur. J. Biochem.* 190, 155–159.

38. Verhaert, R.M.D., Hilhorst, R., Vermüe, M., Schaafsma, T.J., Veeger, C. 1990. Description of enzyme kinetics in reverse micelles. 1. Theory. *Eur. J. Biochem.* 187, 59–72.

39. Papadimitriou, V., Xenakis, A., Evangelopoulos, A.E. 1993. Proteolytic activity in various water-in-oil microemulsions as related to the polarity of the reaction medium. *Colloids Surf. B. Biointerf.* 1, 295–303.

40. Ruckhenstein, E., Karpe, P. 1990. Enhanced enzymatic activity in reverse micelles. *Eur. J. Biochem.* 139, 408–436.

41. Garti, N. 2003. Microemulsions as microreactors for food applications. *Curr. Opinion Colloid Interface Sci.* 8, 197–211.

42. Stamatis, H., Xenakis, A., Provelegiou, M., Kolisis, F.N. 1993. Esterification reactions catalyzed by lipases in microemulsions. The role of enzyme localization in relation to its selectivity. *Biotechnol. Bioeng.* 42, 103–110.

43. Pavlenko, I.M., Kuptsova, O.S., Klyachko, N.L., Levashov, A.V. 2002. The lipase/lipoxygenase bienzyme system in AOT reversed micelles in octane. *Russ. J. Bioorg. Chem.* 28, 44–49.

44. Lüthi, P., Luisi, P.L. 1984. Enzymatic synthesis of hydrocarbon-soluble peptides with reverse micelles. *J. Am. Chem. Soc.* 106, 7285–7286.

45. Hayes, D.G., Gulari, E. 1990. Esterification reactions of lipase in reverse micelles. *Biotechnol. Bioeng.* 35, 793–801.

46. Stamatis, H., Kolisis, F.N., Xenakis, A., Bornscheuer, U., Scheper, T., Menge, U. 1993. *Pseudomonas cepacia* lipase: Esterification reactions in AOT microemulsion systems. *Biotechnol. Lett.* 15, 703–708.

47. Stamatis, H., Kolisis, F.N., Xenakis, A. 1993. Enantiomeric selectivity of a lipase from *Penicillium simplicissimum* in the esterification of menthol in microemulsions. *Biotechnol. Lett.* 15, 471–476.

48. Stamatis, H., Xenakis, A., Dimitriadis, E., Kolisis, F.N. 1995. Catalytic behavior of *Pseudomonas cepacia* lipase in w/o microemulsions. *Biotechnol. Bioeng.* 45, 33–41.

49. Fadnavis, N.W., Deshpande, A. 2002. Synthetic applications of enzymes entrapped in reverse micelles and organo-gels. *Curr. Org. Chem.* 6, 393–410.

50. Hilhorst, R., Spruijt, R., Laane, C., Verger, C. 1984. Rules for the regulation of enzyme activity in reversed micelles as illustrated by the conversion of apolar steroids by 20-b-hydroxysteroid dehydrogenase. *Eur. J. Biochem.* 144, 459–462.

51. Smolders, A.J.J., Pinheiro, H.M., Noronha, P., Cabral, J.M.S. 1991. Steroid bioconversion in a microemulsion system. *Biotechnol. Bioeng.* 38, 1210–1217.

52. Ichinose, H., Michizoe, J., Maruyama, T., Kamiya, N., Goto, M. 2004. Electron-transfer reactions and functionalization of cytochrome P450cam monooxygenase system in reverse micelles. *Langmuir* 20, 5564–5568.

53. Pandey, A., Pandey, A., Srivastava, P., Pandey, A. 2007. Using reverse micelles as microreactor for hydrogen production by coupled systems of *Nostoc/R. palustris* and *Anabaena/R. palustris*. *World J. Microbiol. Biotechnol.* 23, 269–274.

54. Rao, A.M., John, V.T., Gonzalez, R.D., Akkara, J.A., Kaplan, D.L. 1993. Catalytic and interfacial aspects of enzymatic polymer synthesis in reversed micellar systems. *Biotechnol. Bioeng.* 41, 531–540.

55. Menger, F.M., Yamada, K. 1979. Enzyme catalysis in water pools. *J. Am. Chem. Soc.* 101, 6731–6734.

56. Martinek, K., Levashov, A.V., Klyachko, N.L., Pantin, V.I., Berezin, I.V. 1981. The principles of enzyme stabilization. VI. Catalysis by water-soluble enzymes entrapped into reversed micelles of surfactants in organic solvents. *Biochim. Biophys. Acta* 657, 277–294.

57. Clapes, P., Pera, E., Torres, J.L. 1997. Peptide bond formation by the industrial protease, neutrase, in organic media. *Biotechnol. Lett.* 19, 1023–1026.

58. Kumar, D., Bhalla, T.C. 2005. Microbial proteases in peptide synthesis. Approaches and applications. *Appl. Microbiol. Biotechnol.* 68, 726–736.

59. Chen, Y.X., Zhang, X.Z., Zheng, K., Chen, S.M., Wang, Q.C., Wu, X.X. 1998. Protease catalyzed synthesis of precursor dipeptides of RGD with reverse micelles. *Enzyme Microb. Technol.* 23, 243–248.

60. Chen, Y.X., Zhang, X.Z., Chen, S.M., You, D.L., Wu, X.X., Yang, X.C., Guan, W.Z. 1999. Kinetically controlled syntheses catalyzed by proteases in reverse micelles and separation of precursor dipeptides of RGD. *Enzyme Microb. Technol.* 25, 310–315.

61. Schellenberger, V., Jakube, H.D. 1991. Protease-catalyzed kinetically controlled peptide synthesis. *Angew. Chem. Int. Ed. Engl.* 30, 1437–1449.

62. Hou, R.Z., Zang, N., Li, G., Huang, Y.B., Wang, H., Xiao, Y.P., Liu, Y.J., Yang, Y., Zhao, L., Zhang, X.Z. 2005. Synthesis of tripeptide RGD amide by a combination of chemical and enzymatic methods. *J. Molec. Catal. B: Enzym.* 37, 9–15.

63. Jorba, X., Clapés, P., Xaus, N., Calvet, S., Torres, J.L., Valencia, G., Mata, J. 1992. Optimization and kinetic studies of the enzymatic synthesis of Ac-Phe-Leu-NH$_2$ in reversed micelles. *Enzyme Microb. Technol.* 14, 117–124.

64. Jorba, X., Clapés, P., Torres, J.L., Valencia, G., Mata-Alvarez, J. 1995. Ethyl acetate modified AOT water-in-oil microemulsions for the α-chymotrypsin catalyzed synthesis of a model dipeptide derivative. *Colloids Surf. A* 96, 47–52.

65. Serralheiro, M.L.M., Prazeres, D.M.F., Cabral, J.M.S. 1994. Dipeptide synthesis and separation in a reversed micellar membrane reactor. *Enzyme Microb. Technol.* 16, 1064–1073.

66. Xing, G.W., Liu, D.J., Ye, Y.H., Ma, J.M. 1999. Enzymatic synthesis of oligopeptides in mixed reverse micelles. *Tetrahedron Lett.* 40, 1971–1974.

67. Fan, K.K., Ouyang, P., Wu, X., Lu, Z.J. 2001. Enhancement of the activity of papain in mixed reverse micellar systems in the presence of Tween 80. *Chem. Technol. Biotechnol.* 76, 27–30.

68. Barbaric, S., Luisi, P.L. 1981. Micellar solubilization of biopolymers in organic solvents. Activity and conformation of α-chymotrypsin in isooctane–AOT reverse micelles. *J. Am. Chem. Soc.* 103, 4239–4244.

69. Ishikawa, H., Noda, K., Oka, T. 1990. Kinetic properties of enzymes in AOT–isooctane reversed micelles. *J. Ferment. Bioeng.* 70, 381–385.

70. Mao, Q., Walde, P., Luisi, P.L. 1992. Kinetic behavior of α-chymotrypsin in reverse micelles. A stopped-flow study. *Eur. J. Biochem.* 208, 165–170.

71. Walde, P., Peng, Q., Fednavis, N.W., Battistel, E., Luisi, P.L. 1988. Structure and activity of trypsin in reverse micelles. *Eur. J. Biochem.* 173, 401–409.
72. Peng, Q., Luisi, P.L. 1990. The behavior of proteases in lecithin reverse micelles. *Eur. J. Biochem.* 188, 471–480.
73. Avramiotis, S., Lianos, P., Xenakis, A. 1997. Trypsin in lecithin based (w/o) microemulsions. Fluorescence and enzyme activity studies. *Biocatal. Biotransform.* 14, 299–316.
74. Fadnavis, N.W.; Babu, R.L., Deshpande, A. 1998. Reactivity of trypsin in reverse micelles: pH-effects, on the w_o versus enzyme activity profiles. *Biochimie* 80, 1025–1030.
75. Dasgupta, A., Das, D., Das, P.K. 2005. Reactivity of trypsin in reverse micelles: Neglected role of aggregate size compared to water-pool components. *Biochimie* 87, 1111–1119.
76. Stupishina, E.A., Khamidullin, R.N., Vylegzhanina, N.N., Faizullin, D.A., Zuev, Y.F. 2006. Ethylene glycol and the thermostability of trypsin in reverse micelles system. *Biochemistry (Moscow)* 71, 533–537.
77. Vicente, L.C., Aires-Barros, R., Empis, J.M. 1994. Stability and proteolytic activity of papain in reverse micellar and aqueous media: A kinetic and spectroscopic study. *J. Chem. Tech. Biotechnol.* 60, 291–297.
78. Verger, R. 1980 Enzyme kinetics of lipolysis. In Colowick SP, Kaplan NO (Eds.). *Methods in Enzymology*, Vol. 64B, pp. 340–92. New York: Academic Press.
79. Derewenda, Z.S., Sharp, A.M. 1993 News from the interface: The molecular structures of triacylglyceride lipases. *Trends Biochem. Sci.* 205, 20–25.
80. Marangoni, A.G. 1993 Effects of the interaction of porcine pancreatic lipase with AOT/isooctane reversed micelles on enzyme structure and function follow predictable patterns. *Enzyme Microb. Technol.* 15, 944–949.
81. Tsai, S.W., Lee, K.P., Chiang, C.L. 1995. Surfactant effects on lipase catalyzed hydrolysis of olive oil in AOT/isooctane reverse micelles. *Biocatal. Biotrans.* 13, 89–98.
82. Brown, E.D., Yada, R.Y., Marangoni, A.G. 1993. The dependence of the lipolytic activity of *R.* arrhizus lipase on surfactant concentration in Aerosol-OT/isooctane reverse micelles and its relationship to enzyme structure. *Biochim. Biophys. Acta* 1161, 66–72.
83. Crooks, G.E., Rees, G.D., Robinson, B.H., Svensson, M., Stephenson, G.R. 1995. Comparison of hydrolysis and esterification behavior of *Humicola lanuginosa* and *Rhizomucor miehei* lipases in AOT-stabilized water-in-oil microemulsions: II. Effect of temperature on reaction kinetics and general considerations of stability and productivity. *Biotechnol. Bioeng.* 48, 190–196.
84. Yao, C., Tang, S., Heb, Z., Deng, X. 2005. Kinetics of lipase-catalyzed hydrolysis of olive oil in AOT/isooctane reversed micelles. *J. Mol. Catal. B: Enzym.* 35, 108–112.
85. Han, D., Rhee, J.S. 1986. Characteristics of lipase-catalyzed hydrolysis of olive oil in AOT–isooctane reversed micelles. *Biotechnol. Bioeng.* 28, 1250–1255.
86. Han, D., Rhee, J.S., Lee, S.B. 1987. Lipase reaction in AOT–isooctane reversed micelles: Effect of water on equilibria. *Biotechnol. Bioeng.* 30, 381–388.
87. Stamatis, H., Xenakis, A., Menge, U., Kolisis, F.N. 1993. Kinetic study of lipase catalyzed esterification reactions in microemulsions. *Biotechnol. Bioeng.* 42, 931–937.
88. Otero, C., Rua, M.L., Robledo, L. 1995. Influence of the hydrophobicity of lipase isoenzymes from *Candida rugosa* on its hydrolytic activity in reverse micelles. *FEBS Lett.* 360, 202–206.

89. Rees, G.D., Robinson, B.H., Stephenson, R.G. 1995. Macrocyclic lactone synthesis by lipases in water-in-oil microemulsions. *Biochim. Biophys. Acta* 1257, 239–248.

90. Zaman, M.M., Hayashi, Y., Talukder, M.M.R., Kawanishi, T. 2005. Enhanced activity and stability of *Chromobacterium viscosum* lipase in AOT reverse micellar systems by pretreatment with acetone. *J. Mol. Catal. B: Enzym.* 32, 149–155.

91. Waks, M. 1986. Proteins and peptides in water-restricted environments. *Proteins* 1, 4–12.

92. Walde, P., Han, D., Luisi, P.L. 1993. Spectroscopic and kinetic studies of lipase solubilized in reverse micelles. *Biochemistry* 32, 4029–4034.

93. Malakhova, E.A., Kurganov, B.I., Levashov, A.V., Berezin, I.V., Martinek, K. 1983. A new approach to the study of enzymatic reactions with the participation of water-insoluble substrates. Pancreatic lipase enclosed in inverted micelles of surface-active substances in an organic solvent. *Dokl. Akad. Nauk. SSSR* 270, 474–477.

94. Holmberg, K., Osterberg, E. 1988. Enzymatic preparation of monoglycerides in microemulsion. *J. Am. Oil Chem. Soc.* 65, 1544–1548.

95. Chen, J.P., Chang, K.-C. 1993. Lipase catalyzed hydrolysis of milk fat in lecithin reverse micelles. *J. Ferment. Bioeng.* 76, 98–104.

96. Fletcher, P.D.I., Freedman, R.B., Oldfield, C. 1985. Activity of lipase in water-in-oil microemulsions. *J. Chem. Soc. Faraday Trans. I* 81, 2667–2679.

97. Stark, M., Scagerlind, P., Holmberg, K., Carlfors, J. 1990. Depedence of the activity of *Rhizopus* lipase on microemulsion composition. *Colloid Polym. Sci.* 268, 384–388.

98. Valis, T.P., Xenakis, A., Kolisis, F.N. 1992. Comparative studies of lipase from *Rhizopus delemar* in various microemulsion systems. *Biocatalysis* 6, 267–279.

99. Yamada, Y., Kuboi, R., Komasawa, I. 1993. Increased activity of *Chromobacterium viscosum* lipase in aerosol OT reverse micelles in the presence of nonionic surfactants. *Biotechnol. Prog.* 9, 468–472.

100. Hayes, D.G., Kleiman, R. 1993. 1,3-Specific lipolysis of *Lesquerella fendleri* oil by immobilized and reverse-micellar encapsulated enzymes. *J. Am. Oil Chem. Soc.* 70, 1121–1127.

101. Miyake, Y., Owari, T., Matsuura, K., Teramoto, 1993. M. Enzymatic reaction in water-in-oil microemulsions. Part 1. Rate of hydrolysis of a hydrophilic substrate: Acetylsalicylic acid. *J. Chem. Soc. Faraday Trans.* 89, 1993–1999.

102. Talukder, M.M.R., Hayashi, Y., Takeyama, T., Zamam, M.M., Wu, J.C., Kawanishi, T., Shimizu, N. 2003. Activity and stability of *Chromobacterium viscosum* lipase in modified AOT reverse micelles. *J. Mol. Catal. B: Enzym.* 22, 203–209.

103. Fernandes, M.L.M., Krieger, N., Baron, A.M., Zamora, P.P., Ramos, L.P., Mitchell, D.A. 2004. Hydrolysis and synthesis reactions catalysed by *Thermomyces lanuginosa* lipase in the AOT/isooctane reversed micellar system. *J. Mol. Catal. B: Enzym.* 30, 43–49.

104. Shome, A., Roy, S., Das, P.K. 2007. Nonionic surfactants: A key to enhance the enzyme activity at cationic reverse micellar interface. *Langmuir* 23, 4130–4136.

105. Morgado, M.A.P., Cabral, J.M., Prazeres, D.M.F. 1996. Phospholipase A2-catalyzed hydrolysis of lecithin in a continuous reversed-micellar membrane bioreactor. *J. Am. Oil Chem. Soc.* 73, 337–346.

106. Prazeres, D.M., Garcia, F.A.P., Cabral, J.M.S. 1993. An ultrafiltration membrane bioreactor for the lipolysis of olive oil in reversed micellar media. *Biotechnol. Bioeng.* 41, 761–770.

107. Prazeres, D.M., Lemos, F., Garcia, F.A.P., Cabral, J.M.S. 1993. Modeling lipolysis in a reversed micellar system: Part II. Membrane reactor. *Biotechnol. Bioeng.* 42, 765–771.

108. Holmberg, K. 1994. Organic and bioorganic reactions in microemulsions. *Adv. Colloid Interf. Sci.* 51, 137–174.
109. Schmidli, P.K., Luisi, P.L. 1990. Lipase-catalyzed reactions in reverse micelles formed by soybean lecithin. *Biocatalysis* 3, 367–376.
110. Chang, P.S., Rhee, J.S., Kim, J.-J. 1991. Continuous glycerolysis of olive oil by *Chromobacterium viscosum* lipase immobilized on liposome in reversed micelles. *Biotechnol. Bioeng.* 38, 1159–1165.
111. Hayes, D.G., Gulari, E. 1991. 1-Monoglyceride production from lipase-catalyzed esterification of glycerol and fatty acid in reverse micelles. *Biotechnol. Bioeng.* 38, 507–517.
112. Hayes, D.G., Gulari, E. 1992. Formation of polyol-fatty acid esters by lipases in reverse micellar media. *Biotechnol. Bioeng.* 40, 110–118.
113. Ayyagari, M.S., John, V.T. 1995. Substrate-induced stability of the lipase from *Candida cylindracea* in reversed micelles. *Biotechnol. Lett.* 17, 177–182.
114. Holmberg, K., Osterberg, E. 1987. Enzymatic transesterification of a triglyceride in microemulsions. *Progr. Colloid Polym. Sci.* 74, 98–102.
115. Hedstrom, G., Backlund, M., Slotte, P.J. 1993. Enantioselective synthesis of ibuprofen esters in AOT/isooctane microemulsions by *Candida cylindracea* lipase. *Biotechnol. Bioeng.* 42, 618–624.
116. Bello, M., Thomas, D., Legoy, M.D. 1987. Interesterification and synthesis by *Candida cylindracea* lipase in microemulsions. *Biochim. Biophys. Res. Commun.* 146, 361–367.
117. Lu, Z., Chu, C., Han, Y., Wang, Y., Liu, J. 2005. Enzymatic esterification of DL-menthol with propionic acid by lipase from *Candida cylindracea*. *J. Chem. Technol. Biotechnol.* 80, 1365–1370.
118. Bozreix, F., Monot, F., Vandecasteele, J.P. 1992. Strategies for enzymatic esterification in organic solvents: Comparison of microaqueous, biphasic and micellar media. *Enzyme Microb. Technol.* 14, 791–797.
119. Crooks, G.E., Rees, G.D., Robinson, B.H., Svensson, M., Stephenson, G.R. 1995. Comparison of hydrolysis and esterification behavior of *Humicola lanuginosa* and *Rhizomucor miehei* lipases in AOT-stabilized water-in-oil microemulsions: I. Effect of pH and water content on reaction kinetics. *Biotechnol. Bioeng.* 48, 78–88.
120. Oliveira, A.C., Cabral, J.M.S. 1993. Kinetic studies of *Mucor miehei* lipase in phosphatidylcholine microemulsions. *J. Chem. Technol. Biotechnol.* 56, 247–252.
121. Avramiotis, S., Stamatis, H., Kolisis, F.N., Lianos, P., Xenakis, A. 1996. Structural studies of lecithin- and AOT-based water-in-oil microemulsions, in the presence of lipase. *Langmuir* 12, 6320–6328.
122. Stamatis, H., Macris, J., Kolisis, F.N. 1996. Esterification of hydrophilic diols catalysed by lipases in microemulsions. *Biotechnol. Lett.* 18, 541–546.
123. Baron, A., Sarquis, M.I.M., Baigori, M., Mitchell, D.A., Krieger, N. 2005. A comparative study of the synthesis of *n*-butyl-oleate using a crude lipolytic extract of *Penicillum coryophilum* in water-restricted environments. *J. Mol. Catal. B: Enzym.* 34, 25–32.
124. Backlund, S., Eriksson, F., Karlsson, S., Lundsten, G. 1995. Enzymatic esterification and phase behavior in ionic microemulsions with different alcohols. *Coloid Polym. Sci.* 273, 533–538.
125. Singh, C.P., Shah, D.O., Holmberg, K. 1994. Synthesis of mono- and diglycerides in water-in-oil microemulsions. *J. Am. Oil Chem. Soc.* 71, 583–587.
126. Singh, C.P., Skagerlind, P., Holmberg, K., Shah, D.O. 1994. A comparison between lipase-catalyzed esterification of oleic acid with glycerol in monolayer and microemulsion systems. *J. Am. Oil Chem. Soc.* 71, 1405–1409.

127. Stamatis, H., Xenakis, A., Kolisis, F.N. 1995. Studies on enzyme reuse and product recovery in lipase-catalyzed reactions in microemulsions. *Ann. NY Acad. Sci.* 750, 237–241.
128. Bornscheuer, U., Stamatis, H., Xenakis, A., Yamane, T., Kolisis, F.N. 1994. A comparison of different strategies for lipase-catalyzed synthesis of partial glycerides. *Biotechnol. Lett.* 16, 697–702.
129. Holmberg, K. 1989. Lipase catalyzed processes and reactions in microemulsions. *J. Surface Sci. Technol.* 5, 209–222.
130. Stamatis, H., Xenakis, A. 1999. Biocatalysis using microemulsion-based polymer gels containing lipase. *J. Mol. Catal. B: Enzym.* 6, 399–406.
131. Macris, J.B., Stamatis, H., Kolisis, F.N. 1996. Microemulsions as a tool for the regioselective lipase-catalysed esterification of aliphatic diols. *Appl. Microbiol. Biotechnol.* 46, 521–523.
132. Stamatis, H. 1996. Enzymatic conversion of amphiphilic substrates in microemulsions. PhD Thesis, University of Patras.
133. Nagayama, K., Matsura, S., Doi, T., Imai, M. 1998. Kinetic characterization of esterification catalyzed by *Rhizopus delemar* lipase in lecithin–AOT microemulsion systems. *J. Mol. Catal. B: Enzym.* 4, 25–32.
134. Martinek, K., Levashov, A.V., Klyachko, N.L., Berezin, I.V. 1977. Catalysis of water soluble enzymes in organic solvents. Stabilization of enzyme against denaturation (inactivation) when they are included in inversed micelles of surface active substance. *Dokl. Akad. Nauk. SSSR* 236, 920.
135. Mahiuddin, S., Renoncourt, A., Bauduin, P., Touraud, D., Kunz, W. 2005. Horseradish peroxidase activity in a reverse catanionic microemulsion. *Langmuir* 21, 5259–5262.
136. Bru, R., Sanchez-Ferrer, A., Garcia-Carmona. 1989. Characteristics of tyrosinase in AOT–isooctane reverse micelles. *Biotechnol. Bioeng.* 34, 304–308.
137. Rodakiewicz-Novak, J. 2000. Phenols oxidizing enzymes in water-restricted media. *Top. Catal.* 11/12, 419–434.
138. Gazaryan, I.G., Klyachko, N.L., Dulkis, Y.K., Ouporov, I.V., Levashov, A.V. 1997. Formation and properties of dimeric recombinant horseradish peroxidase in a system of reversed micelles. *Biochem. J.* 328, 643–647.
139. Rojo, M., Gomez, M., Estrada, P. 2001. Polyphenol oxidase in reverse micelles of AOT/cyclohexane: A thermostability study. *J. Chem. Technol. Biotechnol.* 76, 69–77.
140. Rodakiewicz-Novak, J., Pozdnyakova, N.N., Turkovskaya, O.V. 2005. Water-in-oil microemulsions as the reaction medium for the solvent sensitive yellow laccases. *Biocat. Biotransform. 2005* 23, 271–279.
141. Jurgas-Grudzinska, M., Gebicka, L. 2005. Influence of Ipegal reverse micellar and micellar systems on activity and stability of heme peroxidases. *Biocat. Biotransform.* 23, 293–298.
142. Papadimitriou, V., Sotiroudis, T.G., Xenakis, A. 2007. Olive oil microemulsions: Enzymatic activities and structural characteristics. *Langmuir* 23, 2071–2077.
143. Douzou, P., Keh, E., Balny, C. 1979. Cryoenzymology in aqueous media: Micellar solubilized water clusters. *Proc. Natl. Acad. Sci. USA* 76, 681–684.
144. Visser, A.J.W.G., Fendler, J.H. 1982. Reduction of reversed micelle entrapped cytochrome c and cytochrome c_3 by electrons generated by pulse radiolysis or by pyrene photoionization. *J. Phys. Chem.* 86, 947–950.
145. Hilhorst, R., Laane, C., Veeger, C. 1982. Photosensitized production of hydrogen by hydrogenase in reversed micelles. *Proc. Natl. Acad. Sci. USA* 79, 3927–3930.
146. Hilhorst, R., Laane, C., Veeger, C. 1983. Enzymatic conversion of apolar compounds in organic media using an NADH-regenerating system and dihydrogen as reductant. *FEBS Lett.* 159, 225–228.

147. Larsson, K.M., Adlercreutz, P., Mattiasson, B. 1987. Activity and stability of horse liver alcohol dehydrogenase in sodium dioctylsulfosuccinate/cyclohexane reverse micelles. *Eur. J. Biochem.* 166, 157–161.

148. Yang, F., Russell, A.J. 1994. Two-step biocatalytic conversion of an ester to an alde-hyde in reverse micelles. *Biotech. Bioeng.* 43, 232–241.

149. Bommarius, A.S., Hatton, T.A., Wang, D.I.C. 1995. Xanthine oxidase reactivity in reversed micellar systems: A contribution to the prediction of enzymatic activity in organized media. *J. Am. Chem. Soc.* 117, 4515–4523.

150. Orlich, B., Schomaecker, R. 1999. Enzymatic reduction of a less water soluble ketone in reverse micelles with NADH regeneration. *Biotech. Bioeng.* 65, 357–362.

151. Orlich, B., Berger, H., Lade, M., Schomacker, R. 2000. Stability and activity of alcohol dehydrogenases in w/o-microemulsions: Enantioselective reduction includ-ing cofactor regeneration. *Biotech. Bioeng.* 70, 638–646.

152. Das, S., Mozumdar, S., Maitra, A.J. 2000. Activity and conformation of yeast alco-hol dehydrogenase (YADH) entrapped in reverse micelles. *Colloid Interf. Sci.* 230, 328–333.

153. Andrade, S.L., Kamenskaya, O., Levashov, A.V., Moura, J.J.G. 1997. Encapsulation of flavodoxin in reverse micelles. *Biochem. Biophys. Res. Commun.* 234, 651–654.

154. Andrade, S., Moura, J.J.G. 2002. Hydrogen evolution and consumption in AOT-isooctane reverse micelles by *Desulfovibrio gigas* hydrogenase. *Enzyme Microb. Technol.* 31, 398–402.

155. Shinoda, K., Kaneko, T.J. 1988. Characteristic properties of lecithin as a surfactant. *J. Dispers. Sci. Technol.* 9, 555–559.

156. Shinoda, K., Araki, M., Sadaghiani, A., Khan, A., Lindman, B. 1991. Lecitihn-based microemulsions: Phase behavior and microstructure. *J. Phys. Chem.* 95, 989–993.

157. Kahlweit, M., Busse, G., Faulhaber, B. 1995. Preparing microemulsions with leci-thins. *Langmuir* 11, 1576–1583.

158. Kahlweit, M., Busse, G., Faulhaber, B., Eibl, H. 1995. Preparing nontoxic micro-emulsions. *Langmuir* 11, 4185–4187.

159. Avramiotis, S., Bekiari, V., Lianos, P., Xenakis, A. 1997. Structural and dynamic properties of lecithin–alcohol based (w/o) microemulsions. A luminescence quench-ing study. *J. Colloid Interface Sci.* 194, 326–331.

160. Wormuth, K.R., Kaler, E.W. 1987. Micromechanics of surfactant microstructures. *J. Phys. Chem., 1987* 91, 611–617.

161. Aboofazeli, R., Lawrence, C.B., Wicks, S.R., Lawrence, M. 1994. Investigations into the formation and characterization of phospholipid microemulsions. III. Pseudo-ternary phase diagrams of systems containing water-lecithin-isopropyl myristate and either an alkanoic acid, amine, alkanediol, polyethylene glycol alkyl ether or alcohol as cosurfactant. *Int. J. Pharm.* 111, 63–72.

162. Shinoda, K., Kunieda, H., Arai, T., Saijo, H. 1984. Principles of attaining very large solubilization (microemulsion): Inclusive understanding of the solubilization of oil and water in aqueous and hydrocarbon media. *J. Phys. Chem.* 88, 5126–5129.

163. Bachhav, Y.G., Date, A.A., Patravale, V.B. 2006. Exploring the potential of *N*-methyl pyrrolidone as a cosurfactant in the microemulsion systems. *Int. J. Pharm.* 326, 186–189.

164. Ayala, G.A., Kamat, S., Beckman, E.J., Russel, A.G. 1992. Protein extraction and activity in reverse micelles of a nonionic detergent. *Biotech. Bioeng.* 39, 806–814.

165. Constantinides, P.P., Scalart, J.P. 1997. Formulation and physical characterization of water-in-oil microemulsions containing long versus medium chain glycerides. *Int. J. Pharm.* 158, 57–68.

166. Nakamura, N., Yamaguchi, Y., Häkansson, B., Olsson, U., Tagawa, T., Kunieda, H. 1999. Formation of microemulsion and liquid crystal in biocompatible sucrose alkanoate systems. *J. Dispersion Sci. Technol.* 20, 535–557.

167. Garti, N., Clement, V., Fanun, M., Leser, M.E. 2000. Some characteristics of sugar ester non-ionic microemulsions in view of possible food applications. *J. Agric. Food Chem.* 48, 3945–3956.

168. Wang, Y., Guo, R., Guo, X. 2007. The self-organization properties of n-dodecylammonium α-glutamate/n-$C_5H_{11}OH$/water system. *Colloid Polym. Sci.* 285, 1423–1431.

169. Watnasirichaikul, S., Davies, N.M., Rades, T., Tucker, I.G. 2000. Preparation of biodegradable insulin nanocapsules from biocompatible microemulsions. *Pharmac. Res.* 17, 684–689.

170. Mele, S., Murgia, S., Caboi, F., Monduzzi, M. 2004. Biocompatible lipidic formulations: Phase behavior and microstructure. *Langmuir* 20, 5241–5246.

171. Formariz, T.P., Chiavacci, L.A., Sarmento, V.H.V., Santilli, C.V., Tabosa do Egito, E.S., Oliveira, A.G. 2007. Relationship between structural features and in vitro release of doxorubicin from biocompatible anionic microemulsion. *Colloids Surf. B: Biointerf.* 60, 28–35.

172. Ichikawa, S., Sugiura, S., Nakajima, M., Sano, Y., Minoru Seki, M., Furusaki, S. 2000. Formation of biocompatible reversed micellar systems using phospholipids. *Biochem. Engin. J.* 6, 193–199.

173. Sugiura, S., Ichikawa, S., Sano, Y., Nakajima, M., Liu, X.Q., Seki, M., Furusaki, S. 2001. Formation and characterization of reversed micelles composed of phospholipids and fatty acids. *J. Colloid Interface Sci.* 240, 566–572.

174. Malcolmson, C., Lawrence, M.J. 1995. Three-component non-ionic oil-in-water microemulsions using polyethylene ether surfactants. *Colloids Surf. B: Biointerfaces* 4, 97–109.

175. Warisnoicharoen, W., Lansley, A.B., Lawrence, M.J. 2000. Nonionic oil-in-water microemulsions: The effect of oil type on phase behaviour. *Int. J. Pharm.* 198, 7–27.

176. Podlogar, F., Gasperlin, M., Tomsic, M., Jamnik, A., Rogac, M.B. 2004. Structural characterization of water–Tween 40/Imwitor 308–isopropyl myristate microemulsions using different experimental methods. *Int. J. Pharm.* 276, 115–128.

177. Shevachman, M., Shani, A., Garti, N. 2004. Formation and investigation of microemulsiuons based on jojoba oil and nonionic surfactants. *J. Am. Oil Chem. Soc.* 81, 1143–1152.

178. Papadimitriou, V., Sotiroudis, T.G., Xenakis, A. 2005. Olive oil microemulsions as a biomimetic medium for enzymatic studies. Oxidation of oleuropein. *J. Am. Oil Chem. Soc.* 82, 1–6.

179. Spernath, A., Yaghmur, A., Aserin, A., Hoffman, R.E., Garti, N. 2002. Food-grade microemulsions based on nonionic emulsifiers: Media to enhance lycopene solubilization. *J. Agric. Food Chem.* 50, 6917–6922.

180. Spernath, A., Yaghmur, A., Aserin, A., Hoffman, R.E., Garti, N. 2003. Self-diffusion nuclear magnetic resonance microstructure transitions a solubilization capacity of phytosterols and cholesterol in Winsor IV food-grade microemulsions. *J. Agric. Food Chem.* 51, 2359–2364.

181. Papadimitriou, V., Pispas, S., Syriou, S., Pournara, A., Zoumpanioti, M., Sotiroudis, T.G., Xenakis, A. 2008. Biocompatible microemulsions based on limonene: Formulation, structure and applications. *Langmuir*, 24, 3380–3386.

182. Nascimento, M.G., Rezende, M.C., Vecchia, R.D. 1992. Enzyme-catalyzed esterifications in microemulsion-based organo gels. *Tetrahedron Let.* 33, 5891–5894.

183. Rees, G.D., Nascimento, M.G., Jenta, T.R.J., Robinson, B.H. 1991. Reverse enzyme synthesis in microemulsion-based organo-gels. *Biochim. Biophys. Acta* 1073, 493–501.

184. Rees, G.D., Robinson, B.H., Stephenson, G.R. 1995. Preparative-scale kinetic resolutions catalysed by microbial lipases immobilised in AOT-stabilised microemulsion-based organo-gels: Cryoenzymology as a tool for improving enantioselectivity. *Biochim. Biophys. Acta* 1259, 73–81.

185. Luisi, P.L., Laane, C. 1986. Solubilization of enzymes in apolar solvents via reverse micelles. *Trends Biotechnol.* 4, 153–161.

186. Serralheiro, M.L.M., Prazeres, D.M.F., Cabral, J.M.S. 1999. Continuous production and simultaneous precipitation of a dipeptide in a reversed micellar membrane reactor. *Enzyme Microb. Technol.* 24, 507–513.

187. Hakoda, M., Enomoto, A., Hoshino, T., Shiragami, N. 1996. Electroultrafiltration bioreactor for enzymatic reaction in reversed micelles. *J. Ferment. Bioeng.* 82, 361–365.

188. Larsson, K.M., Adlercreutz, P., Mattiasson, B. 1990. Enzymatic catalysis in microemulsions: Enzyme reuse and product recovery. *Biotechnol. Bioeng.* 36, 135–141.

189. Haering, G., Luisi, P.L. 1986. Hydrocarbon gels from water-in-oil microemulsions. *J. Phys. Chem.* 90, 5892–5895.

190. Quellet, C., Eicke, H.-F. 1986. Mutual gelation of gelatin and water-in-oil microemulsions. *Chimia* 40, 233–238.

191. Atkinson, P.J., Grimson, J., Heenan, R.K., Howe, A.M., Robinson, B.H. 1989. Structure of microemulsion-based organogels. *J. Chem. Soc. Chem Commun.* 23, 1807–1809.

192. Atkinson, P.J., Robinson, B.H., Howe, A.M., Heenan, R.K. 1991. Structure and stability of microemulsion-based organo-gels. *J. Chem. Soc. Faraday Trans.* 87, 3389–3397.

193. Jenta, T.R.-J., Batts, G., Rees, G.D., Robinson, B.H. 1997. Biocatalysis using gelatine microemulsion-based organogels containing immobilized *Chromobacterium viscosum* lipase. *Biotech. Bioeng.* 53, 121–131.

194. Lopez, F., Venditti, F., Giuseppe Cinelli, G., Ceglie, A. 2006. The novel hexadecyltrimethylammonium bromide (CTAB) based organogel as reactor for ester synthesis by entrapped *Candida rugosa* lipase. *Proc. Biochem.* 41, 114–119.

195. Fadnavis, N.W., Koteshwar, K. 1999. An unusual reversible sol–gel transition phenomenon in organogels and its application for enzyme immobilization in gelatin membranes. *Biotechnol. Prog.* 15, 98–104.

196. Schuleit, M., Luisi, P.L. 2001. Enzyme immobilization in silica-hardened organogels. *Biotechnol. Bioeng.* 72, 249–253.

197. Pastou, A., Stamatis, H., Xenakis, A. 2000. Microemulsion-based organogels containing lipase: Application in the synthesis of esters. *Prog. Colloid Polym. Sci.* 115, 192–195.

198. Delimitsou, C., Zoumpanioti, M., Xenakis, A., Stamatis, H. 2002. Activity and stability studies of *mucor miehei* lipase immobilized in novel microemulsion-based organogels. *Biocatal. Biotransf.* 20, 319–327.

14 Microemulsions as Decontamination Media for Chemical Weapons and Toxic Industrial Chemicals

Thomas Hellweg, Stefan Wellert,
Hans-Juergen Altmann, and André Richardt

CONTENTS

14.1   INTRODUCTION

After the "Cold War," the rise of the asymmetric warfare and the emergence of chemical terrorism (e.g., the Sarin attack in Japan) show the necessity for the development of new efficient decontamination media. The danger arising from the accidental release of toxic chemicals after the destruction of industrial plants also has to be considered. Hence, effective decontaminants are very much needed for military decontamination forces but also for civilian first responders and firefighters [1].

One major component in the decontaminant is the carrier system which has to solubilize the toxic compounds and the active component simultaneously. For this purpose macroscopic emulsions are often used at present [2]. However, these systems are only kinetically stabilized and due to the resulting limited lifetime they are only temporary applicable. Hence, in recent years several groups were working on the replacement of these carriers by microemulsions [3,4].

For a deeper understanding of the processes related to decontamination, it is necessary to consider the important physicochemical properties of the different toxics. Therefore, in Section 14.1.1, these properties will be presented and discussed followed by a brief review of the desired properties of a decontamination system and of the different active components used in such a process. Microemulsion-based decontaminants were already tested and are still developed at the moment [4]. Section 14.2 discusses the different strategies of decontamination and Section 14.3 gives an overview of the most important properties of microemulsions and the principles which govern their behavior. In Section 14.1.1 some examples for different microemulsion-based decontamination systems are discussed. Finally, a summary is given and the perspectives of microemulsions as carriers for decontamination applications are examined.

14.1.1   PROPERTIES OF CHEMICAL WARFARE AGENTS

Chemicals used as weapons in warfare are called chemical warfare (CW) agents. About 70 different chemicals have been developed and partly used or stockpiled as CW agents during the last 90 years. At normal conditions most of these agents are in the liquid state (Table 14.1), but some are in the gaseous or solid form. Figure 14.1 shows the Chemical structures of the most frequently discussed CW agents. Liquid agents are generally designed to be disseminated by spray or explosive force. Some of the substances have fairly high vapor pressures and therefore a low persistency and others have extremely low vapor pressures and very high persistencies [5].

Vapor pressure in combination with molecular weight is the physical property, which has a strong influence on the persistence and the volatility of the different CW agents. Agents classified as nonpersistent loose effectiveness already after a few minutes or hours. Persistent agents like VX are useful for defensive operations, which means blocking of areas and infrastructure over long periods of time.

Mustard is still a major problem for the decontamination systems since it penetrates varnished metal surfaces and is enriched at the metal/paint interface. This means it can hardly be removed, for example, from vehicle surfaces.

TABLE 14.1
Assortment of Physical Properties of CW Agents

CW Agent	Lost(HD)	Tabun (GA)	Sarin (GB)	Soman (GD)	VX
Density (g/cm³)	1.27	1.073	1.102	1.0222	1.008
Volatility (mg/m³)	610[+]	328[+]	16,091[+]	3900	10.5
Freezing point (K)	287.45	223	217	231	≥222
Saturation concentration (mg/m³)	625[+]	612	21,900	3,000[+]	$7 \times 10^{-4+}$
Vapor pressure (Pa)	14.0	1.8	289.3	53.3	0.03
Water solubility (%)	0.06[+]	9.8	Any ratio	2.1[+]	miscible at ≤9.4°C and at pH ≤6

Notes: Only properties of importance for the mentioned application were chosen and all values are taken from Refs. [6,7]. All values are given for 298 K except those denoted with symbol ([+]) are given for 293 K.

All what applies to CW agents is basically also true for toxic industrial compounds (TIC) like pesticides, etc. Organophosphorus pesticides are used extensively to control agricultural pests. Similary to nerve agents, they inhibit the enzyme acetylcholinesterase (AChE) [8]. Expositions mainly might occur in pesticide production plants and during their application in forestry, agriculture, and horticulture. Chemically similar compounds are used as flexibilizers or additives in lubrication solvents.

14.1.2   Requirements for Decontamination Media

Decontamination is the process of removal or neutralization of surface hazards resulting from a chemical attack or an accidental release of toxic chemicals. Enhanced and modern decontamination systems must be

- Noncorrosive
- Able to extract penetrating chemical agents and industrial chemicals out of sorptive surfaces
- Nontoxic and environmentally safe
- Suitable for a timely clean-up of chemical agents on all materials and surfaces
- Producible with off-the-shelf chemicals wherever possible
- Cost effective
- Storable in conventional warehouses

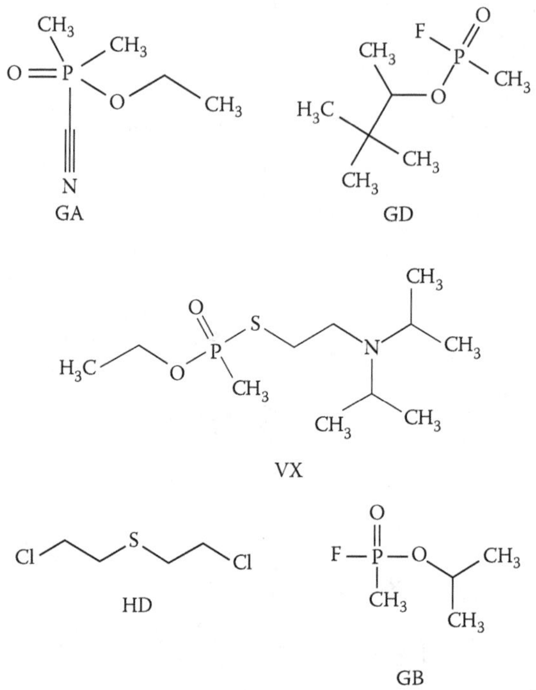

FIGURE 14.1 Chemical structure of the most frequently discussed CW agents VX, soman (GD), tabun (GA), sarin (GB), and lost or mustard (HD).

The requirements include decontaminants and decontamination procedures that both remove and neutralize these toxic chemicals as well as techniques that help prevent the spread of chemical contamination. Modern decontamination media, which match these requirements, will allow to reconstitute personnel and equipment rapidly [1].

14.2　DIFFERENT DECONTAMINATION PROCESSES

14.2.1　PHYSICAL DECONTAMINATION

Physical decontamination methods remove toxic chemicals from contaminated surfaces and reduce the level of contamination without a chemical deactivation of the toxic agent. Examples of physical decontaminants are hot air, pressurized hot or cold water, and weathering. These methods are suitable for the first rough removal of large and fast removal of large quantities of chemical contamination from surfaces. Hot air, for example, can be used for contaminants distributed over nonporous and nonabsorbing surfaces, but is much less efficient in case of surface penetrating CW agents and toxic chemicals with long persistence times [1].

Water is commonly used with the addition of detergents and soaps. The contamination on the surface to be decontaminated is mainly reduced due to the ability of aqueous surfactant solutions to solubilize the contaminants in micelles. Hydrolysis efficiency of this type of decontaminant is low due to slow hydrolysis reactions. Contaminated waste water must be collected during the decontamination process and must be treated for detoxification. Weathering is a passive form, which is mainly based on the natural sources to induce and perform evaporation of the contamination from large areas of terrain, large structures, or high numbers of vehicles, and usually is a very slow process.

Physical methods are not suitable for a rapid decontamination because they are time consuming and the process does not normally end with a detoxified agent. The application of physical methods can only prevent a further spreading of the contaminants and their penetration into nonsealed or chemically hardened surfaces.

14.2.2 CHEMICAL DECONTAMINATION

The most effective way to decontaminate contaminated surfaces is the degradation of CW agents and toxic industrial chemicals by reactive compounds. To do so, reactive chemicals and ideally extractively acting solvents have to be part of the decontamination media used.

Decontamination mechanisms used in this method are mainly oxidation, hydrolysis, chlorination, and combined forms, for example, alkaline catalyzed hydrolysis [6,9]. Typical oxidizing agents are hypochlorites, chloramine-B ($C_6H_5SO_2NCl_2$) and chloramine-T ($CH_3C_6H_4SO_4NCl_2$), sodium permanganate, hydrogen peroxide and peracetic acid, and several inorganic and organic peroxides. Hypochlorites in several forms and combinations are constantly available, and cheap and due to their effectiveness commonly used for the detoxification. The hypochlorite ion is effective against most of the CW agents [1]. Chloramin-B and other organic hypochlorite/chlorine generating compounds (e.g., trichloroisocyanuric acid and its sodium and potassium salts) can be used as milder decontaminants based on the oxidation and chlorination reaction to detoxify CW agents [6,9,10]. Figure 14.2 shows schematically the situation during a microemulsion based decontamination process.

14.3 MICROEMULSION PHASE BEHAVIOR AND PROPERTIES

14.3.1 PHASE BEHAVIOR

In the simplest form, a microemulsion consists of the two immiscible liquids oil and water, which can form stable mixtures in the presence of a surfactant. Well-known systems belonging to this group often are stabilized by a nonionic surfactant of the $C_i E_i$ type.

An instructive way to study the phase behavior is the variation of temperature and surfactant content at equal and constant fractions α of oil and water. Doing so, a so-called "Kahlweit-fish" diagram is obtained [11–15]. Figure 14.3 schematically shows the typical sequence of phases observed. This behavior was already verified for a large variety of microemulsion forming systems. Crossing this cut through the phase diagram either at a constant temperature or a constant surfactant content, different phase sequences are observed. Starting at a properly chosen

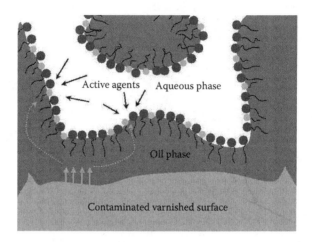

FIGURE 14.2 Simplified scheme of the microemulsion-based decontamination process on varnished surfaces. The contaminant is solubilized in the oil-continuous phase of a microemulsion wetting the contaminated surfaces. At the interface between oil and water, molecules of the toxic compound react with the active agent dissolved in the aqueous phase. Since the reaction takes place at the interfacial film, a large internal surface is expected to influence the decontamination process.

temperature, a two-phase region turns into the three-phase body which then with higher surfactant concentration γ changes into the bicontinuous single phase region starting at the "fish-tail" point [16]. At even higher surfactant content one ends up in the birefringent lamellar phase [17–19]. At a constant surfactant content the increase of temperature leads to different phase sequences, e.g., $\overline{2}$–3–$\underline{2}$ or $\overline{2}$–1–$\underline{2}$. Temperature changes are related to changes in the hydration of the ethoxylene headgroup of the surfactant molecules. This alters the headgroup area of the surfactant molecules resulting in changes of the packing parameter and therefore the curvature of the interfacial film (for more information see Ref. [20]

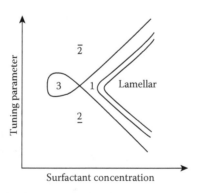

FIGURE 14.3 Schematic of phase behavior of a ternary phase system at constant oil-to-water plus oil ratio α as a function of surfactant concentration and temperature. The position and sequence of typical phase regions are indicated.

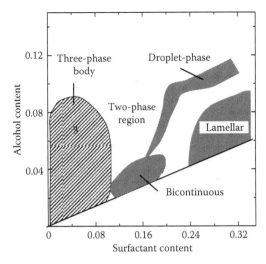

FIGURE 14.4 Kahlweit fish as obtained for the IHF–PCE-2-propanol–water. The technical surfactant can be used to form microemulsions, which exhibit only a rather small temperature dependence. Here, the phase behavior is controlled via the 2-propanol content. (Data from Wellert, S., Imhof, H., Altmann, H.J., Dolle, M., Richardt, A., and Hellweg, T., *Colloid Polym. Sci.*, 285, 2008.)

and references therein). The curvature of the interfacial film is one of the most important parameters controlling the free energy in microemulsion forming systems [21,22]. The phase behavior of ionic surfactant-based [23,27,28] and alkylpolyglucoside-based microemulsions [24] is less sensitive with respect to temperature variations.

In these systems the curvature of the film is modified by the addition of short or medium chain alcohols, which partially partition between oil and water phase [25] and incorporate in the interfacial surfactant film. It has also been shown that the addition of salt can alter the curvature of the surfactant film and therefore significantly influence the phase behavior of microemulsion systems [26,28]. Figure 14.4 shows a fish-like phase diagram, obtained for the Marlowet IHF–perchloroethylene2-propanol–water system. At present these components are employed in the preparation of a macroscopic emulsion applied by the German armed forces for decontamination of HD and VX on varnished surfaces.

Surprisingly, microemulsion systems based on technical grade components show the typical sequence of phases observed in many model systems [4]. This demonstrates the universality and robustness of the principles of microemulsion formation. Although derived from the studies of the phase behavior and the physical properties in microemulsion systems containing highly purified surfactants and hydrocarbons, the found systematics and empirical laws can be applied to systems based on technical grade components. Hence, the systematic study of the phase behavior is crucial with respect to practical applications.

14.3.2 IMPORTANT REQUIREMENTS FOR NEW MICROEMULSION-BASED DECONTAMINATION MEDIA

In abundant literature, a huge number of different microemulsion systems are presented. Some of them were tested for decontamination and reported to be suitable for the application as carrier systems for decontamination. Other microemulsion formulations were developed by titration experiments, where one component is gradually added while the portions of the other components remain constant. A microemulsion is found, when an isotropic and homogeneous one-phase sample is achieved. This method is rather straightforward and fast but it does not provide any further information concerning the phase behavior of the system. Although measuring phase diagrams is more time-consuming it reveals all details and properties concerning the phase structures of the respected system. For this reason it is necessary to carefully choose the components of the prospective microemulsion system.

14.3.2.1 Oil Component

The oil component of all microemulsions formulated for decontamination of CW agents serves as the reservoir for the toxic compound. Usually, it is assumed that hydrocarbons with good solubilization properties for large organic molecules are suitable for this application. Since a large solubilization capability is essential for a fast decontamination a high affinity between oil and contaminant is necessary and, hence, the choice of the oil component demands a lot of attention. In principle, the solubilization capacity of an organic solvent can experimentally be determined. Since in the case of CW agents such a measurement is remarkably a difficult and extensive procedure, a more simple and straightforward method is desirable to estimate the applicability of a solvent.

The semiempirical concept of solubility parameters is a physical approach that takes the interaction forces between the respected liquids into account [29–31]. The basic assumption is the existence of a correlation between the solubility and the cohesive energy density c, expressed as potential energy E per volume V. The cohesive energy is the energy cost, needed to transfer a molecule from the liquid phase into the vapor phase, assuming ideal gas behavior. This potential energy is directly correlated to the intermolecular forces, acting in the liquid phase. The cohesive energy density c is approximately equal to the square of the solubility parameter δ, so the solubility parameter is defined as

$$\delta = \sqrt{\frac{-E}{V}}$$

This is the so-called one-component solubility parameter or also Hildebrand solubility parameter [29,30]. Other definitions of solubility parameters extend this concept and include the various contributions to intermolecular interactions in multicomponent polar and nonpolar liquids. Expressed in the experimentally

TABLE 14.2
Hildebrand Parameters of the Most Important CW Agents and Some Selected Organic Solvents

Substance	δ (MPa$^{1/2}$)
HD	21.9
VX	18.0
GA	18.8
GD	17.2
GB	18.6
Water	47.4
Ethanol	26.0
Cyclohexane	16.8
PCE	19.0
n-Pentane	14.3
n-Hexane	14.9
n-Heptane	15.1

Sources: Barton, A.F.M., *Chem. Rev.*, 75, 731, 1975; Barton, A.F.M., in *CRC Handbook of Solubility Parameters and other Cohesion Parameters*, CRC Press, Boca Raton, 1991.

accessible heat of vaporization ΔH_{vap}, the gas constant R, temperature T at the boiling point, and the molar volume v_m, the definition takes the form

$$\delta = \sqrt{\frac{\Delta H_{vap} - RT}{v_m}}$$

Typically, the van der Waals interaction of nonpolar liquids like hydrocarbons is reflected by low values of δ (smaller than 20 MPa$^{1/2}$), while higher values are found for polar liquids. Due to hydrogen bonding, water has a very high value of 47 MPa$^{1/2}$. Two liquids are ideally miscible, if the difference of their solubility parameters is zero, which means the intermolecular forces in both liquids are equal [29,30]. Table 14.2 summarizes solubility parameters of the most relevant CW agents and typical polar and nonpolar liquids. These numbers reveal the low solubility of the CW agents in water and the need of a nonpolar solvent with appropriate solubility properties. Note that also the hydrocarbons commonly used for studies of the fundamentals of microemulsion systems fail with respect to this application.

Further, important but more practical aspects are environmental compatibility, health risk, and flammability. A nonpolluting and nonharmful decontamination medium is universally applicable to vehicles, exterior technical equipment, as

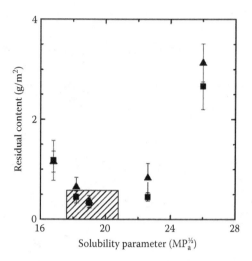

FIGURE 14.5 Residual content of HD on painted metal test sheets for solvents with different solubility parameters. The hatched range marks the region of typical solubility parameters of CW agents and gives the acceptable level of residual contamination. (Based on data from Wellert, S., Imhof, H., Altmann, H.J., Dolle, M., Richardt, A., and Hellweg, T., *Colloid Polym. Sci.*, 285, 2008.)

well as for indoor decontamination. Inflammability is also of great importance, since a decontamination medium might be stored in large containers and is usually sprayed over the contaminated surface and remains there for a time between a few minutes up to hours.

Figure 14.5 shows the result of decontamination of HD from two sorts of painted metal test sheets for several solvents. The residual content, reflecting the residual contamination after decontamination, is plotted as a function of the solubility parameter of the used solvents. For solvents with solubility parameters comparable to values of CW agents the lowest and tolerable residual contents were observed [4].

14.3.2.2 Surfactants

In general, new solubilization problems and important aspects motivate to focus on special surfactant properties, e.g., reactivity, redox-activity, photo-activity, biological functionality, and also environmental compatibility. This manifold of developments is reviewed in a series of articles, Refs. [24,32] provide examples. Currently, the most important, commonly used and intensively studied class of surfactants is the alkyl oligo ethyleneoxide group, abbreviated by C_iE_j. However, due to their temperature dependence with respect to micro-emulsion formation, these surfactants are not well suited for decontamination applications.

Decontamination media might undergo several temperature changes without time for relaxation of the phase structure during the decontamination process. For that reason, surfactants, forming temperature insensitive microemulsion systems, should be preferred. This can be achieved with ionic surfactants or alkyl polyglucosides, the so-called sugar surfactants abbreviated by C_mG_n. Here m means the number of carbons in the alkyl chain and n the mean number of sugar units per surfactant molecule. Here, the addition of a short or medium alkyl chain alcohol as cosurfactant is used to tune the curvature of the interfacial film.

14.3.2.3   Aqueous Phase

It can be deduced from the remarks concerning chemical decontamination (see Section 14.2) that the water phase will usually contain the active component since all relevant oxidizing agents are hydrophilic and at the appropriate pH value the water phase serves as a buffer medium for enzymes. Moreover, the aqueous phase also might contain glycols to prevent the system from freezing or gelators leading to longer contact with the surface.

14.4   MICROEMULSIONS AS DECONTAMINATION MEDIA

In order to replace established but disadvantageous and obsolete decontamination media, studies on microemulsions are of growing importance at present [4,33]. Since this complex topic was beyond the scope of fundamental research on microemulsions, only a few groups worked in this field [3,34–40,43]. Related to this, the number of publications devoted to this subject is limited but reveals basic findings, which are important for further investigations and improvements. These findings will be reviewed further.

Since 1980s, the use of a macroscopic emulsion was established for the decontamination of mustard and nerve agents by the German armed forces. This emulsion is composed of 15 wt % perchloroethylene (PCE), 76.5 wt % water, 7.5 wt % calcium hypochlorite, and 1 wt % emulsifier Marlowet IHF [2]. Although this medium is only a kinetically stable macroscopic emulsion, it shows the main features, which are also relevant for microemulsions with respect to solubilization of CW agents.

The common approach of all reported studies is the solubilization of the usually water insoluble toxic compound in the oil phase, while the water-soluble active agent is concentrated in the aqueous subphase and the chemical deactivation takes place after transfer through the internal interface (surfactant film). Hence, macroscopic emulsions and microemulsions both are suitable to overcome the problem of reagent incompatibility which is, for example, commonly known from hydrolysis of lipophilic organic compounds.

14.4.1   APPLICATIONS OF MICROEMULSIONS

The use of microemulsions as reaction media is an alternative to a two-phase system using a phase transfer catalyst [41]. The latter anyway is not applicable in

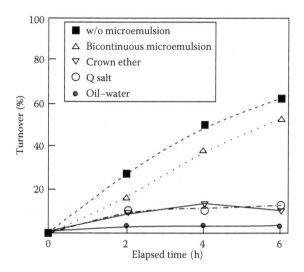

FIGURE 14.6 Kinetics of the synthesis of sodium decyl sulfonate from decyl bromide in different reaction media. (Redrawn from Holmberg, K., *Curr. Opin. Colloid Interface Sci.*, 6, 148, 2001.)

the decontamination of large surfaces of exterior and interior equipment. Furthermore, the reaction speed can be significantly enhanced in a microemulsion-based reaction medium [42]. Figure 14.6 compares the reaction profiles of the synthesis of sodium decyl sulfonate from decyl bromide and sodium sulfite in water-in-oil (w/o) microemulsion, a bicontinuous microemulsion, and a two-phase system with and without an additional phase transfer agent. The reaction yields of the microemulsions used are much higher than the yields achievable with a phase transfer catalyzed by two-phase system [42]. Besides the yield finally obtained, also the reaction speed itself is increased significantly. Both results are of great relevance with regard to the decontamination application.

Menger and Elrington found these principles to be valid too in the case of mustard detoxification [34]. Their oil-in-water (o/w)-type microemulsion was formed by 3 wt % cyclohexane, 82 wt % water, 5 wt % SDS, and 10 wt % butanol. The addition of 0.18 mL half-mustard, a nontoxic substance and chemically similar to mustard, and 5 wt % aqueous hypochlorite to 15 mL microemulsion initiates a fast reaction to sulfoxides, which was finished within 15 s. Figure 14.7 shows the proposed mechanism for the oxidation. An alkyl hypochlorite is assumed to be formed at the oil–water interface and the oxidation of half-mustard to sulfoxide only takes place close to the interfacial region or inside the oil phase. This interfacial reaction is orders of magnitude faster than a sulfide oxidation in a two-phase phase transfer catalysis system [41].

Hypochlorite is one of the cheapest oxidizing agents, but its hygroscopic nature results in a limited shelf life and the need to periodically refresh the hypochlorite stockpile. Furthermore, it is corrosive to painted and varnished surfaces

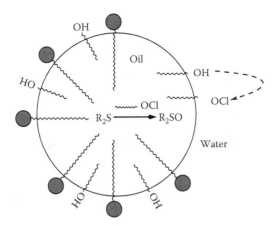

FIGURE 14.7 Scheme of the mechanism for the oxidation of half-mustard. Alkyl hypochlorite formation at the oil–water interface enables an oxidation close to the interface or inside the oil subphase. (From Menger, F. M. and Elrington, A. R., *J. Am. Chem. Soc.*, 112, 8201, 1990. With permission)

and environmentally questionable due to the formation of chlorinated hydrocarbons in aqueous media. Therefore, Wagner et al. developed a more environmental friendly and noncorrosive mildly basic oxidant system [43]. The surfactants SDS and Triton X-100 were used together with isopropanol and butanol as cosurfactants and hexane serves as an oil component in the microemulsion carrier system, which was applied by Wagner et al. [43]. The oxidizing system was formed by 50% H_2O_2 and K_2MoO_4 and oxidizes HD to the corresponding sulfoxides. A few milliliter of HD, GD, or VX was injected into a few milliliter of microemulsion and the reaction was monitored by ^1H NMR (HD) and ^{31}P (GD, VX) NMR. This system serves as a broad spectrum decontaminant but the HD reaction is orders of magnitude faster than the GD and VX detoxification.

Using the same system, the generation of singlet oxygen $^1O_2(^1\Delta_g)$ was applied to oxidize organic compounds. It is known that $^1O_2(^1\Delta_g)$ undergoes selective reactions with a series of electron-rich molecules like heterocycles or polycyclic aromatic hydrocarbons. Aubry and Bouttemy used the disproportionation of H_2O_2 catalyzed by MoO_4^{2-} as a mild chemical source of singlet oxygen [44]. Like the other chemical decontamination methods described here, this reaction proceeds efficiently only in water. A microemulsion, containing at least the seven components, water, surfactant (SDS), cosurfactant (butanol), organic solvent (CH_2Cl_2), substrate, catalyst (Na_2MoO_4), and oxidant H_2O_2, serves as a reservoir for the substrate and generates the singlet oxygen (Figure 14.8). Figure 14.9a and b shows two pseudoternary phase diagrams of this system. The ratio of the surfactant and cosurfactant in the binary mixture was kept constant, while the solution of the catalyst forms the aqueous phase in the system. Methylene chloride (CH_2Cl_2) is an excellent solvent for most organic compounds. The upper phase diagram (Figure 9a)

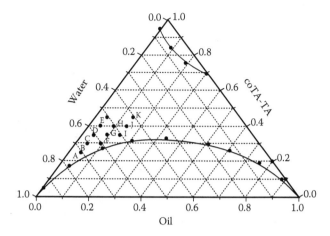

FIGURE 14.8 Pseudoternary phase diagram of a water–benzyl chloride–surfactant-plus-cosurfactant system for on-site decontamination. (Reproduced from Gonzaga, F., Perez, E., Rico-Lattes, I., and Lattes, A., *New J. Chem.*, 25, 151, 2001. With permission.)

was obtained for a weight ratio of butanol/SDS of 2.0 and the lower one (Figure 9b) for a ratio of 1.0 [44].

From these phase diagrams the influence of the catalyst concentration on the phase behavior can be observed. The increase of the molybdate concentration leads to a thinning of the single phase region, which results from changes of the surfactant film curvature due to electrostatic effects. The components were chosen to allow for an easy separation of the products simplifying the analysis of the oxidation products. Rotary evaporation of the microemulsion at 40°C removes methylene chloride, water, and butanol from molybdate, SDS, and oxidized product. The nonpolar oxidized product is selectively solubilized in methylene chloride. The surfactant and the catalyst can be recovered by simple filtration.

The CW agent solubilization capacity of microemulsions is limited due to the oil content of the microemulsion and the saturation concentration of CW agents in solvents. This restricts the application of this kind of microemulsions to the in-depth decontamination of surface regions. Gonzaga et al. [36] replaced the organic solvent by the CW agent, namely mustard analogues and half-mustard of increased viscosity due to the addition of polymers. For the decontamination of large quantities of stock-piled mustard, the continuous phase was water plus formamide in equal amounts, cetylpyridinium chloride as surfactant, and butanone as cosurfactant. Figure 14.10 shows the pseudoternary phase diagram for this system, where up to 30% of oilphase can be solubilized. This allows the neutralization of large quantities of toxins. The oxidizing agent magnesium monoperoxyphtalate (MMPP) was added to the microemulsion as an active component with a molar

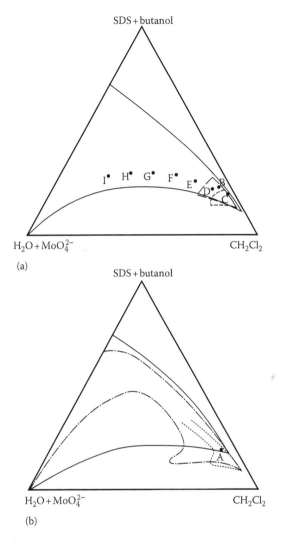

FIGURE 14.9 Phase diagram of the system (a) water–SDS + butanol–methylene chloride. The aqueous phase in (b) contains sodium molybdate in different concentrations. With increasing salt content from 0.2 up to 0.8 M, the single phase region becomes smaller. (Reprinted from Aubry, J.M. and Bouttemy, S., *J. Am. Chem. Soc.*, 119, 5286, 1997. With permission.)

ratio simulant-MMPP of 0.55. For the used mustard simulants, thioanisole, methyl sulfide, and 2-chloroethyl phenyl sulfide, the minimal conversion of simulant into nontoxic sulfoxide is about 98% and, for example, this oxidizing system shows a good chemical selectivity for the production of nontoxic sulfoxides at less than

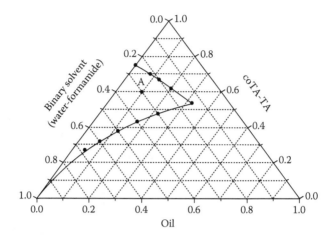

FIGURE 14.10 Pseudoternary phase diagram of a microemulsion system for decomposition of stock-piled mustard agent. The CW agent forms the oil phase, a solubilization in an appropriate organic solvent can be avoided. The aqueous component is a mixture of equal amounts of water and formamide and the surfactant component is a fixed mixture of the surfactant CPCl and the cosurfactant butanone. The diagram was measured using mustard simulants. (From Gonzaga, F., Perez, E., Rico-Lattes, I. and Lattes, A., *New J. Chem*, 25, 151–155, 2001–Reproduced with permission.)

6% of the toxic product sulfone was observed. By using benzyl as oil component, this formulation can be used to efficiently solubilize and detoxify polymer-thickened half-mustard in an on-site decontamination. The corresponding phase diagram is shown in Figure 14.10.

14.4.2 MICROEMULSIONS FOR LOW TEMPERATURE DECONTAMINATION

The volatility of CW agents at ambient conditions (see, e.g., vapor pressures given in Table 14.1) is strongly reduced with decreasing temperature. These physical conditions and the resulting long persistence time make the low temperature decontamination an extremely difficult task. A decontaminant usable in temperature ranges below the freezing point has to fulfill several additional requirements. The decontamination effectiveness has to equal the effectiveness achieved at temperatures above the freezing point and an easy handling has to allow for mixing and disseminating without the need of changing the decontamination equipment and process. In order to prevent freezing of the aqueous component of the decontamination medium, Menger and Rourk used an equally portioned mixture of propylene glycol and water as aqueous phase [45]. Such microemulsions of o/w type can resist freezing and phase separation down to −18°C. Figures 14.11 and 14.12 show HPLC chromatograms of a combination of

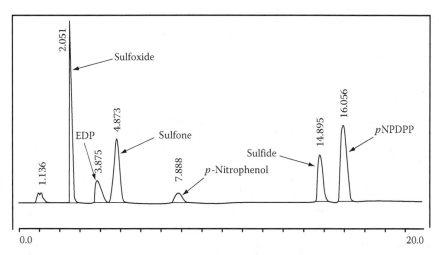

FIGURE 14.11   HPLC chromatogram of CW agent simulants (NPDPP, EDP, *p*NFES) in a microemulsion. The oxidation products sulfoxide, sulfone, and *p*-nitrophenol are added to demonstrate the peak identification. (Reprinted from Menger, F.M. and Rourk, M.J., *Langmuir*, 15, 309, 1999. With permission.)

CW agent simulants, *p*-nitrophenyl diphenyl phosphate (NPDPP), *S*-ethyl diphenyl phosphothioate (EDP), and 4-nitrophenethyl ethyl sulfide (NFES) in a microemulsion, formed by the glycol water mixture, hexanol, heptane, and $C_{12}E_4$ prior and after the reaction at 25°C. From monitoring the HPLC signal it was found that, for example, EDP a simulant for the nerve agent VX was completely hydrolyzed after 8 min at 25°C, while this time was increased up to 90 min at 0°C. This

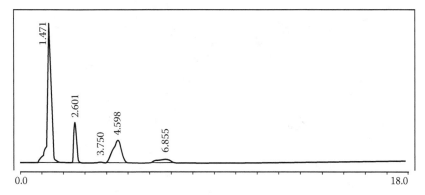

FIGURE 14.12   HPLC chromatogram after an oxidation at 25°C inside the microemulsion. The observed peaks belong to the reaction products. (Reprinted from Menger, F.M. and Rourk, M.J., *Langmuir*, 15, 309, 1999. With permission.)

demonstrates clearly the necessity for a decontamination medium that stays liquid with a low viscosity for longer times and has the capability to detoxify toxic compounds at low temperatures.

14.4.3 ENZYMATIC DECONTAMINATION

Several enzymes with activities against the G-type neurotoxic organophosphorus compounds are reported [1,38,46]. The enzyme quantities needed for a decontamination application can only be expressed for the enzymes organophosphorus hydrolase (OPH) from the bacterium *Pseudomonas diminuta*, organophosphorus acid anhydrase (OPAA) [47] from the bacterium *Alteromonas*, and diisopropyl fluorophosphatase (DFPase) from the head ganglion of the squid *Loligo vulgaris* [48]. For example, the DFPase detoxifies nerve agents by hydrolysis of the bond between phosphorus and the fluoride or cyanide leaving group. Due to its stability and the easy production in large quantities, DFPase is a promising candidate for enzymatic decontamination. The degradation of soman (GD) was studied in the presence of the enzyme in various microemulsions [49]. The aim of these investigations was the determination of the optimal parameters of microemulsions, which can be used as carrier media for decontaminating enzymes. Quantitative Fourier transform infrared (FTIR) spectroscopy was used to follow the decontamination process. Figure 14.13 shows the kinetics of the decomposition

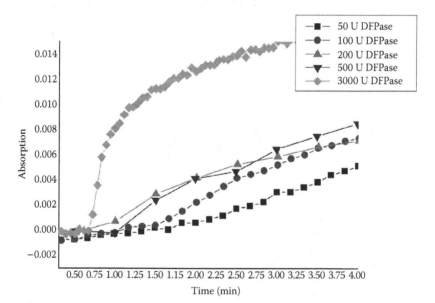

FIGURE 14.13 Enzymatic degradation kinetics of soman (GD) with different concentrations of the enzyme DFPase in o/w microemulsion. Note, instead of pure water a buffer system containing 100 mM $(NH_4)HCO_3$ and 2 mM $CaCl_2$ was used to ensure a pH value of 9.

of 1 vol % GD in 20 mL of o/w microemulsion with different concentrations of the enzyme DFPase. The reaction kinetics were monitored by following the absorbance of the P=O stretching band of the G-agents and their resulting acids produced by the hydrolysis reaction. Based on the environment friendly chemicals and designed for the physical removal (solubilization) of different CW agents as the velocity limiting step in the whole decontamination process, several investigated microemulsions showed promising results as a potential carrier medium for enzymes.

These results also indicate that the reaction rate of the hydrolysis was reduced to 20% compared with the activity of the enzyme in an aqueous solution. However, this still is high enough to justify the intensification of the efforts of an optimization of the decontamination media [49].

14.4.4 NANOPARTICLE-BASED DECONTAMINATION

Catalytic nanoparticles offer a very high surface-to-volume ratio, which is responsible for a variety of unique physical and chemical properties. This makes them interesting for many applications ranging from optical technologies to pharmaceutical formulations [50]. Their surface-to-volume ratio is correlated with a high chemical reactivity. Hence, they are promising with respect to the catalytic decomposition of CW agents. The degradation of toxic chemicals by zero-valent metal nanoparticles (ZVMs) was already investigated and shows some promising potential [51]. Also, the potential of titanate nanoscrolls, a variant of TiO_2 nanocrystals, was employed as a possible decontaminant for CW agents [37]. Furthermore, nanosized CaO (AP-CaO) was found to hydrolyze VX and GD to surface-bound complexes of nontoxic ethyl methylphosphonate and pinacolyl methylphosphonate observed in a solid-state MAS NMR study [52]. Similar behavior was observed for HD on dried or hydrated CaO. The reaction profiles for the detoxification of VX, GD, and HD by hydrolysis on the surface of highly reactive, nanosized MgO showed a dependence of the steady-state half-life on the vapor pressure of the CW agent [53]. For example, for the hydrolysis of GD at room temperature a $t_{1/2} = 28$ min was found. In contrast to this, in the case of VX, $t_{1/2} = 68$ h was observed.

Since the surface of such particles can erode as a result of the catalysis, the bulk material of the particles can participate in the reaction process. This interesting effect leads to a high reaction capacity of such nanoparticles [50]. It is known that metal nanoparticles can be in-situ synthesized in microemulsions [54–59]. However, microemulsions with nanoparticles for decontamination applications still remain to be studied.

14.5 SUMMARY

The decontamination of toxic industrial agents and CW agents is a rather difficult task since most of these compounds are lipophilic, while the reasonable decontaminating agents are hydrophilic. Only kinetically stable emulsion systems or thermodynamically stable microemulsions can be used to overcome this difficulty.

Additionally, a fast and maximal degradation of a large spectrum of toxic chemicals is desired. Moreover, decontamination systems should leave the treated surfaces intact and should not harm the environment.

First results show clearly that a microemulsion designed especially for the extraction of entrapped toxic compounds in surfaces can also be used as a nano-reactor for the decomposition of these chemicals by oxidizing agents and enzymes. Different degradation methods were already demonstrated to work in model microemulsion systems and can probably be applied also in practically important microemulsion systems. Here, technical grade components suitable for extraction and solubilization have to be used instead of purified surfactants and oils.

It was already shown that this is possible and the understanding of the behavior of such technical grade components benefits from the great knowledge of the fundamental general principles of the phase behavior in amphiphilic systems obtained in numerous studies. Hence, for the future new decontaminants media as a hard surface cleaner based on microemulsion formulation can be expected for both military decontamination forces and civilian first responders.

REFERENCES

1. Blum, M.-M. and Richardt, A. 2008. *Decontamination of Warfare Agents: Enzymatic Methods for the Removal of B/C Weapons*. Weinheim, Wiley VCH.
2. Altmann, H.-J. 1989. German patent DE 3638625 c2. Technical report, Munich, Deutsches Patentamt.
3. Yang, Y.-C., Baker, J. A., and Ward, J. R. 1992. Decontamination of chemical warfare agents. *Chem Rev*, **92**, 1729–1743.
4. Wellert, S., Imhof, H., Altmann, H. J., Dolle, M., Richardt, A., and Hellweg, T. 2008. Decontamination of chemical warfare agents using perchloroethylene/marlowet IHF/H$_2$O based microemulsions: Wetting and extraction properties on realistic surfaces. *Colloid Polym Sci*, **286**, 417–426.
5. Love, A. H., Vance, A. L., Reynolds, J. G., and Davisson, M. L. 2004. Investigating the affinities and persistence of VX nerve agent in environmental matrices. *Chemosphere*, **57**, 1257–1264.
6. US Army. 1990. *Field Maunal 3–9: Potential Military Chemical/Biological Agents and Compounds*. New York, Wiley.
7. R. E. Kirk and Othmer D. F. 1978. *Encyclopedia of Chemical Technology*. New York, Wiley.
8. Mileson, B. E., Chambers, J. E., Chen, W. L., Dettbarn, W., Enrich, M., Eldefrawi, A. T., Gaylor, D. W., Hamernik, K., Hodgson, E., Karczmar, A. G., Padilla, S., Pope, C. N., Richardson, R. J., Saunders, D. R., Sheets, L. P., Sultatos, L. G., and Wallace, K. B. 1998. Common mechanism of toxicity: A case study of organophosphorus pesticides. *Toxicol Sci*, **41**, 8–20.
9. Franke, S. 1977. *Lehrbuch der Militärchemie*. 2nd edition, Militärverlag der DDR, Berlin.
10. US Department of Justice. 2001. *Guide for the Selection of Chemical and Biological Equipment for Emergency First Responders*. NIJ-guide, Vol. I.
11. Kahlweit, M. and Strey, R. 1985. Phasenverhalten ternärer systeme des typs H$_2$O-Öl-nichtionisches amphiphil (mikroemulsionen). *Angew Chem*, **97**, 655–669.
12. Strey, R. 1994. Microemulsion microstructure and interfacial curvature. *Colloid Polym Sci*, **272**, 1005–1019.

13. Strey, R. 1996. Phase behavior and interfacial curvature in water-oil-surfactant systems. *Curr Opin Colloid Interface Sci*, **1**, 402–410.

14. Burauer, S., Sachert, T., Sottmann, T., and Strey, R. 1999. On microemulsion phase behavior and monomeric solubility of surfactant. *PCCP*, **1**, 4299–4306.

15. Hellweg, T. 2002. Phase structures of microemulsions. *Curr Opin Colloid Interface Sci*, **7**, 50–56.

16. Jahn, W. and Strey, R. 1988. Microstructure of microemulsions by freeze-fracture electron-microscopy. *J Phys Chem*, **92**, 2294–2301.

17. Olsson, U., Würz, U., and Strey, R. 1993. Cylinders and bilayers in a ternary nonionic surfactant system. *J Phys Chem*, **97**, 4535–4539.

18. Farago, B., Monkenbusch, M., Göcking, K. D., Richter, D., and Huang, J. S. 1995. Dynamics of microemulsions as seen by neutron spin echo. *Physica B*, **213/214**, 712–717.

19. Nagao, M., Seto, H., Okuhara, D., Okabayashi, H., Takeda, T., and Hikosaka, M. 1998. A small-angle neutron-scattering study of the effect of pressure on structures in a ternary microemulsion system. *Physica B*, **241–243**, 970–972.

20. Langevin, D. 1992. Micelles and microemulsions. *Annu Rev Phys Chem*, **43**, 341–369.

21. Helfrich, W. 1973. Elastic properties of lipid bilayers: Theory and possible experiments. *Z Naturforschung*, **28c**, 693–703.

22. de Gennes, P. G. and C. Taupin, C. 1982. Microemulsions and the flexibility of oil/water interfaces. *J Phys Chem*, **86**, 2294–2304.

23. Mehta, S. K. and Kawaljit. 2002. Phase diagram and physical properties of a waterless sodium bis(2-ethylhexylsulfosuccinate)-ethylbenzene-ethyleneglycol microemulsion: An insight into percolation. *Phys Rev E*, **65**(2), 1502–1511.

24. Stubenrauch, C. 2001. Sugar surfactants—aggregation, interfacial, and adsorption phenomena. *Curr Opin Colloid Interface Sci*, **6**, 160–170.

25. Penders, M. and Strey, R. 1995. Phase behavior of the quaternary system H₂O/
n-octane/C₈E₅/n-octanol: Role of the alcohol in microemulsions. *J Phys Chem*, **99**, 10313–10318.

26. Kahlweit, M. and Strey, R. 1988. Phase behavior of quinary mixtures of the type H₂O-oil-nonionic amphiphile-ionic amphiphile-salt. *J Phys Chem*, **92**, 1557–1563.

27. Eastoe, J., Sharpe, D., Heenan, R. K., and Egelhaaf, S. 1997. Rigidities of cationic surfactant films in microemulsions. *J Phys Chem B*, **101**, 944–948.

28. Olsson, U., Ström, P., Söderman, O., and Wennerström, H. 1989. Phase behavior, self-diffusion, and ²H NMR relaxation studies in an ionic surfactant system containing cosurfactant and salt. A comparison with nonionic surfactant systems. *J Phys Chem*, **93**, 4572–4580.

29. Barton, A. F. M. 1975. Solubility parameters. *Chem Rev*, **75**, 731–753.

30. Barton, A. F. M. 1991. *CRC Handbook of Solubility Parameters and Other Cohesion Parameters*. 2nd edition, Boca Raton, CRC Press.

31. Charlesworth, J. M., Riddell, S. Z., and Mathews, R. J. 1993. Determination of polymer-solvent interaction parameters using piezoelectric crystals, with reference to the sorption of chemical warfare agents. *J Appl Polymer Sci*, **47**, 653–665.

32. Holmberg, K. 2001. Natural surfactants. *Curr Opin Colloid Interface Sci*, **6**, 148–159.

33. Wellert, S., Tiersch, B., Koetz, J., Imhof, H., Altmann, H.-J., Dolle, M., Richardt, A., and Hellweg, T. 2008. *J Colloid Interf Sci*, **325**, 250–258.

34. Menger, F. M. and Elrington, A. R. 1990. Rapid deactivation of mustard via microemulsion technology. *J Am Chem Soc*, **112**, 8201–8203.

35. Zhang, S. and Busling, J. F. 1995. Dechlorination of polychlorinated biphenyls on soil and clay by electrolysis in a bicontinuous microemulsion. *Environ Sci Technol*, **29**, 1185–1189.

36. Gonzaga, F., Perez, E., Rico-Lattes, I., and Lattes, A. 2001. New microemulsions for oxidative decontamination of mustard gas analogues and polymer-thickened half-mustard. *New J Chem*, **25**, 151–155.

37. Kleinhammes, A., Wagner, G. W., Kulkarni, H., Jia, Y. Y., Zhang, Q., Qin, L. C., and Wu, Y. 2005. Decontamination of 2-chloroethyl ethylsulfide using titanate nanoscrolls. *Chem Phys Lett*, **411**, 81–85.

38. Richardt, A., Blum, M.-M., and Mitchell, S. 2006. Enzymatische Dekonatmination von Nervenkampfstoffen—Was wissen Calamari über Sarin? *Chem in Unserer Zeit*, **40**, 252–259.

39. Tafesse, F. and Mndubu, Y. 2007. Iron promoted decontamination studies of nitrophenylphosphate in aqueous and microemulsion media: A model for phosphate ester decontamination. *Water Air Soil Pollut*, **183**, 107–113.

40. Talmage, S. S., Watson, A. P., Hauschild, V., Munro, N. B., and King, J. 2007. Chemical warfare agent degradation and decontamination. *Curr Org Chem*, **11**(3), 285–298.

41. Ramsden, J. H., Drago, R. S., and Riley, R. 1989. A kinetic study of sulfide oxidation by sodium hypochlorite using phase-transfer catalysis. *J Am Chem Soc*, **151**, 2958–2961.

42. Haeger, M., Currie, F., and Holmberg, K. 2003. *Colloid Chemistry II*, Vol. 227 of *Topics in current chemistry*, Chapter: Organic Reactions in Microemulsions, pp. 53–74. Berlin, Springer Verlag.

43. Wagner, G. W., Procell, L. R., Yang, Y.-C., and Bunton, C. A. 2001. Molybdate/peroxide oxidation of mustard in microemulsions. *Langmuir*, **17**, 4809–4811.

44. Aubry, J. M. and S. Bouttemy, S. 1997. Preparative oxidation of organic compounds in microemulsions with singlet oxygen generated chemically by the sodium molybdate/hydrogen peroxide system. *J Am Chem Soc*, **119**, 5286–5294.

45. Menger, F. M. and Rourk, M. J. 1999. Deactivation of mustard and nerve agent models via low-temperature microemulsions. *Langmuir*, **15**, 309–313.

46. Richardt, A. and Mitchell, S. 2006. Enzymes for environmentally friendly decontamination of sensitive equipment. *J. Defence Sci*, **10**, 261–265.

47. Raushel, F. M. 2002. Bacterial detoxification of organophosphate nerve agents. *Curr Opin Microbiol*, **5**, 288–295.

48. Blum, M.-M., Löhr, F., Richardt, A., Rüterjans, H., and Chen, J. C.-H. 2006. Binding of a designed substrate analogue to diisopropyl fluorophosphatase: Implications for the phosphotriesterase mechanism. *J Am Chem Soc*, **128**, 12750–12757.

49. Hellweg, T., Wellert, S., and Mitchell, S. 2008., Microemulsions as carriers for decontamination agents in *Decontamination of Warfare Agents: Enzymatic Methods for the Removal of B/C Weapons*, Blum, Chapter 12, pp. 223–242, M.-M. and Richardt, A. (eds.). Weinheim, Wiley VCH.

50. Schwarz, J. A., Contescu, C., and Putyera, K. 2004. *Dekker Encyclopedia of Nanosience and Nanotechnology*. 1st edition, Boca Raton, CRC Press.

51. Mackenzie, K., Hildebrand, H., and Kopinke, F. D. 2007. Nano-catalysts and colloidal suspensions of carbo-iron for environmental application. *NSTI-Nanotech*, **2**, 639–642.

52. Wagner, G. W., Koper, O. B., Lucas, E., Decker, S., and Klabunde, K. J. 2000. Reactions of VX, GD, and HD with nanosize CaO: Autocatalytic dehydrohalogenation of HD. *J Phys Chem B*, **104**, 5118–5123.

53. Wagner, G. W., Bartram, P. W., Koper, O., and Klabunde, K. J. 1999. Reactions of VX, GD, and HD with nanosize MgO. *J Phys Chem B*, **103**, 3225–3228.

54. Osseo-Asare, K. and Arriagada, F. J. 1990. Preparation of SiO_2 nanoparticles in a non-ionic reverse micellar system. *Colloid Surf*, **50**, 321–339.

55. Petit, C., Lixon, P., and Pileni, M. 1993. In situ synthesis of silver nanocluster in aot reverse micelles. *J Phys Chem*, **97**, 12974–12983.
56. Karayigitoglu, C. F., Tata, M., John, V. T., and McPherson, G. L. 1994. Modification of *CdS* nanoparticle characteristics through synthesis in reversed micelles and exposure to enhanced gas pressures and reduced temperatures. *Colloid Surf A: Physicochem Eng Aspects*, **82**, 151–162.
57. Lee, M.-H., Oh, S.-G., and Yi, S.-C. 2000. Preparation of Eu-doped Y_2O_3 luminescent nanoparticles in nonionic reverse microemulsions. *J Colloid Interface Sci*, **226**(1), 65–70.
58. Panda, A. K., Bhowmik, B. B., Das, A. R., and Moulik, S. P. 2001. Dispersed molecular aggregates. 3. Synthesis and characterization of colloidal lead chromate in water/sodium bis(2-ethylhexyl)sulfosuccinate/*n*-hepatane water-in-oil microemulsion medium. *Langmuir*, **17**, 1811–1816.
59. Wu, M.-L., Dong, D.-H., and Huang, T.-C. 2001. Preparation of Pd/Pt bimetallic nanoparticles in water/aot/isooctane microemulsions. *J Colloid Interface Sci*, **243**(1), 102–108.

15 Microemulsions as Potential Interfacial Chemical Systems Applied in the Petroleum Industry

Afonso Avelino Dantas Neto, Tereza Neuma de Castro Dantas, Maria Carlenise Paiva de Alencar Moura, Eduardo Lins de Barros Neto, and Alexandre Gurgel

CONTENTS

15.1   INTRODUCTION: AN OVERVIEW OF MICROEMULSIONS

Since the first scientific description in 1943, by Hoar and Schulman [1], on typically transparent oil–water chemical systems, innumerous academic and technological investigations on microemulsions have been reported and research activities in this area are continuously growing. The term "microemulsion" itself, however, was suggested by Schulman and coworkers only in 1959 [2], and has been used ever since to refer to dispersed, macroscopically homogeneous, thermodynamically stable, optically transparent, single-phase systems, formed by the spontaneous solubilization of two otherwise immiscible liquids, in the presence of surfactants. In some cases, short-chain alcohols or amines are added to the mixture as cosurfactants, which act by reducing interfacial tensions to very low levels, promoting distinct interactions at the interface and improving the fluidity of the interfacial film. The microemulsion can then be formed by the intimate dispersion of one liquid in the other, as droplets.

At the molecular level, microemulsions are heterogeneous systems, in that they comprise water and oil regions with a common interface onto which the surfactant molecules are adsorbed. As a result of interfacial adsorption phenomena, microemulsions have particular properties that allow their use in many commercial and industrial applications. Their high stability, low interfacial tension at low surfactant concentration, ability to stabilize large amounts of two immiscible liquids in a single macroscopically homogeneous phase, and large interfacial area between the microheterogeneous phases are fundamental properties that are exploited in the design of such applications.

More basic structural models for microemulsions involve the dispersion of approximately spherical droplets with diameter ranging between 10 and 100 nm, featuring a monolayer of surfactant molecules at the interface. However, this is highly dependent on their composition and structure of the surfactant molecules that stabilize them.

The potential of microemulsions in technological applications, however, has not been fully exploited. There are many established industrial activities effectively being carried out with emulsions, for instance polymerization reactions in micellar, emulsion, and miniemulsion environments. Technological aspects involved in this type of activity have been discussed in a series of review articles published by Capek [3–6], but the role of microemulsions has yet been restricted to academical investigations, of which works reported by Kaler et al. are a good example [7–10].

It is important to consider the main differences between emulsions and microemulsions to better design their applications. This is briefly shown in Table 15.1.

One interesting aspect featured by microemulsions and listed in Table 15.1 is the fact that they are highly dynamic systems. By that, not only does one refer to

TABLE 15.1

Main Structural Differences between Emulsions and Microemulsions

Emulsions	Microemulsions
Unstable, with eventual phase separation	Thermodynamically stable
Relatively large-sized droplets (1–10 μm)	Small aggregates (around a few tens of nanometers)
Relatively static systems	Highly dynamic systems
Moderately large interfacial area	Very high interfacial area
Small amount of surfactant required for stabilization	Large amount of surfactant required for stabilization
Low curvature of the water–oil interface	Interfacial film may be highly curved

the constant chemical exchange of surfactant molecules from the bulk microphases to the interfaces and vice versa, but also their ability to solubilize compounds with opposite polarity and the high values of mass diffusivity constants attained in such environment. It is thus evident that technological processes where mass transfer phenomena are the main steps may be substantially enhanced with the use of such systems.

When dissolved at sufficiently high concentrations, surfactant aggregates tend to assume specific configurations that are directly related to the shape and geometry of the individual molecules. This is complementary to the concept of hydrophilic–lipophilic balance (HLB), useful to provide a preliminary indication regarding the final use of the surfactants. However, the way by which surfactant structure controls the orientation of the molecules at interfaces is more quantitatively expressed by the packing parameter (P_p) [11,12]. It is defined by Equation 15.1, where V is the volume of the hydrocarbon chain, a_0 is the effective head-group area, and L is the effective hydrocarbon chain length, chiefly between 80% and 90% of the fully extended hydrocarbon chain [13].

$$P_p = \frac{V}{a_0 L} = \frac{S_h}{a_0} \tag{15.1}$$

In essence, the packing parameter is a measure of the ratio between the effective areas occupied by the hydrophobic (S_h) and hydrophilic (a_0) parts of the surfactant. This model is ideal when considering aqueous systems, but may also be applied to dispersions in oil. Depending on the surfactant structure, P_p assumes specific values and the packing constraints in the medium allow for the formation of a preferred aggregate shape configuration. This theoretical approach is also considered in microemulsion systems, as discussed below.

The stability of the microemulsion is determined by how intermolecular cohesive interactions affect the preferred curvature of the interfacial film. The energies of interaction between the different species present in the medium (surfactant, oil, and water molecules, etc.) may be taken by unit interfacial area. Equilibrium is therefore established with regard to the degree of solvation of the surfactant molecule with oil (A_{so}) compared with the degree of interaction with water (A_{sw}). Winsor [14] proposed that the ratio A_{so}/A_{sw} as a good indicator of what to expect, by introducing the "R-theory" of interfacial curvature, which is a more elaborate representation of the HLB concept applied to microemulsions. If the simplest case of a microemulsion formed by water, oil, and surfactant is to be considered, different energy interactions (A_{ij}) arise between its components (i, j), as schematically depicted in Figure 15.1, where the indices stand for (o) oil, (w) water, (t) surfactant tail-group, and (h) surfactant head-group.

A dimensionless parameter, denoted as R and given by Equation 15.2, was thus suggested, to account for the curvature of the interfacial film and, ultimately, the stability of the microemulsion system. The interface, therefore, bends according to the interaction energy established between the water and oil phases, thus defining its preferred curvature. It is worth reminding that, at times, interactions across the interface (A_{tw} and A_{ho}) are negligible compared to the ones developed in the bulk phases (see Figure 15.1).

$$R = \frac{A_{so}}{A_{sw}} = \frac{A_{to} + A_{ho} - A_{tt} - A_{oo}}{A_{tw} + A_{hw} - A_{hh} - A_{ww}} \cong \frac{A_{to} - A_{tt} - A_{oo}}{A_{hw} - A_{hh} - A_{ww}} \qquad (15.2)$$

The cohesive energy balance is established from both sides of the interface. When solvation forces on the water side are stronger than in the oil side, the surfactant film is curved towards the oil phase, assuming a configuration conventionally denominated as having a *positive curvature*. In this case, $R < 1$ and oil-in-water microemulsions are formed, either as a single phase or in equilibrium with excess oil (commonly referred to as a Winsor I system). The opposite effect occurs when the solvation forces are stronger in the oil side. In this case, the film is curved towards the water phase ($R > 1$, *negative curvature*) and water-in-oil microemulsions are

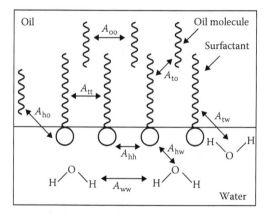

FIGURE 15.1 Interaction energies in the amphiphilic membrane at the water–oil interface of a microemulsion.

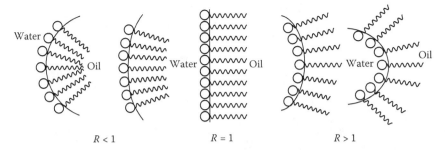

FIGURE 15.2 Evolution of interfacial curvature of microemulsion systems according to parameter R.

favored, either as a single microemulsion phase or in equilibrium with excess water (Winsor II system). Intermediary situations correspond to $R = 1$ and the interface assumes a planar configuration. Winsor III systems (microemulsion with excess water and oil phases) tend to be formed in this case. This is pictorially shown in Figure 15.2.

Winsor's R-theory is important when examining changes in microemulsified systems that cause a variation in the radius (denoted as r_d) of the dispersed droplets. This is particularly useful when studying two-phase systems (Winsor I and II). The droplets are formed according to their preferred curvature, and their sizes tend to decrease as the value of R varies, in both directions, from the point where it equals one. This is important when examining the process mechanisms in any technological application for these systems. For example, changes in particle size and shape may affect the extent of a reaction or alter the partitioning coefficient of a given solute in the medium (important in separation science and technology, which involve mass transfer).

The reader is encouraged to consult the vast literature available on surfactant aggregation and microemulsion formation for further details. In particular, a few more representative works are cited in the references section, which may help guide this study [13,15–20].

15.2 RECENT ADVANCES IN THE USE OF MICROEMULSION SYSTEMS IN THE PETROLEUM INDUSTRY

The focus of this article is based on experimental results obtained and reported by members of our research group in recent years, on the use of self-assembled chemical systems, particularly microemulsions, in some areas involved in modern petroleum industry [21–39]. After initial presentation on some historical aspects that led to the development of this work, we aim to cover the following main subject areas:

• Removal of humidity from natural gas
• Inhibition of corrosion on metallic surfaces

- Microemulsion formulations for cycle-diesel engines
- Use of microemulsions in enhanced oil recovery

The results are discussed in the light of the typical properties and characteristics of self-assembled or microemulsified systems, emphasizing its structural and dynamic aspects.

15.2.1 EARLY DEVELOPMENTS

Engineers have always faced the question of process optimization in the industry, and it is not different in petrochemical activities. The search for novel compounds or operations that increase the efficiency of a certain process is constant, justifying the importance of microemulsion systems in these investigations.

15.2.1.1 Problem of Natural Gas Dehydration

Natural gas is generally exploited in conjunction with petroleum, and can contain significant amounts of water in its composition. It is considered as more environmentally friendly than traditional fossil fuels, because of cleaner combustion. However, water can be the main cause of corrosion problems. During natural gas processing, therefore, water removal is a fundamental step, since the combination of hydrocarbons and moisture enables the formation of a corrosive medium and hydrates. In view of this, drying of natural gas has been enhanced with the discovery of different water-removal fluids or alteration of old drying processes. Back in 1977, Fowler and Protz [40] reported on the use of organic glycol and its diethylene and triethylene derivatives, which reduce the dew point of natural gas, via an absorption process effected in proper columns.

Subsequent changes have been continuously reported. These comprised for example regeneration and recycling of the desiccant fluid [41], incorporation of polymeric membranes that are highly selective to water transport and sufficiently porous to allow gas permeation [42] or physical adaptations to the original dehydration operation [43,44]. However, in terms of cost, the search for novel dehydration agents seems to be more advantageous. In that respect, surfactant-based fluids have proven to be promising new systems, especially when commercial compounds or structures easily synthesized from local raw materials can be used. The purpose is to form microemulsions with the water contained in the natural gas, which can be separated, and enhance the efficiency of normal absorption packed columns by increasing the mass transfer coefficients.

15.2.1.2 Problem of Corrosion on Metallic Surfaces

A lot of attention has been paid to corrosion on metallic surfaces, which leads to material loss, especially in technological activities. Mineral acids and saline media are regarded as the main agents that lead to corrosion in some geographical areas (e.g., seaside developments). In oil industry, salt-induced corrosion in oil pipelines promoted by highly saline water and oil mixtures is a very representative example. Thus, it is not surprising that much research has focused on the use of inhibitors for industrial and technological activities. The more common example of corrosion inhibiting agents is paints, colloidal particles deposited onto metallic surfaces in general. However, organic compounds and, in particular, surfactant molecules have been reported as possessing important corrosion inhibition abilities.

Early works demonstrated the potential of amine compounds as corrosion inhibitors [45,46], by impairing the acid attack to steel via an anodic inhibition phenomenon. As early as 1936, n-alkylamine molecules with more than 5 carbon atoms in the alkyl chain were reported as good inhibiting agents, although this impairs their solubility in an acidic aqueous medium [47]. It has been discovered more recently that nitrogenated surfactant-based micellar systems can be advantageously used as iron corrosion inhibitors in acidic medium. These comprise aromatic amines [48], n-alkylbetaines [49], 2-(alkyldimethyl ammonium) alkanol bromides [50], aminated anionic surfactants [51], among others, and regardless of the corrosive medium, the efficiency is augmented with increasing surfactant concentration. The literature therefore demonstrates the potential of surfactant-based systems as corrosion inhibitors, but, in spite of their unique properties, microemulsions have not been thoroughly investigated as corrosion inhibitors in metallic surfaces.

15.2.1.3   Search for Microemulsion-Based Fuels

In the twentieth century, atmospheric pollution was aggravated with the increase in the number of automotive vehicles and the limited availability of petroleum started to become evident. This has been the motivation to pursue several investigations worldwide on the search for novel energy sources, particularly with regard to low-emission, renewable fuel formulations, without hampering engines performance.

Frequently, phase separation effected upon incorporation of additives or cosolvents is a major concern when developing novel fuel formulations. Microemulsion-based mixtures can overcome this problem, and has been the focus of more recent works. In that respect, Friberg and Force, in 1976, patented a diesel-based microemulsion formulation that could be used as fuel, with reduced NO_x emissions when compared to pure diesel [52]. Subsequent works have focused on phase behavior, stability, and performance of different mixtures, most of which involving surfactant-based mixtures [26,39], but it is worth considering the recent advances in the use of microemulsified systems incorporating other fluids like vegetable oils and alcohols.

We have reported on mixtures comprising diesel and soy oil, water, and surfactants that provide an economy of 25% in fuel consumption compared to pure diesel [32]. Incorporation of additives to ethanol–diesel mixtures can ultimately enhance the solubility of the final product, featuring typical microemulsion properties like thermodynamical stability and macroscopical transparency [53].

The use of emulsion and microemulsion systems as alternative fuels is therefore justified in terms of the reduced pollutants emission rates and improved engine performance. Addition of microemulsified water to diesel can potentially diminish emissions of organic volatile compounds, particulate matter, and noxious gaseous substances. Furthermore, engine performance is enhanced because of high-pressure water vapor formation in the combustion chamber [26]. On that account, microemulsions have been advantageously and continuously investigated in view of their unique properties and constitute strategic research programs especially devised by international agencies and governmental institutions.

An additional advantage of microemulsion systems is discussed in terms of potential environmental applications, although not being exactly related with the development of fuel formulations. For example, in the remediation of soils contaminated with diesel, formation of microemulsions can be the mechanism by which the cleaning operation is effected. Recently, we have reported on the applicability of an anionic surfactant derived from coconut oil (saponified coconut oil, SCO) to remove diesel from contaminated soil [54]. This surfactant was tested in aqueous solutions and in a microemulsion-precursory solution (surfactant + cosurfactant + water). Bench-scale assays were carried out using both column and batch setups with artificially contaminated soil. Parameters like the cosurfactant nature, the cosurfactant/surfactant ratio (C/S) and the effect of electrolyte dissolved in the aqueous phase (NaCl) were examined. It was observed that the formation of diesel-in-oil microemulsion makes easy the removal of contaminants from soil samples. Depending on the technique used (column or batch experiments), up to 75% contaminant removal could be achieved, showing the potential applicability of SCO in microemulsion systems for cleaning up contaminated sandy soils. The microemulsions provided the best results because of enhanced oil solubilization and inhibition of soil pores clogging.

15.2.1.4   Use of Microemulsions in Enhanced Oil Recovery

The first attempt to displace petroleum from rock reservoirs upon microemulsion injection was carried out in 1963 by the *Marathon Oil Company*, following a process denominated *Maraflood*. Later, from 1973, Healy and Reed [55] reported on some fundamental aspects involved in microemulsion flooding, such as viscosity, interfacial tension and salinity, referring the results of phase behavior of self-assembled systems to the Winsor's concepts.

Further development of recovery methods contemplated physical and chemical procedures, such as application of pressure, water injection, or implementation of techniques that alter system miscibility. All aspects that impact on the final recovery yield must be considered, for example: capillary forces; oil viscosity; contact angle between the adsorbed oil and the solid surface; permeability, wettability, and porosity of the solid reservoir; among others. Hence, it is obvious that huge perturbations are provoked in the oil reservoirs when surfactant-based systems are used in enhanced oil recovery operations. The potential role of microemulsions in such activities is once again highlighted.

15.2.2 REMOVAL OF HUMIDITY FROM NATURAL GAS

The problem of water removal from natural gas was efficiently accounted for with the development of an experimental apparatus and methodology specifically applied to local raw material. In particular, the gas produced in the Potiguar Basin (Northeastern Brazil) possesses high water content (about 4000 ppm), and common absorption and adsorption processes have been operated to promote dehydration in a gas treatment unit (GTU). These processes are very expensive and inefficient when stringent regulations, requiring 0.5 ppm water at the exit of the GTU, are to be complied with. In this context, microemulsions have been effectively used in absorption processes as an alternative to dehydrate natural gas. The higher-solubilization capacity, low-interfacial tension, and large interfacial area between continuous and dispersed microphases are important parameters to ensure efficient operation of adapted absorption columns, employing microemulsions as desiccant fluid.

We therefrom report on natural gas dehydration prompted by microemulsion extraction. To do this, phase diagrams were constructed, and compositions that favor the formation of water-in-oil microemulsion systems were selected, since it is interesting and even required that water be accommodated within the cores of reversed droplets in the self-assembled systems generated.

The experimental approach comprised two parts: gas umidification system and adaptation of an absorption column apparatus. The umidification step is required to calibrate humidity probes, based on standard water contents, according to the ASTM D 4178/82 norm, revised in 1999. During the project of an absorption column, the main parameters to be considered are the internal diameter, the choice of the packing material, and the flooding velocity. Gas phase distribution, pressure drop, flow rate measurement, feeding, and collection points must also be accounted for. Figure 15.3 is a schematic diagram depicting the whole experimental water-absorption apparatus designed for use with natural gas, controlled by a series of valves (V) and manometers (M).

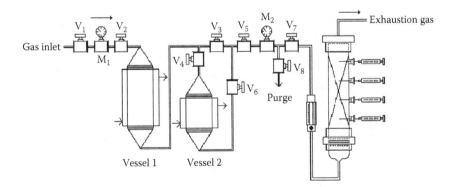

FIGURE 15.3 Experimental apparatus designed for natural gas dehydration via absorption operations.

Two adsorption vessels packed with molecular sieves constitute the initial part of the apparatus, where gas saturation takes place. The main absorption column is packed with Raschig rings and is operated batchwise to assess the absorption capacity of the desiccant fluid. Four collection points were inserted along the main column and secured with rubber rings. Samples are collected in glass syringes with stainless steel needles and taken for analysis in laboratory.

15.2.2.1 Components of the Microemulsion Systems

Surfactants are the principal constituents in the absorption process presented here, since their ability to reduce interfacial tensions between polar and nonpolar phases effectively induces the formation of microemulsions, which are the desiccant fluids in the dehydration system. It was decided to use nonionic surfactants to avoid alcohols as cosurfactants, which may hinder subsequent water analyses with the samples collected. The following low-cost commercial hydrocarbon nonionics have been selected for this study: 1-Renex, Amide 60-PBC, and 2-Amide 60 Henkel. These comprise a group of structures featuring the polyethoxylated moiety as polar headgroup, with different polydispersity ranges.

Common hydrocarbon solvents derived from petroleum (*n*-hexane, *n*-heptane, and naphtha formed by higher-molecular weight liquid hydrocarbons, with distillation temperature range between 150°C and 250°C) and local vegetable oils (*babaçu*, coconut, sunflower, castor oils) have been tested as nonpolar phases. The more promising surfactants were chosen based on their solubility in these nonpolar solvents.

Information on maximal solubilization of the dispersed phase is acquired upon construction of phase diagrams. When dealing with microemulsions, (pseudo)ternary diagrams are very useful to represent the phase behavior established when varying the composition of a chemical system that comprises a certain number of components.

15.2.2.2 Water Absorption and Stripping

The first step of the absorption process is preparatory. The main absorption column is mounted and the injection syringes are inserted for sampling. Random packing is carried out with the Raschig rings, aiming to avoid empty spaces that may interfere in the flow of fluids. The basic operation performed in the apparatus illustrated in Figure 15.3 consists in pumping wet gas through the packed bed that is soaked with the desiccant fluid (nonpolar solvent + surfactant). The wet gas enters the main column after being humidified along the adsorption vessels. The mechanism whereby the gas dehydration takes place involves water withdrawal from the bulk gas current via formation of reversed microemulsion droplets, as illustrated in Figure 15.4.

When saturation of the desiccant fluid is reached, mass transfer from gas to liquid is interrupted. To reuse it, a regenerative stripping operation is performed, whereby heated compressed air is pumped along the column to promote water

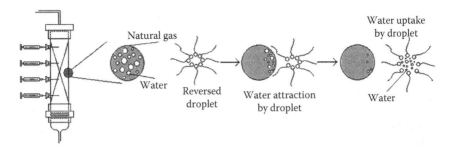

FIGURE 15.4 Mechanism of water transfer from bulk natural gas to the core of reversed microemulsion droplets.

diffusion from the packed bed. The microemulsion solution is therefrom demixed and the initial surfactant–oil solution is formed, which can be recycled to be used in other batches. A special apparatus was devised to perform this operation. It features a coiled tubing through which the compressed air flows, disposed on top of a Bunsen burner that can be adapted to the main column inlet valve.

15.2.2.3 Implementation of the Absorption System

In the phase behavior experiments, several systems have been examined, but two have been selected for giving better results in terms of broader microemulsion regions obtained in ternary diagrams. System I was formed by water, naphtha, and a mixture of two nonionic surfactants in a 1:1 mass ratio (1-Renex + Amide 60-PBC); System II comprised water, naphtha, and a different nonionic surfactant (2-Amide 60 Henkel). This conclusion was based in the diagrams constructed with these components, as shown in Figure 15.5.

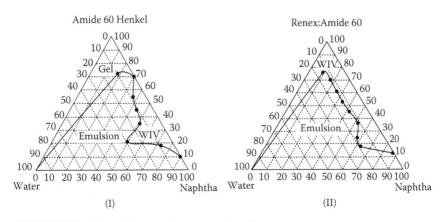

FIGURE 15.5 Phase diagrams for systems I and II.

It must be pointed out that systems favoring the formation of more extensive water-in-oil microemulsion regions (WIV regions in the diagrams of Figure 15.5) are more advantageous. Even broader microemulsion regions could be prepared with addition of alcohols as cosurfactants. However, as mentioned in Section 15.2.2.1, incorporation of alcohols in the mixture would impair the results of water analyses performed later. Systems I and II, therefore, may be suggested as potential desiccant fluids in the dehydration of natural gas. This is also corroborated by the results below.

The umidification system devised specifically for this purpose effectively met the requirements of the assays. By keeping constant low temperatures in the vessels, the concentration of water in natural gas was adjusted accordingly, which is essential to the final dehydration step.

No axial concentration gradients should be detected in the absorption column. For this reason, a sufficiently low tower must be used. However, if higher columns are available, studies on concentration as a function of column height must be undertaken to optimize the absorption process.

Microemulsions formed with components of Systems I and II were used as desiccant fluids in water absorption operations. The amount of surfactant in the samples prepared was set at 50% to avoid turbidity that would be formed below that concentration level. In Figure 15.6, the amount of water that can be transferred from natural gas to the desiccant fluid (C, in ppm) is given as a function of time for System I, at three different gas flow rates: 300, 700, and 1200 mL min^{-1}. It was observed that 240 min was required to attain the maximal dehydration capacity of the system at the lower flow rate. The concentration of water in the desiccant fluid decreases with increasing gas flow rate, because of the relatively high viscosity of the fluid and the reduced contact time between phases at higher flow rates. Viscosity is therefore an important parameter to be considered in such mass transfer applications. On the other hand, higher gas flow rates are favorable

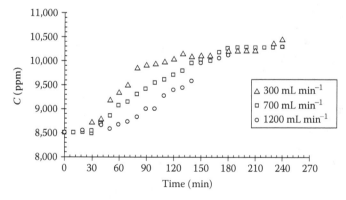

FIGURE 15.6   Absorption assays for system I as a function of time and gas flow rate.

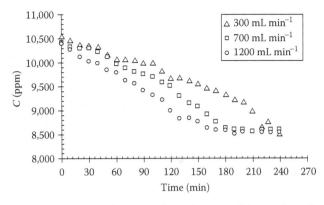

FIGURE 15.7   Stripping assays for system I as a function of time and gas flow rate.

in the subsequent stripping operations. This can be demonstrated by the lower water concentrations detected in the desiccant fluid as the flow rate increases (see Figure 15.7, for System I). Similar results were obtained with System II, as shown in Figures 15.8 (absorption assays) and 15.9 (stripping assays).

In summary, the specific implementation of this absorption system provides another approach to the problem of water removal from natural gas streams. Humidity levels as low as 300 ppm were detected in the treated gas stream. This is still above the levels established by the regulations, but corresponds to a yield of 92.5% in the absorption capacity of the microemulsions. The surfactants used in this investigation do contain a certain amount of water in their original constitution, and no previous treatment was performed with them. Interestingly, their potential as good components in desiccant fluid formulations was demonstrated in the results presented herein, since high water absorption capacity was effected by the system

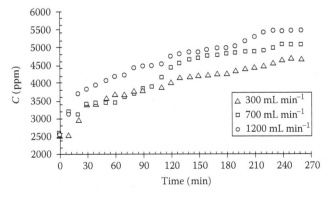

FIGURE 15.8   Absorption assays for system II as a function of time and gas flow rate.

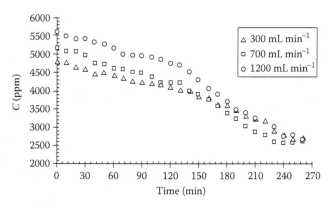

FIGURE 15.9   Stripping assays for system II as a function of time and gas flow rate.

devised. Parameters like viscosity and gas flow rate seem to significantly affect the overall system efficiency. Microemulsions are therefore presented as good options in activities that require natural gas dehydration. Their enhanced water uptake capacity is the property that distinguishes them among other common procedures, e.g., those employing molecular sieves. Experimental facilities could then be projected to contemplate either the sole use of microemulsion techniques or their combinations with common separation processes already in use, resulting from typical optimization procedures.

15.2.3   INHIBITION OF CORROSION ON METALLIC SURFACES

The ability of aminated compounds to inhibit corrosion on metallic surfaces via adsorption phenomena has been already certified. Since operations taking place at interfaces are greatly affected by variations in surface tension, aminated surfactant molecules are expected to provide even better results. This has been the case, when self-assembled micellar or microemulsion systems are used as corrosion inhibitors. In that aspect, surfactants may be used as organic corrosion inhibitors, and act by forming a protective film onto surfaces which are exposed to corrosive media, like oxygen and saline or acidic solutions. When microemulsions are used, an oil film is also adsorbed onto the surface with the surfactants' tails oriented towards it, in view of the usually positive character of the surface. In the petroleum industry, the oil itself may be the nonpolar component of such systems. Figure 15.10 is a schematic of these types of films.

When surfaces are covered with oil, water is repelled and any potential corrosion cell is broken via the effective generation of a barrier between electrolytes and the surface. However, the film is not definitive, and must be replaced or reconstructed regularly. This determines the film's persistence over time.

A good example of this mechanism is illustrated by the use of a commercial mixture comprising alcohol, water, imidazoline derivatives, and quaternary

Surfactant molecules   Oil

Metallic surface

FIGURE 15.10   Monolayers formed by adsorbed surfactant molecules or surfactant + oil onto metallic surfaces.

ammonium salts (surfactant mixture "*A*"). Its anticorrosion capacity can be tested by constructing cylindrical pyrex glass electrochemical cells with a Teflon lid that is adjustable to allow insertion of gas bubblers and electrodes. Three electrodes are normally used in such assays: one silver/silver chloride reference electrode (Ag/AgCl), a platinum counterelectrode, and a work electrode made of API5LX Gr X42 steel. They are immersed in the corrosive medium, which can be a saline or acidic solution, with or without inhibitor. The reference electrode is directly involved in the corrosion potential measurements, from which anodic and cathodic polarizations are effected. The counterelectrode is used as an auxiliary to complete the cell and balance charges. A typical apparatus is represented in Figure 15.11.

The work electrode was designed by inserting a cylindrical piece of the API5LX steel in epoxy resin in such a way to expose only a $1\,cm^2$ metal surface to the corrosive medium. This surface must be polished before each measurement. An insulated copper support is adapted to provide electrical contact for the work electrode. API5LX steel is much used in the construction of ducts used in the transportation of crude oil.

Potentiokinetic assays are carried out in equipments known as potentiostats, which work by inducing higher tensions than the nominal corrosion potentials of

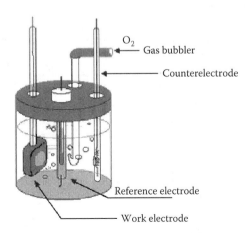

FIGURE 15.11   Schematic of an electrochemical cell used in corrosion inhibition assays.

the electrodes, in both anodic and cathodic directions. The metal is thus polarized and the potential is controlled at a linear scanning velocity. Gas must be bubbled during all measurements to avoid any diffusion phenomena within the corrosive medium. Oxygen and nitrogen can be used, depending on the purpose of the experiment. Oxygen, for instance, may constitute another corrosive medium, whereas nitrogen is inert. In each measurement, the following parameters are quantified as a function of inhibitor concentration: corrosion current (I_{corr}), corrosion potential (E_{corr}), and Tafel's anodic (β_a) and cathodic (β_c) curvatures.

In real petroleum applications, surfactant-based systems are applied as corrosion inhibitors in the beginning of pipeline flows, together with the petroleum emulsion (crude oil + brine). The surfactant may be used either as a micellar solution or in the microemulsified form. Its maximal solubility must be known in advance; therefore, phase behavior studies must be previously made. If the right conditions are promoted, dilution of the inhibitor occurs spontaneously along the emulsion flow. When microemulsions are used, this phenomenon is more effective, since more surfactant molecules can be solubilized in the medium.

Since crude oil itself may contain a significant amount of water, the mixture is pumped in such a way that all excess, nonemulsified water forms a bottom layer in the pipelines. It is exactly this part of the ducts that are more susceptible to corrosion. If more hydrophilic surfactants are used, their corrosion inhibiting properties will be enhanced as a result of more effective adsorption phenomena.

In a typical electrochemical assay, the efficiency of the system in inhibiting corrosion ($E\%$) is estimated by means of Equation 15.3:

$$E\% = \frac{100 \times (I_{corr} - I'_{corr})}{I_{corr}} \tag{15.3}$$

where I_{corr} and I'_{corr} denote the corrosion current densities in the absence and presence of inhibitor, respectively.

Since the commercial mixture "A" mentioned above is a complex chemical system, its critical micelle concentration (CMC) can be determined in weight/volume percentage when the product is dissolved. Table 15.2 lists the CMC of this compound in some specific situations that represent potential sources of corrosion in the industry, that is, saline solutions and high temperatures.

It is noticed that the CMC is practically independent on temperature. Interestingly, when electrolytes are present and high temperatures are detected, lower amounts of inhibitor are necessary to form a protective monolayer onto the surfaces, which implies lower costs. In Figure 15.12, the calculated efficiencies of this surfactant when dissolved in some corrosive solutions are shown as a function of surfactant concentration.

It can be seen that the highest efficiency was reached with only 0.8% of surfactant. Even below the CMC values, the efficiency increases. However, there is a decrease in performance at higher temperatures, since enhanced corrosion velocities and ionic mobility are promoted at such conditions. Pitting corrosion is mostly expected to occur when chloride-rich solutions are affected by temperature.

TABLE 15.2
CMC of a Commercial Mixture Formed by Alcohol, Water, Imidazoline Derivatives, and Quaternary Ammonium Salts in Some Solvents

Solvent	CMC (% w/v)
Water at 27°C	0.4352
Water at 60°C	0.4360
Aqueous NaCl 0.5 M at 27°C	0.2269
Aqueous NaCl 1.0 M at 27°C	0.2226
Aqueous NaCl 0.5 M at 60°C	0.2156
Aqueous NaCl 1.0 M at 60°C	0.2262

Since nonaggregated surfactant molecules are responsible for interfacial tension effects, any interfacial tension variations will be hindered by self-assembly above CMC. This is pictorially represented in Figure 15.13 for the particular case of commercial mixture "A" in 0.5 M NaCl aqueous solutions at 27°C (CMC = 0.2269%).

Our research group has endeavored to employ regional raw materials in surfactant applications. Particularly with regard to the line of activity involving discovery of novel corrosion inhibitors, cashew nut shell liquid, extracted from

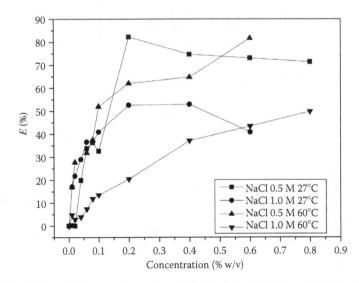

FIGURE 15.12 Efficiency of corrosion inhibition on API5LX Gr X42 steel with a commercial mixture formed by alcohol, water, imidazoline derivatives, and quaternary ammonium salts as inhibitor in different NaCl solutions at 27°C or 60°C.

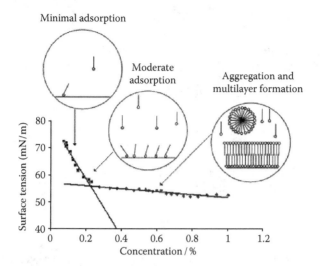

FIGURE 15.13 Surface tension versus inhibitor concentration for a 0.5 M aqueous NaCl solution at 27°C. Note how the metal surface is affected as the surfactant concentration increases.

the cashew nut tree (*Anacardium occidentale*), and castor oil (*Euphorbiacea ricinus communis*) are potential materials used to synthesize novel surfactant molecules. For example, the anionic structures (named as AR1S, AE2S, and AE1S) shown in Figure 15.14 have been fully characterized and their properties as corrosion inhibitors either in micellar or microemulsion environment are testified [21,25,51]. Note the presence of the diethylamino group in all three molecules. Their efficiency as corrosion inhibitors in metal surfaces has been partly attributed by this moiety.

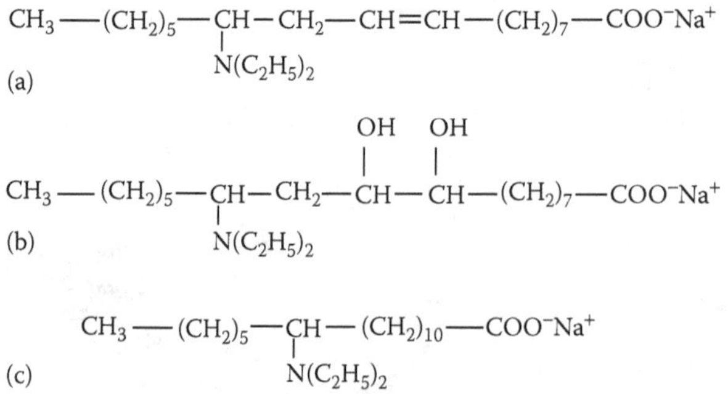

$$CH_3-(CH_2)_5-\underset{\underset{N(C_2H_5)_2}{|}}{CH}-CH_2-CH=CH-(CH_2)_7-COO^-Na^+$$
(a)

$$CH_3-(CH_2)_5-\underset{\underset{N(C_2H_5)_2}{|}}{CH}-CH_2-\overset{\overset{OH}{|}}{CH}-\overset{\overset{OH}{|}}{CH}-(CH_2)_7-COO^-Na^+$$
(b)

$$CH_3-(CH_2)_5-\underset{\underset{N(C_2H_5)_2}{|}}{CH}-(CH_2)_{10}-COO^-Na^+$$
(c)

FIGURE 15.14 Molecular structures of synthesized aminated surfactants: (a) AR1S, (b) AE2S, and (c) AE1S. (From Dantas, T.N.C., Moura, E.F., Scatena Jr., H., Dantas Neto, A.A., and Gurgel, A., *Coll. Surf. A*, 207, 243, 2002. With permission.)

In Figure 15.15, corrosion inhibition efficiency data are particularly shown for systems containing the surfactant AR1S in structurally distinct media. The metal surface affected was API5LX Gr X42 steel. The surfactant was previously dissolved in NaCl aqueous solutions, at 0.5 M or 1.0 M salt concentrations. The tests were carried out either using micellar or microemulsion systems, at 30°C and 60°C. The microemulsion system was prepared with the surfactant NaCl aqueous solutions, butan-1-ol as cosurfactant (C/S ratio = 1.0) and kerosene as oil phase.

In general, quite high efficiencies are reached, but microemulsified structures bring about an important effect on the inhibition ability of the surfactant molecules. In general, higher inhibition levels are attained with microemulsion systems. The deleterious effect of both salinity and temperature is also observed.

It is important to understand the mechanism of corrosion inhibition promoted by surfactant-based systems. The transition of the metal–solution interface from an active dissolution state to a passivation state is highly important in petroleum fields. Normally, surfactants are added to aqueous media to occupy the interface, hence reducing corrosion of the pipelines. It is known that increasing surfactant concentrations reduce interfacial tensions, as a result of enhanced aggregation and physical adsorption upon micelle formation at concentrations above the CMC.

Additional surface coverage in the form of multilayers and increasing viscosity at concentrations above the CMC is thought to further enhance corrosion inhibition. The explanation is simple: with narrow coverage (less than a monolayer, for instance), surfactants molecules might inhibit cathodic and anodic reactions upon occupation of active sites or ordinarily by hampering the supply of oxidation agents and blocking the transport of reaction products. At higher surfactant concentrations, additional molecules may have difficult access to the surface and corrosion is thereby inhibited.

In summary, the role of surfactant-based systems as corrosion inhibitors is highlighted. Microemulsions are particularly better systems, with the advantage

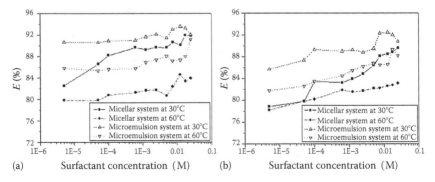

FIGURE 15.15   Efficiency of corrosion inhibition on API5LX Gr X42 steel with surfactant AR1S as inhibitor at 30°C and 60°C, dissolved in micellar or microemulsion systems. The surfactant had been previously dissolved in (a) 0.5 M NaCl aqueous solution or (b) 1.0 M NaCl aqueous solution.

over micellar solutions of providing larger interfacial contact and higher surfactant solubilization capacity in several corrosive media. The protective layers formed by such systems are more homogeneous, and the possibility of oil layers formation on top of the original surfactant films further enhances their performance. This has been the motivation to apply these systems in the petroleum industry.

15.2.4 MICROEMULSION FORMULATIONS FOR CYCLE-DIESEL ENGINES

In this topic, particular discussions are made on the development of alternatives to operate cycle-diesel engines in view of the need to rationalize the use of conventional, existing fuels, improvements on the performance of modern engines, application of more rigorous environmental legislation, and continuously increasing energy demands. Some alternatives consist in changing the current diesel oil compositions, for example by adding vegetable oils and/or alcohols, and forming emulsions or microemulsions. It is interesting that microemulsion formulations can be introduced as potential candidates to help solving these problems. As is the case when investigating novel microemulsion systems, a complete understanding of their phase behavior is required.

As an example of experimental investigation, a selection of commercial non-ionic surfactants was tested in microemulsion systems with diesel oil and water, to determine the best conditions that provide the formation of microemulsions. The surfactants are typical polyethoxylated compounds with various HLB degrees, and were tested either on their own or as binary surfactant mixtures (to adjust the HLB value and benefit from any synergistic effects). In Table 15.3, a list of the surfactants and mixtures tested is presented.

Surfactant AE 1, with low HLB, is easily dissolved in diesel oil but not in water. Surfactants AE 4 through AE 9 have HLB higher than 10 and do not solubilize satisfactory amounts of water in diesel (only emulsions with less than 1% of water can be formed). Since the main objective of this special type of investigation is the generation of large microemulsion areas in ternary diagrams, the best results are obtained when surfactants with appropriate HLB values are used. For example, in Figure 15.16, a series of phase diagrams is presented for systems comprising surfactant or surfactant mixture + diesel oil + water, prepared at 28°C. In all cases, a gel region is formed, which seems to be a common feature with nonionics. Although the emulsion regions are large, quite good WIV microemulsion regions are also obtained. The fact that these systems are formed in the oil-rich part of the diagrams is advantageous, since the aim is to disperse water droplets in a continuous oil phase to prepare the novel fuel formulations.

Based on these results, several formulations can be tested and compared with commercial diesel. For example, the viscosity of a microemulsified diesel sample formed by 6% water + 5% surfactant AE 3 + 89% diesel is more affected by temperature than commercial diesel. For instance, it is shown that the viscosity of this microemulsified diesel decreases from over 8 cSt at 25°C and stops at 3.5 cSt from 40°C, complying with specific Brazilian regulations. Within that temperature range, the viscosity of commercial diesel falls below 3 cSt. When 50 cm^3 of each

TABLE 15.3

HLB of Some Commercial Nonionic Surfactants and Their Mixtures Used in Fuel Formulations

Surfactant Mixture	HLB[a]	Surfactant Mixture	HLB[a]
Pure AE 1	5.30	AE 3/AE 4 1:1	10.45
Pure AE 2	8.90	AE 2/AE 6 1:1	10.95
Pure AE 3	10.00	AE 1/AE 6 1:4	11.46
Pure AE 4	10.90	AE 1/AE 7 1:4	11.70
Pure AE 5	11.70	AE 1/AE 2 1:1	7.10
Pure AE 6	13.00	AE 1/AE 3 1:1	7.65
Pure AE 7	13.30	AE 1/AE 4 1:1	8.10
Pure AE 8	11.00	AE 1/AE 5 1:1	8.50
Pure AE 9	16.70	AE 2/AE 3 1:1	9.45
AE 1/AE 5 4:1	6.58	AE 2/AE 4 1:1	9.90
AE 1/AE 6 4:1	6.84	AE 2/AE 5 1:1	10.30
AE 1/AE 7 4:1	6.90	AE 2/AE 6 1:1	10.95
AE 1/AE 5 1:1	8.50	AE 3/AE 4 1:1	10.45
AE 1/AE 6 1:1	9.15	AE 3/AE 5 1:1	10.85
AE 2/AE 5 4:1	9.46	AE 3/AE 6 1:1	11.50
AE 2/AE 6 4:1	9.72	AE 4/AE 5 1:1	11.30
AE 3/AE 4 4:1	10.18	AE 4/AE 6 1:1	11.95
AE 1/AE 5 1:4	10.42	—	—

[a] The HLB of the surfactant mixture is a ponderal average of the individual components.

fuel is tested in the operation of a cycle-diesel M 790 B engine, between 1500 and 3000 rpm, the results presented in Table 15.4, for microemulsified diesel, and Table 15.5, for commercial diesel, are obtained [t represents the time for fuel consumption; R_{pa} is the number of revolutions per axis; pressure P was applied to calculate the torque τ, the mean angular speed (ω), the maximal power (P_t) and the specific fuel consumption (C)]. The initial combustion and air temperatures are 32°C. The values of C versus $\%P_{max}$, the percentage of the maximal power developed by the engine in each assay, have been plotted and are shown in Figure 15.17.

Two important conclusions are drawn from these experiments:

1. Difference between fuel consumption for microemulsified and commercial diesel samples decreases with increasing power percentage. At higher $\%P_{max}$ values, the rotation frequency is lower as a result of increasing axis load. Hence, water droplets dispersed in the microemulsion formulation are vaporized at high pressure upon fuel combustion, decreasing fuel consumption.

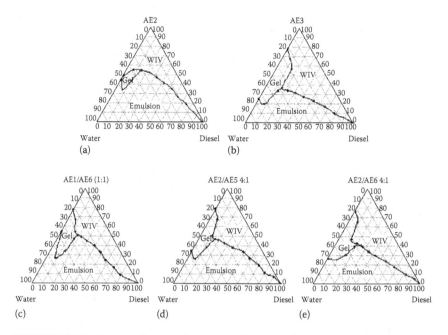

FIGURE 15.16 Ternary diagrams for water + diesel + surfactant or surfactant mixtures at 28°C. WIV denotes microemulsion regions. Surfactants tested: (a) AE2; (b) AE3; (c) AE1/AE6 mixture at 1:1; (d) AE2/AE5 mixture at 4:1; (e) AE2/AE6 mixture at 4:1.

2. At lower-power values, however, the consumption of the microemulsion fuel is augmented in view of higher-engine rotation frequencies. More fuel must be therefore injected into the combustion chamber and the mean cylinder temperature is reduced because of the water contained in the formulation. This apparent problem can be overcome since lower-pollutant emission levels are expected with these novel formulations.

TABLE 15.4
Results of Assays with Microemulsified Diesel at 2000 rpm

Assay	t (s)	R_{pa}	m (kg^{-1})	P (psi)	τ (N·m)	ω (rps)	P_t (kW)	% P_{max}	C (kg kW^{-1})
1	66.0	2192	4.350	62	24.75	33.21	5.16	100	0.000126
2	67.1	2248	4.000	62	22.76	33.50	4.79	86	0.000134
3	71.5	2408	3.500	62	19.91	33.68	4.21	75	0.000143
4	78.3	2605	3.000	65	17.07	33.27	3.57	65	0.000154
5	83.9	2803	2.500	65	14.22	33.41	2.99	54	0.000172
6	87.2	2916	2.250	65	12.80	33.44	2.69	49	0.000183
7	100.6	3338	2.000	65	11.38	33.18	2.37	43	0.000180
8	118.1	3922	1.700	68	9.67	33.21	2.02	36	0.000180

TABLE 15.5
Results of Assays with Commercial Diesel at 2000 rpm

Assay	t (s)	R_{pa}	m (kg^{-1})	P (psi)	τ (N·m)	ω (rps)	P_t (kW)	% P_{max}	C (kg kW^{-1})
1	65.3	2177	4.700	61	26.74	33.34	5.60	100	0.000118
2	70.5	2372	4.000	62	22.76	33.65	4.81	86	0.000127
3	77.5	2617	3.500	62	19.91	33.77	4.23	75	0.000131
4	79.1	2677	3.000	62	17.07	33.84	3.63	65	0.000150
5	95.9	3241	2.500	62	14.22	33.80	3.02	54	0.000148
6	102.5	3470	2.250	68	12.80	33.85	2.72	49	0.000154
7	111.7	3765	2.000	70	11.38	33.71	2.41	43	0.000160
8	131.8	4404	1.700	68	9.67	33.41	2.03	36	0.000161

Moreover, when these microemulsion systems are used as fuels, no starting problems are detected during engine operation. These results are useful to motivate further studies on novel, optimized formulations, requiring continuous adaptations to physical apparatus and application of proper analytical techniques [53].

15.2.5 USE OF MICROEMULSIONS IN ENHANCED OIL RECOVERY

The properties of microemulsions can be exploited with many advantages in enhanced oil recovery applications. In this topic, two basic approaches are discussed:

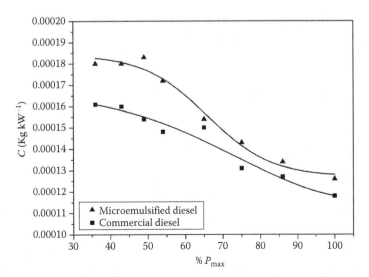

FIGURE 15.17 Specific fuel consumption as a function of percentage of maximal power developed by engines operated with commercial and microemulsion-based diesel.

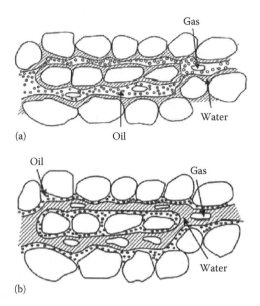

FIGURE 15.18 Distribution of fluids in a reservoir wetted by (a) water and (b) oil.

the use of microemulsion in oil reservoirs drilling operations and the use of surfactant-based systems as fracturing gels. In both cases, it is important to consider the distribution of fluids in a reservoir (water, oil, and even gas). Contact angle measurements are important in directing the applications, since they affect adhesion tensions inside the formation, which ultimately enables removal of oil via an appropriate extraction process.

The wettability must also be accounted for. This is explained if one considers that oil, water, and gas are accommodated in a rock reservoir. It is expected that water preferably wets the solid surfaces, as depicted in Figure 15.18a. Since gas is the least likely fluid to wet the solid surface, it is stored in the cores of the pores, with oil occupying intermediary positions. This is the case in the majority of oil reservoirs, but in some cases, where the formation characteristics are changed during prospection, the rock surface is coated with oil, as shown in Figure 15.18b. This situation requires more attention from engineers and operation personnel in the extraction activities [29,30].

15.2.5.1 Microemulsions as Injection Fluids

During prospection of oil wells, injection fluids are used to alter the pressure equilibrium originally established. Their composition is based on mixtures of liquids and solids and greatly varies according to each specific application. The choice of the most appropriate fluid depends on economical factors, contamination levels, pressure and temperature, to name a few.

In conventional oil recovery activities via waterflooding, low yields are normally observed, basically as a result of high oil viscosity and high interfacial tensions developed when water is injected. When the viscosity of the injected fluid is lower than that of the fluid to be displaced, the previous one flows more swiftly than the latter across the porous medium, often finding preferred paths. In view of high-interfacial tensions, the capacity of the injected fluid to displace the oil is rather impaired, resulting in high contents of residual oil in the wells.

The stability of the final mixture must be guaranteed to maintain high extraction yields. Since petroleum naturally exists in certain underground formations where it is adsorbed onto the rocks' pores, conventional recovery methods are usually capable of removing only 30% of the oil. This can be credited, basically, to three aspects: high viscosity of the oil, geology of the formation, and high interfacial tensions between the reservoir's fluids.

Microemulsions can be formed when appropriate surfactant mixtures are injected, and, to devise enhanced oil recovery operations, constant investigations on optimized chemical systems and physical apparatus must be undertaken. Microemulsion flooding is therefore considered an enhanced method, and has the purposes of desorbing the oil from the rock formation and attaining good efficiency levels in the displacement of the microemulsified oil formed. Comparisons can then be made with the conventional techniques already in use by oil companies. Figure 15.19 illustrates the main enhanced oil receovery methods available [30].

An initial laboratory approach involving adsorption and petroleum recovery assays may comprise the following steps:

- Assemblage of a fluid injection prototype
- Insulation of the porous medium simulating the original well formations
- Determination of CMC of each surfactant tested

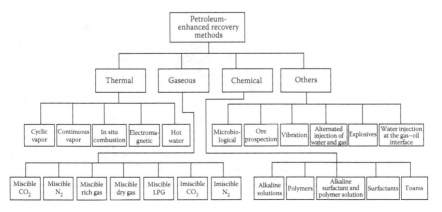

FIGURE 15.19   Main enhanced petroleum recovery methods available.

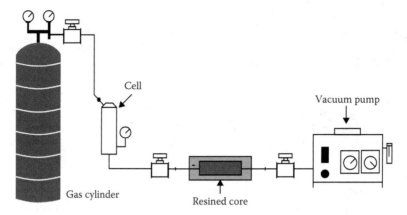

FIGURE 15.20 Schematic representation of a system of fluid injection into resined cores at constant pressure and temperature.

- Saturation of cores with reservoir fluids (brine and petroleum)
- Determination of porosity and permeability of the medium, relative to oil and water

Figure 15.20 shows schematics of a suggested apparatus designed for fluid injection assays at constant pressure and temperature, comprising a nitrogen cylinder, a stainless steel cell and a cylindrical core or plug, previously calcined and resined to mimic a geological formation. To the edges of the core, acrylic disks were attached to allow for uniform fluid distribution.

The same procedure may be carried out at constant flow rate and constant temperature, whereby a pressure gradient is applied across the core. The corresponding apparatus for this system is shown in Figure 15.21.

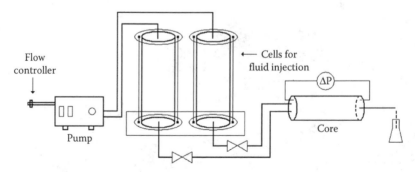

FIGURE 15.21 Schematic representation of a system of fluid injection into resined cores at constant flow rate and temperature.

A representative experiment is made with 2% KCl solution, in weight, and a 2:3 mixture of petroleum and kerosene, in volume. Crude petroleum samples, free from demulsifying agents, were supplied by Petrobras (Guamare, Brazil). The investigations were performed with five commercial nonionic surfactants belonging to a homologous series, with different ethoxylation degrees, two anionics (one of which synthesized in laboratory) and one cationic (also synthesized), and mixtures of them [27]. The selection was based on desired recovery factors provided by surfactants in solution. Surfactant solutions, in water or brine, at concentrations above their CMCs were used as injection fluids. The improved performance of the recovery operation results from the adsorption and flow mechanism occurring across the model apparatus. Single or multilayers of surfactant molecules are adsorbed onto the surface of the petroleum-saturated cores, and the oil is displaced and extracted via formation of microemulsion droplets that flow along the system. This is depicted in Figure 15.22.

Surfactant adsorption onto the core surfaces is evaluated in terms of estimation of material loss in the rock during fluid flow. Previous knowledge of the surfactants phase behavior is required, since it is necessary to ensure that the extracting mixture will remain stable within the ranges of water content and salinity levels observed during operation.

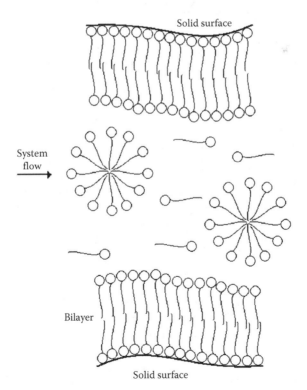

FIGURE 15.22 Surfactant adsorption as bilayers onto solid surfaces.

The performance of the surfactant systems is testified in oil recovery assays involving permeability measurements. Several surfactant solutions at increasing concentrations above the CMC are injected in a core and the amount of oil displaced is evaluated in terms of recovery percentage. This is shown in Figure 15.23 for some surfactant systems in terms of percentage of oil recovered from the model reservoirs as a function of injected porous volume (measured after porosity determination; the injected volume correspond to a proportion of the actual porous volume). The results are shown for four different nonionics, two anionics, the synthesized cationic, and a mixture of one nonionic and one anionic. In all cases, the yields obtained are higher than those promoted when brine is used as injection fluid, with no surfactant content. There seems to be a direct relationship between the beginning of surfactant layer formation in the internal surface of the oil reservoir, the injected porous volume, and the ethoxylation degree for the nonionics. It could be concluded that surfactant absorption is effected via bilayer formation (see Figure 15.22), with an important function being developed by the surfactants' polar headgroups. Higher recovery percentages are reached with anionics, which have a more pronounced adsorption capacity. With the surfactants tested, there seem to be no relevant synergistic effects when mixtures are prepared (see graph (h) in Figure 15.23).

It is also suggested that seawater be used as brine or aqueous phase in microemulsion formulations, because high salinity levels are observed in marine reservoirs. The enhanced recovery methods employ different injection fluids, such as microemulsions, to act in areas where the conventional process cannot provide good recovery rates. With this in mind, it is important to determine all parameters that affect microemulsion formation and stability, like surfactant and cosurfactant types, C/S ratio, and salinity.

The evaluation is made in terms of displacement efficiency with microemulsions that remain stable over a wide concentration range, within the water-rich domain (higher water content). In a typical experiment, core plugs constituted of regional *Assu* (Northeastern Brazil) or *Botucatu* (Southern Brazil) sandstones were assessed as to porosity and permeability and then submitted to the steps of saturation with seawater and oil, conventional recovery with water and enhanced recovery with the selected microemulsions [29]. The *Botucatu* sandstone presented better recovery rates, and a microemulsion system prepared with a nonionic surfactant presented the highest recovery efficiency (26.88% of the original oil "in-place"), among all surfactants tested.

When constructing phase diagrams, we are interested in systems that remain stable over a wider composition region, which ultimately defines their application limits. In the particular case of the salinity study, filtered seawater was used as part of the water phase in the formulations. In Figure 15.24, the effect of amount of seawater added as aqueous phase is demonstrated with the phase diagrams constructed. Kerosene was used as oil phase and butan-1-ol as cosurfactant. An anionic surfactant (C/S ratio = 2) was synthesized in laboratory from regional raw material and used to stabilize the systems. Salt addition usually reduces the CMC of anionic surfactants upon destabilization of the interactions between surfactant

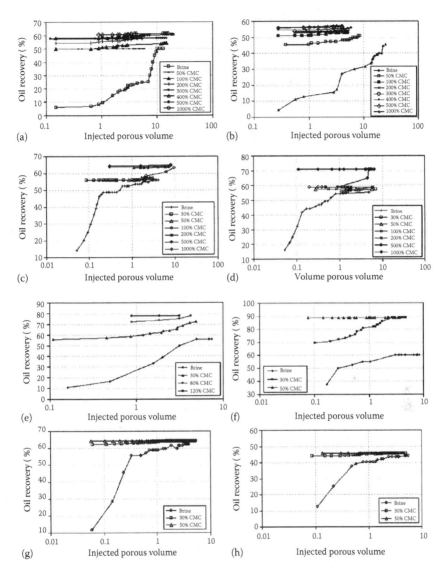

FIGURE 15.23 Oil recovery percentages obtained with some surfactant systems. The results were determined as a function of injected porous volume for surfactant systems with various concentrations above the CMC. Graphs (a), (b), (c), and (d) refer to different nonionics; graphs (e) and (f) correspond to the anionics; graph (g) shows the behavior of the cationic surfactant system; graph (h) is the result of a nonionic and anionic surfactant mixture.

headgroups. The results in Figure 15.24 confirm that, as the salt content in the formulation increases, the area of the microemulsion region is reduced, which represents a limitation to the applicability of the final mixture. Sodium is the counterion of the surfactant, which is also present in seawater. In view of the

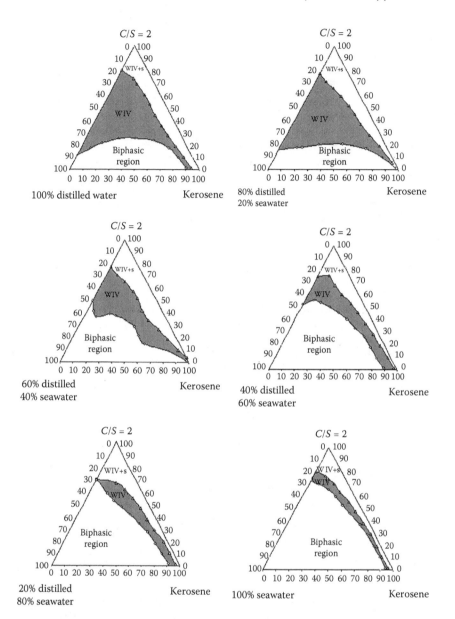

FIGURE 15.24 Effect of seawater content in microemulsion formation (WIV) for the system kerosene + aqueous phase + surfactant + butan-1-ol. C/S ratio = 2; all diagrams show a WIV region, a biphasic region and a region containing excess solid surfactant (S) with microemulsion.

common salt effect, the surfactant solubility in water is impaired and the extent of the WIV regions in the diagrams progressively diminishes.

The viability of the enhanced oil recovery technique is therefore dependent on the amount of microemulsion injected and the salt content in the medium. It is important to establish conditions that allow the formation of stable water-in-oil microemulsions to use their full potential in this application.

15.2.5.2   Use of Surfactant-Based Fracturing Gels

Insoluble residues left in the fractures of oil reservoirs by fracturing fluids have been the objective of many studies that focus on the discovery of novel, less harmful agents. Surfactant-based fracturing gels are considered as clean gels due to the fact that they do not leave insoluble residues in the rock formation after application. This kind of fluid has been developed to minimize or eliminate damages to fractures. In view of this, several investigations on gel rheological properties become more and more required. Viscosity, therefore, is considered as the most important property of fracturing gels.

Hydraulic fracturing is a technique that aims to increase well productivity. The fracturing fluid is applied against the reservoir rock under high differential pressure to create fractures. A proppant (sand, bauxite, or ceramic) is pumped into the well with the fracturing fluid with the purpose of maintaining the fracture open, creating a high-conductivity way that eases the flow of fluids between the formation point and the well. Damage caused by insoluble residues is effected by permeability reduction in the proppant pack or on the surfaces of the fractures themselves. Hence, novel surfactant-based fracturing fluids are thought to reduce the extent of damage [24,35]. Depending on the surfactant concentration and temperature, micelles of different geometries are formed, such as spheres or rods. The determination of the inner structure of surfactant solutions, for example by performing optical measurements, is only possible when the surfactant concentration is very low and the solution is at rest. Rheological measurements, on the other hand, offer an indirect way to determine the inner structure of surfactant solutions over a wide concentration range, and can often provide evidence of structural changes.

A representative example of this kind of approach involves examinations on new anionic surfactant-based gels. Steady and oscillatory shear experiments can be carried out to evaluate the inner structure of the gel prepared with variable composition, within the gel region in a pseudoternary diagram. In Figure 15.25, we present the pseudoternary diagram constructed at 26°C for a system formed by a commercial anionic surfactant derived from fatty acids (20%–30% vegetable oils and 70%–80% animal oils), from the Unilever Group. Isoamyl alcohol was used as cosurfactant, with a fixed C/S ratio equal to 0.5. Distilled water and pine oil as organic phase complete the system composition. It could be observed that small variations in gel composition resulted in changes in the micellar structure (points shown in the gel region).

There is an experimental evidence that gel samples are more commonly characterized as featuring lamellar microstructure, especially in high-surfactant concentration regions [56,57]. The structures thus obtained consist of bilayers with alternated surfactant/cosurfactant layers and aqueous phase. Typically translucid gels are obtained, many presenting some birefringence pattern under cross-polarized light. Another aspect of similar systems, which is also exhibited in Figure 15.25, is the formation of a complex collection of samples showing some

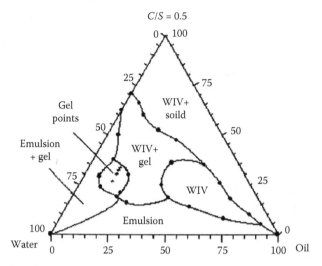

FIGURE 15.25 Pseudoternary diagram showing the gel region used in the formulation of fracturing fluids (temperature: 26°C). The gel points shown correspond to the specific compositions tested in rheological assays. (From Dantas, T.N.C., Santanna, V.C., Dantas Neto, A.A., Barros Neto, E.L., and Moura, M.C.P.A., *Coll. Surf. A*, 225, 129, 2003. With permission.)

kind of equilibrium, e.g., multiphase systems or WIV samples coexisting with a different species, like gels or excess solid surfactant.

In fracturing applications, it is important to employ translucid gels, in the appropriate consistency, to enable smooth pumping and reduced pressure drop levels. Plus, economical viability may be effected if the gels are localized in the water-rich region of the diagram. For example, with the system presented in Figure 15.25, one gel sample made up of 18% surfactant, 9% cosurfactant, 14% organic phase, and 59% aqueous phase may be used in fracturing experiments. For this particular sample, it is seen that temperature does not affect much its viscosity, which remains stable at around 100 mPa·s within a wide temperature range (30°–90°C), as compared to other samples, a direct effect of the branched structure of the cosurfactant used (isoamyl alcohol).

Oscillatory and stationary rheological measurements have been performed with three other gel samples (see points in the gel region of Figure 15.25) with different *C/S* concentrations and constant oil content. Table 15.6 resumes the compositions of the points tested. Stationary measurements were carried out by varying the temperature between 26°C and 86°C, at a constant shear rate of 100 s^{-1}, during 7 min at each temperature. Oscillatory measurements were done at constant pressure of 1 Pa and temperature of 66°C, by varying the frequency between 0.01 and 100 rad·s^{-1}. The temperature of 66°C is the average condition of on-shore wells in the Potiguar Basin (northeastern Brazil), where such studies were performed. These measurements are useful in detecting indirect changes in

TABLE 15.6
**Compositions of the Gel Samples Tested
in Rheological Assays**

Sample	Surfactant (%)	Cosurfactant (%)	C/S (%)	Oil (%)	Water (%)
1	18	9	27	14	59
2	20	10	30	14	56
3	21.3	10.7	32	14	54
4	22.7	11.3	34	14	52

the internal structure of the gels prepared. The viscosity of the samples varies markedly with temperature and C/S concentration, as indicated by stationary measurements. In contrast, viscoelastic gels are characterized in terms of loss and storage moduli which are slightly dependent on frequency. As a result, gels with 27% active matter (sample 1 in Table 15.6) feature a mild elastic behavior. As the C/S concentration increases, this behavior is more pronounced, to the point that highly elastic samples are obtained (sample 4 in Table 15.6). This is shown in Figure 15.26.

Another aspect of the structural changes observed in these systems is elucidated by the stationary measurements. The surfactant-based gels behave as non-Newtonian mixtures, and their activation energies can be determined with the well-known Arrhenius equation. If a positive activation energy is acquired, there

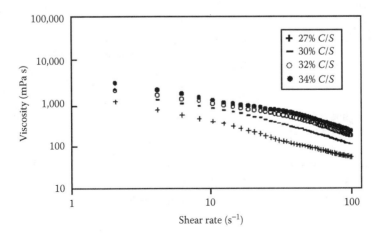

FIGURE 15.26 Rheological behavior of gel samples as a function of shear rate. (From Dantas, T.N.C., Santanna, V.C., Dantas Neto, A.A., Barros Neto, E.L., and Moura, M.C.P.A., *Coll. Surf. A*, 225, 129, 2003. With permission.)

is little variation in the volume of the micelles as temperature changes; when a negative activation energy results, the effect of volume variation of the micelles prevails in the system, as a result of intense water–surfactant interactions. The analogy with the model proposed by Winsor in his R-theory is notorious [14].

These results motivate continuous searches for alternatives in petroleum prospection activities, either by promoting complete changes in the physical apparatus used, or by improving the common methods with the application of novel chemical systems. Surfactant- and microemulsion-based systems are therefore presented as innovative and advantageous substitutes in such developments.

15.3 CONCLUSION

In this chapter, we have focused on some peculiarities of microemulsion systems, namely the internal structure, balance of intermolecular forces, and potentialities in mass transfer operations, which are advantageously exploited in applications in the petroleum industry. Several microemulsion systems may be used as alternatives to conventional techniques, for example, in dehydration of natural gas, corrosion inhibition on metallic surfaces, development of novel fuel formulations, and enhanced oil recovery.

The versatility of the microemulsion systems has been highlighted in terms of the mechanisms whereby several technological applications were effected. In particular, we discussed on how water absorption methods can be enhanced when microemulsions are used as desiccant fluids, and how mass transfer coefficients are affected by composition and variable thermodynamical conditions—especially temperature and pressure. Also, the ability of some molecules featuring specific moieties, such as aminated surfactants, to inhibit corrosion on metallic surfaces was pointed out, and the optimization studies involve previous, comprehensive information on the phase behavior of the system, since adsorption phenomena dominate in such applications. On a more recent departure, strategical projects have been conducted on the preparation of fuel alternatives, the main contribution of which being a reduction in the worldwide energy crisis. Microemulsion-based formulations are interesting particularly with regard to the lower emission levels of pollutants observed. However, continuous investigations on improved engine performance and fuel consumption are required to foster their use as compared to conventional petroleum-based fuels. Furthermore, the discovery of different surfactant-based fracturing gels is an important contribution to the petroleum industry, with direct implications on material processability and recovery yields. Finally, in enhanced oil recovery, the full engineering potential of microemulsion systems can alter the way by which the operations are carried out.

In summary, a general overview of microemulsion applications has been illustrated in petroleum-processing technologies, where dynamic mass transfer phenomena across interfaces and enhanced solubility levels are requirements that are perfectly fulfilled by the physicochemical nature of these self-assembled systems. This is further corroborated by the possibility of microemulsions recycling, profiting from their high load capacity, and ability to coexist in equilibrium with other phases.

In any case, all parameters that interfere with microemulsion formation and stability must be carefully examined when attempting to develop a particular application. Systems showing superior efficiencies may therefore be designed for each situation. We hope to have once again emphasized the technological importance of both micellar and microemulsion systems, especially in the petroleum industry, a line of work which has directed a relevant part of our research efforts in more recent times.

ACKNOWLEDGMENTS

The authors are indebted to the numerous students and coworkers who have joined the research group throughout the years, helping consolidate it as an important group in the colloid science area. The authors also acknowledge financial support from some Brazilian institutions: CNPq, CAPES, FINEP, and ANP-PRH 14, and all companies that have kindly supplied our group with surfactants and oil samples.

SYMBOLS AND TERMINOLOGIES

a_0 effective area of a surfactant headgroup
A_{ij} energy interactions between components i and j in a multicomponent system
C amount of water removed from natural gas via absorption, in ppm
CMC   critical micelle concentration
C/S cosurfactant/surfactant concentration ratio
cSt   viscosity unity *centistoke*
$E\%$ efficiency of the system in inhibiting corrosion
E_{corr} corrosion potential
HLB   hydrophilic–lipophilic balance
I_{corr} corrosion current density in the absence of inhibitor
I'_{corr} corrosion current density in the presence of inhibitor
L effective hydrocarbon chain length
LPG   liquefied petroleum gas
NO_x general denomination of nitrogen oxides
P_P packing parameter
SCO   saponified coconut oil
S_h effective area occupied by the hydrophobic part of a surfactant molecule
V volume of the hydrocarbon chain of a surfactant molecule
WIV   abbreviation given for a general microemulsion sample (Winsor IV system)
β_a Tafel's anodic curvature
β_c Tafel's cathodic curvature

REFERENCES

Observation: Although we have suggested a selection of representative international references to support this work, some dissertations and theses produced by a number of postgraduate students in our research group must be reported. These are numbered below from 21 to 30, listed with their original titles freely translated into English, and are further indicated by the term "in Portuguese" in brackets. Some scientific articles reporting on these results are also listed, and resulted from the works reported in the dissertations and theses. The full content of these works is available at http://acessolivre.capes.gov.br.

1. Hoar, T. P. and Schulman, J. H. 1943. Transparent water-in-oil dispersions: The oleopathic hydro-micelle. *Nature*, 152, 102–103.
2. Schulman, J. H., Stoeckenius, W., and Prince, L. M. 1959. Mechanism of formation and structure of micro emulsions by electron microscopy. *J. Phys. Chem.*, 63, 1677–1680.
3. Capek, I. 1999. Radical polymerization of polar unsaturated monomers in direct microemulsion systems. *Adv. Coll. Int. Sci.*, 80, 85–149.
4. Capek, I. 1999. Microemulsion polymerization of styrene in the presence of anionic emulsifier. *Adv. Coll. Int. Sci.*, 82, 253–273.
5. Capek, I. 2001. On the role of oil-soluble initiators in the radical polymerization of micellar systems. *Adv. Coll. Int. Sci.*, 91, 295–334.
6. Capek, I. 2001. Microemulsion polymerization of styrene in the presence of a cationic emulsifier. *Adv. Coll. Int. Sci.*, 92, 195–233.
7. Co, C. C., de Vries, R., and Kaler, E. W. 2001. Microemulsion polymerization. 1. Small-angle neutron scattering study of monomer partitioning. *Macromolecules*, 34, 3224–3232.
8. de Vries, R., Co, C. C., and Kaler, E. W. 2001. Microemulsion polymerization. 2. Influence of monomer partitioning, termination, and diffusion limitations on polymerization kinetics. *Macromolecules*, 34, 3233–3244.
9. Co, C. C., Cotts, P., Burauer, S., de Vries, R., and Kaler, E. W. 2001. Microemulsion polymerization. 3. Molecular weight and particle size distributions. *Macromolecules*, 34, 3245–3254.
10. Hentzer, H.-P. and Kaler, E. W. 2003. Polymerization of and within self-organized media. *Curr. Opin. Coll. Int. Sci.*, 8, 164–178.
11. Israelachvili, J. N., Mitchell, D. J., and Ninham, B. W. 1976. Theory of self-assembly of hydrocarbon amphiphiles into micelles and bilayers. *J. Chem. Soc. Faraday Trans. II*, 72, 1525–1568.
12. Mitchell, D. J. and Ninham, B. W. 1981. Micelles, vesicles and microemulsions. *J. Chem. Soc. Faraday Trans. II*, 77, 601–629.
13. Sjöblom, J., Lindberg, R., and Friberg, S. E. 1996. Microemulsions—phase equilibria characterization, structures, applications and chemical reactions. *Adv. Coll. Int. Sci.*, 65, 125–287.
14. Winsor, P. A. 1954. *Solvent Properties of Amphiphilic Compounds*. London: Butterworth.
15. Dantas Neto, A. A., Dantas, T. N. C., Moura, M. C. P. A., Gurgel, A., and Barros Neto, E. L. 2007. Extraction of heavy metals by microemulsions: A novel approach. In: Phillip B. Warey (Ed.), *New research on hazardous materials* (pp. 311–346). Hauppauge, NY: Nova Science Publishers.
16. Clint, J. H. 1992. *Surfactant aggregation*. New York: Chapman & Hall.

17. Rees, G. D. and Robinson, B. H. 1993. Microemulsions and organogels: Properties and novel applications. *Adv. Mater.*, 5, 608–619.
18. Solans, C. and García-Celma, M. J. 1997. Surfactants for microemulsions. *Curr. Opin. Coll. Int. Sci.*, 2, 464–471.
19. Moulik, S. P. and Paul, B. K. 1998. Structure, dynamics and transport properties of microemulsions. *Adv. Coll. Int. Sci.*, 78, 99–195.
20. Paul, B. K. and Moulik, S. P. 2001. Uses and applications of microemulsions. *Curr. Sci.*, 80, 990–1001.
21. Moura, E. F. 2002. Synthesis of novel aminated surfactants derived from castor oil and applications of micellar and microemulsion systems on carbon-steel corrosion inhibition, D.Sc. thesis, PPGEQ, Chemical Engineering Department, UFRN, Brazil (in Portuguese).
22. Silva Neto, M. A. 2002. Technical contribution of a vegetable oil-based reversed emulsion system for as drilling fluids, M.Sc. dissertation, PPGEQ, Chemical Engineering Department, UFRN, Brazil (in Portuguese).
23. Nóbrega, G. A. S. 2003. Water removal from natural gas via absorption using microemulsion systems, M.Sc. dissertation, PPGEQ, Chemical Engineering Department, UFRN, Brazil (in Portuguese).
24. Santanna, V. C. 2003. Preparation and study on the properties of a novel biocompatible hydraulic fracturing fluid, D.Sc. thesis, PPGEQ, Chemical Engineering Department, UFRN, Brazil (in Portuguese).
25. Wanderley Neto, A. O. 2004. Study on novel corrosion inhibitors for oil pipelines, M.Sc. dissertation, PPGQ, Chemistry Department, UFRN, Brazil (in Portuguese).
26. Fernandes, M. R. 2005. Development of a new diesel-based microemulsified fuel, M.Sc. dissertation, PPGEQ, Chemical Engineering Department, UFRN, Brazil (in Portuguese).
27. Curbelo, F. D. S. 2006. Petroleum enhanced recovery using surfactants, D.Sc. thesis, PPGEQ, Chemical Engineering Department, UFRN, Brazil (in Portuguese).
28. Moura, J. I. P. 2006. performance of a commercial inhibitor in the corrosion inhibition of API 5LX Gr X42 steel within chloride and oxygen media, M.Sc. dissertation, PPGEQ, Chemical Engineering Department, UFRN, Brazil (in Portuguese).
29. Paulino, L. C. 2007. Study on seawater-based microemulsion systems used in petroleum enhanced recovery, M.Sc. dissertation, PPGEQ, Chemical Engineering Department, UFRN, Brazil (in Portuguese).
30. Ribeiro Neto, V. C. 2007. Development of surfactant-based systems for petroleum enhanced recovery, M.Sc. dissertation, PPGEQ, Chemical Engineering Department, UFRN, Brazil (in Portuguese).
31. Dantas, T. N. C., Dantas Neto, A. A., and Moura, E. F. 2001. Microemulsion systems applied to breakdown petroleum emulsions. *J. Petr. Sci. Eng.*, 32, 145–149.
32. Dantas, T. N. C., Silva, A. C., and Dantas Neto, A. A. 2001. New microemulsion systems using diesel and vegetable oils. *Fuel*, 80, 75–81.
33. Dantas, T. N. C., Moura, E. F., Scatena Jr., H., and Dantas Neto, A. A. 2002. Microemulsion system as a steel corrosion inhibitor. *Corrosion*, 58, 723–727.
34. Dantas, T. N. C., Santanna, V. C., Dantas Neto, A. A., and Barros Neto, E. L. 2003. Application of surfactants for obtaining hydraulic fracturing gel. *Petr. Sci. Tech.*, 21, 1145–1157.
35. Dantas, T. N. C., Santanna, V. C., Dantas Neto, A. A., Barros Neto, E. L., and Moura, M. C. P. A. 2003. Rheological properties of a new surfactant-based fracturing gel. *Coll. Surf. A*, 225, 129–135.
36. Dantas, T. N. C., Santanna, V. C., Dantas Neto, A. A., and Moura, M. C. P. A. 2005. Hydraulic gel fracturing. *J. Disp. Sci. Tech.*, 26, 1–4.

37. Dantas, T. N. C., Santanna, V. C., Dantas Neto, A. A., Curbelo, F. D. S., and Garnica, A. I. C. 2006. Methodology to break test for surfactant-based fracturing gel. *J. Petr. Sci. Eng.*, 50, 293–298.

38. Dantas Neto, A. A., Dantas, T. N. C., Barros Neto, E. L., and Nóbrega, G. A. S. 2004. Process of natural gas dehydration with microemulsion, Brazilian Patent PI0401240.

39. Barros Neto, E. L., Dantas Neto, A. A., Fernandes, M. R., Dantas, T. N. C., and Moura, M. C. P. A. 2006. Diesel-based microemulsified fuel, Brazilian Patent PI0600616.

40. Fowler, A. E. and Protz, J. E. 1977. Gas dehydration with liquid desiccants and regeneration thereof, United States Patent 4,005,997.

41. Honerkamp, J. D. and Ebeling, H. O. 1983. System of gas dehydration using liquid desiccants, United States Patent 4,375,977.

42. Rice, A. W. and Murphy, M. K. 1988. Gas dehydration membrane apparatus, United States Patent 4,783,201.

43. Schievelbein, V. H. and Piglia, T. J. 1992. Glycol dehydration apparatus for natural gas, United States Patent 5,141,536.

44. Jullian, S., Lebas, E., and Thomas, M. 2001. Process for adsorbing and desorbing a solvent contained in a natural gas from a dehydration process, United States Patent 6,251,165.

45. Hoar, T. P. and Holliday, R. D. 1953. The inhibition by quinolines and thioureas of the acid dissolution of mild steel. *J. Appl. Chem.*, 3, 502–513.

46. Kaesche, H. and Hackerman, N. 1958. Corrosion inhibition by organic amines. *J. Electrochem. Soc.*, 105, 191–198.

47. Mann, C. A., Lauer, B. E., and Hultin, C. T. 1936. Organic inhibitors of corrosion: aliphatic amines. *Ind. Eng. Chem.*, 28, 159–163.

48. Banerjee, G. and Malhotra, A. T. 1992. Contribution to adsorption of aromatic-amines on mild-steel surface from HCl solutions by impedance, UV, and Raman-spectroscopy. *Corrosion*, 48, 10–15.

49. Hajjaji, N., Rico, I., Srhiri, A., Lattes, A., Soufiaoui, M., and Benbachir, A. 1993. effect of *n*-alkylbetaines on the corrosion of iron in 1-M HCl solution. *Corrosion*, 49, 326–334.

50. El-Achouri, M., Hajji, M.S., Kertit, S., Essassi, E.M., Salem, M., and Coudert, R. 1995. Some surfactants in the series of 2-(alkyldimethylammonio) alkanol bromides as inhibitors of the corrosion of iron in acid chloride solution. *Corros. Sci.*, 37, 381–389.

51. Dantas, T. N. C., Moura, E. F., Scatena Júnior, H., Dantas Neto, A. A., and Gurgel, A. 2002. Micellization and adsorption thermodynamics of novel ionic surfactants at fluid interfaces. *Coll. Surf. A*, 207, 243–252.

52. Friberg, S. E. and Force, E. G. 1976. Diesel fuel, Germany Patent 2526814 GE.

53. Silva, E. J., Zaniquelli, M. E. D., and Loh, W. 2007. Light-scattering investigation on microemulsion formation in mixtures of diesel oil (or hydrocarbons) plus ethanol plus additives. *Energy Fuels*, 21, 222–226.

54. Dantas, T. N. C., Moura, M. C. P. A., Dantas Neto, A. A., Pinheiro, F. S. H. T., and Barros Neto, E. L. 2007. The use of microemulsion and flushing solutions to remediate diesel-polluted soil. *Braz. J. Petr. Gas*, 1, 26–33.

55. Healy, R. N. and Reed, R. L. 1973. Physicochemical aspects of microemulsion flooding. SPE 4583 presented at the SPE-AIME 48th Annual Meeting, Las Vegas, USA; Healy, R. N. and Reed, R. L. 1974. Physicochemical aspects of microemulsion flooding. *Soc. Petr. Eng. J.*, 14, 491–501.

56. Gurgel, A., Ferreira, M. S., and Loh, W. 2007, Phase equilibria and elucidation of self-assembly structures in systems containing silicone oil, water and silicone surfactant. In: *Proceedings of the 21st Conference of the European Colloid and Interface Society, ECIS 2007,* Geneva, Switzerland, v. 1. p. 474.
57. Berni, M. G., Lawrence, C. J., and Machin, D. 2002. A review of the rheology of the lamellar phase in surfactant systems. *Adv. Coll. Int. Sci.*, 98, 217–243.

16 Nanoparticle Formation in Microemulsions: Mechanism and Monte Carlo Simulations

M. de Dios, F. Barroso, and C. Tojo

CONTENTS

16.1 INTRODUCTION

Microemulsions are colloidal nanodispersions of water in oil (or oil in water) stabilized by a surfactant film. These thermodynamically stable dispersions can be considered as true nanoreactors, which can be used to synthesize nanomaterials. The main idea behind this technique is that by appropriate control of the synthesis

451

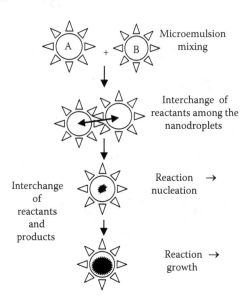

Microemulsion mixing

Interchange of reactants among the nanodroplets

Interchange of reactants and products

Reaction → nucleation

Reaction → growth

FIGURE 16.1   Microemulsion technique scheme.

parameters one can use these nanoreactors to produce tailor-made products down to a nanoscale level. The technique is very simple, as can be seen in Figure 16.1. Two reactants (A and B) are introduced in two identical microemulsions. After mixing both emulsions, droplets collide and interchange the reactants. Then the reaction can take place inside the nanoreactors.

Why in microemulsions? The surfactant-stabilized nanodroplets provide a cage-like effect that limits particle nucleation, growth, and agglomeration. It is a powerful method for controlling particle sizes and size distribution. The nanoparticle formation in microemulsions is controlled by the reactant distribution in the nanodroplets and by the dynamics of intermicellar exchange [1]. Since the synthesis of metal nanoparticles in microemulsions was first reported by Boutonnet et al. [2], the microemulsion technique has been successfully used for the preparation of a number of nanoparticles. Different metallic and bimetallic particles have been prepared using just the same microemulsion and the same reductor agent, the metal being the only difference [3–7]. Although experimental results are not completely consistent, they were all ascribed to a difference in the nucleation process, which could be affected by the material, the reaction media, and the preparation conditions. It was proposed that when the reduction rate was so large that most ions were reduced before the formation of nuclei, particle size would be uniform and determined by the number of nuclei formed at the very beginning of the reaction [5]. That is, nucleation determined the final nanoparticle sizes.

Nucleation is a much-studied phenomenon, although the rate at which it occurs remains difficult to predict. Small nuclei form spontaneously in supersaturated solutions, but unless their size exceeds a critical limit they will redissolve rather than grow. The nucleation rate depends on the very small probability that a critical nucleus forms spontaneously, and on a kinetic factor that measures the rate at which critical nuclei subsequently grow [8]. Considering the formation of a nucleus, the total Gibbs energy might vary with the compositions due to the different bond enthalpies or interactions between metals, and hence lead to the number of atoms required for the formation of a nucleus. Experimental results show that, using the same microemulsion, different metals need different time to reach the final sizes, a fast formation rate being related to a smaller final nanoparticle size [3–7]. Given the absence of a priori knowledge of either quantity, classical nucleation theory is commonly used, in spite of the compartmentalization of the microemulsion media. But previous simulation results [9,10] suggest that the usual classical crystallization theory used to explain the crystallization process in homogeneous media has to be modified to include the material interchange among the nanodroplets. A number of parameters influence this interchange [9–23]. Due to Brownian motion, droplets collide forming a transient dimer and interchange reactants with a rate constant k_{ex}. This interchange is strongly determined by the elasticity of the surfactant film (see later). Because a lot of parameters enter into consideration, different algorithms have been developed to study nanoparticle formation in microemulsions [9–10,17–23]. The objective of this chapter is to study the role of two parameters which strongly affects nucleation: the critical nucleus size and the chemical reaction rate. This study allows us to understand how these parameters affect final nanoparticle size. A comparison study between experimental and simulation results supports the conclusions of this study.

16.2 SIMULATION PROCEDURE

Each simulation begins with the microemulsion droplets randomly located on a lattice (see Ref. [24] for details). In this study, $\varphi = 10\%$ portion of the space is occupied by droplets. The reactant species were distributed throughout the droplets using a Poisson distribution, $\langle c_A \rangle$ and $\langle c_B \rangle$ being the mean number of reactants inside each droplet. Initially, droplets performed random walks to nearest neighbor sites, colliding when they occupy contiguous lattice sites. The algorithm was improved to save computation time, and now the two colliding droplets are chosen at random at each step.

One of the most important points in the algorithm was how to simulate the microemulsion dynamics: The droplets are not static, but are in continuous movement and collide with each other, forming a fused dimer. At this stage, material interchange can take place if a channel is opened through the surfactant bilayer. The ease with which channels can form to communicate colliding droplets and the size of these channels are also governed by the surfactant film flexibility. The dimer formation is a slow process because it requires the opening of the micellar walls and the inversion of the surfactant film curvature (see Figure. 16.2). Dimer stability depends on intermicellar attractive potential, which depends on surfactant, solvent, and cosurfactant. In addition, the rate permeation through the channel would be highly material dependent: a voluminous and highly charged ion

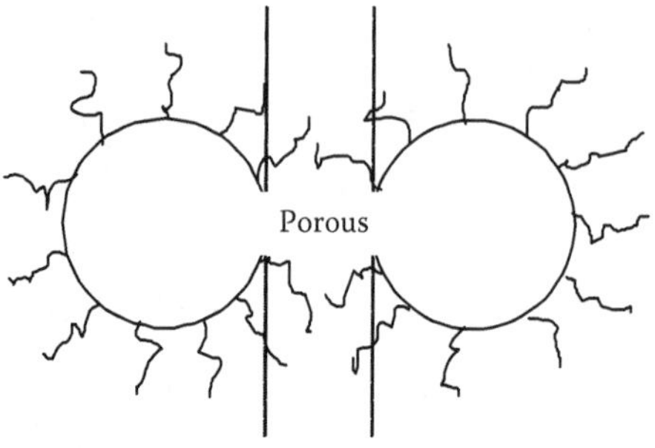

FIGURE 16.2   Channel through surfactant bilayer.

will hardly cross the channel, whereas a small, neutral molecule will cross more easily. A highly flexible film will allow the interchange of larger particles than a rigid film. From this picture, we can simulate the exchange of reactants: The larger the dimer stability, the longer the two water pools remain together, and more reactants can be transferred during a collision. The reactants are redistributed in agreement with the concentration gradient principle: Mass transfer takes place from the droplet containing more reactants to the droplet with fewer reactants. The reactant exchange parameters, $k_{ex,A}$ and $k_{ex,B}$, determine how many units of reactant (A or B) could be transferred during a collision.

16.2.1 CHEMICAL REACTION

When both reactants are located inside the same water pool, chemical reaction can take place ($A + B \rightarrow P$). By changing the reaction rate, we can study not only the reactions controlled by the interdroplet exchange, but also the reactions in which both parameters (chemical rate and exchange) can play an important role. Only a percentage, v_r, of reactants inside the colliding droplets gives rise to products. The reactants (which did not react) remain in the water pool. These can be exchanged and will react during a later collision. The fastest reaction corresponds to $v_r = 1$ (100% reactants transform into products).

16.2.2 NUCLEATION

Nucleation is the process by which atoms (or ions) that are free in solution come together to produce a thermodynamically stable cluster. The cluster must exceed the critical nucleus size, n^*, and then it becomes a supercritical nucleus capable of further growth. If the nucleus is smaller than the critical size, spontaneous dissolution can occur. In the simulation, if the number of products inside the same droplet is smaller than the n^* parameter, they are considered free inside the droplet. When the number of products is equal or greater than n^*, they come together forming a stable nucleus. This nucleus has to be exchanged as a whole.

16.2.3 EXCHANGE

To discuss the exchange of products, it is important to take into account Ostwald ripening. Ripening theory assumes that the largest particles will grow by condensation of material coming from the smallest particles, which solubilize more than the larger ones. This mass transfer is limited by the intermicellar channel size in the dimer, which depends on film flexibility. We can relate the film flexibility around the droplets and the ease with which channels communicating colliding droplets can form. Surfactant film flexibility therefore places a limit on the size of the particles traversing the droplet–droplet channels. From this picture, two different cases can be found. First, aggregates of products (growing particles): The surfactant film flexibility is simulated by varying a parameter f specifying a maximum particle size for transfer between droplets. In this way, a highly flexible film will allow the interchange of larger aggregates than a rigid film. Second, free products: This interchange is governed by the $k_{ex,P}$ parameter.

16.2.4 AUTOCATALYSIS

As the reaction takes place, more droplets contain reactants and the growing nucleus simultaneously. The interchange of reactants between two colliding droplets in the presence of a growing particle allows us to simulate an autocatalytic reaction, which will be catalyzed by the surface of a growing particle. When both droplets are carrying aggregates, autocatalysis takes place on the bigger one. (A larger aggregate has a greater probability of playing as a catalyst because of its bigger surface.) Autocatalysis and ripening favor the growth of the biggest nanoparticles, and can be considered as possible ways of growth. The intermicellar exchange plays an important role in both process, but an autocatalytic growth depends on the intermicellar exchange of reactants. On the contrary, a growth by ripening depends on the intermicellar exchange of small growing nanoparticles.

Finally, the simulation allows the modification of the droplet size with the q parameter, which restricts the maximum number of products which can be carried by a droplet. In this work, droplets do not restrict nanoparticles growth.

16.3 RESULTS AND DISCUSSION

16.3.1 CRITICAL NUCLEUS SIZE

According to classical crystallization theory, the final particle size depends mainly on the ratio of nucleation to growth rates. A slow nucleation leads to a low number of nuclei, which can grow and reach large polydisperse sizes. On the contrary, if nucleation is quick, a large number of nuclei are formed so that the final particle size will be small and monodisperse.

To study the influence of the critical nucleus size on nucleation, Figure 16.3 shows the initial stage of nanoparticle formation. The number of small particles (aggregates constituted by less than five P units), which simulate nuclei, is plotted versus the initial time (the first 120 steps), at different critical

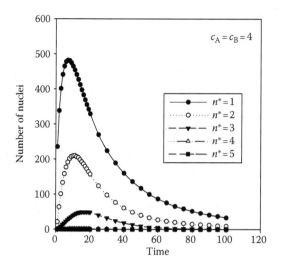

FIGURE 16.3 Time evolution of the number of droplets carrying nuclei ($c_A = c_B = 4$, $f = 5$, $k_{ex,A} = k_{ex,B} = k_{ex,P} = 1$, and $v_r = 1$). (From Tojo, C., Barroso, F., and de Dios, M., *J. Colloid Interface Sci.*, 296, 591, 2006. With permission.)

nucleus size n^*. One can observe the existence of two well-defined processes: nucleation and growth. Nucleation is associated with an increase in the number of nuclei and growth is associated with a decrease. As expected, the nucleation rate is faster as the critical nucleus is smaller. As increases n^*, more time is needed to locate inside the same droplet the minimum number of products needed to reach a stable nucleus. Consequently, the nucleation rate is slowing down and less nuclei are formed.

The critical nucleus number effect on growth can be observed in Figure 16.4, which shows the average nanoparticle size during the whole process. It is interesting to point out the inversion of the curves: an increase of critical nucleus leads to a slowdown of the nucleation, but it leads also to an acceleration of the growth. This faster growth at high critical nucleus gives rise to larger sizes. To compare the simulation results with experimental data, Figure 16.4B represents the variation of the mean diameter with reaction time for Au and Pt nanoparticles (data taken from Ref. [4]). The fast formation process associated with Au nanoparticles could be related to a small critical nucleus number n^*, and a value $n^* = 1$ could be assigned to Au. The slower Pt nucleation rate allows us to suggest a higher critical nucleus number $n^* = 3$. By comparing both kinds of experiments, one can conclude that a more difficult nucleation leads to a growth during more time, giving rise to larger particles. This behavior is explained by taking into account that a small value of n^* implies the quick formation of a large number of seed nuclei, which will compete for the available reactant to grow. On the contrary, a larger critical nucleus

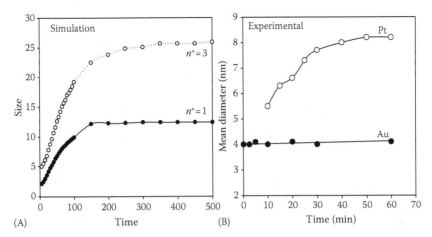

FIGURE 16.4 Variation of mean nanoparticle size versus reaction time. (A) Simulation results for $n^* = 1$ (black symbols) and $n^* = 3$ (white symbols) ($c_A = c_B = 4$, $f = 5$, $k_{ex,A} = k_{ex,B} = k_{ex,P} = 1$, and $v_r = 1$). (B) Au (black symbols) and Pt (white symbols) nanoparticles synthesized in water/AOT/isooctane. (From Tojo, C., Barroso, F., and de Dios, M., *J. Colloid Interface Sci.*, 296, 591, 2006. With permission.)

size will lead to a slow formation of less number of nuclei, which grow easier because the reactants/nuclei ratio is higher. In this case, the critical nucleus number effect would be attenuated if concentration increases, as one can observe in Figure 16.5.

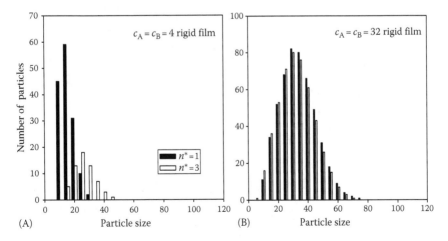

FIGURE 16.5 Final nanoparticle size distribution for different n^*: $n^* = 1$ (black) and $n^* = 3$ (white). (A) Low and (B) high concentration ($f = 5$, $k_{ex,A} = k_{ex,B} = k_{ex,P} = 1$, and $v_r = 1$).

16.3.2 CHEMICAL REACTION RATE

Figure 16.6 shows nucleation. Each curve corresponds to a different value of the reaction rate, v_r, keeping constant the other synthesis variables ($\varphi = 10\%$, initial reactant concentration inside the droplets $c_A = c_B = 4$, $n^* = 1$). To compare with experimental results obtained in an AOT microemulsion, a rigid film was used ($f = 5$ and $k_{ex} = 1$). Similar qualitative simulation results have been obtained for different film flexibilities and for different reactant concentrations. As expected, the nucleation process depends on chemical reaction rate: the maximum decreases and it appears longer when the chemical reaction rate is slower. As v_r decreases, the probability of the reaction between A and B is smaller and fewer nuclei are formed at the beginning of the process. Consequently, the nucleation is slower and the maximum is reached at longer times.

To study nanoparticle growth, Figure 16.7 shows the size distribution during the course of the synthesis, for fast and slow reactions. The last histograms correspond to the equilibrium distribution, and the first and second histograms correspond to 3% and 50% conversion, respectively. At the beginning of the process, large quantity of droplets carrying very small particles appears. At 50% conversion, a very different behavior appears depending on the reaction rate. When the reaction is fast, most of the nanoparticles have already reached the final size, and a very low number of the smallest particles are appearing. However, a high number of nuclei are still forming at this stage when the reaction is slow. At the same time, some particles have already grown to the final value, i.e., nucleation and growth take place simultaneously. This overlapping of both processes, which is more pronounced as the reaction is slower, leads to larger and more polydisperse final sizes. Similar qualitative behaviors are found at different concentrations and film flexibilities. A slow chemical reaction favors a continuous nuclei production,

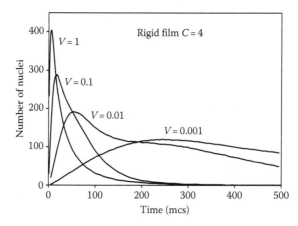

FIGURE 16.6 Temporal evolution of the number of nuclei at different values of reaction rate v_r. Synthesis conditions: $c_A = c_B = 4$, $f = 5$, $k_{ex} = 1$, $\varphi = 10\%$, and $n^* = 1$.

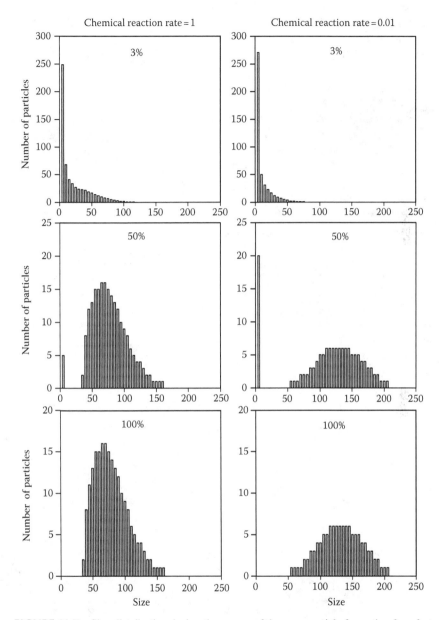

FIGURE 16.7 Size distribution during the course of the nanoparticle formation for a fast (left column, $v_r = 1$) and a slow (right column, $v_r = 0.01$) chemical reactions. Synthesis conditions: $c_A = c_B = 4, f = 5, k_{ex,} = 1, \varphi = 10\%$, and $n^* = 1$. (From de Dios, M., Barroso, F., Tojo, C., Blanco, M.C., and Lopez-Quintela, M.A., *Colloid. Surface. A. Physiochem. Eng. Aspect.*, 270, 83–87, 2005. With permission.)

keeping always a certain number of nuclei in the system. As a result, growth by ripening can take place during the whole process, and a bigger particle size is obtained for a slow reaction. Therefore, a slow chemical reaction rate can be associated with a more important ripening contribution (see Ref. [25] for details).

One can conclude that chemical reaction rate affects both kinds of growth. If the reaction is slow, growth by autocatalysis takes place for a longer time because there are available reactants, contributing to an increase in the biggest nanoparticles. Growth by ripening also takes place longer, because of the continuous nuclei production associated to a slow reaction. Both factors lead to a bigger particle size if the reaction is slower. This fact is reflected in Figure 16.8, which shows the time evolution of the average nanoparticle size at different reaction rates, keeping constant the synthesis conditions.

These results allow us to explain experimental data concerning the formation rates of Au and Ag nanoparticles by reduction with hydrazine in an AOT–isooctane–water microemulsion [6]. The Ag nanoparticle formation was completed in 30 min, and the final Ag nanoparticle size was 13 nm. The final size of Au nanoparticles was reached in 4 s, and the diameter of Au nanoparticles was 5 nm. Therefore, using exactly the same synthesis conditions, Ag nanoparticles are bigger and the formation rate is slower than the corresponding values for Au nanoparticles. Since the concentration and the film flexibility were fixed, the factor affecting nucleation rate might be the critical nucleus number. But both metals crystallize in an fcc lattice with very close lattice constants (4.078 Å for Au, 4.086 Å for Ag), so it can be expected to have a similar n^* value. Although these results have been explained by assuming that both chemical reaction rates are instantaneous, but the nucleation rate of Au is much faster than of Ag [6], Manna et al. [26] explained

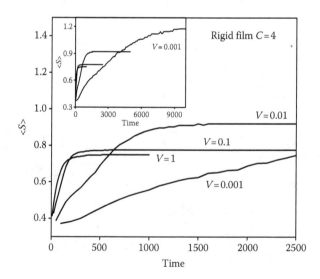

FIGURE 16.8 Time evolution of the average nanoparticle size at different values of reaction rate v_r. Synthesis conditions: $c_A = c_B = 4$, $f = 5$, $k_{ex} = 1$, $\varphi = 10\%$, and $n^* = 1$.

the size difference between Au and Ag on the basis of the standard redox potentials (1.002 eV for Au^{3+}/Au and 0.79 eV for Ag^+/Ag [27]). The higher potential of Au^{3+}/Au pair favors the faster rate of reduction, resulting in the formation of smaller metal particles. By comparing experimental and simulation data, and in the light of previous discussion, the difference in the Au and Ag nanoparticle formation can be explained on the basis of different chemical reaction rates. The faster formation rate and the smaller nanoparticle sizes of Au allow us to associate Au to a high reaction rate. On the contrary, the slower Ag formation rate and the larger size suggest a low reaction rate. The reduction rate extremely affects nucleation and growth, making easier the material intermicellar exchange as the reaction rate decreases. Consequently, longer growth is observed and bigger nanoparticles are obtained.

16.4   CONCLUSION

Critical nucleus value strongly affects the nucleation and growth rates of nanoparticles synthesized in microemulsions. An increase of critical nucleus leads to a slowdown of the nucleation rate and to an acceleration of the growth. As a consequence, larger nanoparticle sizes are obtained for a higher critical nucleus. It is predicted that this dependence will be less pronounced at high concentrations. On the other hand, a slow chemical reaction makes possible the material exchange in later stages, favoring autocatalysis and ripening. Both factors contribute to an increase in the biggest nanoparticles and to a longer formation process. Consistency between experimental and simulation results supports the conclusions of this study.

ACKNOWLEDGMENTS

The authors are grateful for the financial support of the Ministerio de Educación y Ciencia (CTQ2006-04085/BQU) and Xunta de Galicia (PGIDIT06PXIB383004PR).

SYMBOLS AND TERMINOLOGIES

$\langle c_A \rangle$	mean number of A reactants inside each droplet at the beginning
$\langle c_B \rangle$	mean number of B reactants inside each droplet at the beginning
f	maximum particle size for transfer between droplets (surfactant film flexibility)
$k_{ex,A}$	number of units of reactant A which could be transferred during a collision (interdroplet exchange constant of the A reactant)
$k_{ex,B}$	number of units of reactant B which could be transferred during a collision (interdroplet exchange constant of the B reactant)
$k_{ex,P}$	number of units of product P which could be transferred during a collision (interdroplet exchange constant of the P product)
v_r	chemical reaction rate
n^*	critical nucleus size
φ	portion of the space occupied by droplets

REFERENCES

1. López-Quintela, M.A. 2003. Synthesis of nanomaterials in microemulsions: Formation mechanisms and growth control. *Current Opinion in Colloid and Interface Science*, 8, 137–144.
2. Boutonnet, M., Kizling, J., Stenius, P., and Maire, G. 1982. The preparation of monodisperse colloidal metal particles from microemulsions. *Colloids and Surfaces*, 5, 209–225.
3. Wu, M.L., Chen, D.H., and Huang, T.C. 2001. Synthesis of Au/Pd bimetallic nanoparticles in reverse micelles. *Langmuir*, 17, 3877–3883.
4. Wu, M.L., Chen, D.H., and Huang, T.C. 2001. Preparation of Au/Pt bimetallic nanoparticles in water-in-oil microemulsions. *Chemistry of Materials*, 13, 599–606.
5. Wu, M.L. and Lai, L.B. 2004. Synthesis of Pt/Ag bimetallic nanoparticles in water-in-oil microemulsions. *Colloids and Surfaces, A: Physicochemical and Engineering Aspects*, 244, 149–157.
6. Chen, D.H. and Chen, C.J. 2002. Formation and characterization of Au–Ag bimetallic nanoparticles in water-in-oil microemulsions. *Journal of Materials Chemistry*, 12, 1557–1562.
7. Wu, M.L., Chen, D.H., and Huang, T.C. 2001. Preparation of Pd/Pt bimetallic nanoparticles in water/AOT/isooctane microemulsions. *Journal of Colloid and Interface Science*, 243, 102–108.
8. Auer, S. and Frenkel, D. 2001. Prediction of absolute crystal-nucleation rate in hard-sphere colloids. *Nature*, 409, 1020–1023.
9. Tojo, C., Blanco, M.C., and López-Quintela, M.A. 1998. The influence of reactant excess and film flexibility on the mechanism of nanoparticle formation in microemulsions: A Monte Carlo simulation. *Langmuir*, 14, 6835–6839.
10. Kumar, A.R., Hota, G., Mehra, A., and Khilar, K. 2004. Modeling of nanoparticles formation by mixing of two reactive microemulsions. *AIChE Journal*, 50, 1556–1567.
11. Bagwe, R.P. and Khilar, K.C. 1997. Effects of the intermicellar exchange rate and cations on the size of silver chloride nanoparticles formed in reverse micelles of AOT. *Langmuir*, 13, 6432–6438.
12. Bagwe, R.P. and Khilar, K.C. 2000. Effects of intermicellar exchange rate on the formation of silver nanoparticles in reverse microemulsions of AOT. *Langmuir*, 16, 905–910.
13. Curri, M.L., Agostiano, A., Manna, L., Della Monica, M., Catalano, M., Chiavarone, L., Spagnolo, V., and Lugarà, M. 2000. Synthesis and characterization of CdS nanoclusters in a quaternary microemulsion: The role of the cosurfactant. *Journal of Physical Chemistry B*, 104, 8391–8397.
14. Kitchens, C.L., McLeod, M.C., and Roberts, C.B. 2003. Solvent effects on the growth and steric stabilization of copper metallic nanoparticles in AOT reverse micelle systems. *Journal of Physical Chemistry B*, 107, 11331–11338.
15. Cason, J., Miller, M.E., Thompson, J.B., and Roberts, C.B. 2001. Solvent effects on copper nanoparticle growth behavior in AOT reverse micelle systems. *Journal of Physical Chemistry B*, 105, 2297–2302.
16. Eastoe, J. and Warne, B. 1996. Nanoparticle and polymer synthesis in microemulsions. *Current Opinion in Colloid and Interface Science*, 1, 800–805.
17. Tojo, C., Blanco, M.C., and López-Quintela, M.A. 1997. Preparation of nanoparticles in microemulsions: A Monte Carlo study of the influence of the synthesis variables. *Langmuir*, 13, 4527–4534.

18. Quintillán, S., Tojo, C., Blanco, M.C., and López-Quintela, M.A. 2001. Effects of the intermicellar exchange on the size control of nanoparticles synthesized in microemulsions. *Langmuir*, 17, 7251–7254.

19. Tojo, C., Blanco, M.C., and López-Quintela, M.A. 1998. Microemulsions as microreactors: A Monte Carlo simulation on the synthesis of particles. *Journal of Non-Crystalline Solids*, 235–237, 688–691.

20. López-Quintela, M.A., Rivas, J., Blanco, M.C., and Tojo, C. 2003. Synthesis of nanoparticles in microemulsions. In L.M. Liz-Marzán, P.V. Kamat (Eds.), *Nanoscale Materials*, Kluwer Academic Publishers, Boston, pp. 135–155.

21. Li, Y. and Park, C.-W. 1999. Particle size distribution in the synthesis of nanoparticles using microemulsions. *Langmuir*, 15, 952–956.

22. Bandyopadhyaya, R., Kumar, R., and Gandhi, K.S. 2000. Simulation of precipitation reactions in reverse micelles. *Langmuir*, 16, 7139–7149.

23. Jain, R. and Mehra, A. 2004. Monte Carlo models for nanoparticle formation in two microemulsion systems. *Langmuir*, 20, 6507–6513.

24. Tojo, C., Barroso, F., and de Dios, M. 2006. Critical nucleus size effects on nanoparticle formation in microemulsions: A comparison study between experimental and simulation results. *Journal of Colloid and Interface Science*, 296, 591–598.

25. de Dios, M., Barroso, F., Tojo, C., Blanco, M.C., López-Quintela, M.A., Colloid. Surface A: Physicochem. Eng. Aspect, 270, 83–87, 2005.

26. Manna, A., Imae, T., Yogo, T., Aoi, K., and Okazaki, M. 2002. Synthesis of gold nanoparticles in a Winsor II type microemulsion and their characterization. *Journal of Colloid and Interface Science*, 256, 297–303.

27. Emsley, J. 1998. *The Elements*, 3rd edn. Clarendon, Oxford, p. 87.

17 Nanoparticle Uptake by (W/O) Microemulsions

Maen M. Husein and Nashaat N. Nassar

CONTENTS

17.1 INTRODUCTION

Nanoparticles meet the need for materials with specific physical, chemical, and electronic properties by virtue of their size-dependent properties [1–5]. Consequently, ability to control particle size is considered an essential aspect of a nanoparticle preparation technique. (w/o) Microemulsions have been extensively used to prepare different types of nanoparticles [4–13]. These systems provide good control over particle size and produce highly homogeneous particles, due to their efficient mixing at the molecular level [14–18]. Moreover, (w/o) microemulsions were employed to form core-shell and onion-structured nanoparticles [19], which find applications in catalysis. Particle size manipulation in these systems is achieved by controlling some microemulsion and operating variables [4,8,20–28]. The effect of a given variable is, however, dependent on the reactant addition scheme [20,21,26–28]. Two reactant addition schemes have been identified in the literature: the mixing of two microemulsions and the single microemulsion schemes. The first scheme involves adding the precursors to two identical (w/o) microemulsions followed by mixing the microemulsions. This mode of preparation typically limits rapid reactions of nanoparticle formation to the rate of opening of the surfactant surface layer [7,8,29,30]. Slow rate of successful collisions leads to simultaneous nucleation and growth and produces large and polydispersed particles [29]. This, in turn, limits the ability to control particle size. The review by Eastoe et al. [20] reported on the sometimes conflicting trends found in the literature on the effect of a given microemulsion or operating variable on the final particle size. The single microemulsion scheme, on the other hand, involves sequential addition of precursors to the same microemulsion. This mode of reactant addition allows for *intramicellar* nucleation and growth, thus reduces the impact of intermicellar nucleation and growth on the final particle size. A review of trends in nanoparticle size in response to changes in microemulsion and operating variables has been recently communicated by Husein and Nassar [26]. It is worth noting that the use of microemulsions with reactive surfactants, functionalized surfactants, is more common in the single microemulsion scheme. The single microemulsion scheme requires the use of lesser microemulsion volumes and is more suited for in situ preparation of nanoparticles [21,22,31–33].

In general, microemulsion systems suffer from low-reactant solubilization and product stabilization capacities [34]. Nonetheless, for some applications, including nanofluids, dip coating, and ultradispersed catalysis in organic media, (w/o) microemulsions serve as an ideal preparation medium. Maintaining the highest-possible concentration of stabilized colloidal nanoparticles is essential for these applications. Husein and coworkers introduced the term nanoparticle uptake to refer to the time-independent maximum possible concentration of colloidal nanoparticles in the (w/o) microemulsions for given values of microemulsion and operating variables [24,25,35]. For ultradispersed catalysis in organic media, the high concentration of stable colloidal nanoparticles should be coupled with small particle size to achieve high-surface area per liter of the reactive system.

This chapter reviews the work pertaining to nanoparticle uptake and discusses the early experiments that led to introducing this concept and its evolution. Single (w/o) microemulsions with reactive and nonreactive surfactants were employed during these investigations.

17.2   NANOPARTICLE UPTAKE IN REACTIVE SURFACTANT SYSTEMS

The work on nanoparticle uptake stemmed from an investigation on silver halide nanoparticle preparation in single microemulsions formed with surfactants having the halide counterion [27,28,36]. Three different surfactants were employed, namely, dioctyldimethylammonium chloride ($(C_8H_{17})_2(CH_3)_2N^+Cl^-$) [28], dioctyldimethylammonium bromide ($(C_8H_{17})_2(CH_3)_2N^+Br^-$) [35], and cetyltrimethylammonium bromide (CTAB) [27]. In these experiments, when excess amount of the $AgNO_{3(aq)}$ precursor was added, bulk precipitate of silver halide formed. This precipitate disappeared after 24 h. UV-spectroscopy of the samples after 24 h revealed a peak specific to colloidal silver halide nanoparticles. In light of this observation, separate experiments involving mixing bulk powders of AgCl and AgBr with the microemulsions formed with surfactants having the corresponding halide counterion were undertaken. These experiments confirmed the formation of colloidal silver halide nanoparticles from their powder precursors. More detailed experimental procedure can be found elsewhere [24,25,35]. The resultant particles were analyzed for their size and colloidal concentration. The concentration of the nanoparticles increased at early stages of mixing and leveled off after sometime. The time–invariant concentration of the colloidal silver halide nanoparticles corresponded to the maximum concentration that can be maintained stable in the (w/o) microemulsions was termed the nanoparticle uptake. The following mechanism was put forward to explain the formation of the silver halide nanoparticles from their bulk powder precursors.

17.2.1   Particle Formation Mechanism

First, surfactant molecules together with their water of hydration adsorb at the surface of the powder. In this environment, the concentration of the halide counterion is very high due to the limited amount of water, and as a result soluble higher halides formed [37]. Surfactant molecules at the solid surface and the reverse-micellar surface exchange by virtue of the dynamic nature of the (w/o) microemulsions allowing the transfer of the soluble halide to the reverse-micelles interface. Once at the interface, the soluble halides diffuse to the bulk water pool where they precipitate as solid monomers due to the accompanying dilution. *Intramicellar* nucleation commences in water pools containing critical number of the halide monomers at a rate governed by the degree of supersaturation [38–40]. Particle growth follows by adding more of the silver halide monomers. Intermicellar nucleation and growth also contribute to particle formation when conditions leading to low-surfactant layer rigidity dominate [7,8,29,30]. Further growth and agglomeration are prevented by the surfactant protective layer, and the concentration of the colloidal nanoparticles and their size distribution become time-independent. Particles growing beyond that range precipitate under gravity effect.

The discussion below details the effect of some microemulsion and operating variables on the nanoparticle uptake and particle size for the reactive surfactant systems.

17.2.2 EFFECT OF MIXING

Since, and in accordance with the proposed mechanism, particle formation comprises of reaction and mass transfer steps, it was essential to explore the effect of mixing on the rate of particle formation. Figure 17.1 depicts typical trends obtained for mixed and unmixed samples. This figure belongs to AgCl nanoparticle formation from bulk AgCl powder in $(C_8H_{17})_2(CH_3)_2N^+Cl^-$ microemulsions. Figure 17.1 clearly shows that mixing improves the rate of nanoparticle formation and shortens the time required to attain nanoparticle uptake. These experiments, however, could not provide an answer to what is the controlling step, since the rate of mass transfer and the surface area for reaction were allowed to simultaneously vary during the experiment. Particle size distribution histograms collected from transition electron microscopy (TEM), photographs of the nanoparticles were almost identical [24,25,35]. This suggests the existence of a rigid surfactant layer, which limited particle aggregation even with increasing rate of collision at higher rates of mixing. Rigid surfactant layer favors *intramicellar* nucleation and growth, and limits the role of intermicellar nucleation and growth.

17.2.3 EFFECT OF TEMPERATURE

Temperature affects the stability of (w/o) microemulsion systems [41,42], and consequently, the stability of the colloidal nanoparticles. Moreover, temperature

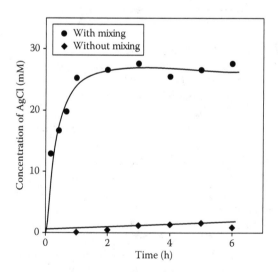

FIGURE 17.1 Effect of mixing time on the concentration of AgCl nanoparticles in $(C_8H_{17})_2(CH_3)_2N^+Cl^-$ (w/o) microemulsions. (Reprinted from Husein, M., Rodil, E., and Vera, J.H., *J. Colloid Interface Sci.*, 288, 457, 2005. With permission.)

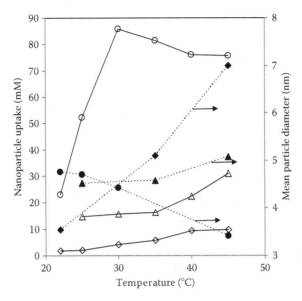

FIGURE 17.2 Variation of nanoparticle uptake (open symbols) and mean particle diameter (solid symbols) upon increasing temperature for ($\triangle$,$\blacktriangle$) $(C_8H_{17})_2(CH_3)_2N^+Br^-$ [35], ($\circ$,$\bullet$) $(C_8H_{17})_2(CH_3)_2N^+Cl^-$ [25], and ($\diamond$,$\blacklozenge$) CTAB [26] (w/o) microemulsions.

increases rates of reaction and influences rates of mass transfer. Figure 17.2 is a plot of the silver halide nanoparticle uptake and mean size as a function of temperature for the dioctyldimethylammonium chloride and bromide and CTAB. It is worth noting that in the temperature range presented, stable (w/o) microemulsion systems were maintained.

Figure 17.2 shows a monotonous increase in nanoparticle uptake with temperature, with only AgCl in $(C_8H_{17})_2(CH_3)_2N^+Cl^-$ displaying a decrease at high temperatures. Moreover, AgCl uptake by $(C_8H_{17})_2(CH_3)_2N^+Cl^-$ system displayed significantly higher uptake compared to the other systems. The increase in the nanoparticle uptake suggested that the particle formation reaction follows an endothermic equilibrium reaction scheme, which shifted towards more product equilibrium concentration as the temperature increased. It should be, nonetheless, stated that reactions involving formation of nanoparticles should not be simply viewed as endothermic equilibrium reactions. This statement finds support from observations on particle size variation with temperature shown in Figure 17.2. In the case of AgCl uptake, the temperature increase was accompanied by a reduction in the particle size. Smaller particles possess larger surface areas and require more surfactant molecules to maintain their colloidal stability. Consequently, nanoparticle uptake decreased at high temperatures. On the other hand, temperature significantly increased the nanoparticle size for AgBr/CTAB system. Therefore monotonous, however moderate, increase in nanoparticle uptake was observed. For the AgBr/$(C_8H_{17})_2(CH_3)_2N^+Br^-$ system, temperature had a limited effect on particle size and more appreciable increase in nanoparticle uptake was observed than CTAB system.

17.2.4 Effect of Surfactant Concentration

The surfactant layer limits growth and aggregation of nanoparticles prepared in (w/o) microemulsions and maintains their sizes within the nanodomain. Furthermore, increasing the surfactant concentration at constant water to surfactant mole ratio, R, is accompanied by an increase in the population of reverse micelles [43]. In addition, reactive surfactants participate in product formation and shift equilibrium reactions towards more product concentration. All these facts suggest that an increase of the surfactant concentration favors higher-nanoparticle uptake. Higher uptake, on the other hand, may lead to particle aggregation due to the increase in probability of collision between nanoparticle-populated reverse micelles.

Figure 17.3 shows that, within the range considered, nanoparticle uptake increased linearly with the surfactant concentration. The resultant straight lines displayed system-specific slopes and intercepts, which led to the conclusion that $(C_8H_{17})_2(CH_3)_2N^+Cl^-/AgCl$ system has the greatest ability to form and stabilized silver halide nanoparticles, followed by $(C_8H_{17})_2(CH_3)_2N^+Br^-/AgBr$ and lastly CTAB/AgBr system. These findings are significant, since CTAB is widely employed in the literature due to its ability to form and stabilized reverse-micellar systems at high concentrations of ionic precursors.

Figure 17.3 reveals an increase in the nanoparticle size in response to the increase in the surfactant concentration in all the systems. This is attributed to particle aggregation at higher uptake as noted above.

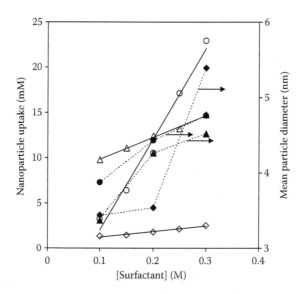

FIGURE 17.3 Variation of nanoparticle uptake (open symbols) and mean particle diameter (solid symbols) upon increasing surfactant concentration at constant R for ($\triangle,\blacktriangle$) $(C_8H_{17})_2(CH_3)_2N^+Br^-$ [35], ($\circ,\bullet$) $(C_8H_{17})_2(CH_3)_2N^+Cl^-$ [25], and ($\diamond,\blacklozenge$) CTAB [26] (w/o) microemulsions.

17.2.5   EFFECT OF WATER TO SURFACTANT MOLE RATIO, *R*

The water pools of the (w/o) microemulsions mediate reactions, and in this work dilute the higher halides and precipitate the halide monomers in the water pools. High amount of water at a constant surfactant concentration, however, compromises the stability of the (w/o) microemulsions [44]. Furthermore, at high-water content, the surfactant protective layer becomes less rigid [29]. This, in turn, promotes the role of intermicellar nucleation and growth and increases the probability of nanoparticle aggregation.

Figure 17.4 portrays a decrease in nanoparticle uptake with increasing *R* at a constant surfactant concentration. This decrease was sharp for the dioctyldimethylammonium chloride, and less dramatic for the rest of the surfactants. It should be noted here that no common range of *R* where stable (w/o) microemulsions could be found for the three surfactant systems. The dioctyldimethylammonium chloride and bromide surfactants could form stable (w/o) microemulsions at lower *R*, and hence were able to uptake more nanoparticles. Figure 17.4 shows an increase in the nanoparticle size, which resulted from particle aggregation as stated above.

17.2.6   EFFECT OF COSURFACTANT

The ammonium salt surfactants are not capable of forming stable reverse-micellar structures in the absence of cosurfactants [45]. High concentrations of cosurfactants, on the other hand, may reduce the surfactant participation at the interface of

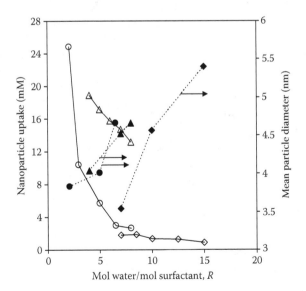

FIGURE 17.4   Variation of nanoparticle uptake (open symbols) and mean particle diameter (solid symbols) upon increasing *R* at constant surfactant concentration for (Δ,▲) $(C_8H_{17})_2(CH_3)_2N^+Br^-$ [35], (○,●) $(C_8H_{17})_2(CH_3)_2N^+Cl^-$ [25], and (◇,◆) CTAB [26] (w/o) microemulsions.

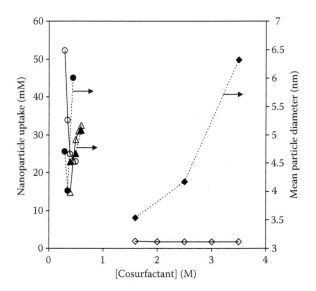

FIGURE 17.5 Variation of nanoparticle uptake (open symbols) and mean particle diameter (solid symbols) upon increasing cosurfactant concentration for (Δ,$\blacktriangle$) $(C_8H_{17})_2(CH_3)_2N^+Br^-$ [35], ($\bigcirc$,$\bullet$) $(C_8H_{17})_2(CH_3)_2N^+Cl^-$ [25], and ($\diamond$,$\blacklozenge$) CTAB [26] (w/o) microemulsions.

reverse micelles. Lower-water uptake by Winsor type II microemulsions was obtained as the cosurfactant concentration increased, which was attributed to lower participation of surfactant molecules at the reverse-micellar interfac in the present of high concentration of cosurfactant [45,46].

Figure 17.5 shows that nanoparticle uptake of the silver halides decreased as the cosurfactant concentration increased for $(C_8H_{17})_2(CH_3)_2N^+Cl^-$ and CTAB systems. However, for $(C_8H_{17})_2(CH_3)_2N^+Br^-$ system, the uptake increased. The drop in the nanoparticle uptake was attributed to a poor participation of the surfactant molecules at the reverse-micellar interface. The declining population of surfactant molecules at the reverse-micellar interface contributed to lower-surface layer rigidity and eventually larger particle size, as shown on Figure 17.5. In the case of $(C_8H_{17})_2(CH_3)_2N^+Br^-$, it seems that, within the range considered, the cosurfactant was still contributing to the stability of the (w/o) microemulsions, and hence, more nanoparticles could be stabilized. The relatively limited increase in particle size for $(C_8H_{17})_2(CH_3)_2N^+Br^-/AgBr$ system is attributed to particle aggregation at higher rate of collision between nanoparticle-populated reverse-micelles at higher uptake.

17.3 NANOPARTICLE UPTAKE IN NONREACTIVE SURFACTANT SYSTEMS

This investigation considered the uptake of metal oxide/hydroxide nanoparticles in single microemulsions formed with AOT surfactant as model systems. Aerosol OT is one of the few surfactants that can stabilize (w/o) microemulsion systems without the need for a cosurfactant. Colloidal metal oxides/hydroxides find application as heterogeneous catalysts for organic reactions [47–50]. As such, their effectiveness relies heavily on their surface area per unit volume of the reactive medium. High-surface area per volume of the organic phase can be attained by maintaining small colloidal particles at high-uptake values. The highest-possible concentration of colloidal nanoparticles, nanoparticle uptake, was experimentally realized by sequential addition of water-soluble metal salts followed by sodium hydroxide solution to the microemulsions. Even though exchange between the metal ion and the surfactant sodium counterion may take place, this reaction is easily reversed upon the precipitation of the metal hydroxide. Therefore, these microemulsions were considered as the nonreactive surfactant systems. More details on nanoparticle formation mechanism are given below.

17.3.1 Particle Formation Mechanism

The formation of metal oxide/hydroxide nanoparticles started with the dissolution of the corresponding solid metal precursor salt, typically chloride or nitrate, in the water pools. The dissolution is believed to proceed as follows. Reverse micelles from the bulk oil phase diffuse to the surface of the metal chloride or nitrate powder. At the surface of the solid, reverse micelles collapse at a rate governed by the rigidity of surfactant surface layer. Exposure of the solid to the bulk water pool allows dissolution of the precursor salt. In addition, some surfactant molecules in the presence of moisture at the surface of the solid facilitate the dissolution of the salt via ion exchange between the surfactant counterion, Na^+, and the corresponding metal ion, including Fe^{3+}, Cu^{2+}, or Ni^{2+} [4,45,51–54]. The reverse micelles re-form near the surface and migrate back to the bulk oil phase. The ions, Fe^{3+}, Cu^{2+} or Ni^{2+} and Na^+, redistribute between the bulk water pool and the stern layer in accordance with their corresponding ion exchange constants and their initial concentrations in the two regions [55]. The dissolution of the precursor salt is mass transfer limited, since the rate of disappearance of the precursor salt increased with shaking [21,22,31]. The ion exchange reactions, on the other hand, proceed at very high rates [34,55], and equilibrium is established the moment the reverse-micelles re-form near the solid surface. This mechanism of the precursor dissolution is supported by the fact that no salt could be solubilized in the absence of free water, $R < 2$, as will be explained later.

The formation of the metal oxide/hydroxide nanoparticles commenced upon the addition of the precipitating agent, the NaOH solution, to reverse micelles containing the dissolved salt. The addition of the aqueous solution disrupts the equilibrium momentarily as the reverse micelles re-group to accommodate the added solution [27,28]. This disruption is manifested by the appearance of immediate cloudiness. Reaction R3.1 proceeds in the bulk water pools, upon re-grouping of the reverse micelles, and metal hydroxide nanoparticles form by means of *intramicellar* [27,28,56,57] and intermicellar [57,58] nucleation and growth.

$$M^{z+}_{(aq)} + z\ OH^-_{(aq)} \rightarrow M(OH)_{z(s)} \quad\quad (R3.1)$$

where $M^{z+}_{(aq)}$ is the metal cation with z oxidation state. Nucleation takes place in reverse micelles containing the critical number of product monomers at a rate governed by the degree of supersaturation, while growth on an existing particle is instantaneous [38–40]. Metal ions migrate instantaneously from the stern layer to the bulk water pool, where they react, due to the shift of the ion exchange equilibrium [34]. The formation of metal oxide in the Stern layer is unlikely, since OH⁻ ion is repelled by the negatively charged surfactant head groups [59,60]. Particles that grow or agglomerate to sizes beyond the stabilization capacity of the reverse micelles aggregate and settle under gravitational force. The stabilization capacity of reverse micelles for a given material and temperature is dictated by the ionic strength of the water pools and the water content of the system. For given values of these variables, the reverse-micellar system stabilizes a specific concentration of the nanoparticles with specific particle size distribution [21,22,31].

17.3.2 Effect of Reactant Addition Sequence

To ensure that the experimental procedure adopted results in the highest-possible colloidal nanoparticle concentration, the sequence of precursor addition was reversed. A second scheme which involved mixing the metal salt powder with microemulsions already containing the stoichiometric amount of NaOH solution was tested. Both schemes succeeded in forming stabilized colloidal nanoparticles; however, higher uptake was obtained when scheme 1, solubilizing the metal salt before adding NaOH, was employed. The lower uptake associated with scheme 2 was attributed to the formation of a mass transfer barrier of the metal oxide/hydroxide at the surface of the salt powder, which prevented further dissolution.

17.3.3 Effect of Mixing Time

As noted in the reactive surfactant systems, mixing reduced the time needed to reach the value of nanoparticle uptake. Figure 17.6 shows that for the nonreactive system, a decrease in the nanoparticle concentration with time was observed followed by a plateau after around 1 h of mixing at 300 rpm. The decrease in nanoparticle concentration resulted from bulk precipitation of particles with sizes exceeding the stabilization capacity of the reverse micelles. Mixing improves the rate of aggregation of these particles, and hence they leave the colloidal suspension in shorter period.

17.3.4 Effect of Surfactant Concentration

Increasing surfactant concentration at constant R corresponds to a higher population of reverse micelles, nanoreactors, in the continuous oil phase [43]. Despite the fact that the surfactant is nonreactive, it is anticipated that larger population of reverse micelles will accommodate higher concentration of stabilized nanoparticles leading to higher uptake.

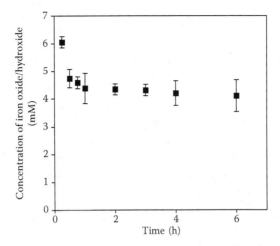

FIGURE 17.6 Effect of mixing time on the concentration of iron oxide/hydroxide nanoparticles in AOT-microemulsions. (Reprinted from Nassar, N.N. and Husein, M.M., *Langmuir*, 23, 13093, 2007a. With permission.)

Figure 17.7a shows a linear increase in nanoparticle uptake with the surfactant concentration for iron, copper, and nickel oxides. Comparison between Figures 17.3 and 17.7a reveals that the trend in nanoparticle uptake is independent of whether the surfactant is reactive or nonreactive. The increase in nanoparticle uptake was coupled with an increase in the particle size. Again, the same trend was reported for the reactive surfactant case. The increase in particle size is attributed to the higher

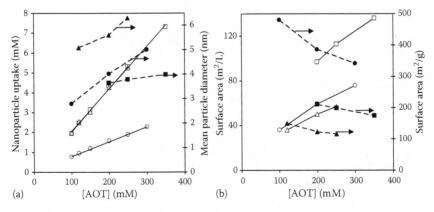

FIGURE 17.7 (a) Variation of nanoparticle uptake (open symbols) and mean particle diameter (solid symbols) upon increasing AOT concentration, (b) variation of surface area per liter (open symbols), and surface area per gram (solid symbols) upon increasing AOT concentration. (△,▲) nickel oxide [31], (○,●) cupper oxide [22], and (□,■) iron oxide [21].

rate of collision among nanoparticle-populated reverse micelles at higher uptake, which led to higher rate of aggregation [24,25,35]. It is worth noting that the value of nanoparticle uptake was nanoparticle specific. Iron oxide displayed higher uptake than nickel and copper oxides. This fact is attributed to a probably higher interaction between iron oxide nanoparticles and AOT head groups.

Figure 17.7b presents the variation of surface area per liter and surface area per gram upon increasing surfactant concentration. The decrease in the surface area per gram as a result of particle aggregation was balanced by an increase in the surface area per liter as a result of uptake increase. The surface area per gram of the stabilized nanoparticles and the surface area per liter of the microemulsions are very relevant to ultradispersed catalysis. To provide high-ultradispersed catalyst surface area, it is crucial to increase the concentration of stable colloidal nanoparticles, in addition to maintaining small particle size.

17.3.5 Effect of Water to Surfactant Mole Ratio, R

R plays an important role in controlling the size of the nanoparticle in the microemulsions. Increasing R, at constant surfactant concentration, leads to larger reverse micelles. Literature investigations showed a linear relationship between the diameter of reverse micelle and R [4,56,61,62]. Since reverse micelles limit the growth of nanoparticles to within the size of their water pools, an increase in the nanoparticle size is expected as R increases at a constant surfactant concentration. Literature has, nonetheless, showed that this simple rule does not necessarily apply to all cases. Some researchers have reported a decrease in particle size in response to an increase in R at constant surfactant concentration [7,8]. This observation was explained on the basis of an increase in intermicellar nucleation, owing to a decrease in surfactant layer rigidity, followed by growth taking place on more nuclei.

Following the experimental procedure outlined for nonreactive surfactants, it was found that a range of R between 2 and 8, at fixed values of other microemulsion variables, successfully maintained stable colloidal particles of the different metal oxides. At $R < 2$, all the water was bounded to the surfactant heads and no water was available to solubilize the metal salt precursor. For $R > 8$, on the other hand, cloudiness appeared indicating a shift towards Winsor type II microemulsions.

Figure 17.8a shows a maximum nanoparticle uptake with varying R. Interestingly, this maximum is metal oxide specific. For instance, the maximum uptake of nickel oxide occurred at $R = 3$, for copper oxide the maximum uptake occurred at $R = 5$, and for iron oxide the maximum occurred at $R = 6.5$. The difference in the values of maximum R is attributed to a different degree of interaction between the surfactant head groups and the colloidal nanoparticles. Iron oxide nanoparticles appear to have the strongest interaction, therefore sustained appreciable uptake at higher values of R. Moreover, Figure 17.8a shows an increase in particle size as R increased for the different metal oxides considered. It is believed that the increase in particle size resulted from the increase in the size of reverse micelles and the decrease in the rigidity of surfactant layer, which promoted particle aggregation.

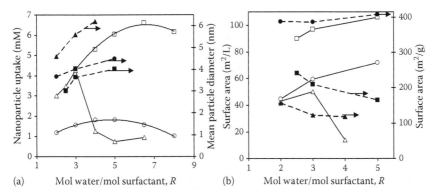

FIGURE 17.8 (a) Variation of nanoparticle uptake (open symbols) and mean particle diameter (solid symbols) upon increasing R, (b) variation of surface area per liter (open symbols), and surface area per gram (solid symbols) upon increasing R. ($\triangle$,$\blacktriangle$) nickel oxide [31], ($\circ$,$\bullet$) cupper oxide [22], and ($\square$,$\blacksquare$) iron oxide [21].

Figure 17.8b shows a decrease in the surface area per gram of nickel and iron oxide nanoparticles upon increasing R, whereas a slight increase was observed for copper oxide nanoparticle, despite the increase in the mean particle size. It seems that the mean particle size of copper oxide did not represent particle size distribution to a good extent. The figure shows an increase in the surface area per liter of iron and copper oxide, only because R was limited to the portion belonging to the increase in iron and copper oxide uptake. For nickel oxide, on the other hand, a sharp decrease in surface area per liter for $R > 3$ resulted from the sharp decrease in nanoparticle uptake at these values of R.

17.3.6 EFFECT OF INITIAL PRECURSOR SALT CONCENTRATION

Increasing precursor salt concentration increases the ion occupancy number per reverse micelle, which promotes *intramicellar* nucleation and growth and leads to higher uptake. On the other hand, high-ionic strength inside water pools of reverse micelles impacts the stability of the (w/o) microemulsions [63,64]. Moreover, the stability of the (w/o) microemulsions depends on the surfactant counterion, which results from ion exchange between the solubilized metal salt and sodium. Since the focus of this work was on nanoparticle uptake by (w/o) microemulsions, the concentrations of the metal precursors were limited to values which maintained stable reverse-micellar structures.

Figure 17.9a shows an increase in the nanoparticle uptake and the mean particle size as the initial concentration of the precursor salt in the microemulsions increased. This increase is attributed to the increase in the metal ion occupancy number and the subsequent improvement of *intramicellar* nucleation and growth. Figure 17.9b shows an increase in surface area per liter for every type of metal oxide in response to the increase in nanoparticle uptake. Again, surface area per gram declined or did not change significantly depending on the change in particle size.

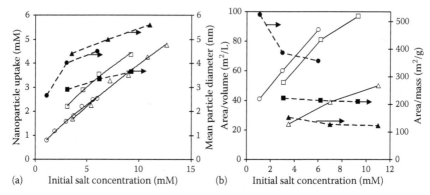

FIGURE 17.9 (a) Variation of nanoparticle uptake (open symbols) and mean particle diameter (solid symbols) upon increasing initial salt concentration, (b) variation of surface area per liter (open symbols) and surface area per gram (solid symbols) upon increasing initial salt concentration. (△,▲) nickel oxide [31], (○,●) copper oxide [22], and (□,■) iron oxide [21].

17.4   CONCLUSIONS

Water-in-oil microemulsions are attractive media for nanoparticle preparation due to their ability to form wide variety of nanoparticles with controllable sizes. These systems, however, suffer from low-reactant solubilization and product stabilization capacities. Nonetheless, for some applications, including heterogeneous catalysis in organic media, nanofluids, dip coating, etc., (w/o) microemulsions serve as ideal preparation route. For ultradispersed catalysis, as an example, it is essential to maintain high-catalyst surface area per liter of the reaction phase. High-surface area of colloidal particles can be achieved when small particle size is coupled with high-colloidal concentration. Nanoparticle uptake refers to the highest-possible concentration of stable colloidal particles in (w/o) microemulsions at given microemulsion and operation variables. Therefore, nanoparticle uptake becomes as pivotal as particle size for these applications.

This review summarizes our findings on the effect of some operating and microemulsion variables on nanoparticle uptake in reactive and nonreactive microemulsion systems. Mixing was found to increase the rate of nanoparticle formation and shorten the time needed to reach the nanoparticle uptake for reactive and nonreactive systems. Temperature increased nanoparticle uptake for exothermic nanoparticle formation reactions in reactive microemulsions, given stable reverse-micellar structure is maintained. The overall effect of temperature, nevertheless, was dependent on the final particle size. Nanoparticle uptake increased linearly with the surfactant concentration, for reactive and nonreactive surfactant systems, most probably due to the increase in the population of reverse micelles, nanoreactors. The effect of surfactant counterion and water to surfactant mole ratio, R, was dependent on their effect on the stability of the reverse micellar structure and the rigidity of the surfactant surface layer. Structure shift towards other Winsor microemulsion equilibria as a result of high-cosurfactant concentration, ionic strength, or R led to lower uptake. Furthermore, low-surfactant layer rigidity at high-cosurfactant concentration or R promoted particle aggregation. Particle aggregation was also encountered at high uptake due to higher probability of collision between nanoparticle-populated reverse micelles. Particle aggregation resulted in lower-surface area per gram of the colloidal nanoparticles. However, this trend was balanced by higher-surface area per liter of the microemulsions encountered at higher uptake. Increasing the precursor concentrations led to higher uptake for reactive and nonreactive microemulsion, provided stable reverse-micellar structures are maintained.

SYMBOLS AND TERMINOLOGIES

AOT   aerosol OT, dioctylsulfosuccinate sodium salt
CTAB   cetyltrimethylammonium bromide
R mol water per mol surfactant
TEM   transmission electron microscopy
w/o   water in oil

REFERENCES

1. Feltin, N. and Pileni, M.P. 1997. New technique for synthesizing iron ferrite magnetic nanosized particles. *Langmuir*, 13, 3927–3933.
2. Lisiecki, I. and Pileni, M.P. 1995. Copper metallic particles synthesized "in situ" in reverse micelles: Influence of various parameters on the size of the particles. *J. Phys. Chem.*, 99, 5077–5082.
3. Petit, C., Lixon, M.P., and Pileni, M.P. 1993. In situ synthesis of silver nanocluster in AOT reverse micelles. *J. Phys. Chem.*, 97, 12974–12983.
4. Pileni, M.P. 1993. Reverse micelles as microreactors. *J. Phys. Chem.*, 97, 6961–6973.
5. Pileni, M.P., Motte, L., and Petit, C. 1992. Synthesis of cadmium sulfide in situ in reverse micelles: Influence of the preparation modes on size, polydispersity, and photochemical reactions. *Chem. Mater.*, 4, 338–345.
6. Debuigne, F., Jeunieau, L., Wiame, M., and Nagy, J.B. 2000. Synthesis of organic nanoparticles in different W/O microemulsions. *Langmuir*, 16, 7605–7611.
7. Bagwe, R.P. and Khilar, K.C. 2000. Effects of intermicellar exchange rate on the formation of silver nanoparticles in reverse microemulsions of AOT. *Langmuir*, 16, 905–910.
8. Bagwe, R.P. and Khilar, K.C. 1997. Effects of the intermicellar exchange rate and cations on the size of silver chloride nanoparticles formed in reverse micelles of AOT. *Langmuir*, 13, 6432–6438.
9. Hirai, T., Sato, H., and Komasawa, I. 1994. Mechanism of formation of CdS and ZnS ultrafine particles in reverse micelles. *Ind. Eng. Chem. Res.*, 33, 3262–3266.
10. Hingorani, S., Pillai, V., Kumar, P., Multani, M.S., and Shah, D.O. 1993. Micro-emulsion mediated synthesis of zinc-oxide nanoparticles for varistor studies. *Mater. Res. Bull.*, 28, 1303–1310.
11. Petit, C., Lixon, P., and Pileni, M.P. 1990. Synthesis of cadmium sulfide *in-situ* in reverse micelles. 2. Influence of the interface on the growth of the particles. *J. Phys. Chem.*, 94, 1598–1603.

12. Osseo-Asare, K. and Arriagada, F.J. 1990. Preparation of SiO_2 nanoparticles in a nonionic reverse micellar system. *Colloids Surf.*, 50, 321–339.

13. Boutonnet, M., Kizling, I., Stenius, P., and Maire, G. 1982. The preparation of monodisperse colloidal metal particles from microemulsions. *Colloids Surf.*, 5, 209–225.

14. Zhang, R., Liu, J., Han, B., He, J., Liu, Z., and Zhang, J. 2003. Recovery of nanoparticles from $(EO)_8(PO)_{50}(EO)_8$/p-Xylene/H2O microemulsions by tuning the temperature. *Langmuir*, 19, 8612–8614.

15. Weidenkaff, A., Ebbinghaus, S.G., and Lippert, T. 2002. $Ln_{1-x}A_xCoO_3$ (Ln = Er, La; A = Ca, Sr)/carbon nanotube composite materials applied for rechargeable Zn/air batteries. *Chem. Mater.*, 14, 1797–1805.

16. Palla, B.J., Shah, D.O., Garcia-Casillas, P., and Matutes-Aquino, J. 1999. Preparation of nanoparticles of barium ferrite from precipitation in microemulsions. *J. Nanopart. Res.*, 1, 215–221.

17. Vaucher, S., Fielden, J., Dujardin, E., and Mann, S. 2002. Molecule-based magnetic nanoparticles: Synthesis of cobalt hexacyanoferrate, cobalt pentacyanonitrosylferrate, and chromium hexacyanochromate coordination polymers in water-in-oil microemulsions. *Nano Lett.*, 2, 225–229.

18. Kumar, P., Pillai, V., and Shah, D. 1993. Preparation of Bi-Pb-Sr-Ca-Cu-O oxide superconductors by coprecipitation of nanosize oxalate precursor powders in the aqueous core of water-in-oil microemulsions. *Appl. Phys. Lett.*, 62, 765–768.

19. Tan, W., Santra, S., Zhang, P., Tapec, R., and Dobson, J. 2002. Method for identifying cells, U.S. Patent, Patent serial No. 010807.

20. Eastoe, J., Hollamby, M.J., and Hudson, L., Recent advances in nanoparticle synthesis with reversed micelles, *Adv. Colloid Interface Sci.*, 128–130, 5–15, 2006.

21. Nassar, N.N. and Husein, M.M. 2007a. Study and modeling of iron hydroxide nanoparticle uptake by AOT (w/o) microemulsions. *Langmuir*, 23, 13093–13103.

22. Nassar, N.N. and Husein, M.M. 2007b. Effect of microemulsion variables on copper oxide nanoparticle uptake by AOT microemulsions, *J. Colloid Interface Sci.*, 316, 442–450.

23. Nassar, N. and Husein, M. 2006. Preparation of iron oxide nanoparticles in (w/o) microemulsions starting from $FeCl_3$ solid powder. *Phys. Status Solidi A*, 203, 1324–1328.

24. Husein, M., Rodil, E., and Vera, J.H. 2006. A novel approach for the preparation of AgBr nanoparticles from their bulk solid precursor using CTAB microemulsions. *Langmuir*, 22, 2264–2272.

25. Husein, M., Rodil, E., and Vera, J.H. 2005. A novel method for the preparation of silver chloride nanoparticles starting from their solid powder using microemulsions. *J. Colloid Interface Sci.*, 288, 457–467.

26. Husein, M.M. and Nassar, N.N. 2008. Nanoparticle preparation using the single microemulsions scheme. *Curr. Nanosci.* (In press).

27. Husein, M., Rodil, E., and Vera, J.H. 2003. Formation of silver chloride nanoparticles in microemulsions by direct precipitation with the surfactant counterion. *Langmuir*, 19, 8467–8474.

28. Husein, M., Rodil, E., and Vera, J.H. 2004. Formation of silver bromide precipitate of nanoparticles in a single microemulsion utilizing the surfactant counterion. *J. Colloid Interface Sci.*, 273, 426–434.

29. Chew C.H., Gan, L.M., and Shah, D.O. 1990. The effect of alkanes on the formation of ultrafine silver bromide particles in ionic w/o microemulsions. *J. Disper. Sci. Technol.*, 11, 593–609.

30. Bommarius, A.S., Holzwarth, J.F., Wang, D.I.C., and Hatton, T.A. 1990. Coalescence and solubilizate exchange in a cationic four-component reversed micellar system. *J. Phys. Chem.*, 94, 7232–7239.

31. Nassar, N.N. and Husein, M.M. 2008. *Maximizing the Uptake of Nickel Oxide Nanoparticles in AOT (w/o) Microemulsions*. In Recent Trends in Surface and Colloid science; Paul, P.K., Ed., Special Issue, World Scientific Publishing Co. Pvt. Ltd., Singapore. (In Press).

32. Nassar, N. 2007. A (w/o) Microemulsion approach for in-situ preparation of high concentration of colloidal metal oxide nanoparticles. PhD thesis, University of Calgary, Calgary, Alberta, Canada.

33. Patruyo, L. 2008. Removal of $H_2S_{(g)}$ using ultradispersed iron oxide nanoparticles. M Eng thesis, University of Calgary, Calgary, Alberta, Canada.

34. Husein, M.M., Weber, M.E., and Vera, J.H. 2000. Nucleophilic substitution sulfonation in microemulsions and emulsions. *Langmuir*, 16, 9159–9167.

35. Husein, M.M., Rodil, E., and Vera, J.H. 2007a. Formation of colloidal AgBr nanoparticles starting from their powder precursor in reactive dioctyldimethylammonium bromide microemulsions. *WJCE*, 2007, 13–25.

36. Husein, M.M., Rodil, E., and Vera, J.H. 2007b. Preparation of AgBr nanoparticles in microemulsions via reaction of $AgNO_3$ with CTA. *J. Nanopart. Res.*, 9, 787–796.

37. Ottewill, R.H. and Woodbridge, R.F. 1961. The preparation of monodisperse silver bromide and silver iodide sols. *J. Colloid Sci.*, 16, 581–594.

38. Kumar, A.R., Hota, G., Mehra, A., and Khilar, K.C. 2004. Modeling of nanoparticles formation by mixing of two reactive microemulsions. *AIChE J.*, 50, 1556–1567.

39. Bandyopadhaya, R., Kumar, R., and Gandhi, K.S. 2000. Simulation of precipitation reactions in reverse micelles. *Langmuir*, 16, 7139–7149.

40. Bandyopadhaya, R., Kumar, R., Gandhi, K.S., and Ramkrishna, D. 1997. Modeling of precipitation in reverse micellar systems. *Langmuir*, 13, 3610–3620.

41. Kahlweit, M., Strey, R., and Busse, G. 1990. Microemulsions: A qualitative thermodynamic approach. *J. Phys. Chem.*, 94, 3881–3894.

42. Kahlweit, M., Strey, R., Schomacker, R., and Hasse, D. 1989. General patterns of the phase behavior of mixtures of water, nonpolar solvents, amphiphiles, and electrolytes. 2. *Langmuir*, 5, 305–315.

43. De, T.K. and Maitra, A. 1995. Solution behaviour of aerosol OT in non-polar solvents. *Adv. Colloid Interface Sci.*, 59, 95–193.

44. Winsor, P.A. 1948. Hydrotropy, solubilisation and related emulsification processes. *Trans. Faraday Soc.*, 44, 376–398.

45. Wang, W., Weber, M.E., and Vera, J.H. 1995. Reverse micellar extraction of amino acids using dioctyldimethylammonium chloride. *Ind. Eng. Chem. Res.*, 34, 599–606.

46. Zabaloy, M.S. and Vera, J.H. 1996. Water uptake and dioctyldimethylammonium chloride distribution in water + hydrocarbon microemulsions containing an alcohol and sodium chloride at 296.1 K. *J. Chem. Eng. Data*, 41, 1499–1504.

47. Bianco, A.D., Panariti, N., Carlo, S.D., Beltrame, P.L., and Carnit, P. 1994. New developments in deep hydroconversion of heavy oil residues with dispersed catalysts. 2. kinetic aspects of reaction. *Energy Fuels*, 8, 593–597.

48. Bianco, A.D., Panarit, N., Carlo, S.D., Elmouchnino, J., Fixari, B., and Perchec, P.L. 1993 Thermocatalytic hydroconversion of heavy petroleum cuts with dispersed catalysts. *Appl. Catal., A*, 94, 1–16.

49. Rymes, J., Ehret, G., Hilaire, L., Boutonnet, M., and Jiratova, K. 2002. Microemulsions in the preparation of highly active combustion catalysts. *Catal. Today*, 75, 297–303.

50. Eriksson, S., Nylén, U., Rojas, S., and Boutonnet, M. 2004. Preparation of catalysts from microemulsions and their applications in heterogeneous catalysis. *Appl. Catal. A: Gen.*, 265, 207–219.

51. Rabie, H.R., Weber, M.E., and Vera, J.H. 1995. Effects of surfactant purity and concentration, of surfactant counterion, and of different ions on water uptake of dioctyldimethyl ammonium salt-decanol-isooctane reverse micellar systems. *J. Colloid Interface Sci.*, 174, 1–9.

52. Rabie, H.R. and Vera, J.H. 1996. Generalized water uptake modeling of water-in-oil microemulsions. New experimental results for Aerosol-ot-isooctane-water-salts systems. *Fluid Phase Equilib.*, 122, 169–186.

53. Concalves, S.A.P., De Pauli, S.H., Tedesco, A.C., Quina, F.H., Okano, L.T., and Bonilla, J.B.S. 2003. Counterion exchange selectivity coefficients at water-in-oil microemulsion interface. *J. Colloid Interface Sci.*, 267, 494–499.

54. Pal, S., Vishal, G., Gandhi, K.S., and Ayappa, K.G. 2005. Ion exchange in reverse micelles. *Langmuir*, 21, 767–778.

55. Romested, L. 1977. A general kinetic theory of rate enhancement for reactions between organic substrates and hydrophilic ions in micelles systems. In *Micellization, Solubilization and Microemulsions*; Mittal, K., Ed.; Plenum Press: New York, p. 509.

56. Pileni, M.P., Zemb, T., and Petit, C. 1985. Solubilization by reverse micelles: Solute location and structure perturbation. *Chem. Phys. Lett.*, 118, 414–420.

57. Natarajan, U., Handique, K., Mehra, A., Bellare, J.R., and Khilar, K.C. 1996. Ultra-fine metal particle formation in reverse micellar systems: Effects of intermicellar exchange on the formation of particles. *Langmuir*, 12, 2670–2678.

58. Herrera, A.P., Resto, O., Briano, J.G., and Rinaldi, C. 2005. Synthesis and agglomeration of gold nanoparticles in reverse micelles. *Nanotechnology*, 16, S618–S625.

59. Stigter, D. 1964. On the adsorption of counterions at the surface of detergent micelles. *J. Phys. Chem.*, 68, 3603–3611.

60. Stigter, D. and Mysles, K. 1955. Tracer electrophoresis. II. The mobility of the micelle of sodium lauryl sulfate and its interpretation in terms of zeta potential and charge. *J. Phys. Chem.*, 59, 45–51.

61. Eastoe, J., Robinson, B.H., Visser, A.J.W.G., and Steytler, D.C. 1991. Rotational dynamics of AOT reversed micelles in near-critical and supercritical alkanes. *J. Chem. Soc. Faraday Trans.*, 87, 1899–1903.

62. Zulauf, M. and Eicke, H.-F. 1979. Inverted micelles and microemulsions in the ternary system H_2O/aerosol-OT/isooctane as studied by photon correlation spectroscopy. *J. Phys. Chem.*, 83, 480–486.

63. Sjoblom, J., Skurtveit, R., Saeten, J.O., and Gestblom, B. 1991. Structural changes in the microemulsion system didodecyldimethylammonium bromide/water/dodecane as investigated by means of dielectric spectroscopy. *J. Colloid Interface Sci.*, 141, 329–337.

64. Guering, P. and Lindman, B. 1985. Droplet and bicontinuous structures in microemulsions from multicomponent self-diffusion measurements. *Langmuir*, 1, 464–468.

18 TiO$_2$ Nanoparticles in Microemulsion: Photophysical Properties and Interfacial Electron Transfer Dynamics

Hirendra N. Ghosh

CONTENTS

18.1   INTRODUCTION

Nanocrystalline semiconductor materials [1] exhibit a wide range of novel chemical and physical properties that are finding applications in devices such as solar cells [2], waste water treatment [3], and nano-electronic devices [4]. Dye sensitization of wide band-gap semiconductor electrodes has gained sufficient attention in recent years, largely owing to the demonstration of dye-sensitized solar cells with a conversion efficiency as high as 10% [2]. In nanocrystalline materials, a significant fraction of the atoms reside on the nanocluster surface. These surface atoms having "dangling bonds" that may act as electron and hole traps that can dominate electron/hole recombination and other processes. It is often possible (and desirable) to passivate the surface traps. Several studies have shown that passivation of surface traps has large effects on the nanocluster photophysics [5–7]. These surface states also can take part in the interfacial electron transfer (IET) reaction. We have reported earlier that surface states of wide band-gap materials like ZrO_2 can take part in the IET processes [8–10]. Presence of surface states in the nanostructured materials actually brings down the efficiency of the devices. To gain higher efficiency in the devices it is very important to pacify these surface states. Modification of these states is possible using suitable modifier molecules. By this process it is possible to remove most of the lower lying surface states. Surface modification of semiconductor NPs changes their optical, chemical, and photo-catalytic properties significantly [11]. It can lead to the following effects: (1) it may enhance their excitonic and defect emission by blocking nonradiative electron/hole (e^-/h^+) recombination at the defect sites (traps) on the surface of the semiconductor NPs [5]; (2) it may enhance the photostability of semiconductor NPs [5]; (3) it may create new traps on the surface of the NPs leading to the appearance of new emission bands [12]; and (4) it may enhance the selectivity and efficiency of light-induced reactions occurring on the surface of semiconductor NPs [11,13]. On surface modification, density of surface states (lower lying states below the conduction band) can be changed drastically. We have observed that on surface modification the optical and photochemical properties of NP changed [14]. Electron injection dynamics has been found unaffected by surface modification; however, back electron transfer (BET) dynamics is found to be slow on modified surface as compared to that on bare surface [15,16], which in turn can increase the efficiency of solar cell.

Water/surfactant/oil reversed micelle is one such nanometer-sized confined water system and has been extensively studied as a model for water molecules on proteins or biological membranes to understand the chemical and biological processes in such systems [17]. These studies have shown that a reversed micelle is spherical, and the radius of water droplet r is proportional to the water–surfactant molar ratio w_0. It is well known that the structure, dynamics, and physicochemical properties of water molecules confined in a nanometer-sized space are greatly different from those of bulk water. It will be interesting to monitor the photophysical properties of NPs dispersed in the water core of the reverse micelles. We have observed that surfactant molecules can modify the NPs surface effectively which in turn change the photophysical properties of the NPs [14,18]. In the present article we have shown the optical and photophysical properties of TiO$_2$ NPs after dispersing the particles in the water pool of a microemulsion. Again water-in-oil microemulsions (or reverse micelles) have been used as microreactors to synthesize ultrafine powder with a narrow distributed particle size by controlling the growth inside [19]. Typically, they are used to synthesize nanometer-sized TiO$_2$ by using microemulsion-mediated processing [20]. It is possible to synthesize desired size NPs by changing water-to-surfactant ratio. In this review article, we report the synthesis of very small size TiO$_2$ particles (called subnanoparticles, SNPs) in AOT/n-heptane/ water microemulsion with $w_0 = 1$ ($w_0 = $ [H$_2$O]/[surfactant]). Size-quantization behavior of small size TiO$_2$ particles has been confirmed from the blueshift in the absorption band edge. After synthesizing we have sensitized the TiO$_2$ NPs which resides in the water core of the microemulsion with catechol (Cat) molecules. Steady-state optical measurements show that catechols interact strongly with the NPs with the formation of charge transfer (CT) complex. Femtosecond transient absorption spectroscopy has been carried out in Cat-sensitized TiO$_2$ NPs in microemulsion, exciting with 400 nm laser pulse. Electron injection has been confirmed by direct detection of injected electron in the conduction band of TiO$_2$. We have also monitored BET dynamics by monitoring the kinetic decay trace of the injected electron. To find out the effect of microenvironment and effect of particle size on IET dynamics we have also carried experiments by sensitizing bigger size TiO$_2$ particles as prepared by solgel process by the same sensitizer (Cat).

18.2 MATERIALS AND METHODS

18.2.1 MATERIALS

Sodium lauryl sulfate (NaLS) was obtained from Fluka and purified from ether. Aerosol-OT (AOT) and n-heptane were also obtained from Aldrich. titanium(IV) tetraisopropoxide {Ti[OCH(CH$_3$)$_2$]$_4$} (Aldrich, 97%), cyclohexane (Aldrich), 1-butanol (Aldrich), and isopropyl alcohol (Aldrich) were purified by distillation. Nanopure water (Barnsted System, USA) was used for making aqueous solutions.

18.2.2 SYNTHESIS OF TiO₂ NANOPARTICLES IN SOLGEL METHODS

Nanometer-sized TiO_2 was prepared by controlled hydrolysis of titanium(IV) tetraisopropoxide [21]. A solution of 5 mL $Ti[OCH(CH_3)_2]_4$ (Aldrich, 97%) dissolved with 95 mL isopropyl alcohol (Aldrich) was added dropwise (1 mL/min) to 900 mL of nanopure water (2°C) at pH 1.5 (adjusted with HNO_3). The solution was continuously stirred for 10–12 h until a transparent colloid was formed. Both titanium(IV) tetraisopropoxide and isopropyl alcohol were purified by distillation. The colloidal solution was concentrated at 35°C–40°C with a rotary evaporator and then dried with nitrogen stream to yield a white powder.

18.2.3 FEMTOSECOND VISIBLE SPECTROMETER

The femtosecond tunable visible spectrometer has been described earlier and developed based on a multipass amplified femtosecond Ti:sapphire laser system from Avesta, Russia (1 kHz repetition rate at 800 nm, 50 fs, 800 μJ/pulse) [15]. The 800 nm output pulse from the multipass amplifier is split into two parts to generate pump and probe pulses. One part, with 200 μJ/pulse, is frequency doubled and tripled in BBO crystals to generate pump pulses at 800, 400, or 267 nm. In the present investigation we have used 400 nm (frequency doubled in BBO crystal) laser light as pump pulse to excite the samples. To generate visible probe pulses, about 3 μJ of the 800 nm beam is focused onto a 1.5 mm thick sapphire window. The intensity of the 800 nm beam is adjusted by iris size and ND filters to obtain a stable white light continuum from 400 to over 1000 nm region. The probe pulses are split into the signal and reference beams and are detected by two matched photodiodes with variable gain. To monitor kinetics at different wavelength we have used interference filter for different wavelength before both the photodiodes. The noise level of the white light is ~0.5% with occasional spikes due to oscillator fluctuation. We have noticed that most laser noise is low frequency noise and can be eliminated by comparing the adjacent probe laser pulses (pump blocked vs unblocked using a mechanical chopper). The typical noise in the measured absorbance change is about <0.3%.

18.3 PHOTOPHYSICAL PROPERTIES OF TiO₂ NANOPARTICLES IN MICROEMULSION

18.3.1 PREPARATION OF MICROEMULSION

To study the optical and photophysical properties of in microemulsion we have used TiO_2 NPs which are synthesized by solgel method as described earlier. The microemulsion was prepared using the following composition (2 g NaLS/1.8 cc water/46.5 cc cyclohexane/9.5 cc 1-butanol). The diameter of this water pool is around 4–5 nm. Pileni et al. [22] have shown an empirical relation between w_0 and r as $r = 1.5\,w_0$ (where r is the radius of the water pool for microemulsion). According to Adhikari et al. [23], the water droplet size is ~4.2 nm in the present microemulsion studied. TiO_2 NPs in this study can just fit in the water pool (Figure 18.1).

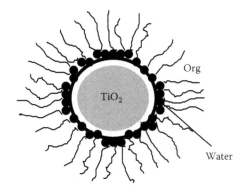

FIGURE 18.1 Schematic diagram of the location of semiconductor NP in a microemulsion (NaLS/water/cyclohexane/1 butanol with $w_0 = 14$). Water droplet size is ~4–5 nm.

For making the experimental solution, first dissolve NaLS in cyclohexane and then add 1-butanol in it. Dry TiO$_2$ powder is dissolved separately in water and then certain volume of this solution is added to the surfactant solution to get the desired w_0. After adding TiO$_2$/water the solution is shaken vigorously for 10–15 min to obtain a transparent microemulsion solution. This part is very crucial because emission can only be obtained when TiO$_2$ particles just fit in the water pool, though slight swelling of the reverse micelle is expected The experiments are also carried out at other w_0 values (e.g., $w_0 = 12$, 16, and 19).

18.3.2 Optical Absorption and Emission of TiO$_2$ Nanoparticles in Microemulsion

UV-Vis spectroscopy has been employed to measure the optical absorbance of the NP in the microemulsion as shown in Figure 18.2. We have shown the optical absorbance of microemulsion (with out NPs), TiO$_2$ NPs in water, and TiO$_2$ NPs in microemulsion. It has been observed that the optical absorbance of the NP does not change by dispersing the NP in the water core of the microemulsion. It is interesting to observe that with incorporating TiO$_2$ NPs in the microemulsion the scattering property of the experimental solution does not change and it is still optically clear to perform optical experiments.

Steady-state emission measurements have been carried on TiO$_2$ NPs in the microemulsion. The emission from the TiO$_2$ NP in microemulsion at room temperature was observed by the excitation of the sample with 350 nm light using a fluorescence spectrometer and is shown in Figure 18.3. The samples were also excited at other wavelengths to find out that emission is not due to other scattering process. It has been observed that emission spectra are unchanged with different exciting wavelengths. At each excitation wavelength, the emission band was the same with a peak at 445 nm and a hump at 550 nm. The emission quantum yield (ϕ) was found to be 0.2%. The emission band matched well with photoluminescence

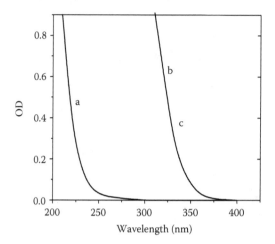

FIGURE 18.2 Optical absorbance spectra of (a) microemulsion, (b) TiO_2 in water, and (c) TiO_2 in microemulsion. The concentration of TiO_2 nanoparticles is kept at 0.1 g/L.

spectra of anatase TiO_2 powder in air [24]. No band-gap emission was observed. Experiments were also carried out for TiO_2 NPs in pure water. No emission was observed from the same TiO_2 NPs as they dispersed in water. Emission from indirect band-gap semiconductors [25] like TiO_2 and GaP is difficult to observe. Except few examples like photo- and electroluminescence, spectra were detected by Nakato et al. [26] from n-TiO_2 and transition metal doped n-TiO_2 on the electrode surface. In situ photoluminescence of commercially available TiO_2 particle at 77 K

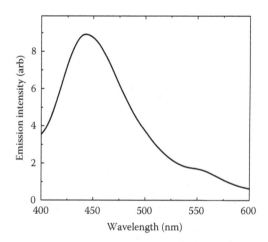

FIGURE 18.3 Emission spectrum (curved) of TiO_2 NP in microemulsion (NaLS/water/ cyclohexane/1 butanol with $w_0 = 14$, where $w_0 = $ [water]/[surfactant]) exciting at 350 nm light at 298°C. $[TiO_2] = 0.10$ g/L ($\phi = 0.002$).

was reported by Anpo et al. [27]. Emission from bare colloidal anatase TiO$_2$ NPs was also observed by Serpone et al. [28] in the visible region. In earlier observations the emission quantum yields were found to be really low. However, in the present investigation clear photoluminescence spectra have been observed as the TiO$_2$ NPs dispersed in the water core of the microemulsion. In our earlier investigation [14] we observed that on surface modification of TiO$_2$ surface with sodium dodecyl benzene sulfonate (SDBS) emission has been observed from the particles. A dramatic enhancement of emission from surface modified CdS colloidal NP has been observed earlier [29]. A simple hypothesis has been proposed that the kinetically stable semiconductor colloids have a high density of surface defect sites. These sites cover a broad range of energies and structures. Many such defect sites exist at mid-band-gap energies. These sites are involved in trapping initially produced electron–hole pairs, as evident from the fact that emission (resulting from recombination of electron–hole pairs) is significantly redshifted relative to the band-gap absorption. The existence of different trap sites provides multiple pathways for radiative and nonradiative recombination. The molecules, which modify the surface of NPs, bind to lower energy trap sites, which are directly involved in nonradiative decay. Such bindings sites can act as efficient traps and can increased the quantum yield of emission. In the present studies also surface modification might be playing a big role for the enhancement of luminescence quantum yield of the particles. As we described earlier, a microemulsion can be defined as a thermodynamically stable, optically isotropic solution of two immiscible liquids (e.g., water and oil) consisting of microdomains of one or both liquids stabilized by an interfacial film of surfactant [30]. The surfactant molecule generally has a polar (hydrophilic) head group and a long-chained aliphatic (hydrophobic) tail. Such molecules optimize their interactions by residing at the oil/water interface, thereby reducing the interfacial tension. In this condition the polar head group can interact efficiently with the TiO$_2$ NP, where the polar head (sulfonic acid) can modify surface states of the NP. Here the TiO$_2$ NPs are dispersed in the water pool of a microemulsion. After dissolving the NP in water, the pH of the colloidal solution becomes ~2.5–3. At this lower pH, the surface of TiO$_2$ NP becomes positively charged. The head group of the surfactant molecules (NaLS) can be attached to the surface of the NP. The sulfonic acid group can act as a surface modifier and can remove the lower lying surface defects. Such interactions can promote the radiative recombination pathway and can enhance the emission intensity.

18.4 ELECTRON TRANSFER DYNAMICS IN CATECHOL-SENSITIZED TiO$_2$ NANOPARTICLES

18.4.1 Preparation of Semiconductor Colloids in Reverse Micelles

To study IET dynamics on NP surface in the water pool of microemulsion TiO$_2$ particles were synthesized. Colloidal TiO$_2$ suspension in reverse micelles was prepared by hydrolysis of titanium(IV) isopropoxide/2-propanol solution as followed by Sant and Kamat [31]. Water-in-oil type of microemulsions of $w_0 = 1$

was prepared by adding 0.22 mL $HClO_4$ (0.02 M) in 30 mL solution containing 0.4 M AOT in *n*-heptane. An aliquot of 0.2 mL titanium isopropoxide (0.136 M) was added dropwise to 20 mL of the above microemulsion under mild stirring at room temperature. The stirring was continued for about 20 min after the addition of the reagent. A clear colorless transparent solution was obtained which contains the TiO_2 NPs.

18.4.2 DYE ADSORPTION ON TiO_2 COLLOIDS IN MICROEMULSION AND WATER

It is very important to choose dye molecules, which will be adsorbed on the surface of TiO_2 colloids in microemulsion, effectively. We have observed that the dye molecules like coumarin 343 and ruthenium polypyridine complexes cannot be adsorbed on the surface of TiO_2 colloids in microemulsion. These kinds of sensitizer molecules have good solubility in *n*-heptane. As a result the dye molecules will prefer to stay in *n*-heptane than adsorbing on the NPs, which are residing in the water pool in microemulsion. However, if the same dye molecules can interact strongly with the NPs then it can sensitize the NPs even in the microemulsion. On the other hand, if the dye molecules are water soluble then it can reside in the water pool and easily adsorb with the NPs. In the present investigation we have chosen Cat which can interact strongly with NPs. After synthesizing the TiO_2 colloids in microemulsion we have added the dye molecules into it and stirred mildly for 1 h till deep red-colored solutions were formed. The dye-NP solutions were found to be stable for a couple of days.

Sensitization of the bare particles was carried out by directly mixing TiO_2 NPs (20 g/L) and the sensitizer molecules in nanopure water and stirred for 1 h. The solution turned deep red and was found to be stable for many days.

18.4.3 OPTICAL ABSORPTION SPECTROSCOPY OF NANOPARTICLES AND INTERACTION WITH DYE MOLECULES

We have carried out optical absorption spectroscopy of TiO_2 colloids in microemulsion with different w_0 (1–5). It has been observed that with decreasing w_0, optical absorption spectrum of TiO_2 colloid gets blueshifted. This is a clear indication of decrement in their sizes. Figure 18.4 shows the optical absorption spectrum of TiO_2 colloid in microemulsion with $w_0 = 1$. In Figure 18.2 we have also shown the optical absorption spectrum of TiO_2 colloid in water as prepared by solgel process. It is seen clearly that the absorption onset of TiO_2 particles in microemulsion is noticeably blueshifted compared to that of the particles prepared in water by the solgel process. The size of the particles in colloids prepared in water is significantly larger (>5 nm), which has absorption onset at ~380 nm. The absorption onset was found to be 320 nm ($E_g = 3.65$) for TiO_2 colloids in microemulsion with $w_0 = 1$, which matches well with that reported by Sant and Kamat [31].

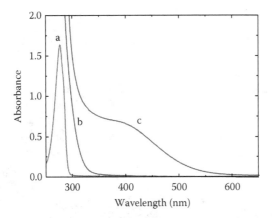

FIGURE 18.4   Absorption spectra of (a) free Cat, (b) SNP TiO$_2$, and (c) Cat-sensitized SNP TiO$_2$.

The aim of this investigation is to study interfacial ET dynamics after sensitizing these particles in microemulsion. For this purpose it is necessary to monitor the dye-NP interactions. Figure 18.4 shows the optical absorption spectra of free Cat (Figure 18.4a), TiO$_2$ QDs in microemulsion (SNP) (Figure 18.4b), and Cat-sensitized TiO$_2$ QDs (Figure 18.4c). It is clearly seen that an additional new absorption band centered at ~390 nm and extending up to 600 nm has appeared. This band has been attributed to a CT complex of Cat/TiO$_2$ system [32,33]. We have also reported earlier [15,34–37] that catecholate molecules bind strongly with TiO$_2$ NPs with the formation of a five-membered chelating ring (Figure 18.5).

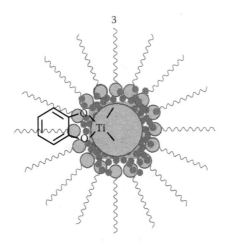

FIGURE 18.5   Molecular structure of Cat coupling with TiO$_2$ NPs in the water pool of microemulsion.

18.4.4 TRANSIENT ABSORPTION MEASUREMENTS IN MICROEMULSION

It has been demonstrated by many workers including us that optical excitation of dye molecules adsorbed on TiO_2 NP surface injects electron into the conduction band of the NP [15,34–37]. In the present investigation, we have carried out femtosecond laser flash photolysis experiments exciting the Cat/TiO_2 system in microemulsion with 400 nm laser light to study the ET dynamics on semiconductor surface where particles are synthesized in the water pool of microemulsion. Figure 18.6 shows the time-resolved transient absorption spectra of Cat/TiO_2 system in microemulsion at 0.2, 0.5, 1, 2, and 10 ps after excitation at 400 nm. The spectrum at each delay time consists of bleach in the 470–500 nm region and a transient absorption band in the 500–900 nm region. The transient absorption and the bleach are separated by a well-defined isosbestic point at ~500 nm. The transient spectra show a transient absorption band in the 500–700 nm region with a peak at 610 nm and a flat absorption band in the 700–900 nm region. The positive absorption band can be attributed to the charge-separated species, i.e., Cat cation radical and injected electron in the NPs. However, Cat cation radical does not have any absorption beyond 500 nm [37]. So the transient absorption signal at 500–900 nm can be attributed to the injected electrons in TiO_2. Lian and coworkers have also assigned this transient band in Cat-sensitized TiO_2 NPs in water as injected electrons in NPs. However, in earlier measurements we [15,21,34–36] have shown that the transient spectra of the injected electron in the visible region are quite broad and structureless. It is also interesting to see that at longer time delay (Figure 18.6), the transient peak at 610 nm gets smeared off with the flat absorption band (700–900 nm). In the case of Cat/TiO_2 SNP system, on excitation by 400 nm laser light, only the CT band is excited. On excitation of the CT band, electron is

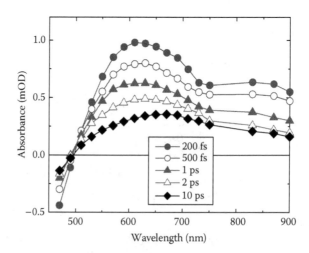

FIGURE 18.6 Transient absorption spectra of Cat/SNP TiO_2 complex in microemulsion at 0.2, 0.5, 1, 2, and 10 ps after 400 nm excitation.

directly injected to the Ti metal ion. From the Ti ion the electron gets delocalized from the metal ion to the conduction band of the NPs [33]. So, at the end of the laser pulse in the transient absorption spectra of Cat/TiO$_2$ SNP system we have observed charge-separated CT state, where electron is localized in the metal ion. Hence, in the present investigation transient absorption band peaking at ~610 nm can be attributed to the excited CT state of Cat/TiO$_2$ SNP system, where electrons are localized in the metal ion. To determine the electron injection we have monitored the appearance signal of the transients at both 610 and 900 nm. We have observed that both the kinetic traces look very similar. The transient signal at 900 nm can be attributed to the injected electron in the conduction band of semiconductor NPs and it is shown in Figure 18.7a for Cat/TiO$_2$ SNP system. Time constant for the appearance signal has been fitted and found to be pulse-width limited, i.e., <50 fs (Figure 18.7a, inset). Electron injection time in a strong coupling dye-NP system is an adiabatic process, which indicate that on photo-excitation electron transfer takes place from the sensitizer to NP very fast. It is reported by Schnadt et al. [38] that electron injection can take place as fast as sub-3 fs from the aromatic adsorbate to TiO$_2$.

Similarly in the present investigation also we have observed that injection time is too fast to measure by our laser pulse (pulse width ~50 fs). In our earlier work [36] we have shown there that pulse-width limited injection means that the data can be fitted with time constant between 1 and 25 fs. We had seen that 50 fs rise time showed bad fitting and we had attributed that injection is pulse-width

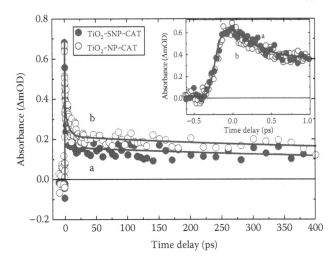

FIGURE 18.7 Normalized transient absorption decay profiles at 900 nm after 400 nm excitation for (a) Cat sensitized SNP TiO$_2$ (●) in microemulsion and (b) Cat sensitized NP TiO$_2$ (O) in water. The solid lines are best fits to the data. Inset: Same kinetic decay trace is plotted in a shorter timescale.

limited. Similarly in the present investigation also we have fitted the data in all the systems with 1–25 fs rise time, which clearly indicate pulse-width limited injection. To determine BET dynamics we have monitored the kinetic decay trace at 900 and it is shown in Figure 18.7a. The transient decay kinetics at 900 nm, which can be fitted multiexponentially with time constants of 300 fs (74.0%), 6 ps (11.3%), and >400 ps (14.7%). Charge recombination dynamics in dye-sensitized TiO_2 NPs is multiexponential in nature. Scheme 18.1 can be used to understand the multiexponential nature of BET dynamics. We have depicted the electron transfer process in Scheme 18.1, where $E_{S/S+}$ and $E_{S^*/S+}$ are the ground and excited state redox potentials of Cat, and $E_{CT*/S+}$ the excited state potential of the CT complex. $E_{S/S+}$ is found to be +1.13 V [33]. $E_{S^*/S+}$ has been determined from ground state redox potential ($E_{S/S+}$) subtracting the $S_1 \leftarrow S_0$ transition (Figure 18.4a) of Cat molecule and found to be −3.01 V. Similarly $E_{CT*/S+}$ has been determined from ground state redox potential ($E_{S/S+}$) subtracting the $S_{CT} \leftarrow S_0$ transition (Figure 18.4c) of CT complex of Cat/TiO_2 e− and found to be −0.94 V. It is interesting to

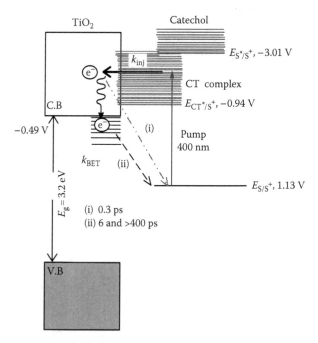

SCHEME 18.1 Mechanistic scheme of electron transfer for the Cat-sensitized TiO_2 NPs. Here S^*/S^+ is the excited-sensitized dye/cation radical couple, and CT^*/S^+ is the excited CT state/cation radical couple. On 400 nm excitation, CT complex of Cat/TiO_2 gets excited and electron injection to the conduction band takes place. $E_{S/S+}$ and $E_{S^*/S+}$ is the ground and excited state redox potential of Cat, respectively, and $E_{CT*/S+}$ the excited state potential of the CT complex.

see that on 400 nm excitation CT complex is excited in Cat/TiO$_2$ system and directly injects electron into the conduction band of TiO$_2$. Cat is a tricky molecule and it absorbs light below 300 nm. So on excitation by 400 nm light pure Cat molecule cannot be excited. We can excite only the CT band by exciting the CT band. Theoretical calculations [39,40] also confirmed that in Cat/TiO$_2$ system electron injection takes place directly into the conduction band. To find out back ET dynamics we have fitted the data in lab-view program that yielded best-fitted three exponential time constants. These data could not be fitted biexponentially rather we could fit three exponentially with very good χ^2. The fastest component, i.e., 300 fs for Cat/TiO$_2$ system is the recombination dynamics of hot electron (process (i) in Scheme 18.1), which recombines with the parent cation just after injection due to strong coupling in nature. The second time constants, 6 ps component, might be the recombination time constant of injected electron in the conduction band edge or shallow trapped electron and parent cation. On the other hand the longer component (>400 ps) is the recombination time constant between deeper trapped electron and parent cation. The longer components have been attributed to process (ii) in Scheme 18.1. We have also carried out intensity dependence measurements on back ET dynamics in the present studies. However, we did not observe and difference in back ET dynamics like our earlier observation [30].

Although interfacial electron dynamics in Cat-sensitized TiO$_2$ NPs (prepared from solgel process) in water has been carried out by Wang et al. [33], for the sake of comparison of the ET dynamics in the above system in two different types of particles in different media we have also carried out sensitization experiments in Cat/TiO$_2$ NP system in water. We have monitored the kinetic decay trace at 900 nm shown in Figure 18.7b. The kinetic decay trace can be fitted multiexponentially with the time constants of 300 fs (67.3%), 6 ps (13.2%), and >400 ps (19.5%). We have also monitored the appearance signal at 900 nm which is due to the injected electron in the conduction band of TiO$_2$ (Figure 18.7b, inset). It is interesting to see that electron injection time in both the particles is similar (pulse-width limited). We tried to fit both colloidal data and microemulsion data for both the dye/TiO$_2$ system with similar time constants as the kinetic decay traces are very similar and also tried to fit to the model. However, we can see clearly the contribution of the components is different.

18.4.5 EFFECT OF MICROENVIRONMENT IN INTERFACIAL ELECTRON TRANSFER DYNAMICS

One of the main aims of this investigation is to monitor interfacial ET dynamics of exciton-like TiO$_2$ particles synthesized in microemulsion. We had to choose to synthesis particles in microemulsion method as it gives good control over size of the particles, which we have easily synthesized by controlling the size of the water pool with $w_0 = 1$. We have observed that these TiO$_2$ particles in the microemulsion can be sensitize with relatively stronger coupling adsorbate like Cat and Qz-6S. Electron injection and BET dynamics have been monitored by directly detecting the electron in the conduction band and/or cation radical of the adsorbates. To see the effect of microenvironment and size quantization of the particles, the above experiments were also carried out using bigger size particles as synthesized by solgel process. Now let us discuss the ET dynamics of Cat/TiO$_2$ system. Earlier, we have discussed that Cat/TiO$_2$ system is a strong coupling system and Cat forms strong CT complex with TiO$_2$ NPs irrespective of the medium and environment. It is interesting to observe that in both the cases, electron injection is pulse-width limited as Lian and coworkers [33] observed in Cat/TiO$_2$ system. Recently, Rego et al. [39] have studied the interfacial ET dynamics in Cat/TiO$_2$ nanoclusters after combining ab initio DFT molecular dynamics simulations and quantum dynamics calculations. They have found that the primary process that localizes the charge in the Ti^{4+} surface ions next to the Cat adsorbate. As in both microemulsion and water, Cat forms strong CT complex and on excitation of this CT complex at 400 nm, the electron will be first localized in Ti(IV) ion, which is basically the electron injection into the NPs. Electron injection time is found to be similar in both the systems. The primary event is followed by charge delocalization (i.e., charge diffusion) through the nanocrystalline material. This process of localization of the electron at the metal center is only possible when the dye molecules form a strong CT complex with the NPs. However, the injected electrons can also rapidly delocalize and diffuse out and get trapped in different trapping positions with a distribution of trap energy and distance from the adsorbate. As a result, multiexponential recombination (BET) dynamics is expected and observed earlier by us [15,34–36] and many other authors [33]. Ultrafast BET reaction in Cat-sensitized TiO$_2$ NPs has been reported by Wang et al. [33]. They have reported that majority of the injected electrons recombine with the parent cation with time constant ~400 fs. Our kinetic data traces for the charge recombination reaction can be fitted multiexponentially with time the constants of 300 fs (74.0%), 6 ps (11.3%), and >400 ps (14.7%) for smaller size nanoparticles (SNP) and 300 fs (67.3%), 6 ps (13.2%), and >400 ps (19.5%) for the bigger size NP. It is clear from the kinetic decay traces in Figure 18.4 that early dynamics (faster components) for both systems is very similar and longer dynamics is marginally slower on NP surface.

18.5   CONCLUSION

We have studied optical and photophysical properties of TiO$_2$ NPs, where the NPs are dispersed in the water core of a microemulsion. We have observed that the optical absorption of NP does not change in the microemulsion; however, the emission quantum yield increases drastically. As an indirect band-gap semiconductor TiO$_2$ NPs are nonemissive but due to the surface modification by surfactant molecules in the microemulsion the particles become emissive.

We have synthesized TiO$_2$ semiconductor particles of very small size nanoparticles (SNP) in AOT/water microemulsion with $w_0 = 1$. Size-quantization behavior of TiO$_2$ particles of small size has been confirmed from blueshift in the absorption band edge. We have also used these particles in microemulsion along with a suitable sensitizer molecule like Cat. We have also sensitized TiO$_2$ NP by Cat molecules as synthesized by solgel process. In the Cat/TiO$_2$ systems 400 nm excitation of the CT bands promotes an electron from the Cat ligand to titanium(IV) centers. Electron injection has been found to be pulse-width limited. In both SNP and NP surfaces, just after injection a major portion of the injected electron recombines with an ultrafast component (300 fs) with multiple slower components. Charge recombination dynamics was found to be very similar on both the particle surfaces for both the dye/NP pairs. Strong coupling between the dye–NP systems dominates the charge recombination dynamics in early timescale and found to be similar in different type particles. However, in longer time domain BET dynamics is marginally faster on TiO$_2$ surface in microemulsion, which may be due to the modification of the surface states by AOT molecules.

ACKNOWLEDGMENTS

The author acknowledges Drs. M. C. Rath, S. Adhikari, and D. K. Palit for fruitful discussions and Drs. S. K. Sarkar and T. Mukherjee for encouragements.

SYMBOLS AND TERMINOLOGIES

IET	interfacial electron transfer
BET	back electron transfer
NP	nanoparticle
SNPs	subnanoparticles
QD	quantum dot
Cat	catechol
CT	charge transfer
AOT	Aerosol-OT
SDBS	sodium dodecyl benzene sulfonate
NaLS	sodium lauryl sulfate
w_0	[H$_2$O]/[surfactant]

REFERENCES

1. J. Z. Zhang, 2000. Interfacial charge carrier dynamics of colloidal semiconductor nanoparticles. *J. Phys. Chem. B* 104, 7239–7253.
2. B. Oregan and M. Gratzel, 1991. A low-cost, high-efficiency solar cell based on dye-sensitized colloidal TiO$_2$ films. *Nature* 353, 737–740.
3. N. Serpone, 1994. A decade of heterogeneous photocatalysis in our laboratory; pure and applied studies in energy production and environmental detoxification. *Res. Chem. Interm.* 20, 953–992.
4. A. P. Alivisatos, 1996. Perspectives on the physical chemistry of semiconductor nanocrystals. *J. Phys. Chem.* 100, 13226–13239.
5. L. Spahel, M. Haase, H. Weller, and A. Henglein, 1987. Photochemistry of colloidal semiconductors. 20. Surface modification and stability of strong luminescing CdS particles. *J. Am. Chem. Soc.* 109, 5649–5655.
6. T. Dannhauser, M. O'Neil, K. Johansson, D. Whitten, and G. McLendon, 1986. Photophysics of quantized colloidal semiconductors. Dramatic luminescence enhancement by binding of simple amines. *J. Phys. Chem.* 90, 6074–6076.

7. M. G. Bawendi, P. J. Carroll, W. L. Wilson, and L. E. Brus, 1992. Luminescence properties of CdSe quantum crystallites: Resonance between interior and surface localized states. *J. Chem. Phys.* 96, 946–954.
8. G. Ramakrishna, A. K. Singh, D. K. Palit, and H. N. Ghosh, 2004. Dynamics of interfacial electron transfer from photo-excited quinizarin (Qz) into the conduction band of TiO_2 and surface states of ZrO_2 nanoparticles. *J. Phys. Chem. B* 108, 4775–4783.
9. G. Ramakrishna and H. N. Ghosh, 2004. Determination of back electron transfer rate from the surface states of quinizarin sensitized ZrO_2 nanoparticles by monitoring charge transfer emission. *Langmuir* 20, 7342–7345.
10. M. C. Rath, G. Ramakrishna, T. Mukherjee, and H. N. Ghosh, 2005. Electron injection into the surface states of ZrO_2 nanoparticles from photo-excited quinizarin and its derivatives: Effect of surface-modification. *J. Phys. Chem. B* 109, 20485–20492.
11. P. V. Kamat, 1993. Photochemistry on nonreactive and reactive (semiconductor) surfaces. *Chem. Rev.* 93, 267–300.
12. L. Spanhel, H. Weller, A. Fojtik, and A. Henglein., 1987. Photochemistry of semiconductor colloids 17. Strong luminescing CdS and $CdS–Ag_2S$ particles. *Ber. Bunsenges. Phys. Chem.* 91, 88–95.
13. M. A. Fox and M. T. Dulay, 1993. Heterogeneous photocatalysis. *Chem. Rev.* 93, 341–357.
14. G. Ramakrishna and H. N. Ghosh, 2003. Optical and photochemical properties of sodium dodecylbenzenesulfonate (DBS)-capped TiO_2 nanoparticles dispersed in nonaqueous solvents. *Langmuir* 19, 505–508.
15. G. Ramakrishna, A. K. Singh, D. K. Palit, and H. N. Ghosh, 2004. Slow back electron transfer in surface-modified TiO_2 nanoparticles sensitized by alizarin. *J. Phys. Chem. B* 108, 1701–1707.
16. G. Ramakrishna, A. Das, and H. N. Ghosh, 2004. Effect of surface modification on back electron transfer dynamics of dibromo fluorescein sensitized TiO_2 nanoparticles. *Langmuir* 20, 1430–1435.
17. P. L. Luisi and B. E. Straub, 1984. *Reverse Micelles: Biological and Technological Relevance of Amphiphilic Structures in Apolar Media*, Plenum, New York.
18. H. N. Ghosh and S. Adhikari, 2001. Trap state emission from TiO_2 nanoparticles in microemulsion solutions. *Langmuir* 17, 4129–4130.
19. J. H. Fendler, 1987. Atomic and molecular clusters in membrane mimetic chemistry. *Chem. Rev.* 87, 877–899.
20. V. Chhabra, V. Pillai, B. K. Mishra, A. Morrone, and D. O. Shah, 1995. Synthesis, characterization, and properties of microemulsion-mediated nanophase TiO_2 particles. *Langmuir* 11, 3307–3311.
21. H. N. Ghosh, 1999. Charge transfer emission in coumarin 343 sensitized TiO_2 nanoparticle: A direct measurement of back electron transfer. *J. Phys. Chem. B* 103, 10382–10387.
22. M. P. Pileni, T. Zemb, and C. Petit, 1985. Solubilization by reverse micelles: Solute localization and structure perturbation. *Chem. Phys. Lett.* 118, 414–420.
23. S. Adhikari, R. Joshi, and C. Gopinathan, 1997. Hydrated electrons in a quaternary microemulsion system: A pulse radiolysis study. *J. Colloid Interface Sci.* 191, 268–271.
24. S. K. Poznyak, V. V. Sviridov, A. I. Kulak, and M. P. Samtsov, 1992. Photoluminescence and electroluminescence at the TiO_2—electrolyte interface. *J. Electroanal. Chem.* 340, 73–97.
25. J. I. Pankove, 1971. *Optical Processes in Semiconductors*, Dover Publications, New York.

26. Y. Nakato, A. Tsumura, and H. Tsubomura, 1982. Electro- and photo-luminescence spectra in various n-type semiconductors in relation with anodic reaction intermediates. *Chem. Phys. Lett.* 85, 387–390.

27. M. Anpo, M. Tomonari, and M. A. Fox, 1989. In situ photoluminescence of titania as a probe of photocatalytic reactions. *J. Phys. Chem.* 93, 7300–7302.

28. N. Serpone, D. Lawless, and R. Khairutdinov, 1995. Size effects on the photophysical properties of colloidal anatase TiO$_2$ particles: Size quantization versus direct transitions in this indirect semiconductor? *J. Phys. Chem.* 99, 16646–16654.

29. T. Dannhauser, M. O'Neil, K. Johansson, D. Whitten, and G. McLendon, 1986. Photophysics of quantized colloidal semiconductors. Dramatic luminescence enhancement by binding of simple amines. *J. Phys. Chem.* 90, 6074–6076.

30. P. G. De Gennes and C. Taupin, 1982. Microemulsions and the flexibility of oil/water interfaces. *J. Phys. Chem.* 86, 2294–2304.

31. P. A. Sant and P. V. Kamat, 2002. Interparticle electron transfer between size-quantized CdS and TiO$_2$ semiconductor nanoclusters. *Phys. Chem. Chem. Phys.* 4, 198–203.

32. J. Moser, S. Punchihewa, P. P. Infelta, and M. Gratzel, 1991. Surface complexation of colloidal semiconductors strongly enhances interfacial electron-transfer rates. *Langmuir* 7, 3012–3018

33. Y. Wang, K. Hang, N. A. Anderson, and T. Lian, 2003. Comparison of electron transfer dynamics in molecule-to-nanoparticle and intramolecular charge transfer complexes. *J. Phys. Chem. B* 107, 9434–9440.

34. G. Ramakrishna, H. N. Ghosh, A. K. Singh, D. K. Palit, and J. P. Mittal, 2001. Dynamics of back-electron transfer processes of strongly coupled triphenyl methane dyes adsorbed on TiO$_2$ nanoparticle surface as studied by fast and ultrafast visible spectroscopy. *J. Phys. Chem. B* 105, 12786–12796.

35. G. Ramakrishna, A. D. Jose, D. Krishnakumar, A. Das, D. K. Palit, and H. N. Ghosh, 2005. Strongly coupled ruthenium-polypyridyl complexes for efficient electron injection in dye-sensitized semiconductor nanoparticles. *J. Phys. Chem. B* 109, 15445–15453.

36. G. Ramakrishna, A. D. Jose, D. Krishnakumar, A. Das, D. K. Palit, and H. N. Ghosh, 2006. Interfacial electron transfer between the photoexcited porphyrin molecule and TiO$_2$ nanoparticles: Effect of catecholate binding. *J. Phys. Chem. B* 110, 9012–9021.

37. E. J. Land, 1993. Preparation of unstable quinones in aqueous solution via pulse radiolytic one-electron oxidation of dihydroxybenzenes. *J. Chem. Soc. Faraday Trans.* 89, 803–810.

38. J. Schnadt, P. A. Bruhwiler, L. Patthey, J. N. O'Shea, S. Sodergren, M. Odelius, R. Ahuja, O. Karis, M. Bassler, P. Persson, H. Siegbahn, S. Lunell, and N. Martensson, 2002. Experimental evidence for sub-3-fs charge transfer from an aromatic adsorbate to a semiconductor. *Nature* 418, 620–623.

39. G. L. S. Rego and V. S. Batista, 2003. Quantum dynamics simulations of interfacial electron transfer in sensitized TiO$_2$ semiconductors. *J. Am. Chem. Soc.* 125, 7989–7997.

40. W. R. Duncan and O. V. Prezhdo, 2005. Electronic structure and spectra of catechol and alizarin in the gas phase and attached to titanium. *J. Phys. Chem. B* 109, 365–373.

19 Microemulsions as Pseudostationary Phases in Electrokinetic Chromatography: I. Estimation of Physicochemical Parameters. II. Analysis of Drugs in Pharmaceutical and Biofluidic Matrices

Valeria Tripodi and Silvia Lucangioli

CONTENTS

19.1 INTRODUCTION

Capillary electrophoresis (CE) is a high-efficiency analytical technique that separates compounds by applying high voltage across buffer-filled capillaries. The CE advantages with respect to other analytical techniques, such as very high resolution in short time of analysis, versatility, small volume of sample, and low cost, have made this technique adequate for the analysis of numerous compounds like biological macromolecules, chiral drugs, inorganic ionics, organic acids, DNA fragments, and even whole cells and virus particles [1–3].

CE has been applied in different modes, one of them is electrokinetic chromatography (EKC) where the mobile phase is normally an aqueous buffer and the pseudostationary phase can be micelles (MEKC), vesicles (VEKC), or microdroplets (MEEKC).

Microemulsions are dispersed systems consisting of nanometer-size droplets of an inmiscible liquid, stabilized by surfactant and cosurfactant molecules. Microemulsions can be prepared as water droplets in an oil phase (W/O) or oil droplets in a water phase (O/W). Due to many potential advantages of microemulsions such as high stability, ability to interact with a wide range of hydrophobic and hydrophilic compounds, transparency, and easy preparation, they have been used in CE as pseudostationary phases in the EKC mode [4–6].

Oil in water (O/W) is the type of microemulsion commonly used in EKC prepared with different surfactants like sodium dodecylsulfate or biosurfactants such as phosphatidylcholine (PC) and/or bile acids (BA), cosurfactants like butanol, oils (octane, hexane), and buffers at different pH values.

MEEKC has become an important field of research in CE offering a large number of applications such as estimation of physicochemical properties and for the analysis of different endogenous and exogenous compounds present in different biological and pharmaceutical samples.

19.2 PRINCIPLES OF CAPILLARY ELECTROPHORESIS

Separation by electrophoresis is based on differences in analyte velocity in an applied electric field within the capillary. The electrophoretic mobility of the analyte (μ_{ep}) depends on the characteristics of the analyte (electrical charge, molecular size, and shape) and the characteristics of the running buffer (type and ionic strength of the electrolyte, pH, viscosity, and properties of the additives) in which the migration takes place [1–3].

The electrophoretic velocity of an ion can be given by

$$v_{ep} = \mu_{ep}E = \frac{q}{6\pi\eta r}\frac{V}{L} \tag{19.1}$$

where
v_{ep} is the analyte velocity
μ_{ep} is the electrophoretic mobility
E is the applied electric field
q is the effective charge
η is the viscosity of the buffer
r is the size of the solute ion
V is the applied voltage
L is the total length of the capillary

The direction and velocity of that movement are determined by the sum of two components: the migration of the ionic solute and the electroosmotic flow (EOF). EOF is a flow generated inside the capillary when an electric field is applied. If Equation 19.1 is modified, it can be considered as

$$v_{ep} = (\mu_{ep} + \mu_{eof})E \tag{19.2}$$

where μ_{eof} is the mobility of the EOF (Figure 19.1).

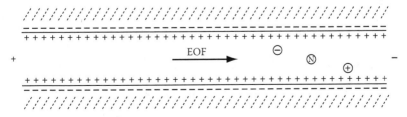

FIGURE 19.1 Development of EOF and order of elution of ionic and neutral solutes.

μ_{ep} is a physical constant specifically related to individual ionic species, the EOF results from the effect of the applied electric field at the interface between the solution and the capillary wall. The magnitude of the EOF can be expressed by

$$V_{eof} = \frac{\varepsilon\varsigma}{4\pi\eta} E \qquad (19.3)$$

where
  V_{eof} is the EOF velocity
  ε is the dielectric constant
  ς is the zeta potential
  η is the viscosity of the buffer
  E is the applied electric field

Migration time, t_m, is the time required for the analyte to migrate toward the point of detection

$$t_m = \frac{l}{v_{ep}} \qquad (19.4)$$

l is the effective capillary length (the distance between the points of injection and detection) and v_{ep} is the analyte velocity.

The apparent mobility of the analytes is determined by their t_m and other experimental parameters:

$$\mu_{app} = \frac{lL}{t_m V} \qquad (19.5)$$

where
  μ_{app} is the apparent mobility
  l is the effective capillary length (to the detector)
  L is the total capillary length
  t_m is the migration time
  V is the applied voltage

The effective mobility, μ_{eff}, is the value corrected by the contribution of the mobility of the EOF (μ_{eof}):

$$\mu_{eff} = \mu_{app} - \mu_{eof} \qquad (19.6)$$

The μ_{eof} can be experimentally determined by using a neutral marker such as methanol, acetone, acetonitrile, dimethylsulphoxide, or some other neutral compounds.

19.3   ELECTROKINETIC CHROMATOGRAPHY

19.3.1   Mobilities and Retention Factors of the Analytes

In EKC systems, the separation of the compounds is carried out by the partition of the analytes between the mobile phase and a pseudostationary phase. The mobile phase is normally an aqueous buffer at different concentration and pH values and the pseudostationary phase may be micelles (MEKC), vesicles (VEKC), and microdroplets (MEEKC).

In MEEKC, the mass partition coefficient is calculated from the retention factor (k') by the determination of the electrophoretic mobilities of the compounds. The mobilities of the microdroplets are calculated by using as marker dodecaphenone or other highly water-insoluble neutral compound [7] (Figure 19.2).

Retention factor (k') is defined as the mass distribution coefficient according to

$$k' = \frac{n_{i,s}}{n_{i,m}} = \frac{c_{i,s} V_s}{c_{i,m} V_m} \tag{19.7}$$

where $n_{i,s}$ and $n_{i,m}$ are the mole numbers of analytes, i, in the pseudostationary phase (s), and mobile phase (m), respectively. c_i are the molar concentrations of the analyte in the respective phases with volumes V_s and V_m. Retention factor of neutral analytes can be calculated from the mobilities according to the following equation:

$$k' = \frac{(\mu_{meas} - \mu_{eof})}{(\mu_{meas}^{mic} - \mu_{eof}^{mic}) - (\mu_{meas} - \mu_{eof})} = \frac{\mu_{eff}}{\mu_{eff}^{mic} - \mu_{eff}} \tag{19.8}$$

μ is the electrophoretic mobility of the neutral analyte
μ_{eof} is the electrophoretic mobility of the EOF
μ_{meas}^{mic} is the electrophoretic mobility of the microdroplets
μ_{eof}^{mic} is the electrophoretic mobility of the EOF (in the presence of the microdroplets marker)
μ_{eff} is the effective mobility of the neutral analyte
μ_{eff}^{mic} is the effective mobility of the microdroplet

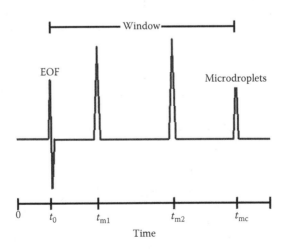

FIGURE 19.2 Schematic representation of the electrophoretic separation principle of MEEKC. An EOF/microdroplet marker and the separation of two analytes.

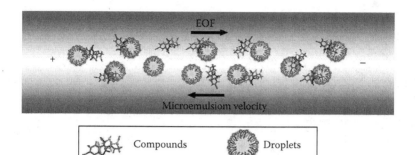

FIGURE 19.3   Scheme of the MEEKC principle.

The values of k' of the ionic analytes can be calculated according to

$$k' = \frac{\mu_{\text{eff}} - \mu_{\text{eff}}^{\text{free}}}{\mu_{\text{eff}}^{\text{mic}} - \mu_{\text{eff}}} \qquad (19.9)$$

where

   μ_{eff} is the effective mobility of the ionic analyte

   $\mu_{\text{eff}}^{\text{free}}$ is the effective mobility of the ionic analyte in free solution (without microdroplets)

   $\mu_{\text{eff}}^{\text{mic}}$ is the effective mobility of the microdroplets (Figure 19.3)

19.4   HYDROPHOBICITY

19.4.1   Estimation of the Hydrophobicity

Hydrophobicity is considered one of the most important physicochemical properties in the design process of a new bioactive substance. It can be related to biological activities such as absorptivity, transportation, bioaccumulation, effectivity, and toxicity, and it is taken as a parameter able to reflect interactions of the drugs with biomembranes [7,8].

The traditional method known as shake-flask method to determine hydrophobicity is usually expressed by the thermodynamic 1-octanol–water partition coefficient (P_{ow}) (or its logarithm, log P_{ow}) defined as the ratio of concentrations of a species in the two phases at equilibrium. This method is time consuming, tedious, requires highly pure compounds, and it is not adequate for compounds with log P_{ow} higher than 4 [8–10]. Many alternative methods have been used for the log P_{ow} measurement, direct methods such as potentiometric method [11] and indirect methods such as thin layer chromatography (TLC) [12], reversed-phase liquid chromatography (RP-LC) [13,14] and different mathematical models [15]. In RP-LC, the estimation of the hydrophobicity is achieved by using the linear correlation between the log P_{ow} and the logarithm of the retention factor (log k'). More recently, EKC methods based on micelles [16,17], vesicles [18], or microdroplets [19–21] have also turned out as reliable, rapid, and economical alternatives for the indirect measurement of the hydrophobicity of the compounds. However, more appropriate models to estimate the hydrophobicity of bioactive compounds require the design and study of systems that may adjust better to biopartitioning. Microemulsions containing biosurfactants like PC and cobiosurfactants such as sodium salts of BAs as cholic (SC) or deoxycholic (SDC) acids may be adequate for such purposes [7,22].

19.4.2 MICROEMULSION ELECTROKINETIC CHROMATOGRAPHY: EXPERIMENTAL

A typical microemulsion in MEEKC consists of sodium dodecyl sulfate (SDS) as surfactant that also implements charges to the microdroplets, 1-butanol as cosurfactant, octane, or hexane as oil and an aqueous buffer at pH 7.50, in order to mimic physiological acid-base conditions and also to establish EOF in the capillary.

Apart from the most commonly used microemulsion, other systems are used to estimate the hydrophobicity of several analytes based on PC as biosurfactant and sodium cholate or sodium deoxycholate as cobiosurfactant, isopropyl myristate as oil and an aqueous phosphate buffer at pH 7.50 (Table 19.1).

TABLE 19.1
MEEKC Systems for the Estimation of log P of Betamethasone and Derivatives

System	Composition of the EKC Systems	Voltage, Temperature Detection
MEEKC-SDS	SDS 1.44% Octane 0.81% 1-Butanol 6.61% Sodium phosphate (20 mM pH 7.5) 91.14%	+22 kV, 22°C, 214 nm
MEEKC-PCSC	PC 3.5% Isopropyl myristate 1.9% SC 2.0% 1-Butanol 7.5% Sodium phosphate (20 mM pH 7.5) 85.1%	+25 kV, 25°C, 214 nm
MEEKC-PCSC	PC 3.5% Isopropyl myristate 1.9% SDC 2.0% 1-Butanol 7.5% Sodium phosphate (20 mM pH 7.5) 85.1%	+25 kV, 25°C, 214 nm

Note: Percentages are based on w/w.

19.4.3   RETENTION FACTOR AND 1-OCTANOL–WATER PARTITION COEFFICIENT

The partition coefficients, P, of the analytes in EKC systems can be calculated from their k' values by using reference compounds whose P_{ow} values are known. The reference compounds for the calibration of the log k' versus log P_{ow} scale were selected taking into account the range of log P_{ow} between 0.55 and 4.50 taken from literature [20,23]. These data allow construction of a calibration curve for each MEEKC system with the measured log k' values according to the following equation:

$$\log P_{ow} = a \log k' + b \qquad (19.10)$$

where
  a is the slope
  b is the intercept.

The log P_{ow} for 10 reference compounds are given in Table 19.2 together with the measured log k' in three MEEKC systems and the Equation 19.10 for each system is included.

TABLE 19.2

Measured log k' of the Reference Compounds Used for the Calibration of the log P Scale

Compounds	log P_{ow}	MEEKC (SDS) log k'	MEEKC (PCSC) log k'	MEEKC (PCSDC) log k'
Hydroquinone	0.55	−0.809	−0.609	−0.257
Resorcinol	0.80	−0.625	−0.078	−0.106
Benzyl alcohol	1.10	−0.349	−0.009	−0.003
Phenol	1.46	−0.258	0.317	0.284
m-Cresol	1.96	0.051	0.637	0.635
Anisole	2.11	0.235	0.878	0.825
2-Naphthol	2.84	0.477	1.180	0.888
Benzophenone	3.18	0.735	1.149	0.919
Naphthalene	3.37	0.887	1.259	1.210
Anthracene	4.45	1.264	2.046	1.720
R		0.9936	0.9802	0.9817
Equation 19.10		$\log P = 1.6 \log k' + 1.7$	$\log P = 1.6 \log k' + 1.1$	$\log P = 2.0 \log k' + 1.0$

Source: Lucangioli S. E., Carducci C. N., Scioscia S. L., Carlucci A., Bregni C., Kenndler E., *Electrophosesis*, 24, 984, 2003. With permission.

Note: Data for log P_{ow} were taken from Refs. [20,23], r, linear correlation coefficient for log k' versus log P_{ow}.

19.4.4 Relation between log P_{ow} from Octanol–Water and log P from MEEKC

The relation between log P_{ow} from octanol–water and log P determined by different MEEKC systems has been studied for different compounds. Typical microemulsions with SDS (MEEKC-SDS) and MEEKC based on PC and bile salts like sodium cholate (MEEKC-PCSC) and deoxycholate (MEEKC-PCSDC) have been studied.

The selected compounds studied were the corticoids, betamethasone, and derivatives. Corticoids are a group of drugs with therapeutic action as anti-inflammatory, antialergic, and antirheumatic properties. Betamethasone and synthetic derivatives have similar chemical structures (Figure 19.4). Esterification in position 17 and/or 21 leads to compounds with different hydrophobicity and they have been used in different pharmaceutical administrations. log k' for betamethasone and derivatives have been determined in those MEEKC systems and the Equation 19.10 applied for calculation of log P. The measured mobilities and retention factors of the analytes in different MEEKC systems are shown in Table 19.3.

The log P calculated in the three MEEKC systems (Table 19.4) agree for bethametasone and bethametasone acetate, both for experimental and software values. In the case of bethametasone phosphate, which is in anionic form, the software does not calculate the log P_{ow} and the log P agree in less grade than the others. For bethametasone valerate and dipropionate, the MEEKC systems with

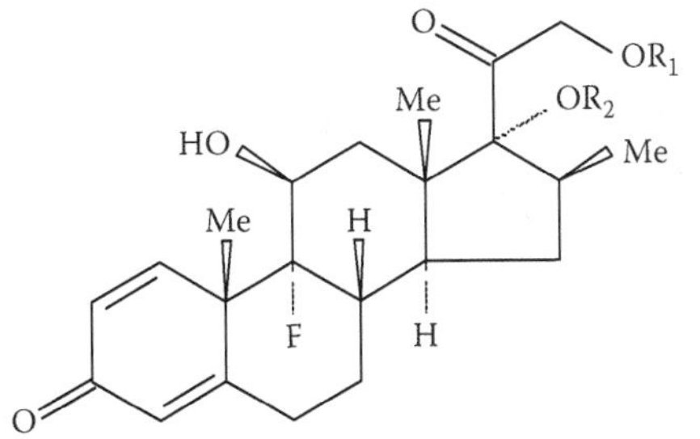

Compounds	R_1	R_2
1 Betamethasone 21–phosphate	PO_2^{2-}	H
2 Betamethasone	H	H
3 Betamethasone 21–acetate	$-COCH_3$	H
4 Betamethasone 17–valerate	H	$-COCH_2CH_2CH_2CH_3$
5 Betamethasone 17–21-dipropionate	$-COCH_2CH_3$	$-COCH_2CH_3$

FIGURE 19.4 Chemical structures of betamethasone and derivatives. (From Lucangioli S. E., Carducci C. N., Scioscia S. L., Carlucci A., Bregni C., Kenndler E., *Electrophosesis*, 24, 984, 2003. With permission.)

TABLE 19.3

Mobilities and Retention Factors for Betamethasone and Derivatives for MEEKC Systems

Compounds	MEEKC-SDS			MEEKC-PCSC			MEC-PCSDC		
	μ_{med}	μ_{eff}	k'	μ_{med}	μ_{eff}	k'	μ_{med}	μ_{eff}	k'
Bethamethasone phosphate	28.6	−29.2[c]	0.43	30.9	−20.3[d]	0.60	31.5	−22.3[e]	1.01
Bethamethasone	22.5	−35.2	3.45	35.3	−17.1	4.10	34.5	−19.3	3.41
Bethamethasone acetate	20.0	−37.7	4.90	32.9	−19.5	11.1	32.2	−21.6	6.36
Bethamethasone valerate	15.2	−42.5	14.3	31.6	−20.9	54.1	30.5	−24.0	26.5
Bethamethasone dipropionate	15.5	−42.3	13.6	31.4	−21.0	90.7	30.2	−24.3	39.1

Source: Lucangioli S. E., Carducci C. N., Scioscia S. L., Carlucci A., Bregni C., Kenndler E., *Electrophosesis*, 24, 984, 2003. With permission.

Note: Mobilities are expressed in 10^{-9} m^2 V^{-1} s^{-1}.

μ_{eff}^{free} of the anion (betamethasone phosphate) is −22.9(MEEKC-SDS); −19.6(MEEKC-PCSC); −19.6(MEC-PCSDC) (10^{-9} m^2 V^{-1} s^{-1}).

μ_{eff} microdroplets −45.4(MEEKC-SDS); −21.3(MEEKC-PCSC); −24.9(MEC-PCSDC) (10^{-9} m^2 V^{-1} s^{-1}).

TABLE 19.4

log P in Different MEEKC Systems from the log k' of the Analytes upon Calibration by log P_{ow} versus log k' from the Reference Compounds Listed in Table 19.2, and log P_{ow} Calculated from the Molecular Structure

Compounds	log Software[a]	P_{ow} Experim[b]	log P MEEKC-SDS	MEEKC-PCSC	MEEKC-PCSDC
Bethamethasone phosphate		0.72[1]	1.22	0.78	1.01
Bethamethasone	2.06	2.01[2] 2.08[1]	2.88	2.08	2.07
Bethamethasone acetate	2.61	2.58[2]	3.15	2.76	2.61
Bethamethasone valerate	3.97	3.60[2]	4.01	3.84	3.85
Bethamethasone dipropionate	4.23	3.85[1]	3.96	4.19	4.18

Source: Lucangioli S. E., Carducci C. N., Scioscia S. L., Carlucci A., Bregni C., Kenndler E., *Electrophosesis*, 24, 984, 2003. With permission.

[a] Data calculated from the molecular structure by Software Solaris Ref. [66].

[b] Data measured experimentally: (1) by own measurements and (2) from Ref. [67].

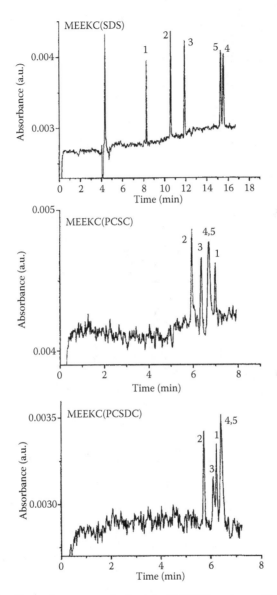

FIGURE 19.5 Electropherograms obtained by MEEKC systems. (1) Betamethasone 21–phosphate, (2) Betamethasone, (3) Betamethasone 21–acetate, (4) Betamethasone 17–valerate, (5) Betamethasone 17–21-dipropionate. (From Lucangioli S. E., Carducci C. N., Scioscia S. L., Carlucci A., Bregni C., Kenndler E., *Electrophosesis*, 24, 984, 2003. With permission.)

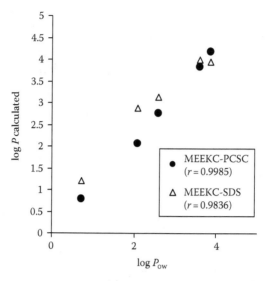

FIGURE 19.6 Calculated log P versus log P_{ow} for betamethasone and derivatives for MEEKC-SDS and MEEKC-PCSC.

biosurfactants like PC and bile salts give the best accordance. Moreover, in the MEEKC-SDS, there is an inverted elution for the most hydrophobic compounds (Figure 19.5) [22].

In conclusion, MEEKCs based on biosurfactants are apparently better models to estimate the hydrophobicity of compounds with P_{ow} higher than a value of 4 (Figure 19.6).

19.5 CRITICAL MICELLAR CONCENTRATION

19.5.1 METHODS FOR THE DETERMINATION OF THE CRITICAL MICELLAR CONCENTRATION

Critical micellar concentration (CMC) of a surfactant molecule is defined as the solute concentration at which micelles firstly appear in solution [24].

Many different techniques are used for determination of CMC such as surface tension [25], small-angle x-ray scattering [26], electron paramagnetic resonance (EPR) [27], nuclear magnetic resonance (NMR) [28], light scattering and refractometry [29], microcalorimetric titration [27], spectrophotometry [30,31], and RP-LC [32]. In the last time, EKC methods are based on the changes in the migration time of a neutral marker compound by using different concentrations of the studied surfactant as part of the electrolyte. At below CMC of the surfactant molecule, the marker compound migrates together with the EOF. However, at above CMC, the migration time of the marker increases with the augment of the surfactant concentration [33,34].

19.5.2 IMPORTANCE OF THE STUDY OF BILE ACIDS CMC

BAs are steroid compounds, hydroxyl-derivatives of 5-β-cholan-24-oic acid. BAs are the final products of the hepatic biotransformation of cholesterol and they are normally present in the bile as mixed micelles. They have different physicochemical properties according to the number, position, and orientation of their hydroxyl groups, and the type of conjugation with glycine and taurine, which forms the glyco- and tauro-derivatives. These factors influence their solubility, hydrophobicity, and detergent properties [35–37].

Ursodeoxycholic acid (UDCA), cholic acid (CA), chenodeoxycholic acid (CDCA), deoxycholic acid (DCA), lithocholic acid (LCA), and their glyco and tauro-derivatives (Figure 19.7) play an important role in biological systems. Their biological functions are principally associated with lipid digestion and absorption,

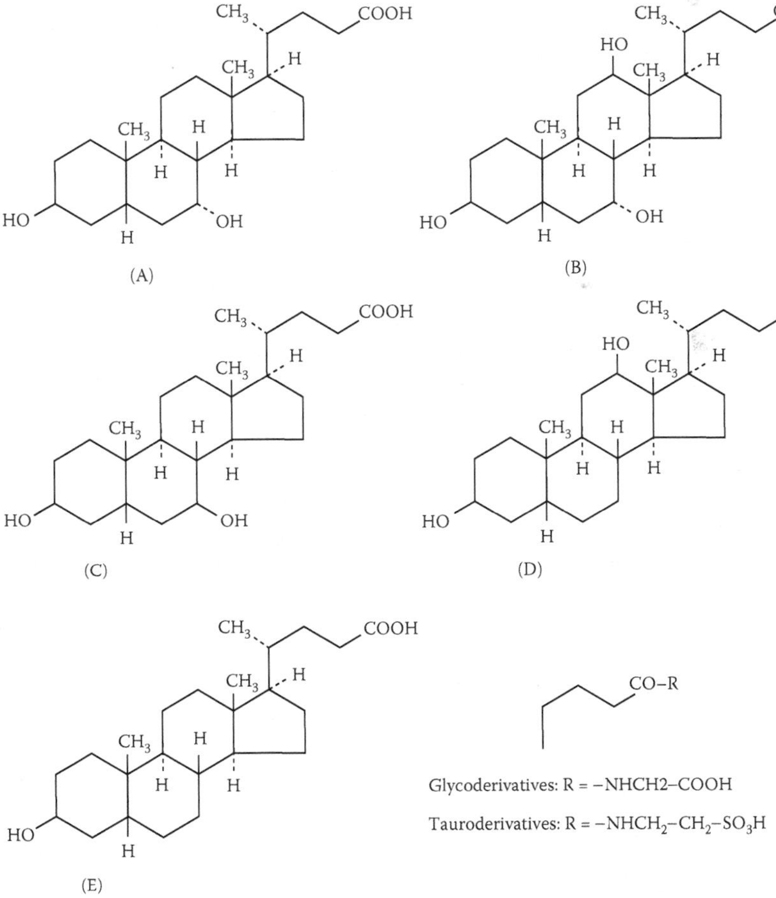

Glycoderivatives: R = −NHCH2−COOH

Tauroderivatives: R = −NHCH$_2$−CH$_2$−SO$_3$H

FIGURE 19.7 Chemical structures of BAs. (A) UDCA, (B) CA, (C) CDCA, (D) DCA, and (E) LCA. (From Tripodi, V.P., Lucangioli, S.E., Scioscia, S.L., and Carducci, C.N., *J. Chromatogr. B*, 785, 147, 2003. With permission.)

solubilization of cholesterol, and bile formation influencing the volume and composition of the bile [38]. Some of the BAs are used as therapeutic agents, UDCA is administered in cholestatic liver disease and cholesterol gallstone dissolution [37]. Moreover, in the study of the relationship between BA detergency and hepatotoxicity, the degree of toxicity is directly related to their CMC values [37,39].

In this context, the study of the CMC of BAs is a fundamental parameter in the evaluation of the biological activity of this type of compounds.

19.5.3 RETENTION FACTOR AND CMC

The microemulsion system to estimate the CMC of free, glycine, and taurine derivative is MEEKC-SDS. This system employs SDS as surfactant, butanol as cosurfactant, octane as oil and an aqueous buffer at pH 7.50. At that pH, the BAs are in the anionic form.

The mobilities of the BA were calculated in the MEEKC system and in free solution (Equation 19.6). Methanol and dodecaphenone were the EOF and microdroplets marker, respectively (Table 19.5).

Table 19.6 shows the log k' values reported for BAs and their CMC values taken from the literature [36,37,40]. In the same table, the position and orientation

---

TABLE 19.5

Electrophoretic Mobilities and k' Values of Free and Conjugate BAs in MEEKC-SDS

Mobilities Expressed as $10^{-9} m^2 V^{-1} s^{-1}$

Abbreviation	Compound	μ_{med}	μ_{eff}	μ_{eff}^{free}	k'
UDCA	Ursodeoxycholate	21.0	−34.7	−13.7	4.78
CA	Cholate	20.5	−35.2	−15.1	5.19
CDCA	Chenodeoxycholate	18.7	−37.1	−15.8	10.54
DCA	Deoxycholate	18.5	−37.3	−16.0	11.61
LCA	Lithocholate	17.7	−38.0	−17.5	19.50
GUDCA	Glycoursodeoxycholate	20.7	−34.7	−15.1	4.49
GCA	Glycocholate	20.2	−35.2	−15.8	5.05
GCDCA	Glycochenodeoxycholate	18.7	−36.7	−16.2	8.64
GDCA	Glycodeoxycholate	18.4	−37.1	−16.3	10.19
GLCA	Glycolithocholate	17.7	−37.7	−17.9	13.80
TUDCA	Tauroursodeoxycholate	20.1	−35.6	−15.7	5.76
TCA	Taurocholate	19.7	−35.9	−16.1	6.32
TCDCA	Taurochenodeoxycholate	18.3	−37.4	−16.7	12.09
TDCA	Taurodeoxycholate	17.9	−37.8	−16.5	16.94
TLCA	Taurolithocholate	17.5	−38.2	−18.0	22.84

μ_{eff} microdroplets −39.1 ($10^{-9} m^2 V^{-1} s^{-1}$)

TABLE 19.6
log k' Values and CMC of Free and Conjugate BAs

BA	Number of OH	Position/ Orientation of OH Group	k'	CMC (mmol/L)
UDCA	2	$3\alpha7\beta$	4.78	7.0
GUDCA	2	$3\alpha7\beta$	4.49	4.0
GCA	3	$3\alpha7\alpha12\alpha$	5.05	10.0
CA	3	$3\alpha7\alpha12\alpha$	5.19	11.0
TUDCA	2	$3\alpha7\beta$	5.76	2.2
TCA	3	$3\alpha7\alpha12\alpha$	6.32	6.0
GCDCA	2	$3\alpha7\alpha$	8.64	2.0
GDCA	2	$3\alpha12\alpha$	10.19	2.0
CDCA	2	$3\alpha7\alpha$	10.54	4.0
DCA	2	$3\alpha12\alpha$	11.61	3.0
TCDCA	2	$3\alpha7\alpha$	12.09	3.0
GLCA	1	3α	13.80	—
TDCA	2	$3\alpha12\alpha$	16.94	2.4
LCA	1	3α	19.50	0.6
TLCA	1	3α	22.84	—

CMC values taken from Refs. [36,37,40]; —, no data available.

of the OH groups of the BAs are indicated. It is observed that BA retention factors decrease when the number of OH groups decreases. Exceptions are UDCA and derivatives (GUDCA and TUDCA). These acids have the same number of OH groups as their epimer (CDCA and derivatives) but one OH group in β orientation and resulting in a smaller k' than trihydroxy acids (CA and derivatives). Moreover, in Table 19.6, it can be seen that the tauro-derivatives in general interact more strongly than the other BAs (free and glycine derivatives) with the microdroplets.

Due to the properties of the BAs of former micelles, it could be assumed that BAs are distributed between aqueous buffer and SDS microdroplets in the same way as the other hydrophobic compounds. Based on NMR data from Ref. [41], BAs are incorporated into SDS micelle not randomly, but under orientation of the ionic group towards the outside of the micelle. The CMC values of each BA show that a high number of OH groups, correlatives with higher CMC values, and lesser detergency. Moreover, derivative BAs have lower-CMC values respect to their free form. Actually, the CMC strongly reflects the influence of the orientation of the OH group in the molecule. As an example, UDCA has two times larger CMC than its epimer CDCA. For this reason, UDCA is less detergent and less toxic and it is used as therapeutic agent in different pathologies [37].

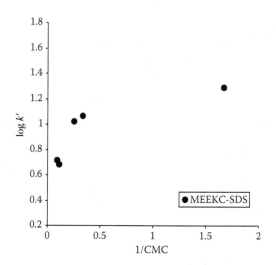

FIGURE 19.8   Relationship of log k' and 1/CMC of free BAs.

Figure 19.8 shows the correlation between log k' versus 1/CMC for free BAs. The retention of BAs in MEEKC system seems to reflect their role in living systems as detergents, and probably might serve as a model for estimation of detergency of hydrophobic compounds [38].

19.6   MEEKC SYSTEMS

19.6.1   Method Development of MEEKC

In the last time, MEEKC has become a powerful tool in the analytical laboratory, specially in the determination of a wide range of compounds present in different biological and pharmaceutical samples [42,43].

The convenience of working with microemulsions is based on their properties: thermodynamic stability, optical transparency, and easy preparation that make possible to quantitate drugs of different hydrophobicity, both neutral and ionic compounds in the same run.

Moreover, MEEKC has been shown to be a highly applicable system for the analysis of complex mixtures such as multicomponent formulations, related substances of active principles and excipients in bulk drugs, chiral analysis, natural products, and bioanalytical separations.

Commonly, microemulsions used in MEEKC are based O/W systems. A carefully selection of the components of a microemulsion should be planned in the development of a MEEKC system.

19.6.2 SURFACTANT: TYPE AND CONCENTRATION

The type and concentration of a surfactant molecule affect the analytical separations of the analytes in MEEKC. High concentrations of a surfactant increase the stability of the microemulsion, but it enlarges the size of microdroplets and lengthen the retention time of the analytes.

SDS, an anionic tensioactive and with single-chain structure, is the surfactant mostly employed in the development of a microemulsion at concentrations in the range between 50 and 100mM. Selectivity can be modified by mixing SDS with another surfactant like a bile salt (sodium cholate) or polyoxyethyleneglycol ethers [43]. Cationic surfactants have been used in MEEKC resulting in the reverse of polarity [19].

It must be noticed that resolution of compounds with different and high hydrophobicity using SDS as surfactant in MEEKC sometimes may be unsuccessful [44]. Double-chain structure surfactants such as phosphatydilcholine achieved a better selectivity compared to the traditional MEEKC-SDS system [7,22].

Another double-chain, anionic, and hydrophobic surfactant is sodium bis(2-ethylhexyl) sulfosuccinate (AOT) used in the development of a novel microemulsion in a MEEKC system for the separation of estrogens with different hydrophobicity in the same run [45].

Chiral surfactants used in MEEKC have also been like dodecoxycarbonylvaline (DDCV) for the separation of different enantiomeric analytes [46,47]. Chiral polymeric surfactant has been also employed in the enantiomeric resolution of barbiturate, binaphthyl, and paveroline [48].

19.6.3 COSURFACTANT

The addition of a cosurfactant to the microemulsion reduces the interfacial tension between the oil and aqueous phases [43]. Commonly, it is used a straight chain alcohol like 1-butanol. Increasing the concentration of a cosurfactant affects the viscosity and changes the EOF rate [42]. A series of alcohols from 1-propanol to 1-hexanol have shown different selectivity in the separation of antiepileptic drugs [49].

A chiral cosurfactant such as series of alcohols ranging from R-(–)-butanol to R-(–)-heptanol has been used in MEEKC to improve chiral resolution of ephedrine and related compounds [50].

19.6.4 OIL PHASE

Octane, heptane, hexane, and octanol are the oils used in MEEKC [2]. They show similar selectivity and retention of the analytes. However, octane is the mostly used oil in a MEEKC system. Another oil is isopropyl myristate, employed as cosurfactant with PC and salts of BAs [7,22].

19.6.5 Buffer: Type, Concentration and pH

Solely phosphate and borate or their mixtures are generally employed as buffer in MEEKC at low concentrations (5–20 mM) to produce EOF without an increment of high-current intensity [42].

Low concentrations of salts generate high-speed microemulsion [51] and higher concentrations suppress the EOF and generate higher current, which limits the applied voltage for operation [42].

Zwiterionic buffers (ACES, CAPS, CAPSO, etc.) have been used in MEEKC to reduce the intensity of current produced during migration, allowing that higher voltages may be applied for faster separation [52]. The use of CAPSO and phosphate with AOT as tensioactive in the microemulsion provides not only an appropriate capacity buffer, but also a higher stabilization of the microemulsion system in comparison to the same buffer without CAPSO [45].

The pH of the buffer has a pronounced effect on the separation selectivity as it affects both analyte ionization and velocity of the EOF generated during the running. In MEEKC, typical buffers have been used in the range of pH 7 and 9. Low pH values below 3 have been applied to reduce or cancel the EOF and negative voltages need to be applied through the capillary resulting that the most hydrophobic compounds are firstly detected [42]. At pH values, above 12 have been used to eliminate ionization in the case of basic compounds or to produce ionic compounds in the case of compounds like estrogens and improve their resolution [43,45].

19.6.6 Additives

Organic solvents such as methanol, acetonitrile, 2-propanol have been added to the microemulsion to improve resolution of the analytes, specially those with strong hydrophobicity [53]. The addition of the organic solvents increases the viscosity of the electrolyte reducing the EOF and thus increasing the migration time of the analytes. The addition of cyclodextrins allows to resolve enantiomeric drugs or analytes with closely related structure [46,47].

19.7 APPLICATIONS

19.7.1 Pharmaceutical Analysis

MEEKC is an interesting alternative to traditional methods of analysis applied in research of new products. Determination of wide range of pharmaceutical and excipients [54], different drugs such as steroids [55], vitamins (fat and hydro-soluble) achieved in the same run [56], antibiotics [57], natural, and synthetic estrogens [45], etc., have been performed by MEEKC systems in pharmaceutical samples. Figure 19.9 shows the separation by MEEKC of estrogens with different hydrophobicity by using AOT as tensioactive and CAPSO-phosphate buffer at pH 12.5.

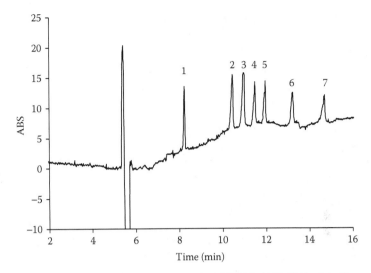

FIGURE 19.9   Electropherogram of estrogens by MEEKC-AOT. (1) Estriol, (2) estradiol-17-hemisuccinate, (3) estrone, (4) etinilestradiol, (5) estradiol, (6) estradiol-3-benzoate, and (7) estradiol 17-valerate. (From Tripodi, V., Flor, S., Carlucci, A., and Lucangioli, S., *Electrophoresis*, 27, 4431, 2006. With permission.)

The determination of related substances of drugs has been reported [45,58] (Figure 19.10). Chiral analysis for the separation of enantiomeric compounds has been achieved by using a chiral surfactant [46,47].

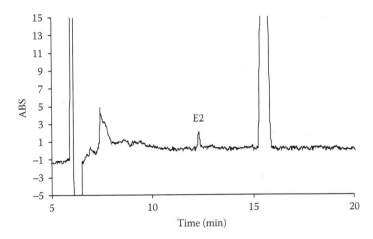

FIGURE 19.10   Electropherogram of estradiol E2 (1% w/w) and estradiol-17-valerate. By MEEKC-AOT. (From Tripodi, V., Flor, S., Carlucci, A., and Lucangioli, S., *Electrophoresis*, 27, 4431, 2006. With permission.)

The method development by MEEKC is useful for the determination of active and excipient compounds as well as related substances of drugs, with very high precision and accuracy. The application of MEEKC in pharmaceutical analysis offers a powerful tool in quality control of pharmaceutical laboratories.

19.7.2 Bioanalysis

Many applications in the bioanalysis require an analytical system capable of handling nanoliter volumes of samples with high efficiency and resolution in a short time of analysis. However, for clinical determinations in which the early detection of the onset of a disease process may be essential to the patient's survival, the analytical technique employed should have an excellent mass sensitivity.

CE is becoming an all-purpose technology since it has been applied to solve analytical problems connected to forensic chemistry, food and clinical chemistry, biochemistry, neuroscience, molecular biology, and environmental science. For this reason, CE has become an established technique comparable to gas chrometography (GC) and High-performance liquid chromatography (HPLC). The most common mode of CE used the bioanalysis are capillary zone electrophoresis (CZE) and MEKC [59] and few reports of MEEKC for the analysis of biological samples. Many authors have compared MEKC and MEEKC systems and they found that MEEKC offered no additional advantage for routine analysis [60]. However, although O/W systems are similar to micelles in the function of solubilization of hydrophobic compounds, microemulsions have a much larger capacity. In that field, Furumoto et al. compared the solubilization capacity between MEKC and MEEKC and they confirmed the advantage of using MEEKC [61].

Moreover, comparing MEKC and MEEKC for steroids analysis, it was found that these analytes are able to penetrate the surface of microdroplets more easily than surface of the micelles. This ability allows MEEKC to extend the analysis to wide range of solutes. In addition, it could allow an easily optimization of the migration time windows that allow to analyze many structurally related compounds as possible in a single run [45].

Although at present there is only a reduced number of publications describing the use of MEEKC, this mode applied to separation of steroids [62], diphenyl hydrazine derivatives [63], cardiac glycosides [64], proteins [65] bases, and nucleosides [61].

In conclusion, it is predicted that the number of applications referred to MEEKC will increase in CE with the possibility of resolving complex mixtures of analytes possessing very different grades of polarity and solubility with optimal selectivity.

ACKNOWLEDGMENTS

The authors wish to thank Prof. Clyde Carducci and Prof. Ernst Kenndler for their direction of the research and also thank Dr. Jorge Muse for assistance.

SYMBOLS AND TERMINOLOGIES

ACES	2-carbamoylmethylaminoethanesulfonic acid
AOT	bis(2-ethylhexyl) sulfosuccinate
BA	bile acids
CA	cholic acid
CAPS	N-cyclohexyl-3-aminopropanesulfonic acid
CAPSO	N-cyclohexyl-2-hydroxyl-3-aminopropanesulfonic acid
CDCA	chenodeoxycholic acid
CE	capillary electrophoresis
CMC	critical micellar concentration
CZE	capillary zone electrophoresis
DCA	deoxycholic acid
DDCV	dodecoxycarbonylvaline
EKC	electrokinetic chromatography
EOF	electroosmotic flow
EPR	electron paramagnetic resonance
GC	gas chromatography
GUDCA	glycoursodeoxycholic acid
HPLC	high-performance liquid chromatography
k'	retention factor
LCA	lithodeoxycholic acid
MEEKC	microemulsion electrokinetic chromatography
MEKC	micellar electrokinetic chromatography
NMR	nuclear magnetic resonance
O/W	oil droplets in a water phase
P	partition coefficients
PC	phosphatidylcholine
P_{OW}	1-octanol-water partition coefficient
RP-LC	reverse phase liquid chromatography
SC	sodium cholate
SDC	sodium deoxycholate
SDS	sodium dodecyl sulfate
TLC	thin layer chromatography
TUDCA	tauroursodeoxycholic acid
UDCA	ursodeoxycholic acid
VEKC	vesicles electrokinetic chromatography
W/O	water droplets in an oil phase

REFERENCES

1. Khaledi, M. 1998, *High Performance Capillary Electrophoresis*, John Wiley & Sons, New York.
2. Neubert, R. and Ruttinger, H. 2003, *Affinity Capillary Electrophoresis in Pharmaceutics and Biopharmaceutics*, Marcel Dekker, New York.

3. United States Pharmacopoeial 24 Revision, United States Pharmacopeial Convention/ Capillary electrophoresis/, *Pharmacopeial Forum* 27, 2353, Rockville, MD, USA, 2001.
4. Altria, K. 2000, Background theory and applications of microemulsion electrokinetic chromatography. *J. Chromatogr. A* 892, 171.
5. Altria, K., Mahuzier, P., and Clark, B. 2003, Background and operating parameters in microemulsion electrokinetic chromatography. *Electrophoresis* 24, 315.
6. Huie, C. 2006, Recent applications of microemulsion electrokinetic chromatography. *Electrophoresis* 27, 60.
7. Lucangioli, S., Kenndler, E., Carlucci, A., Tripodi, V., Scioscia, S., and Carducci, C. 2003, Relation between retention factors of immunosuppressive drugs in microemulsion electrokinetic chromatography with biosurfactants and octanol-water partition coefficients. *J. Pharm. Biomed. Anal.* 33, 871.
8. Jia, Z. 2005, Physicochemical profiling by capillary electrophoresis. *Current Pharm. Anal.* 1, 41.
9. Leo, A., Hansch, C., and Elkins, D. 1971, Partition coefficients and their uses. *Chem. Rev.* 71, 525.
10. Yang, S., Bumgarner, J., Kruk, L., and Khaledi, M. 1996, Quantitative structure–activity relationships studies with micellar electrokinetic chromatography influence of surfactant type and mixed micelles on estimation of hydrophobicity and bioavailability. *J. Chromatogr. A* 721, 323.
11. Takacs-Novak, K. and Avdeef, A. 1996, Interlaboratory study of log p determination by shake-flask and potentiometric methods. *J. Pharm. Biomed Anal.* 14, 1405.
12. Sarbu, C., Kuhajda, K., and Kevresan, S. 2001, Evaluation of the lipophilicity of bile acids and their derivatives by thin-layer chromatography and principal components analysis. *J Chromatogr. A* 917, 361.
13. Dorsey, J. and Khaledi, M. 1993, Hydrophobicity estimation by reversed-phase liquid chromatography. Implications for biological partitioning processes. *J. Chromatogr. A* 656, 485.
14. Valkó, K. 2004, Application of high-performance liquid chromatography based measurements of lipophilicity to model biological distribution. *J. Chromatogr. A* 1037, 299.
15. Meylan, W. and Howard, P. 1995, Atom/fragment contribution method for estimating octanol-water partition coefficients. *J. Pharm. Sci.* 84, 83.
16. Ishihama, Y., Oda, Y., Uchikawa, K., and Asakawa, N. 1994, Correlation of octanol-water partition coefficients with capacity factors measured by micellar electrokinetic chromatography. *Chem. Pharm. Bull.* 42, 1525.
17. Smith, J. and Vinjamoori, D. 1995, Rapid determination of logarithmic partition coefficients between *n*-octanol and water using micellar electrokinetic capillary chromatography. *J. Chromatogr. B* 669, 59.
18. Klotz, W., Schure, M., and Foley, J. 2002, Rapid estimation of octanol-water coefficients using synthesized vesicles in electrokinetic chromatography. *J. Chromatogr. A* 962, 207.
19. Ishihama, Y., Oda, Y., and Asakawa, N. 1996, Hydrophobicity of cationic solutes measured by electrokinetic chromatography with cationic microemulsions. *Anal. Chem.* 68, 4281.
20. Klotz, W., Schure, M., and Foley, J. 2001, Determination of octanol-water partition coefficients of pesticides by microemulsion electrokinetic chromatography. *J. Chromatogr. A* 930, 145.
21. Oszwaldowski, S. and Timerbaev, A. 2007, Development of quantitative structure–activity relationships for interpretation of the migration behavior of neutral platinum (ii) complex in microemulsion electrokinetic chromatography. *J. Chromatogr. A* 1146, 258.

22. Lucangioli, S., Carducci, C., Scioscia, S., Carlucci, A., Bregni, C., and Kenndler, E. 2003, Comparison of the retention characteristics of different pseudostationary phases for microemulsion and micellar electrokinetic chromatography of betamethasone and derivatives. *Electrophoresis* 24, 984.

23. Ishihama, Y., Oda, Y., Uchikawa, K., and Asakawa, N. 1995, Evaluation of solute hydrophobicity by microemulsion electrokinetic chromatography. *Anal. Chem.* 67, 1588.

24. Reis, S., Moutinho, C., Matos, C., Castro, B., Gameiro, P., and Lima, J. 2004, Noninvasive methods to determine the critical micelle concentration of some bile acid salts. *Anal. Biochem.* 334, 117.

25. Roda, A., Cerre, A., Fini, A., Sipahi, A., and Baraldini, M. 1995, Experimental evaluation of model for predicting micellar composition and concentration of monomeric species in bile salt binary mixtures. *J. Pharm. Sci.* 84, 593.

26. Matsuoka, H., Kratohvil, J., and Ise, N. 1987, Small-angle X-ray scattering from solutions of bile salts. Sodium taurodeoxicholate in aqueous electrolyte solutions. *J. Colloid Interface Sci.* 118, 387.

27. Simonovic, B. and Momirovic, M. 1997, Determination of critical micelle concentration of bile salts by micro-calorimetric titration. *Mikrochim. Acta* 127, 101.

28. Gouin, S. and Zhu, X. 1998, Fluorescence and NMR studies of the effect of a bile acid dimer on the micellization of bile salts. *Langmuir* 14, 4025.

29. Paula, S., Sus, W., Tuchtenhagen, J., and Blume, A. 1995, Thermodynamics of micelle formation as a function of temperature: A high sensitivity titration calorimetry study. *J. Phys. Chem.* 99, 11742.

30. Sugihara, G., Yamakawa, K., Murata, Y., and Tanaka, M. 1982, Effects of pH, pNa, and temperature on micelle formation and solubilization of cholesterol in aqueous solutions of bile salts. *J. Phys. Chem.* 86, 2784.

31. Egelhaaf, S. and Schurtenberger, P. 1994, Shape transformations in the lecithin-bile salt system: From cylinders to vesicles. *J. Phys. Chem.* 98, 8560.

32. Shaw, R., Elliott, W., and Barisas, B. 1991, Estimation of critical micelle concentration of bile acids by reversed-phase high performance liquid chromatography. *Mikrochim, Acta* 3, 137.

33. Nakamura, H., Sano, A., and Matsuura, K. 1998, Determination of critical micelle concentration of anionic surfactants by capillary electrophoresis using 2-naphthalenemethanol as a marker for micelle formation. *Anal. Sci.* 14, 379.

34. Mrestani, Y., Marestani, Z., and Neubert, R. 2001, Characterization of micellar solubilization of antibiotics using micellar electrokinetic chromatography. *J. Pharm. Biomed. Anal.* 26, 883.

35. Hofmann, A. and Roda, A. 1984, Physicochemical properties of bile acids and their relationship to biological properties: An overview of the problem. *J. Lipid Res.* 25, 1477.

36. Roda, A., Minutello, A., Angellotti, M., and Fini, A. 1990, Bile acid structure–activity relationship: Evaluation of bile acid lipophilicity using 1-octanol/water partition coefficient and reverse phase HPLC. *J. Lipid Res.* 31, 1433.

37. Roda, A., Gioacchini, A., Manetta, A., Cebre, C., Montagnani, M., and Fini, A. 1995, Bile acids: Physico-chemical properties, function and activity. *Ital. J. Gastroenterol.* 27, 327.

38. Lucangioli, S., Carducci, C., Tripodi, V., Kenndler, E. and 2001, Retention of bile salts in micellar electrokinetic chromatography: Relation of capacity factor to octanol-water partition coefficient and critical micellar concentration. *J. Chromatogr. B* 765, 113.

39. Posa, M., Kevresan, S., Mikov, M., Cirin Novta, V., Sarbu, C., and Kuhajda, K. 2007, Determination of critical micellar concentrations of cholic acid and its keto derivatives. *Colloid Surface B: Biointerface* 59, 179.
40. Roda, A., Hofmann, A., and Mysels, K. 1983, The influence of bile salt structure on self-association in aqueous solutions. *J. Biol. Chem.* 258, 6362.
41. Wiedmer, S. and Riekkola, M. 1997, Mixed micelles of sodium dodecyl sulfate and sodium cholate: Micellar electrokinetic capillary chromatography and nuclear resonance spectroscopy. *Anal. Chem.* 69, 1577.
42. Broderick, M., Donegan, S., Power, J., and Altria, K. 2005, Optimization and use of water-in-oil MEEKC in pharmaceutical analysis. *J. Pharm. Biomed. Anal.* 37, 877.
43. McEvoy, E., Marsh, A., Altria, K., Donegan, S., and Power, J. 2007, Recent advances in the development and application of microemulsion EKC. *Electrophoresis* 28, 193.
44. Lin, J., Nakagawa, M., Uchiyama, K., and Hobo, T. 2002, Comparison of three different anionic surfactants for the separation of hydrophobic compounds by nonaqueous capillary electrophoresis. *Electrophoresis* 23, 421.
45. Tripodi, V., Flor, S., Carlucci, A., and Lucangioli, S. 2006, Simultaneous determination of natural and synthetic estrogens by electrokinetic chromatography using a novel microemulsion. *Electrophoresis* 27, 4431.
46. Mertzman, M. and Foley, J. 2004, Chiral cyclodextrin-modified microemulsion electrokinetic chromatography. *Electrophoresis* 25, 1188.
47. Khle, K. and Foley, J. 2007, Review of aqueous chiral electrokinetic chromatography (EKC) with an emphasis on chiral microemulsion EKC. *Electrophoresis* 28, 2503.
48. Iqbal, R., Rizvi, S., Akbay, C., and Shamsi, S. 2004, Chiral separations in microemulsion electrokinetic chromatography. Use of micelle polymers and microemulsion polymers. *J. Chromatogr. A* 1043, 291.
49. Ivanova, M., Piunti, A., Marziali, E., Komarova, N., Raggi, M., Kenndler, E. 2003, Microemulsion electrokinetic chromatography applied for separation of levetiracetam from other antiepileptic drugs in polypharmacy. *Electrophoresis* 24, 992.
50. Zheng, Z., Lin, J., Chan, W., Lee, A., and Huie, C. 2004, Separation of enantiomers in microemulsion electrokinetic chromatography using chiral alcohols as cosurfactants. *Electrophoresis* 25, 3263.
51. Mahuzier, P., Clark, B., Bryant, S., and Altria, K. 2001, High-speed microemulsion electrokinetic chromatography. *Electrophoresis* 22, 3819.
52. Marsh, A., Clark, B., Broderick, M., Power, J., Donegan, S., and Altria, K. 2004, Recent advances in microemulsion electrokinetic chromatography. *Electrophoresis* 25, 3970.
53. Altria, K., Clark, B., and Mahuzier, P. 2000, The effects of operating variables in microemulsion electrokinetic capillary chromatography. *Chromatographia* 52, 758.
54. Altria, K. 1999, Application of microemulsion electrokinetic chromatography to the analysis of a wide range of pharmaceuticals and excipients. *J. Chromatogr. A* 844, 371.
55. Pomponio, R., Gotti, R., Fiori, J., Cavrini, V. 2005, Microemulsion electrokinetic chromatography of corticosteroids. Effect of surfactants and cyclodextrins on separation selectivity. *J. Chromatogr. A* 1081, 24.
56. Altria, K. 2002, Microemulsion electrokinetic chromatography. *J. Cap. Elec. Microchip Tech.* 7, 11.
57. Mrestani, Y., El-Mokdad, N., Ruttinger, H., and Neubert, R. 1998, Characterization of partitioning behavior of cephalosporins using microemulsion and micellar electrokinetic chromatography. *Electrophoresis* 19, 2895.
58. McEvoy, E., Donegan, S., Power, J. and Altria, K. 2007, Optimization and validation of a rapid and efficient microemulsion liquid chromatography (melc) method for the determination of paracetamol (Acetominophen) content in a suppository formulation. *J. Pharm. Biomed. Anal.* 44, 137.

59. Guzman, N., Park, S., Schaufelberger, D., Hernandez, L., Paez, X., Rada, P., Tomlinson, A., and Naylor, S. 1997, New approaches in clinical chemistry: On-line analyte concentration and microreaction capillary electrophoresis for the determination of drugs, metabolic intermediates, and biopolymers in biological fluids. *J. Chromatogr. B* 697, 37.

60. Melin, V. and Perrett, D. 2004, Comparative study of microemulsion electrokinetic capillary chromatography and micellar electrokinetic capillary chromatography for the analysis of UV-absorbing compounds in human urine. *Electrophoresis* 25, 1503.

61. Furumoto, T., Fukumoto, T., Sekiguchi, M., Sugiyama, T., and Watari, H. 2001, Migration mechanism of bases and nucleosides in oil-in-water microemulsion capillary electrophoresis. *Electrophoresis* 22, 3438.

62. Vomastova, L., Miksik, I., and Deyl, Z. 1996, Microemulsion and micellar electrokinetic chromatography of steroids. *J. Chromatogr. B* 681, 107.

63. Miksik, I., Gabriel, J., and Dehyl, Z. 1997, Microemulsion electrokinetic chromatography of diphenylhydrazones of dicarbonyl sugars. *J. Chromatogr. A* 772, 297.

64. Debusschere, L., Demesmay, C., Rocca, J., Lachatre, G., and Lofti, H. 1997, Separation of cardiac glycosides by micellar electrokinetic chromatography and microemulsion electrokinetic chromatography. *J. Chromatog. A* 779, 227.

65. Zhou, G., Luo, G., and Zhang, X. 1999, Microemulsion electrokinetic chromatography of proteins. *J. Chromatogr. A* 853, 277.

66. Software Solaris, Advanced chemistry developed (ACD), 1994–2001.

67. Pones, M., Kempena, J., Shroot, B., and Caron, J. 1986, Glucocorticoids: Binding affinity and lipophilicity. *J. Pharm. Sci.* 75, 973.

Index